International Symposium on Digital Industrial Radiology and Computed Tomography (DIR 2019)

Berichtsband BB 172

Furth, Germany
2-4 July 2019

ISBN: 978-1-7138-3709-1

Printed from e-media with permission by:

Curran Associates, Inc.
57 Morehouse Lane
Red Hook, NY 12571

Some format issues inherent in the e-media version may also appear in this print version.

Deutsche Gesellschaft Fuer Zerstoerungsfreie Pruefung (DGZfP)
Max-Planck-Str. 6
12489 Berlin
Germany

Phone: +49 30 67807-0
Fax: +49 30 67807-109

mail@dgzfp.de

Additional copies of this publication are available from:

Curran Associates, Inc.
57 Morehouse Lane
Red Hook, NY 12571 USA
Phone: 845-758-0400
Fax: 845-758-2633
Email: curran@proceedings.com
Web: www.proceedings.com

TABLE OF CONTENTS

TU.4.A: MULTI-TECHNIQUES FUSION

WE.1.A: IMAGE PROCESSING AND RECONSTRUCTION

WE.2.A: HIGH-ENERGY APPLICATIONS

RoboCT – Application for in-situ Inspection of Join Technologies of large scale Objects

Wolfgang HOLUB [1], Franziska BRUNNER [1], Tobias SCHÖN [1]
[1] Fraunhofer Development Center X-ray Technology EZRT, Fürth, Germany
wolfgang.holub@iis.fraunhofer.de

Abstract. In sophisticated mechanical engineering like e.g. automobile industry, the reliability of joints like rivets, screws or adhesive bonding is usually being examined by regularly repeated destructive testing of these manufactured parts. At full scale, this implies disassembling whole car bodies so far that e.g. all adhesive bondings are being forcibly opened so that the geometry of these joints can be assessed.

Whilst X-ray CT is one established means of NDT applied to cutout specimens in the laboratories of such manufacturers, it is only beginning to become available to full scale objects since few years.

Over more than one decade, EZRT has developed the technology of RoboCT - the application of cooperating industrial robots as large and highly flexible manipulation system for the X-ray components. In 2018, we have taken into operation the very first such system in the field in BMW's prototype plant. This system of four cooperating robots is designed to address large automobile components, whole car bodies or even fully assembled cars and shall help to reduce the time to market of new models.

In this contribution, we introduce the technology and its first implementation in production environment. We demonstrate the exemplary application of RoboCT to different join technologies.

Introduction

X-ray radioscopy and especially computed tomography (CT) are well established means of NDT in laboratories worldwide and also in some production lines for smaller objects with the necessity for demanding production monitoring. Nonetheless, there is a wide field of production scenarios that have not yet been suitable for monitoring by CT. This is mostly due to the size and geometry of the respective objects, which can hardly be handled by available X-ray equipment. Join connections in car bodies belong to the most prominent examples of such applications. Join connections like bondings, screws, rivets or weldings are being examined by X-ray NDT in the laboratory – but only within small cutouts that fit within such laboratory scale CT systems. RoboCT shall bring computed tomography to those large objects – in the laboratory and also in production.

1. RoboCT Technology

Laboratory CT usually depends on ambitious mechanical engineering to provide the geometric precision necessary for the algorithmic reconstruction of 3-D CT images from a series of 2-D radioscopies. Those designs do often rely on bases of granite to provide highest geometric and temperature stability. Consequently, classic CT systems are very limited to their possible dimensions and also to the trajectories, i.e. the kinds of acquisition procedures that they can perform.

EZRT has spent the past decade on the idea to replace highly precise manipulation systems by rather inaccurate industrial robots. On the one hand motivated by the immediate increase in work space (typical industrial robots reach up to more than 3 m, each) and the extremely high flexibility to move to any possible pose (i.e. 6-D position and orientation) or along arbitrary trajectories, on the other hand.

1.1 General System Setup

In general, we speak of RoboCT, i.e. robot based computed tomography, when at least one pair of cooperating industrial robots is being applied to carry X-ray source and detector, with hardly any limit to additional robots for more than one source or detector, object manipulation or complementary modalities like e.g. photography, stereography, thermography or ultrasonics.

Fig. 1. RoboCT system at BMW's prototype plant, robots carrying X-ray sources rear left and front right, robots carrying X-ray detectors and / or optical 3-D scanner front left and rear right, lifting table in the center

The possible couples are then being used to prescribe different trajectories for 2-D or 3-D inspection along or around some object or region of interest (ROI).The possibilities reach from conventional circular trajectories – freely positioned and oriented within the work space

of the cell – to completely arbitrary trajectories without any restriction to regular geometric paths.
The precision of industrial robots does only reach in the magnitude of ½ mm (average absolute euclidean positioning error in space) for the models with highest accuracy. Since this does not suffice for the application of CT with resolutions of some tens of Microns, the key enabling technology is the developed geometric calibration process that delivers the absolute position of X-ray source and detector in 6-D space – even with rather inaccurate manipulation hardware.
Resulting acquisitions are then being reconstructed to 3-D CT images by means of EZRT firefly reconstruction and correction suite with FBP reconstruction for quasi-regular circular trajectories and with SART reconstruction for laminographic or arbitrary trajectories.

1.2 RoboCT at BMW's Prototype Plant

As result of the past research, we achieved to reach the necessary technology readiness level to implement the first laboratory RoboCT cell with BMW at their prototype plant in 2018. There, the system is helping designers and engineers to speed up ITO time – i.e. idea-to-offer time, the time needed to establish stable production processes of new models before they can be released to the market. That system consists of four Kuka Quantec extra KR90 R3100 HA robots on two linear axis, left and right of a lifting table for the objects like e.g. car bodies or fully assembled cars. The robots carry two Comet VarioFocus X-ray sources XRS-225VF (225 kV, adjustable focal spot size down to 250 µm) one Varex XRD3025 detector (300 mm · 250 mm, 100 µm pixel pitch) and one Perkin Elmer XRD 1621 (409.6 mm squared, 200 µm pixel pitch) and one Gocator optical 3-D sensor. The robots can cooperate in up to four different pairs – on either side of the object table or opposite to each other. (see Figure 1)

Fig. 2. Application example of one pair of cooperating robots positioning X-ray source and detector at one car's A-pillar

After several experimental installations, both in aerospace industry and in the laboratory, this is EZRT's first installed productive system in the field in automotive industry. At BMW Group it is being applied for different questions on large components, car bodies or even fully assembled cars that cannot be addressed with their regular CT systems. Situated in their prototype plant, its main aim is to accompany design and production of pilot-series vehicles in order to speed up ITO (idea-to-offer) times for new car models.
The cell geometry is designed so that the system fits BMW's largest models like the Rolls Royce Phantom with 6 m length. It shall enable accessing most ROIs – even of fully assembled cars, like e.g. the A-pillar of a BMW i8 Roadster (see Figure 2).

1.3 Research Setup at EZRT

RoboCT research and development at EZRT had started with some small Stäubli RX90B robots with which we also executed parts of the experimental basis for the application of Micro-CT on large automotive components [1][2][3]. In 2017, we upgraded to a system of two Kuka KR90 robots. This system has been applied for a full-scale feasibility study as R&D basis for BMW's installation [4]. (see Figure 3)

Fig. 3. RoboCT research setup at EZRT 2017
with one floor-mounted pair of robots at a car body rear end

Today, we work with a setup consisting of the two Kuka KR90 robots on 6 m linear axes, each in a cabin with 10 m · 8 m · 4 m (length, width, height) working space. (see Figure 4) We apply one Varex XRD 3025 detector and a prototype source from Comet which shall come to the market, in near future. This MesoFocus source is Comet's development of a closed source towards the focal spot sizes necessary for Micro-CT applications. The 225kV source has an adjustable focal spot of 50 µm to 200 µm with 1 W/µm power [5]. Apart from the higher achievable resolution with 50 µm instead of 250 µm focal spot size (compared to the VarioFocus tube recently applied), this tube gives us better accessibility in RoboCT applications due to the fact that it allows acquisitions at magnification M=2 where the detector may be equally far away from the object ROI as the source.

Fig. 4. RoboCT Research Setup at EZRT 2019
with one pair of robots on linear axes at an automobile on a lift table

2. RoboCT Application on Join Connections

2.1 State of the Art Join Inspection

In automotive industry, the quality of joints e.g. in the car body is regularly being destructively tested on control samples. Therefore, the car body is being completely disassembled in order to perform e.g. micrographic analysis of rivets and visual inspection of adhesive bondings.
This kind of destructive testing costs both time and money and requires long delay until the results become available which leads to long reaction times in case of some process failure.

2.2 Join Inspection by RoboCT

One of the most relevant applications for automotive industry has been to find a solution to assist the inspection of joints in the car body manufacturing by means of X-ray NDT. Together with BMW, we investigated the application of RoboCT on car body parts [4]. As a result, BMW decided to implement the first RoboCT system in their prototype plant. Not only, but also for the application at join connections on the whole automobile. Figure 5 shows results from BMW's publication [6] on their applications of RoboCT. The shape and geometric features of the punch rivet can be visualized and measured and the adhesive bonding between layers in the aluminium car body can easily be visualized.

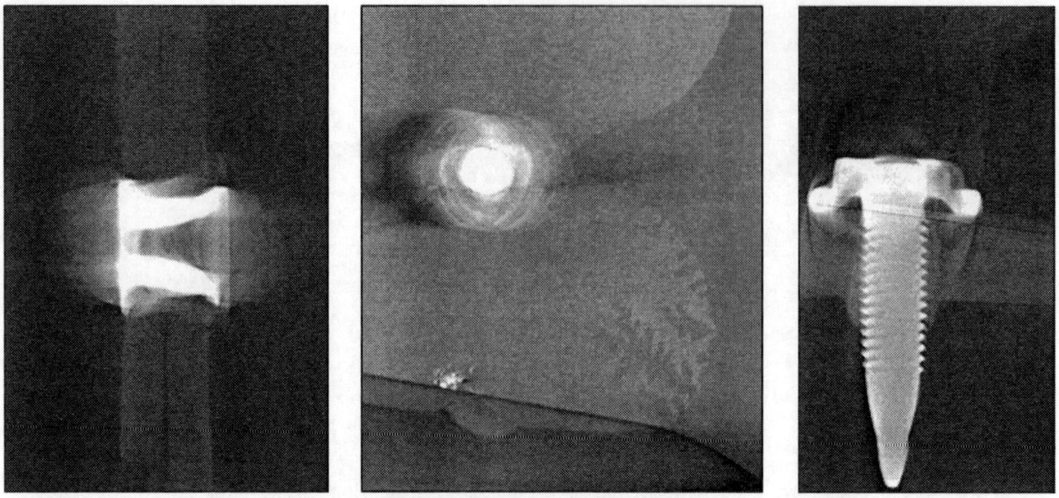

Fig. 5. RoboCT acquisition from a car body front end: punch rivet, adhesive bonding and a self-piercing screw

Figure 6 shows the CT image of one row of different joints within a test specimen of three-dimensionally shaped aluminium sheets of approximately 1 m · 1 m · 40 cm. The shape and deformation geometry both of aluminium and steel pop rivets can easily be seen, as can the insertion of the self piercing screw. The orthogonal slice in between the sheets of aluminium shows the adhesive bonding in between and allows to identify pores and the flow front of the glue.

3. Summary and Outlook

RoboCT proves to more and more become a solution to perform Micro-CT on very large objects that have not yet been accessible by any other CT setup.
For EZRT, RoboCT builds one of the first steps towards cognitive sensor systems, i.e. smart monitoring systems that do intelligently apply and parametrize their multimodal sensors.
RoboCT represents a very powerful, yet difficult to apply technology due to its high dimensional parameter space – especially when it comes to completely arbitrary acquisition trajectories. Smart, AI based acquisition planning will help this technology to become one first cognitive sensor that will assist the user to find the most suitable parameter set for a given task. With RoboCT, the next step will consist of the automatic acquisition planning on safe, collision-free trajectories, followed by optimized acquisitions to provide best possible CT image quality at the given accessibility.
In the future, RoboCT shall become part of a cognitive sensor system – a blackbox of multiple modalities, intelligently combining those either for best image quality or most efficient results given a concrete task with prior information on shape and material of an object and on the monitoring task.

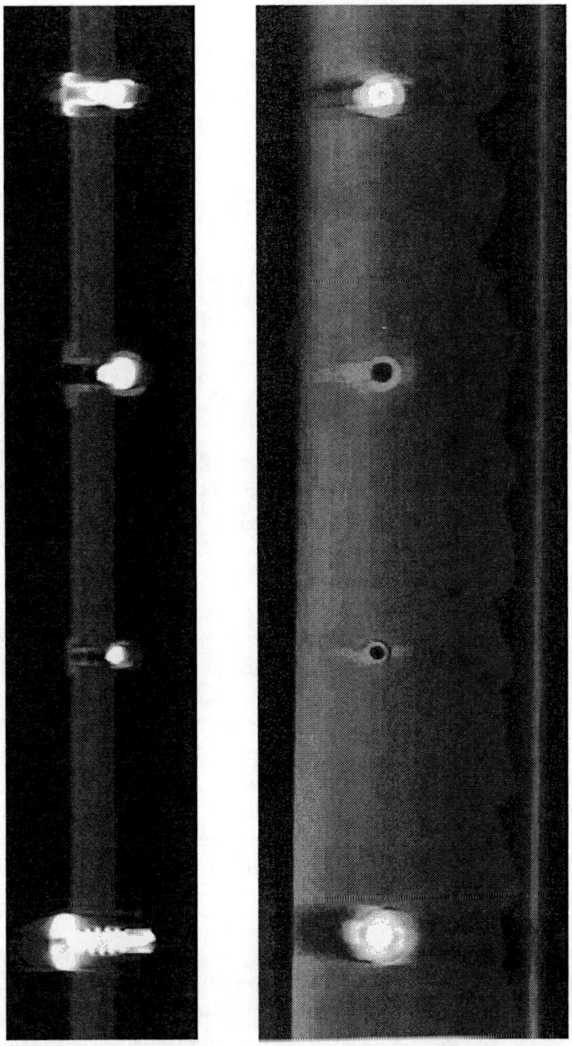

Fig. 6. RoboCT acquisition of sample specimen: one row of steel and aluminium pop rivets, a self-piercing screw and adhesive bonding joining sheets of aluminium

References

[1] Holub, Wolfgang et al.: "XXL-Micro-CT – Comparative Evaluation of Microscopic Computed Tomography for Macroscopic Objects", International Symposium on Digital Industrial Radiology and Computed Tomography Conference DIR 2015

[2] Holub, Wolfgang et al.: "RoboCT – Roboterbasierte XXL-Mikro-Computertomographie an Großbauteilen im Automobilbau", DGZfP Jahrestagung 2017

[3] Holub, Wolfgang et al.: "RoboCT – μCT zur Eskalationsprüfung von UT-Befunden in der Luftfahrt", DGZfP Jahrestagung 2017

[4] Holub, Wolfgang et al.: "RoboCT – Robot based Computed Tomography and its Transition towards the full scale Car Body", 8th International Conference on Industrial Computed Tomography iCT, 2018

[5] Riedo, Adrian: "New MesoFocus NDT X-Ray tube concept", International Symposium on Digital Industrial Radiology and Computed Tomography Conference DIR 2019

[6] Kerscher, Stefan: "Robot-guided, non-destructive computer tomography on whole automobiles", CARAT - International Conference on Applied X-Ray Technologies 2019

[7] Holub, Wolfgang et al.: "RoboCT - High Resolution Robot Based CT on Large Components", CARAT - International Conference on Applied X-Ray Technologies 2019

[8] Holub, Wolfgang et al.: "RoboCT and XXL-CT – X-ray imaging for non-destructive testing of large and complex car components up to the whole car body", AEE Congress – Automotive Engineering Expo 2019

3D Tomogram-less Reconstruction for Industrial Computed Tomography Using a Clustering Approach

Maximilian WATTENBERG [1], Philipp KLEIN [2], Maik STILLE [1], Thorsten M. BUZUG [1]
[1] Institute of Medical Engineering, Universität zu Lübeck, Ratzeburger Allee 160, 23562 Lübeck, Germany, {wattenberg, stille, buzug}@imt.uni-luebeck.de
[2] YXLON International GmbH, Essener Bogen 15, 22419 Hamburg, Germany, klein@hbg.yxlon.com

Contact e-mail: wattenberg@imt.uni-luebeck.de

Abstract. In order to verify specification compliance, some applications in metrology use industrial computed tomography to extract a surface mesh from a reconstructed tomogram. Surface information validity relies on a high-resolution tomogram of good quality. In order to satisfy the sampling theorem and fulfill the tomogram requirements, many images need to be acquired at densely sampled projection angles.

The method proposed in this paper reconstructs an object representation in free 3D space without the use of a tomogram. Instead, the information, which describes the surface of the object, is detected in the projection images by an edge detection algorithm. The detected edges are converted to Cartesian coordinates and transformed from the detector coordinates into the object coordinate system by considering the system geometry at the moment of acquisition for the corresponding projection image. During the reconstruction, two projection images are drawn from an acquired set. For these projection geometries, the x-rays from the source to the coordinates of the edges are calculated. The center point between two rays from different projection geometries is calculated for all rays passing each other in near distance. These coordinates mark folds in the surface and indicate the object's shape.

The proposed method divides into the detection of the projected edge vertices and into the reconstruction of edge vertices by ray-tracing. In this paper the two methods are evaluated separately. It was shown that the edges in the surface of an object can be reconstructed from ideally projected edge vertices. Furthermore, it was shown, that a Canny edge detection algorithm is capable of localizing projected vertices within pixel precision. Therefore, the concept has potential for a tomogram-less reconstruction and dimensional evaluation from few projections.

1 Introduction

Computed tomography (CT) has long since proven its suitability in nondestructive testing and dimensional measurement in industrial context. It is often used to determine if components satisfy or deviate from defined specifications. A high accuracy during this measurement demands – beside a precise knowledge of the system geometry, high image quality, low noise levels, and other factors – for high resolution of the projection image as well as the tomogram. Standard reconstruction approaches like filtered backprojection (FBP) or algebraic reconstruction technique require a densely sampled scan trajectory in order to satisfy sampling theorem and avoid artefacts [1]. High-resolution tomograms therefore lead to an increased scan time and time consuming reconstructions.

Besides the wish for high accuracy, the time for measuring the manufactured product should be as short as possible. Especially in cases where CT will be used for in-line inspection, time is critical. From the delineations above it is easy to see that both demands contradict each other when classical reconstruction algorithms are used. Motivated by this, Jinnouchi et al. introduced a new approach that drastically reduces the number of required projection images compared to ordinary algorithms [2]. The approach takes advantage of the fact that scan objects in industrial CT consist of fewer materials and often have a simpler geometry than their medical counterpart. Furthermore, industrial objects often provide edges or folds in the surface, which characterize parts of the object clearly. Jinnouchi et al. refer to these folds as *creases*. The creases are reconstructed by the following pipeline: 1) performing edge detection on projection images in order to locate the projection of creases and binarize the result, 2) reconstruction of a tomogram using the reconstruction algorithm published by Feldkamp et al. [3], 3) searching for bright voxels in the tomogram utilizing thresholding, and 4) extracting a coordinate for the crease coordinate by interpolating from the grey values of the voxel and its vicinity [2].

The approach described above successfully reduces the number of necessary projections, while increasing the accuracy of the crease position. However, the approach extracts the crease coordinates from the voxels, which is vulnerable to be blurred or to suffer from partial volume artefacts. The present paper aims to overcome these drawbacks by replacing the tomogram-based reconstruction with a clustering approach, in order to directly reconstruct the crease location in the 3D space.

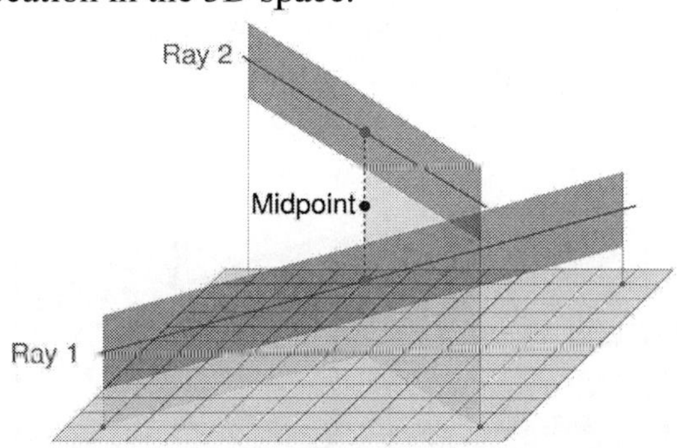

Fig.1. Two rays in 3D space – in the graphic accompanied by their planes, which are orthogonal to the base grid – pass each other rather than they intersect. The dots on the rays mark points on them, which have the closest distance to the other plane. The midpoint marks the center between these two points and has minimum distance to both rays. The midpoint is a good approximation for an intersection of two passing rays.

2 Material and Methods

The subsequent section introduces a vertex reconstruction method, which localizes edge vertices in 3D space. This method uses a raytracing approach and ideally projected vertices from two different scan geometries. Such a reconstruction requires to extract the location of the projected edge vertices from projection images. Section 2.2 describes a methodology to evaluate, if an edge detection algorithm complies with the needs. Furthermore, an experimental setup is described by which the methods are evaluated for two different objects.

2.1 Vertex Reconstruction

This approach calculates the edge vertices from a pair of two projection geometries G_a and G_b and the corresponding projection images. The projection geometries consist of coordinates and orientations of the source s, the detector d, and the scanned object. For this section it is assumed that the location of the projected vertices is ideally extracted from the projection images, resulting in two sets of projected vertices p_{a,n_a} and p_{b,n_b}, with $n_a \in N_a$ and $n_b \in N_b$, where $N_a \in \mathbb{N}$ and $N_b \in \mathbb{N}$ represent the number of projected vertices.

Drawn from the assumption of correctly detected vertices, a vertex of the object's edge must be positioned along the ray path from the source s_a to this projected edge vertex p_{a,n_a}. The same edge vertex will be detected in the projection image acquired under the projection geometry G_b. In an ideal system, a second ray from the second source position s_b to p_{b,n_b} will intersect with the previously mentioned ray at the location of the object's edge vertex. However, this assumption will not hold if the discretization of the detector limits the accuracy of the edge detection. Due to this and additional numerical inaccuracies during the intersection calculation it is unlikely that two rays in 3D space will intersect. Therefore, the algorithm does not check for line intersections, but calculates a midpoint between the lines. This midpoint is located at the center between two points, where both of them are positioned along one of the rays and closest to the other ray (compare Fig. 1). From this definition, it can be derived that the distance between these two points is minimal for any two points from two different rays, as long as the rays are neither parallel nor colinear. The midpoint between these two points has the same distance to each of the rays, indicating the position with the smallest distance to both rays. Thus, it is the point that comes closest to an intersection. Therefore, throughout the text, this coordinate is also referred to as *ray intersection*. Under the assumption that the projected edge vertices $p_{a,0}$ and $p_{b,0}$ originate from the same point on an object edge, the ray from the source s_a to the projected edge vertex $p_{a,0}$ and the ray from the source s_b to the projected edge vertex $p_{b,0}$ are calculated (see Fig. 2). The midpoint between the two rays, which indicates the measured location of an object edge, is calculated.

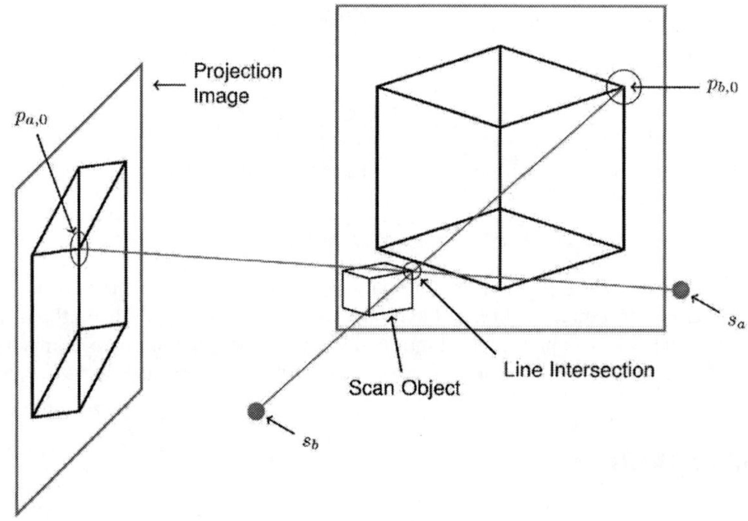

Fig. 2. The cube (5) represents a scan object, which is scanned under two geometries. The midpoint between two closest rays from source (1) to the projection of the edge vertex (3) in the projection (2) marks a potential edge location (6).

Given two different scan geometries, the information on corresponding projected edge vertices is not available. However, it is likely, that the rays of projection vertices that originate from the same edge vertex will pass by closely. Therefore, for each projected vertex of the scan geometry G_a a ray r_{a,n_a} is calculated. Furthermore, for every r_{a,n_a}, its midpoint to every ray $r_{b,n_b}$of the scan geometry G_b is computed. Only the midpoint with the smallest distance to its rays is stored from the whole set of calculated midpoints. This coordinate is assumed to indicate an object edge vertex.

2.2 Detection of Projected Vertices

The positions of the projected vertices need to be extracted from the projection images. Strong changes in the surface curvature result in steep gradients in the gray values of neighboring pixels in the projection image. Therefore, edge detection algorithms might be suitable to localize pixels containing the projection of object edges. The Canny edge detection algorithm is used exemplary for the detection of projected edge vertices [4]. Therefore, a projection image is acquired under a certain scan geometry using the method described in section 2.3.1. The acquired projection image is filtered with the Canny edge algorithm using an upper threshold of 31.25% and a lower threshold of 7.81% based on the initial maximal gray value of the image. Those pixels, which are marked as 'edges' are converted to coordinates. The vertices of the object are ideally projected onto the detector, as described in the section 2.3.2. The coordinates from the edge detection and the coordinates of the projected vertices will be compared for similarity as described in the section 2.4.2.

2.3 Experimental Setup

The underlying methodology of data acquisition for the *vertex reconstruction* and the *detection of projected vertices* are similar and therefore described in this section in detail.

A tomogram-less approach, which aims to reconstruct the edges of an object demands for an accurate localization of these edges in the projection images. In order to evaluate the suitability of an edge detection algorithm to accomplish this task, detected edges must be compared with the location of the ideally projected vertices (see section 2.3.4). Inaccuracies between the scan geometries of the radiographic image and the ideal vertex projection may result in distinct systematic errors. The proposed procedure avoids this error source by relying on simulated radiographic images and projecting the vertices of the digital phantom with the same scan geometry. In this way, the ideally projected vertices can be superimposed on the projection image. The simulation of the CT scan is described in section 2.3.1, whereas the projection of the ideal edge vertices is characterized in section 2.3.2. An explanation of the used scan trajectory concludes the experimental setup.

2.3.1 Simulated CT Scan

A virtual projection image of the step-wedge and the step-cylinder is performed using the program aRTist 2.10 [5]. It calculates the x-ray path from the focus of the source through the object to the detector elements. Based on the material density, the x-ray energy, the exposure time, and by simulating noise as well as scatter the program calculates projection images. In this first study, the simulation complexity was reduced by choosing a very low noise level and a point source in favor of a high-quality image. For each scan geometry defined by the trajectory in section 2.3.3, the source, object and detector were positioned, and a projection image was calculated.

2.3.2 Ideal Vertex Projection

The edge vertices v, which define the structure of the object, are given in 3D homogeneous coordinates. These homogeneous coordinates can be projected onto a plane, or, in this case, onto the detector by multiplication with a projection matrix designed according to the scan geometry. Along the central beam from the source to the detector different vertices can be positioned either closer to the source or to the detector. Due to the cone beam characteristic, these differences in the position lead to different magnifications of the points. The projection matrix takes these facts into account and projects the vertices onto the detector so that the source, the object's vertices and the coordinate of the projected vertices are colinear. For each scan geometry (see section 2.3.3) the vertices of the object are projected onto the detector plane. In Fig. 5 a) and b) the projected vertices of the ideal model are superimposed to the vertices extracted from the projection image acquired using aRTist; as described above.

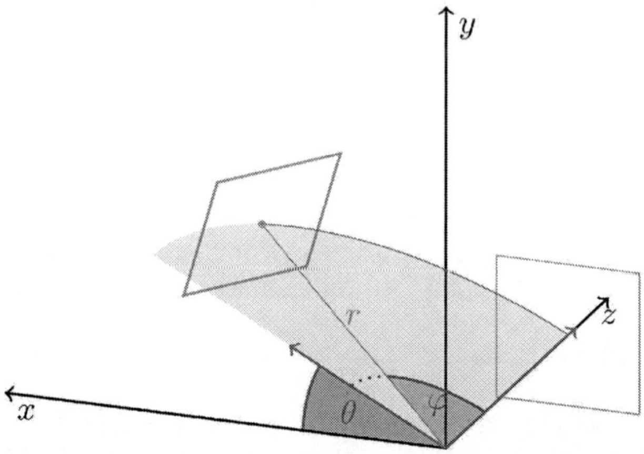

Fig. 3. Spherical coordinates are defined by an azimuthal angle θ, by a polar angle φ and by a radius r. When the centre of the scanned object is located at the origin and r is set to the object-detector-distance these coordinates can be used to rotate a source and a detector about the origin of an object. Like standard circle trajectories the detector is rotated by a read-out angle, or in this case the polar angle. The azimuthal angle however lifts the rotation plane out of the x-z-plane.

2.3.3 Trajectory

Whereas in most industrial CT use-cases the source and detector are steady with a displaceable object or manipulator, the trajectory used in this study positions the source and the detector relatively to a steady object. This definition will support the description of the *vertex reconstruction.* The trajectory is based on a stable system geometry with the same distance for the focus-object-distance and for the object-detector-distance. This results in a magnification of 2. The source and the detector are rotated about the object center based on spherical coordinates and extend a standard circular trajectory to with out of plane rotations. The final orientation of source and detector is determined by the azimuthal and the polar angle. The azimuthal angle lifts the rotation plane out of the x-z-plane. Further, the source and the detector are rotated around the object along a circular trajectory by the polar angle in the plane defined by the azimuthal angle. Fig. 3 depicts, the azimuthal angle $\theta \in [0,180)$ and the polar angle $\varphi \in [0,360)$.

The vertex reconstruction calculates the edge vertex positions based on scan geometry pairs (G_a and G_b), which are defined by the azimuthal and the polar angle. Table 1 lists 5 different parameter sets, for which the reconstruction is evaluated.

Table 1. Randomly chosen azimuthal and polar angles which define distinct scan geometries based on spherical coordinates.

Experiment Number	Scan Geometry			
	G_a		G_b	
	Azimuthal Angle in Degrees	**Polar Angle in Degrees**	**Azimuthal Angle in Degrees**	**Polar Angle in Degrees**
1	138.191	145.50	29.88	177.31
2	19.62	110.41	144.39	80.20
3	40.50	4.03	137.77	5.17
4	1.36	183.61	68.78	50.29
5	126.56	84.72	131.64	16.02

2.3.4 Scanned Objects

The evaluation was performed on two different digital phantoms. A step-wedge and a step-cylinder were designed. Both objects provide areas with a strong surface curvature in a narrow region compared to the rest of the object. The step-wedge (compare Fig. 4) only consists of planar faces and the strong edges in the surface along the corners. The step-cylinder additionally provides areas with in-between curvature along the circular shape. The models were specified in such a way, that vertices are only positioned along the strongest surface curvature. Under the use of a virtual flat-panel CT geometry, the projected positions of these vertices, and thereby the location of the surface-edges, on a projection image can be determined. In order to evaluate the methods described below with sub-pixel precision, the vertices along the surface edges are positioned very densely. The number of vertices was chosen in such a way, that adjacent vertices in the 3D phantom are projected either into the same detector element or into directly adjacent detector elements. The strategic and narrow positioning of the vertices for both test objects is represented in Fig. 4. The vertices are represented by small orange spheres.

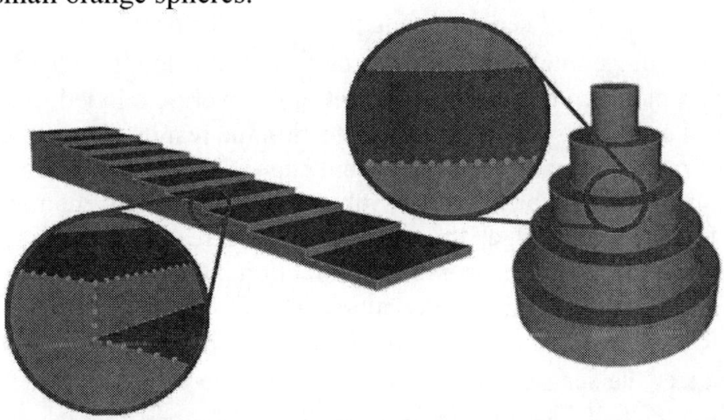

Fig. 4. A digital step-wedge and a step-cylinder are designed. The vertices (orange spheres) are densely positioned at the edges of objects, which is visualized in the magnified areas.

2.4 Evaluation

The evaluation method of the vertex reconstruction and the detection of the projected vertices is characterized in the following sub-sections.

2.4.1 Vertex Reconstruction

The section 2.1 describes a method to reconstruct edge vertices. In order to evaluate this approach under the condition of ideally detected projected vertices, scans of a step-wedge and a step-cylinder are simulated with each projection geometry pair (G_a, G_b) listed in Table 1. Afterwards, the object edges are reconstructed using the vertex reconstruction. The Hausdorff distance is calculated between the point cloud of reconstructed edge vertices and the vertices of the digital reference model [6]. This measure calculates the minimum distance for all points in a point cloud U to any point of a point cloud V. The mean or the maximum distance of all the points in U to all the points in V is a measure for the similarity of the point clouds, whereby a small maximum distance indicates high similarity. As the Hausdorff distance from point cloud U to point cloud V may differ from the results of the Hausdorff distance from point cloud V to point cloud U, both alternatives are evaluated.

2.4.2 *Detection of Projected Vertices*

The methods described in section 2.2 result in two point clouds, of which each point represents a projected edge vertex. The Hausdorff distance is used to evaluate the similarity between ideally projected vertices and the vertices extracted from the projection images. This procedure is performed for the step-wedge and the step-cylinder with the 10 different scan geometries defined in Table 1.

3 Results and Discussion

Fig. 5 shows the mean and maximum values of the Hausdorff distances for the performed experiments. The following sections describe the results for the *vertex reconstruction* and the *detection of projected vertices* separately.

3.1 *Vertex Reconstruction*

The Hausdorff distances of the vertex reconstruction result for the step-wedge are shown in Fig. 5 c). For the scan geometry pairs $G_1^{a,b}$ and $G_5^{a,b}$ the maximum distances from the *reconstructed* point cloud to the reference point cloud were excluded as the reconstructed point cloud showed a single isolated outlier, distorting the results. These outliers most likely emerged from two rays, originating from different edge vertices, bypassing each other in near vicinity. The mean distance, which is more robust to outliers, is in the range of 0.42 pixels and 1.51 pixels for the step-wedge. Fig. 5 d) shows the results for the step-cylinder. The Hausdorff distances for both measurement directions (reference to reconstructed and reconstructed to reference) have the same values. Therefore, all reconstructed vertices were closest to their ideal edge vertex counterpart. Comparing the Hausdorff distances between c) and d) in reference to the scan geometries reveals, that the results of $G_1^{a,b}$ and $G_5^{a,b}$ deviate from the results of $G_2^{a,b}$, $G_3^{a,b}$ and $G_4^{a,b}$ similarly. This indicates, that the pairing of the scan geometries has an impact on the reconstruction quality. Overall, the mean value of the Hausdorff distance from the ideal edge vertices to the reconstructed vertices is within pixel accuracy and therefore comparable to tomogram-based approaches [2].

3.2 *Detection of Projected Vertices*

Fig. 5 e) and f) show the similarity between the ideally projected vertices and the vertices detected by the Canny edge algorithm. For the step-wedge (see Fig.5 e) no significant differences between the distance measurement direction can be observed. The maximum difference is around 2.0 pixels, whereas the mean difference is between 0.72 pixels and 0.82 pixels. These stable results indicate, that in the vicinity of a detected edge a real and ideally projected vertex is located. The demand of the reconstruction for accurately detected vertices is therefore fulfilled.

The maximum distances for the step-cylinder, especially for the distance from the Canny edge-based to the ideally projected vertices is about one order of magnitude higher. The reason for these large distances are the transition areas between the round faces of the cylinder and the background of the projection image. The image processing algorithm detects edges along these faces, whereas the model does not provide vertices here. These additionally found vertices induce the increased error. Future versions of the reconstruction algorithm therefore need to be able to handle false-positive edge vertices.

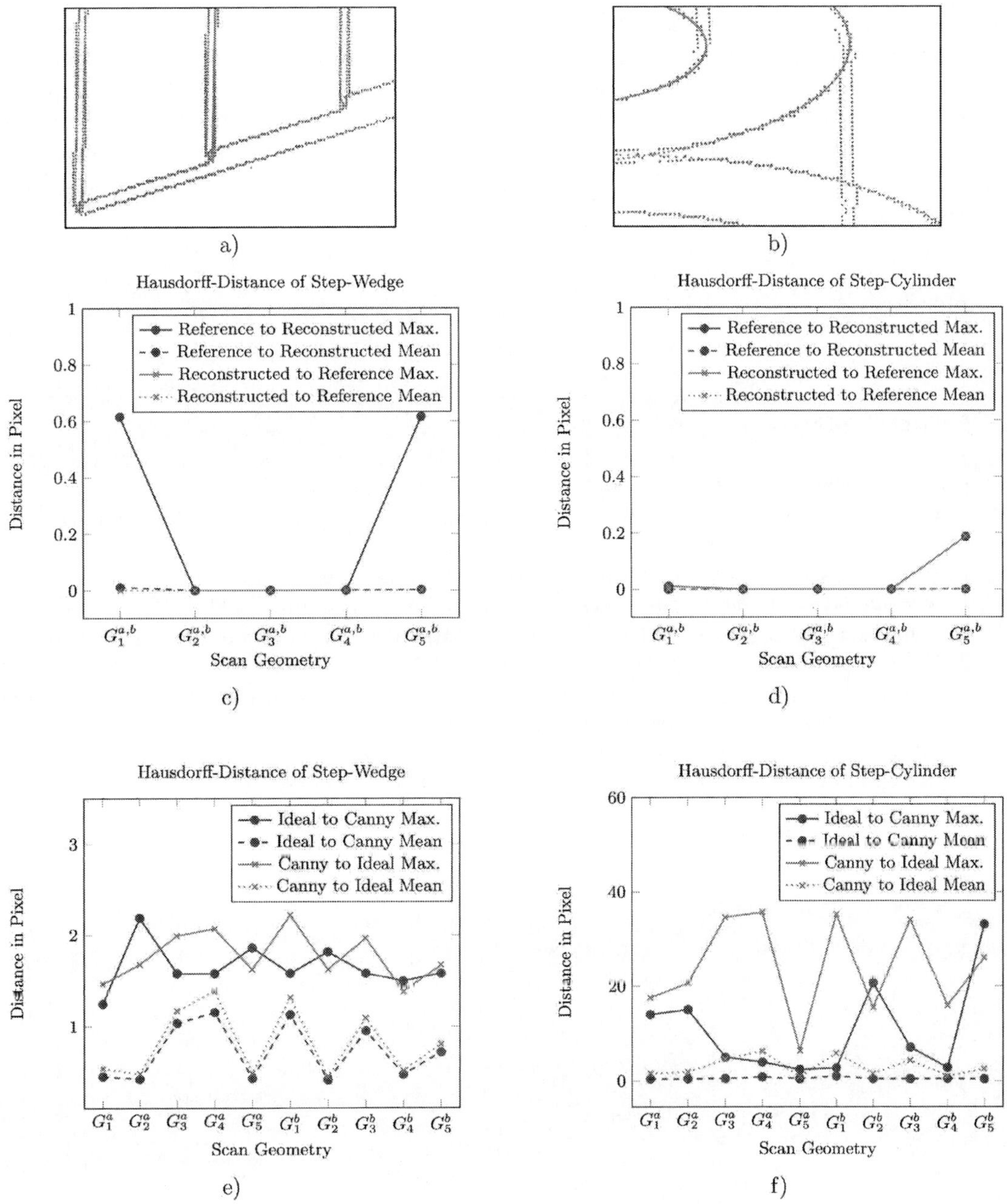

Fig. 5. The figures a, c and e show the results of the step-wedge and the figures b, d and f show the results for the step-cylinder. In a) and b) a section of the ideally projected vertices (orange) and the vertices extracted from the projection image (blue) are superimposed. The plots of c and d show the Hausdorff distances for the edge vertex reconstruction (compare section 2.1). The plots of e and f show the Hausdorff distances between the ideally projected and the extracted vertices (see section 2.2).

4 Conclusion

In this study, it has been shown that the 3D position of an object edge vertex can be reconstructed using a ray-tracing approach with two different scan geometries, if the coordinates of the projected edges are known. Further, it has been demonstrated that edge detection algorithms are suitable to detect the projected vertices within pixel precision. The performance of the projection vertex detection might be improved further by sub-pixel precise edge detection algorithm. The uncertainty in the localization of the projected vertices will lead to inaccuracies in the vertex reconstruction. This drawback might be conquered by reconstructing projection vertices from multiple scan geometries and combining the partial results by a method like the one described in [7]. The two parts of the proposed method showed satisfying results and plausibility for a tomogram-less reconstruction approach. In further studies the two parts will be merged to a fully tomogram-less reconstruction method and compared to state-of-the-art tomogram-based methods.

5 References

[1] T. M. Buzug, Computed Tomography: From Photon Statistics to Modern Cone-Beam CT, Springer, Berlin 2008, pp. 259-261.

[2] R. Jinnouchi, Y. Ohtake, S. Hiromasa and Y. Nagai, "Efficient 3D crease point extraction from 2D crease pixels of sinogram," *Conference on Industrial Computed Tomography,* pp. 439--446, 2014.

[3] L. A. Feldkamp, L. C. Davis and J. W. Kress, "Practical cone-beam algorithm," *Josa A,* vol. 1, no. 6, pp. 612-619, 1984.

[4] P. Bao, L. Zhang and X. Wu, "Canny edge detection enhancement by scale multiplication," *IEEE transactions on pattern analysis and machine intelligence,* vol. 27, pp. 1485-1490, 2005.

[5] C. Bellon and G.-R. Jaenisch, "aRTist - analytical RT inspection simulation tool," *Proc DIR,* pp. 25-27, 2007.

[6] M.-P. Dubuisson and K. Jain, "A modified Hausdorff distance for object matching," *Proceedings of 12th international conference on pattern recognition,* vol. 1, pp. 566-568, 1994.

[7] M. Wattenberg, M. Still, P. Klein and T. M. Buzug, "2D contour reconstruction for industrial computed tomography using crease clusters," *e-Journal of Nondestructive Testing,* vol. 24, 2019.

XRM-II nanoCT – SEM based computed tomography

Fabian LUTTER [1], Kilian DREMEL [2], Jens ENGEL [2], Daniel ALTHOFF [2], Simon ZABLER [2], Nikolas WESTPHAL [3], Randolf HANKE [2]

[1] Chair of X-ray Microscopy LRM, Julius-Maximilians-Universität, Würzburg, Germany

[2] Fraunhofer Development Center X-ray Technology EZRT, Fürth, Germany

[3] ProCon X-Ray GmbH, Sarstedt, Germany

Contact e-mail: fabian.lutter@physik.uni-wuerzburg.de, kilian.dremel@iis.fraunhofer.de, nikolas.westphal@procon-x-ray.de

Abstract. We present the XRM-II nanoCT, a laboratory nano-CT system based on geometric magnification, which is used for material characterization. The prior system, XRM-II, was presented in 2016 and was the starting point for our development. The new system is composed of a scanning electron microscope JEOL-JSM7900, which is used for focusing electrons on a needle shaped tungsten or molybdenum target with a tip diameter of about 80 nm. In this way we generate a X-ray source spot of the same size to accomplish a spatial resolution below 100 nm. In comparison to the prior system the target can be handled independently from the object stage, while maintaining the full functionality of the SEM. Furthermore, we use a photon counting detector with a 1 mm thick CdTe sensorlayer and a resolution of 1290x512 pixels. The XRM-II nanoCT is also capable of element mapping by using a built-in EDS system. Therefore, the system provides multi-modal imaging using three different image-contrasts (SEM, EDS, nano-CT) on the same sample.

The acquisition time for a CT, consisting of 1200 projections, is about 34 hours. By using image correlation to compensate the problem of electron beam drift, the data acquisition is fully automated. Due to the high spatial resolution, sample-drift and positioning inaccuracies need to be compensated during the CT reconstruction. A customized algebraic reconstruction method allows to improve the low signal-to-noise ratio of the CT projections by regularization and to get rid of geometric deviations.

We present details about the experimental setup, high resolution radiographies and reconstruction results of a 3D nano-CT of a metallic alloy.

1. Introduction

Nano-CT resolves structures in the range of a few hundred nanometers to some micrometers. This volumetric imaging technique is important for material scientists, because the characteristics and behaviour of materials is often defined by their microstructure [1]. By the means of electron microscopy, it is possible to examine the surface of a sample with a resolution of a few nanometers, but information from more than few hundreds of nanometers beneath the surface is not accessible. To gain volumetric information of the sample FIB tomography or nano-CT can be used. We presented a laboratory based nano-CT system (XRM-II) in 2016, based on geometric magnification [2]. A spatial resolution in 2D of 3000lp/mm could be achieved with this system, while there was less resolving power in volumetric imaging due to limitations in the signal-to-noise-ratio (SNR). The previous acquisition time of 10-14 days for the computed tomography was quite long and the functional range of the SEM was limited due to the nano-CT setup. In this work we show the improvements made compared to the prior system and give an impression on its functional capabilities by recording an AlCu29 alloy.

2. Materials and Methods

2.1 Principle of geometric magnification

The XRM-II nanoCT system is based on the principle of geometric magnification. The magnification M is given by the source-detector distance z_1 and the source-object distance z_2:

$$M = \frac{z_1}{z_2}$$

The size of the X-ray source is an important point to achieve high resolution [2]. To keep the source spot as small as possible, the electron interaction volume of the reflection target must be reduced to a minimum. A larger spot size will cause a blurring of the image at the same magnification [2].

2.2 X-ray targets

Our approach to reduce the electron interaction volume results in the use of needle shaped reflection targets, which are made of tungsten or molybdenum. We use an electrochemical etching process to create these targets with a tip diameter of less than 100nm [3], [4]. This process is tuned to fulfill the important attributes like shape, stiffness, conductivity and thermal resistance.

A SEM image of a tungsten target is shown in Fig. 1. The electron beam is focused on the very end of the tip, where the diameter is about 80nm. Multiple nano-CTs were performed with this target and it shows no damage caused by the high electron current (~400nA).

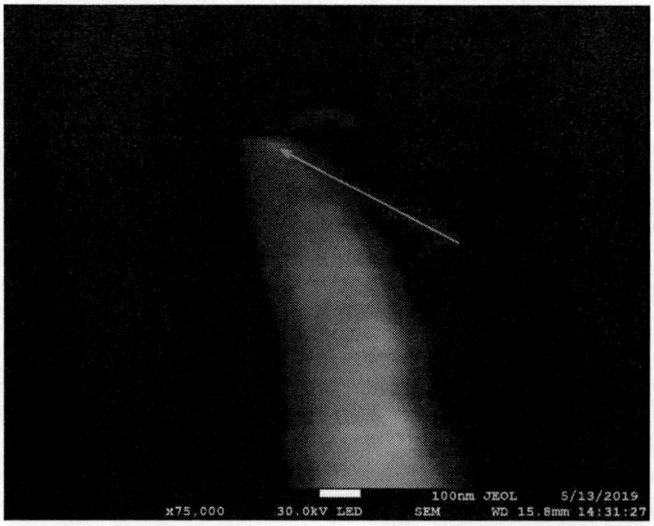

Fig. 1. SEM image of the needle-shaped tungsten target; The green arrow marks the point where the electron beam is focused on to create the X-ray source.

2.3 Experimental setup

The XRM-II nanoCT is based on a field emission scanning electron microscope JEOL JSM-7900. It is capable of high resolution in electron imaging (~0,8nm) and is powerful enough to provide high and stable electron flux (~400nA) to generate X-rays. The most significant difference compared to the prior system is the use of two independent manipulators for target and object. Thus, it is possible to use the stage of the SEM to manipulate the object while maintaining the full functional range of the SEM including the use of the specimen exchange mechanism. For the target manipulator we use a stack of three linear axes, which is attached on the inside of the specimen chamber. As shown in Fig. 2, object and target are inside the vacuum chamber of the SEM, while the X-ray detector is placed outside the microscope behind a 250µm thick Beryllium window. The detector is a photon counting ADVACAM WidePIX with a 1mm thick CdTe sensor layer and a 1290x512 pixel matrix with a pixel size of 55µm. This offers a field of view of 100µm with 80nm sampling. Furthermore, an EDS detector is attached to get information of the sample's elemental composition.

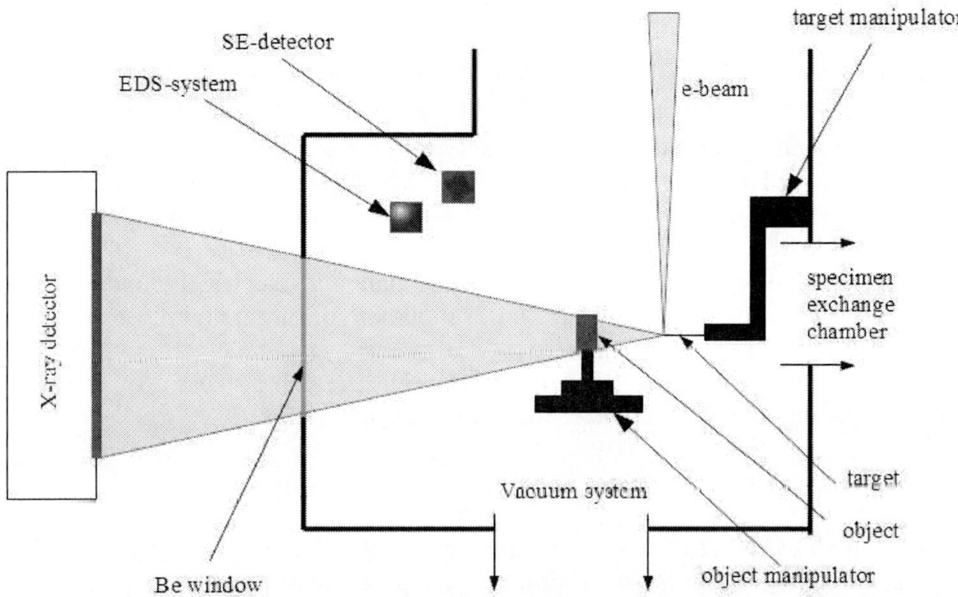

Fig 2. Schematic representation of the specimen chamber; Compared to the prior system, object and target manipulator are now separated, which makes the system easier to use and keeps the complete functional range of the SEM available.

The 30kV electron beam is focused on a needle-shaped tungsten tip with tip diameters less than 100nm (see Fig. 1). This creates an X-ray spot of the same size. The resulting X-ray spectrum, collected with the EDS-system, is shown in Fig. 3. The main peaks are the fluorescence lines of tungsten at $W_{L\alpha}$=8,40keV, $W_{L\beta}$=9,67keV, $W_{L\gamma}$=11,29 keV [5].

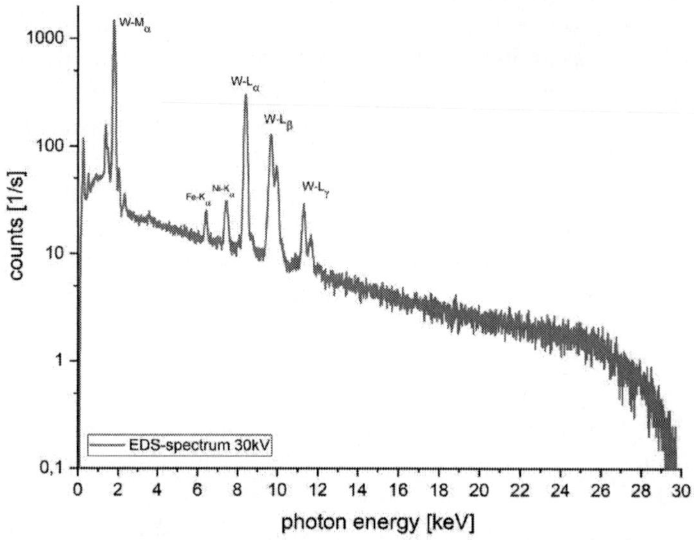

Fig. 3. X-ray Bremsspektrum of the tungsten tip; The peaks originate from the Wolfram fluorescence lines. Photons lower than 5keV don't reach the detector due to the Beryllium window.

The remaining lines of iron and nickel originate from a secondary X-ray spot on the sample stage caused by the transmitted electrons, which were not absorbed by the target. The intensity below 5keV is almost completely cut off by the Be-window, which is why the lower energy threshold of the detector is set to the this energy.

Charging effects of the sample and changing electrical fields during stage movements, especially rotations, cause a misalignment of target and electron beam, which leads to a breakdown of the X-ray intensity. To prevent this, the electron beam is realigned by using image correlation of SEM images during the CT process [6]. Hereby, the CT procedure was fully automated and requires no user interaction.

2.4 Volume data reconstruction

The 3D volume of XRM-II nanoCT data is computed by a custom designed software (see also [2]), which includes a consideration of the typical measurement setup, a compensation for mechanical instabilities and a 3D phase-contrast filter. Since the photon flux is low due to the small focal spot size and the signal-to-noise ratio (SNR) is critical, a regularized reconstruction method is needed to achieve good results. Iterative volume reconstruction gives the possibility to not only include regularization but also deal with the non-equidistant angular setup of the XRM-II nanoCT. We use a total-variation (TV) [10] regularized Simultaneous Algebraic Reconstruction Technique (SART) [7] reconstruction algorithm. Moreover, geometric deviations occur due to the high magnification of the system. Since smallest mechanical instabilities and thermal drift of a few nanometers lead to non-negligible errors in the geometry, an image-correlation based correction compares the forward-projected volume image to the measured data in order to detect sample shifts in the detector plane (see also [11], [8]).

After the reconstruction a Paganin based phase contrast filter [9] is applied in the volume space to get images with better contrast. To deal with the intrinsic blurring by the Paganin Filter, a second deconvolution filter is used.

3. Results

In order to demonstrate the achievable resolution in 2D, we used a Siemens star as a test pattern. The magnification for this measurement was set to 5500x corresponding to a sampling of 10nm per pixel (see Fig.4). The lines of the ring numbered as 3 have a feature width between 176nm and 80nm. These features are clearly resolved, while smaller lines in the following inner ring are also visible.

Fig. 4. Radiography of Siemens star recorded with a Magnification of 5500x or 10 nm pixel sampling and an integration time of 30 minutes. Details down to 80nm are clearly distinguishable, while smaller features are also visible.

We chose an AlCu29 alloy as our sample to analyse with the system and compare the results to our previous measurement of this alloy. Furthermore, its internal structures are in the range of the system's achievable resolution and provide a good contrast for X-ray imaging. First, we recorded a SEM image of the sample (Fig. 5). The grooves on the surface of the sample originate from the preparation with fine sandpaper as well as the light particles on the top.

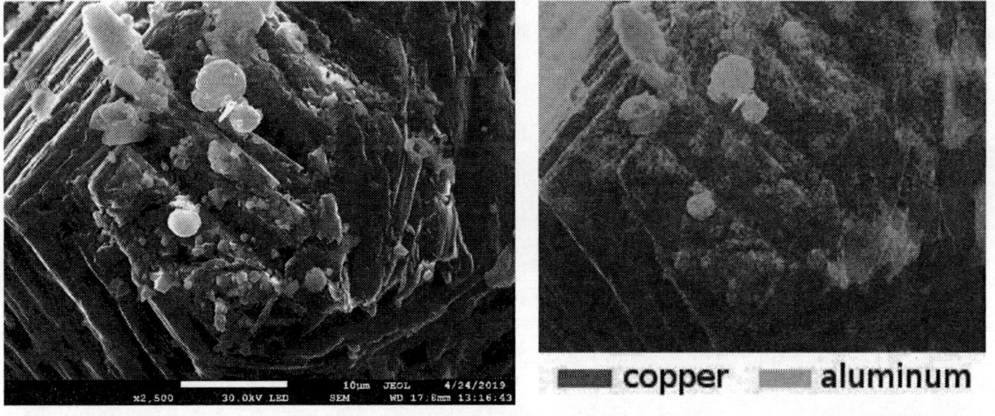

Fig. 5. SEM and EDS image of the sample: (left image) The bright particles are remnants of sandpaper used during the preparation. The light and dark areas on the upper right of the sample are two different metallic phases of the alloy. (right image) In the map, coloured by the EDS-system, aluminum is represented in turquoise, while copper is purple. Both phases are visible on the sample surface. The spheres on top were identified as silicon but excluded from the mapping. The dark areas in the lower left part are caused by the conical shape of the sample shielding the EDS-detector.

We performed an EDS mapping to determine the elemental distribution of the sample. Fig. 5 shows the expected metals (aluminum and copper) and the remnants of sandpaper (silicon dioxide).

The subsequent nanoCT-scan was performed with 1000 projections 100s exposure time each and a magnification of M=585. In total, the acquisition time was about 34 hours.

Slices of the reconstructed and the Paganin-filtered volume (94nm voxel size) as well as line plots of the eutectic structure are shown in Fig. 6.

The Al-Al2Cu eutectic phase with a periodicity of about 1µm is clearly visible. Furthermore, aluminium particles with a size of 20-30µm are present. The finest lamination of the intermetallic phase is about 350nm large. The small sphere on top of the sample, which was identified as Silicon dioxide by the EDS, is also visible in the reconstructed volume as well as the grooves on the surface caused by sample-preparation with sandpaper.

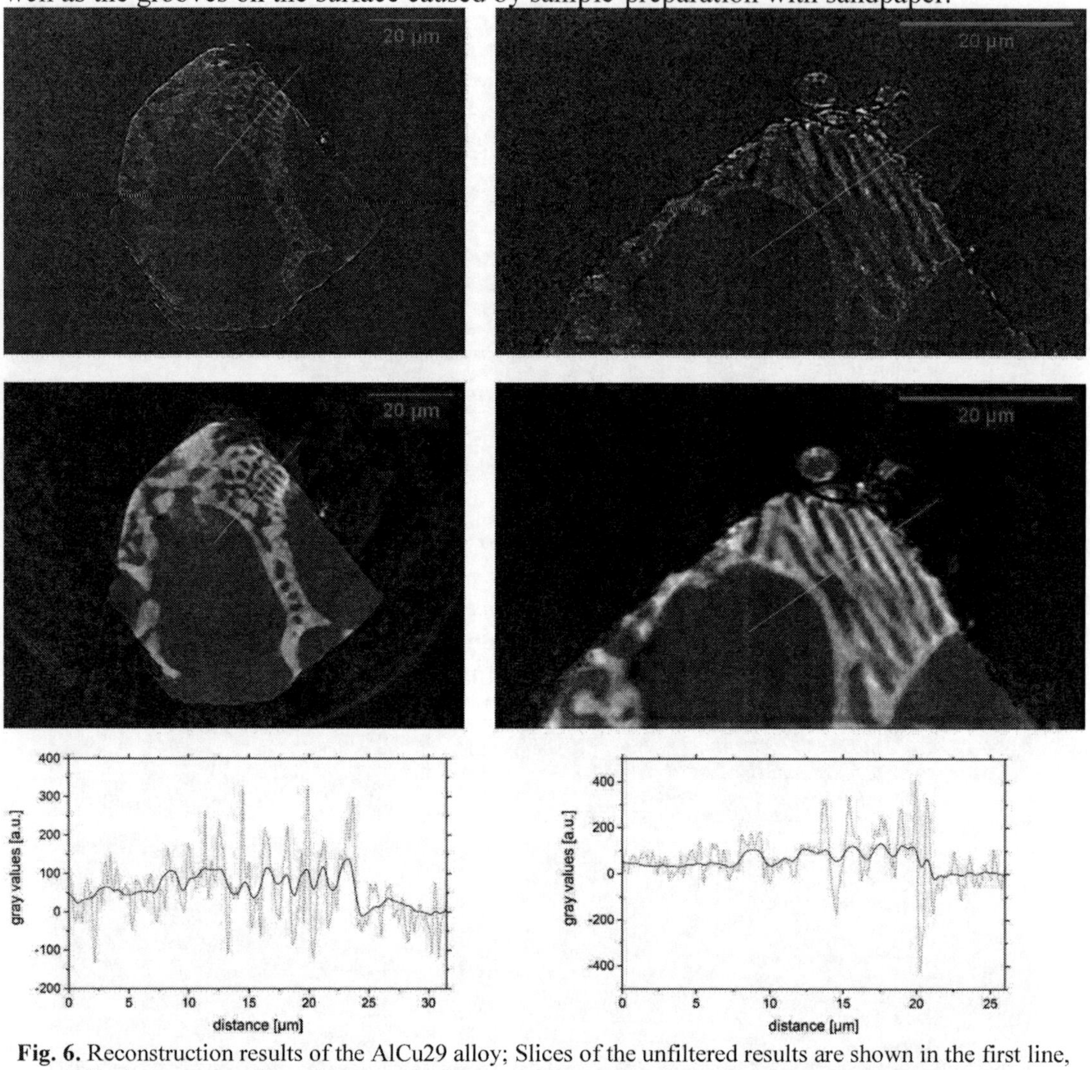

Fig. 6. Reconstruction results of the AlCu29 alloy; Slices of the unfiltered results are shown in the first line, while the Paganing-filtered ones are in the second line. In the line plots along the green lines, phase overshoots can be seen in the unfiltered image (red dashed line), while the filtering smooths the results (black solid line).

We used the FWHM of the derivative of the line profile at the eutectic microstructure to estimate the line spread and the corresponding resolution. For the unfiltered volume, we find a line spread between 1,5 and 2,7 pixels corresponding to a resolution of 135nm – 255nm. In the phase retrieved volume the line spread is between 3 and 4 pixels, which equals to a resolution of 270nm – 360nm. The unfiltered volume offers the better resolution, while the Paganin-filtered one has an improved image contrast.

4. Conclusions and Outlook

The XRM-II nanoCT system provides three different image contrast mechanisms of the same sample. The SEM- and EDS-image can be used for high resolution surface analysis and determination of the constituting elements, while the nano-CT provides additional volume information of the sample.

Due to our improvements in comparison to the prior XRM-II system, it became much faster and easier to use. The total acquisition time for a nano-CT was shortened from 10-14 days to 1,5 days by using a significant higher electron beam current due to optimized parameters of the SEM. By using the alignment between electron beam and target, the acquisition for the computed tomography is now fully automated. The use of two independent motor stages for object and target made the system more user-friendly and maintained the full functionality of the SEM, especially the built-in exchange mechanism. The usable magnification range in X-ray-mode varies from 100x to 1000x (550nm – 55nm voxel size) for computed tomography and up to 5500x (10nm pixel size) for radiographic imaging.

The XRM-II nanoCT system can be used to characterize metallic alloys, but is also capable of examining organic samples or micro electronic devices. Organic samples must be coated with a few nanometer thick layer of gold or carbon to minimize charging effects and electron beam drift.

There are already two systems running and in use in Saarbruecken at the ZRM and in Fuerth at the Fraunhofer EZRT. Both are distributed by ProCon X-Ray GmbH, while a third one is located at the chair of X-ray microscopy in Würzburg using the SEM of the prior system with the new manipulator setup.

In the future it will be possible to further shorten the acquisition time for the CT by increasing the beam current. The currently used JSM-7900 supports currents up to 1µA, which would increase the X-ray intensity by a factor of 2. Furthermore, the design of the target manipulator allows the mounting of up to 4 different targets. In this way, it will possible to switch the target material without venting the specimen chamber of the SEM.

References

[1] C.H.McMurtry, W.D.G.Boecher, S.G.Seshadri, J.S.Zanghi, J.E.Garnier "Microstructure and material properties of SiC-TiB_2 particulate composites. United States: N. p., 1987. Web.

[2] P.Stahlhut, K.Dremel, J.Dittmann, J.M.Engel, S.Zabler, A.Hoelzing, R.Hanke "First results on laboratory nano-CT with a needle reflection target and an adapted toolchain", Proc. SPIE 9967, Developments in X-Ray Tomography X, 996701 (3 October 2016); doi: 10.1117/12.2240561

[3] M.Fotino, "Tip sharpening by normal and reverse electrochemical etching", Rev. Sci. Instrum 64, 159-167 (1992)

[4] W.-T.Chang, I.-S.Hwang, M.-T.Chang, C.-Y.Lin, W.-H.Hsu, J.-L.Hou "Method of electrochemical etching of tungsten tips with controllable profiles," ", Rev. Sci. Instrum 83, 083704 (2012)

[5] A.Thompson, D.Attwood, E.Gullikson, et. Al. "X-ray data booklet", 2009

[6] F.Lutter "Anwendungen für ein Laborbasiertes Nano-CT System," Master Thesis Universität Würzburg (2017)

[7] A.H.Andersen and A.C.Kak, "Simultaneous algebraic reconstruction technique (SART): a superiorimplementation of the ART algorithm," Ultrasonic imaging 6.1, 81-94 (1984)

[8] S.Mayo, P.Miller, D.Gao and J.Sheffield-Parker, "Software image alignment for X-ray microtomography with submicrometre resolution using a SEM-based X-ray microscope," Journal of Microscopy 228.3, 257-263(2007)

[9] D.Paganin, S.C.Mayo, T.E. Gureyev, P.R Miller and S.W.Wilkins, "Simultaneous phase and amplitude extraction from a single defocused image of a homogeneous object," Journal of microscopy 206.1, 33-40 (2002)

[10] E.Y.Sidky, C.M.Kao and X.Pan, "Accurate image reconstruction from few-views and limited-angle data in divergent-beam CT," Journal of X-ray Science and Technology 14.2, 119-139 (2006)

[11] K.Dremel, D.Althoff, and S.Zabler, "CT alignment correction in iterative reconstruction methods," Proc. 4thIntl. Mtg. on image formation in X-ray CT 40, 137 (2016)

Improved Visualization of Polymer Foams using Talbot-Lau Grating Interferometry to Reduce Metal Artifacts

Jonathan GLINZ [1], Sascha SENCK [1], Michael REITER [1], Andreas SCHREMPF [2], David FÜRST [2], Johann KASTNER [1]
[1] University of Applied Sciences Upper Austria, Stelzhamerstraße 23, 4600 Wels, Austria
[2] University of Applied Sciences Upper Austria, Garnisonstraße 21, 4020 Linz, Austria

Contact e-mail: jonathan.glinz@fh-wels.at

Abstract

The sufficiency of bone fixation via ceramic or metallic implants largely depends on the osseointegration at the bone-implant interface and is of major importance in trauma, orthopedic, and maxillofacial surgery. Preclinical imaging via micro computed tomography (XCT) is the state-of-the-art solution to study the ingrowth of mineralized tissue at the bone-implant interface and to investigate the influence of innovative materials on the regeneration of bone tissue.

However, a major challenge in XCT imaging is impaired image quality due to metallic objects such as implants and orthopedic screws as they can cause severe image artefacts characterized by bright and dark streaks due to scattering and beam hardening effects. These artefacts can severely reduce image quality, impeding the quantification of trabecular structures and therefore reducing the diagnostic value of the data. In this paper we present multi-modal image data acquired using Talbot-Lau grating interferometer (TLGI) XCT visualizing the interface between surgical screws and artificial trabecular bone. Our findings reveal that differential phase contrast (DPC) is less prone to metal artefacts improving the visualization of microstructure in bone adjacent to metal components.

For this purpose, we produced polymeric foam structures imitating the microstructure of human bone to systematically investigate the advantages of DPC imaging for multi-component applications. We created a realistic set-up of four synthetic foam cylinders produced with different filler materials representing artificial bone. Two to four titanium screws (thread diameter 2 mm, length 20 mm) were applied into the artificial bone parallel to each other. The specimens were investigated using a TLGI-XCT system (SkyScan 1294) to extract information in relation to absorption contrast (AC), DPC, and dark-field contrast (DFC). In AC, metal streaking artefacts are severe, covering microstructure between the screws completely. DPC images in return show almost no streaking artefacts, which enhances image segmentation and volume rendering between the screws and close to the screw surface.

Improved visualization of polymer foams using Talbot-Lau grating interferometry to reduce metal artifacts

FH OBERÖSTERREICH

Jonathan Glinz, Sascha Senck, Michael Reiter, Andreas Schrempf, David Fürst, Johann Kastner

University of Applied Sciences Upper Austria (Austria)

Motivation

FH OBERÖSTERREICH

❖ **Visualizing metal implants & osseointegration**

- Ingrowth of mineralized tissue at the interface of bone and implant

  → effects of new materials (metal or ceramic), coatings, and surface treatments on the regeneration of bone tissue

- BUT: implants can cause serious **image artifacts**

  → bright and dark streaks due to beam hardening effects and scattering in laboratory XCT

Improving the visualization of microstructural details in artificial bone adjacent to metal components

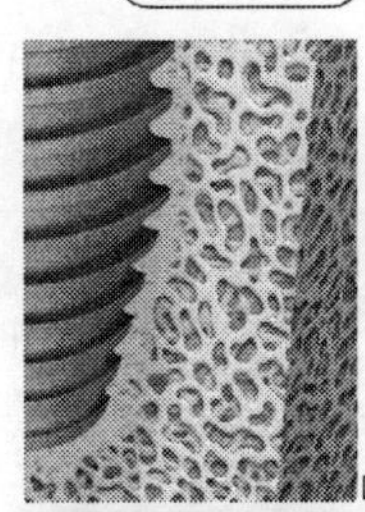

[1]

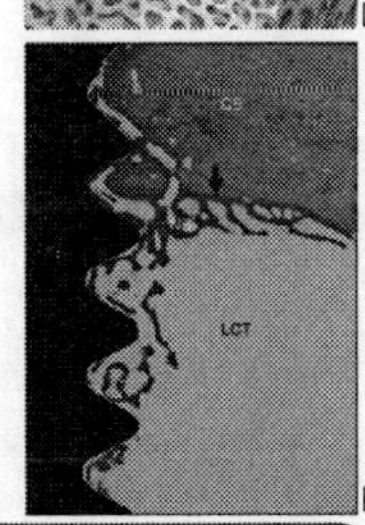

[2]

Methods

❖ Talbot-Lau Grating Interferometer XCT (TLGI-XCT)

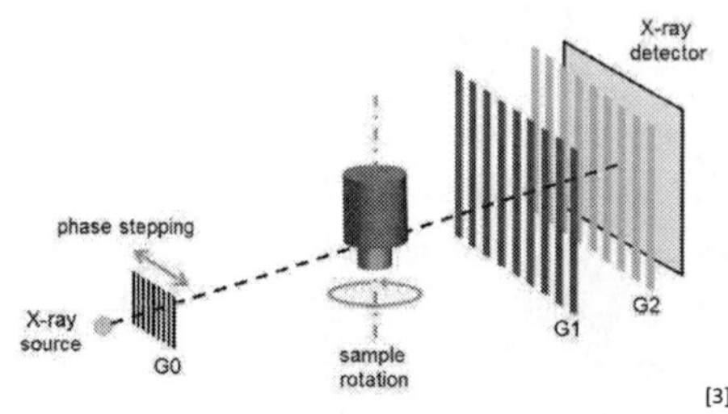

[3]

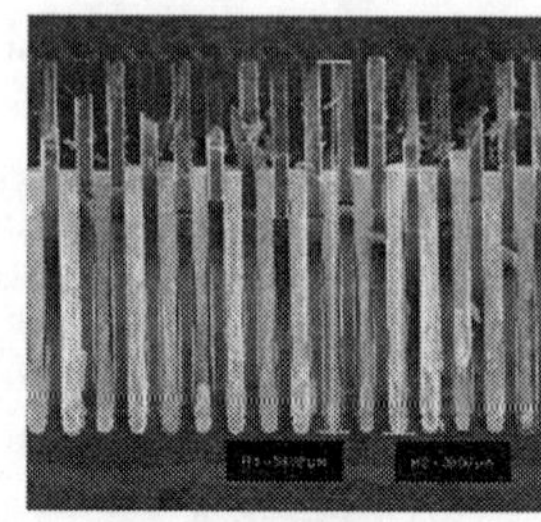

[4]

Absorption grating

- **Source grating (G0)**: creates spatially coherent X-rays & moves sideways for phase stepping
- **Phase grating (G1)**: creates interference pattern
- **Absorption grating (G2)**: converts signal into intensity modulation

Motivation

❖ Talbot-Lau Grating Interferometer XCT (TLGI-XCT)

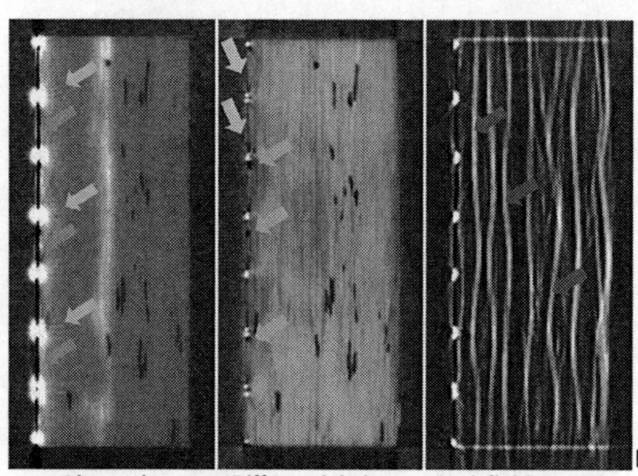

[3]

Absorption contrast (AC) | **Differential phase contrast (DPC)** | **Dark field contrast (DFC)**

Example:

Carbon fiber reinforced polymer (CFRP) sample with a copper mesh

- **AC → artifacts:** Dark bands **between dense objects &** streaking artefacts
- **DPC →** less artifacts**:** details **at metal/polymer interface visible**
- **DFC → visualization of** carbon fibers

Methods

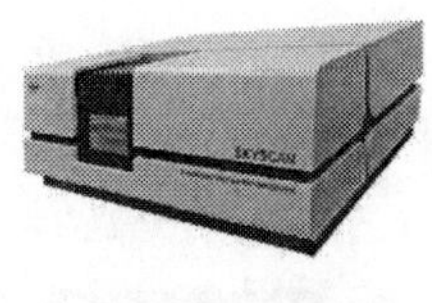

Skyscan 1294

- 60 kV micro-focus X-ray tube
- 4000 x 2672 CCD camera
- Source – camera: 286.0 mm
- Voxel size: 5.7 μm (binning: 11.4 μm, 22.8 μm)
- Sample size: 20 mm (width), 50 mm (height)

GE phoenix Nanotom s

- 180 kV Sub-micro-focus X-ray tube
- 2304 x 2304 pixel detector
- Min. voxel size 0.3 μm
- Sample size: 68 mm (width), 150 mm (height)

DIR 2019 Page 5

Materials

❖ **Samples**

- Polyurethane resin & *mineral filler*
- CMF screws (Stryker): 2 x 20 mm self-tapping screws

1) **Sample 1**: $3Ca_3(PO_4) \bullet CaO$
2) **Sample 2**: $CaCO_3$
3) **Sample 3:** $BaSO_4$
4) **Sample 4:** No filler

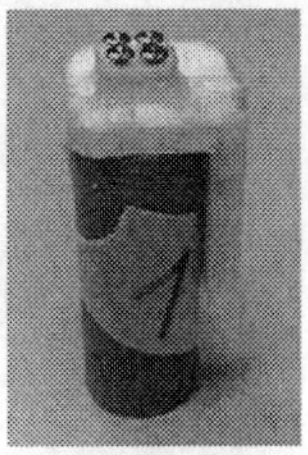

Synthetic foam & screw guide

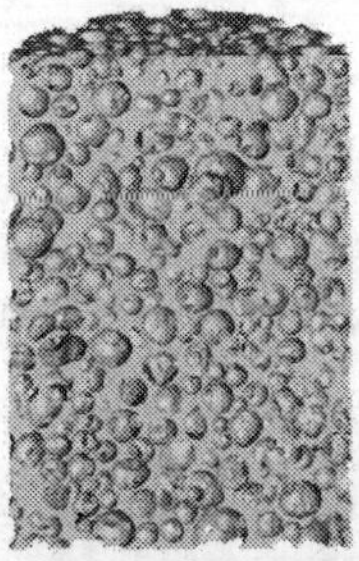

CT volume rendering example

Synthetic foam & Ti screws

DIR 2019 Page 6

Materials

❖ Scan parameters

- Voltage (U) in kV
- Current (I) in µA
- Integration time (t_i) in ms
- Number of projections (N_{proj})
- Averaging (Avg)
- Voxel size (VS) in µm for TLGI-XCT and XCT

	U	I	t_i	N_{proj}	VS	Avg	Filter	Scan time
TLGI-XCT (Skyscan)	50	1000	440	1800	22.8	10	Cu (0.045 mm)	607 min
XCT (Nanotom)	100	110	600	1600	10	5	Cu (0.1 mm)	99 min

❖ Image processing

- Absorption contrast (AC) images: BHC, median filter
- Differential phase contrast images (DPC): non-local means filter
- Dark field contrast (DFC): median filter

DIR 2019 Page 7

Results: Sample 1

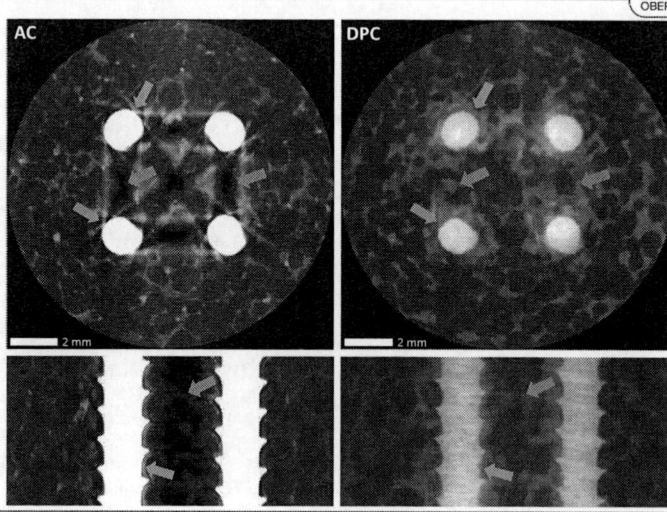

DIR 2019 Page 8

Results: Sample 2

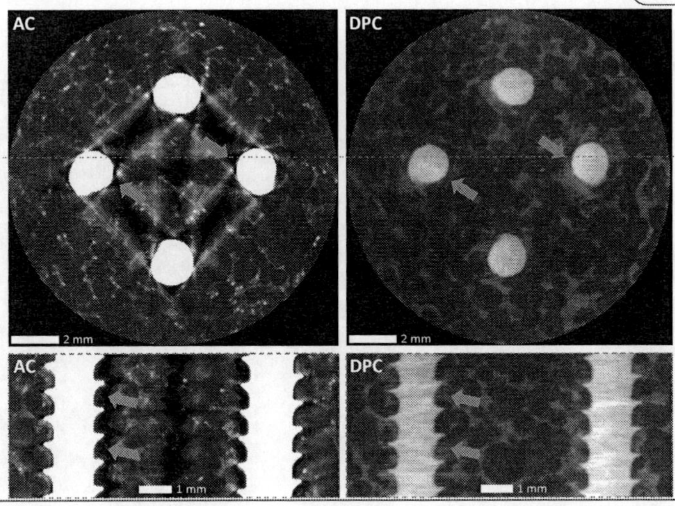

DIR 2019 Page 9

Results: Sample 2

❖ **Comparison Skyscan - Nanotom**

SkyScan – 22.8 µm VS

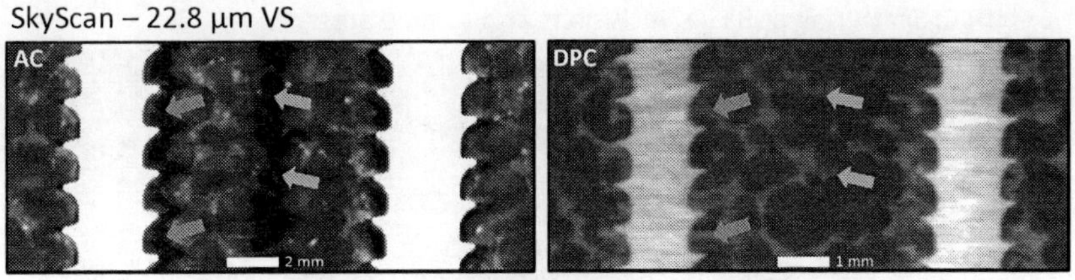

Nanotom – 10 µm VS

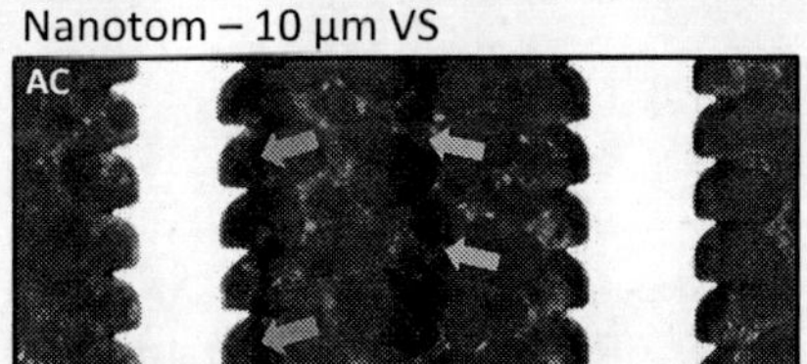

DIR 2019 Page 10

Results: Sample 3

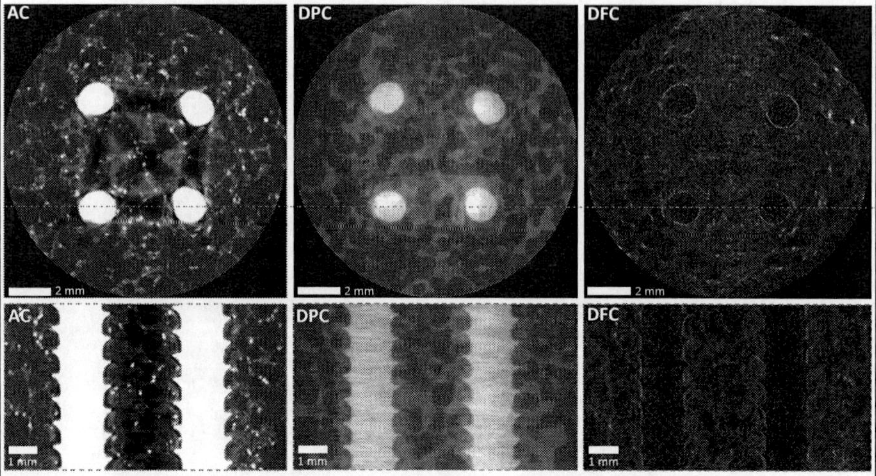

DIR 2019 Page 11

Results: Sample 3

- **DPC:** Foam structure
- **AC:** High absorbing $BaSO_4$ inclusions
- **DFC:** Scattering at $BaSO_4$ inclusions and sample surface

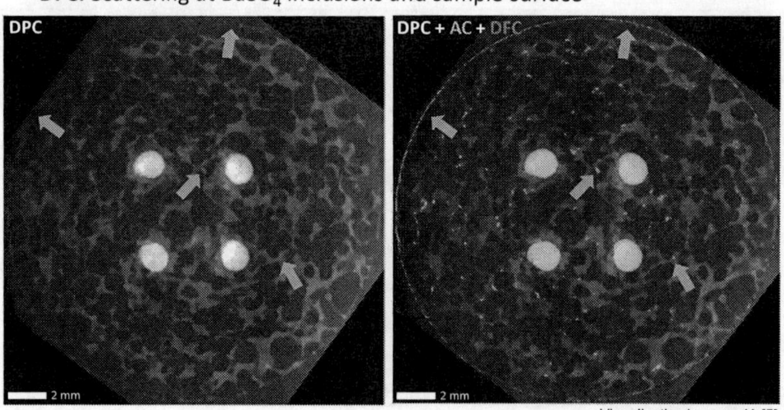

Visualization in open_iA [5]

DIR 2019 Page 12

Results: Sample 4

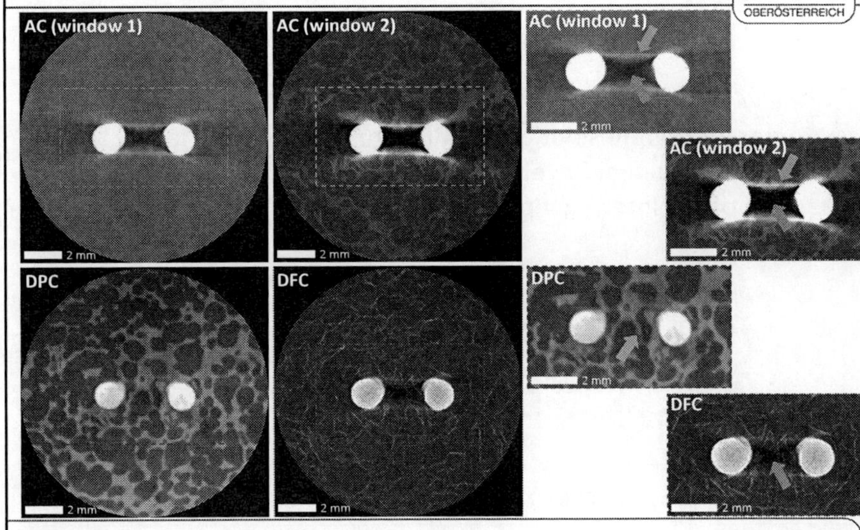

DIR 2019 Page 13

Conclusion

University of Applied Sciences

- **Differential phase contrast is less prone to "metal artifacts"**

  → Improving the visualization of the microstructure in polymer structures adjacent to Ti components

- **TLGI-XCT**

  → Advantages: two additional 3D imaging modalities & larger sample sizes in comparison to XCT

  → Disadvantages: low resolution, long acquisition times & higher level of noise

- **Outlook**

  → Optimized reconstruction & metal artifact reduction, filters, and segmentation

DIR 2019 Page 14

Acknowledgements

FH OBERÖSTERREICH

The work is financed by the project "*Com3d-XCT:* Competence Center for High-Resolution 3D X-ray Imaging" supported by the European Regional Development Fund (EFRE) in the framework of the Interreg V program Austria-Czech Republic

Interreg EUROPÄISCHE UNION
Österreich-Tschechische Republik
Europäischer Fonds für regionale Entwicklung

Europäische Union

Jonathan Glinz, M.Sc. (jonathan.glinz@fh-wels.at)

DIR 2019 Page 15

References

FH OBERÖSTERREICH

[1] P. Neidhart, "Verlagshaus der Ärzte – Gesellschaft für Medienproduktion und Kommunikationsberatung GmbH", www.medizinpopulaer.at, accessed: 20.05.2019

[2] P. Burgos et al., „Early Bone Tissue Responses to Turned and Oxidized Implants in the Rabbit Tibia", *Clinical Implant Dentistry and Related Research*, Vol. 10, No. 3, pp. 181-190, https://doi.org/10.1111/j.1708-8208.2007.00074.x

[3] C. Gusenbauer et al., „Porosity Determination of Carbon and Glass Fibre Reinforced Polymers Using Phase-Contrast Imaging", *Journal of Nondestructive Evaluation*, Vol. 38, No. 1, https://doi.org/10.1007/s10921-018-0529-6

[4] Bruker Corporation, „SkyScan-1294 User Manual", Version 1.1, 01.2015, pp. 24

[5] B. Fröhler et al., „open_iA: A tool for processing and visual analysis of industrial computed tomography datasets", *The Journal of Open Source Software*, 3/2019, https://doi.org/10.21105/joss.01185

DIR 2019 Page 16

Elucidating the distribution of organic consolidants in wood by Neutron Tomography

Amélie NUSSER [1], Stefan RÖHRS [1], Andreas SCHWABE [2], Sonja RADUJKOVIC [3], Thomas BÜCHERL [4], Volker DANGENDORF [5], Uwe ZSCHERPEL [6], Ina REICHE [7], Kurt OSTERLOH [8]

[1] Rathgen-Forschungslabor, Staatliche Museen zu Berlin (SMB), Stiftung Preußischer Kulturbesitz (SPK), Berlin, Germany

[2] Chair of Wood and Fibre Material Technology, Technische Universität, Dresden, Germany

[3] Vorderasiatisches Museum, SMB, SPK, Berlin, Germany

[4] Radiochemie München RCM, Technische Universität München, Garching, Germany

[5] Physikalisch-Technische Bundesanstalt (PTB), Braunschweig, Germany

[6] Bundesanstalt für Materialforschung und –prüfung (BAM), 8.3, Berlin, Germany

[7] PSL University, ENSCP, Institut de recherche de Chimie Paris (IRCP) UMR 8247 CNRS - Centre de recherche et de restauration des musées de France, Paris, France

[8] retired, formerly BAM 8.3, Berlin, Germany

Contact e-mail: kurt.r.s.osterloh@gmail.com

Abstract. While the absorption of X-rays and gamma radiation is determined by the Z-number of the elements a specimen is composed of, it is the hydrogen making an effective contrast with neutron imaging. As a consequence, interrogating with neutrons presents a suitable tool to study the distribution of organic consolidants in materials such as wood as encountered in impregnated wooden artworks.

Four different examples of objects are presented here to demonstrate the potential of neutron CT: 1) small wooden pieces of ship wrecks (< 2 cm thickness) interrogated with cold neutrons (0.5 meV at the ANTARES facility of the FRM II in Garching) to demonstrate the potential and the limitation of using low energy neutrons, 2) a wooden statue soaked with carbolineum (fission neutrons 1.8 MeV at the NECTAR facility of the FRM II), 3) a smaller wooden figure of a skull heavily soaked with carbolineum so it was too tight for the fission neutrons used before with accelerator neutrons (broad range about 5.5 MeV at the PTB in Braunschweig) and 4) pieces of charred wood to study the impregnation with a consolidant (NECTAR, FRM II). With the exception of the last example, all results have been combined with X-ray tomography (BAM 8.3 in Berlin). In the case of the charred wood specimens (example 4) the density histograms of the neutron tomography results were compared with those obtained from untreated references. The observed gain in specific density of the soaked specimens corresponded with an increase of specific weight. All results obtained so far showed distinct distribution patterns attributable to structural peculiarities or organic consolidants providing valuable support for subsequent restoration works.

Introduction

It is a well known fact that hydrogen containing materials are moderating neutron beams raising the problem of the achievable penetration thickness through wooden layers, i.e. the size of specimens that could be interrogated with certain energies of neutron beams. This question has been tackled earlier [1] with the result that cold neutrons are capable to penetrate layers of very few cm, higher energies are necessary for thicker layers as encountered in larger objects. As a consequence, three different neutron radiologic instruments have been used for this study that had neutron beams of various energies from a few meV to several MeV. Since the spatial distribution of the consolidants is of interest, an equipment for computed tomography (CT) was essential.

Due to the fact that wood already is a heterogeneous matrix by itself it is necessary to have some sort of reference or at least a further indication that allows to attribute certain patterns to a distribution of a soaked substance. Ideally, this would be an investigation of the specimen before and after treatment. However, this is usually impossible when interrogating cultural artefact since they have been already treated before in most cases, very often ill documented. The way to identify alternative reference information will be discussed here.

1. Objects and Rationale

This study focusses on wooden objects of the cultural heritage in archaeology and sacral arts. They are unique and with an unknown original state in most cases, but very often in need of conservation measures. Various organic consolidants have been and are used for this purpose to prevent decaying caused by pests, mould and environmental conditions, in some cases the deformation due to dehydration as shown in the first example in Fig. 1. It displays a wooden specimen from a sunken ship wreck [2] that was kindly provided by the Schifffahrtsmuseum in Bremerhaven, Germany. In the course of desiccation it has changed its shape. Based on its small size it was a suitable example to study the interior by interrogating with cold neutrons (~5 meV) that show fairly good spatial resolution in imaging. With higher energies in the MeV range, images generated with neutrons are of inferior resolution than those with X-rays, but show different specific material absorptions, i.e. reveal certain material properties. The radiographic interrogations with X-rays alone may fail show material properties indicating faint traces of a previous treatment, if at all.

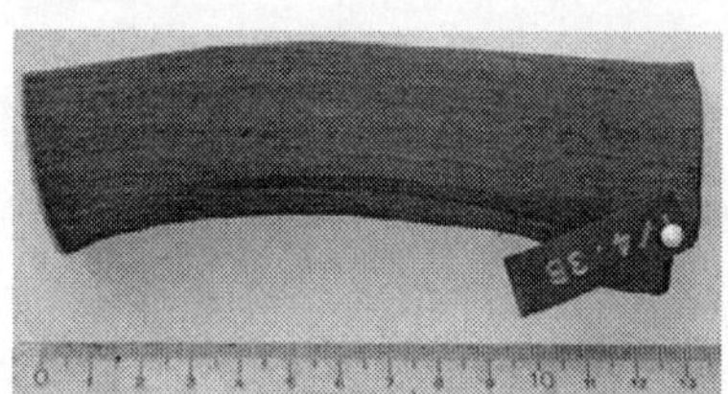

Fig. 1. Wooden specimen from a sunken ship wreck.
Curved shape deformation as a result of dehydration.

A century ago, the problem of biological destruction of wood was fought with organic substances such as carbolineum that was successfully applied to fences and railway sleepers, unfortunately also to wooden artefacts without considering long time effects. Such an example is shown in Fig. 2. The sculpture "Tugend der Hoffung" ("Virtue of Hope") to the left and the skull model to the right were taken from the baroque epitaph Reyer (1704) at the St. Laurentius church in Tönning (Schleswig-Holstein, Germany). The impregnating agent that has been used a century ago has chemically been identified as carbolineum, a mixture of coal tar oil components classified as carcinogenic and harmful to the environment. What an image cannot express is the scent emanating from such an object. A putative protocol to revert this measure has yet to be found but this needs some profound knowledge about the interior of the objects that only can be available by tomographic methods. Interrogating objects in this size require neutrons of higher energies.

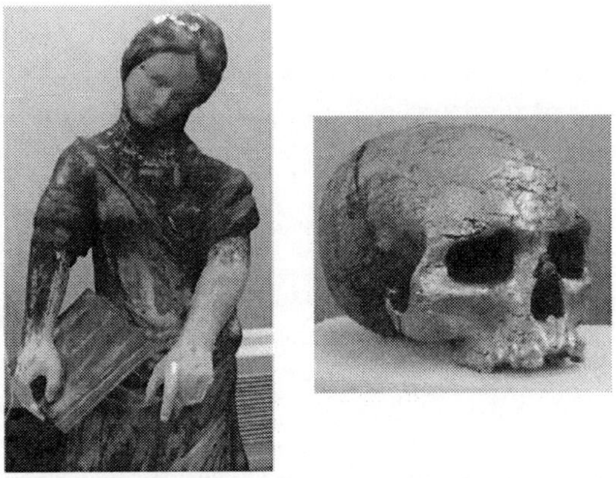

Fig. 2. Wooden artefacts treated with the consolidant carbolineum.
The sculpture "Virtue of Hope" (left) and the sculpture of a skull (right) were taken from an epitaph.

The other problem encountered particularly with archaeological objects are damages caused e.g. by fire and heat as happened to the torso shown in Fig. 3. It was found in the ruins of Uruk in Mesopotamia and served obviously as a cultural statue in Seleucidian times around 300 BC. Owing to the damage, such specimens are rather fragile so it is impossible to transport them over a longer distance. In order to develop an effective protocol for appropriate conservation measures, studies have been performed with charred wooden pieces that have been treated with several consolidants as a replacement (mock-ups) [3]. Examples of typical substances with different uptake and distribution characteristics, linseed oil and polyvinyl butyral, (PVB) are shown in Fig. 3. It should be reminded that this study was focussed on material compositions inside of an object rather than detecting fine structured flaws.

Fig. 3. Heat damaged torso that was not transportable and mock-ups for testing conservation protocols.
The treated specimens (bottom) and the untreated controls (top) were imaged together. The treatment consisted out of soaking with boiled linseed oil (centre) or polyvinyl butyral (PVB, right).

In the course of further conservative measures it is essential to obtain a clue to the current status of the interior of the objects to be treated and, in some cases, about the intermediate stages of a treatment. Since the consolidants used here are organic substances that are distributed in an organic matrix, it might be difficult to detect heterogeneous distribution patterns by X-ray radiography due to the expectingly low contrast. In difference to X-rays, neutrons are efficiently absorbed by hydrogen containing materials so their radiological use promises to detect distribution patterns of organic substances in an organic matrix much more effectively. In order to achieve additional information allowing a clue about the material composition, an X-ray-CT has been produced in addition that also resolves fine structures more in detail. Only in case of using mock-ups to study the effect of various consolidants on damaged objects as shown in Fig. 3 it was possible to compare treated specimens with untreated ones. As a consequence, results from original artefacts are being presented as overlays of neutron- and X-ray-CT images and those obtained from mock-ups by showing areas of increased densities following a treatment as compared to untreated controls.

2. Instruments and Methods

2.1 Facilities

The facilities used in this study have been described previously [2-6]. In brief, the study using cold neutrons (~5 meV) was conducted at the ANTARES instrument of the FRM II neutron source in Garching, Germany [4]. At the same site, the NECTAR instrument [5] was the source of fission neutrons (1.8 MeV average energy). Since both instruments showed a parallel beam geometry, it was sufficient to scan 180° only in 1° steps. Neutrons of still higher energies in the range of 4 – 10 MeV were available at the Physikalisch-Technische Bundesanstalt (PTB) in Braunschweig. Unfortunately, this unique source, where also energy resolved neutron radiography for element sensitive imaging was feasible, has been closed in the meantime. Normally, neutron CT scans were completed within 12 h. X-ray CT scans were conducted at BAM in Berlin, Germany, with a Comet X-ray tube set to 160 kV and 4 mA. The images were recoded with a Perkin-Elmer matrix detector (2048 x 2048 pixels of 200 µm x 200 µm size) and 1000 projections per scan at 15 s per frame.

2.2 Data processing

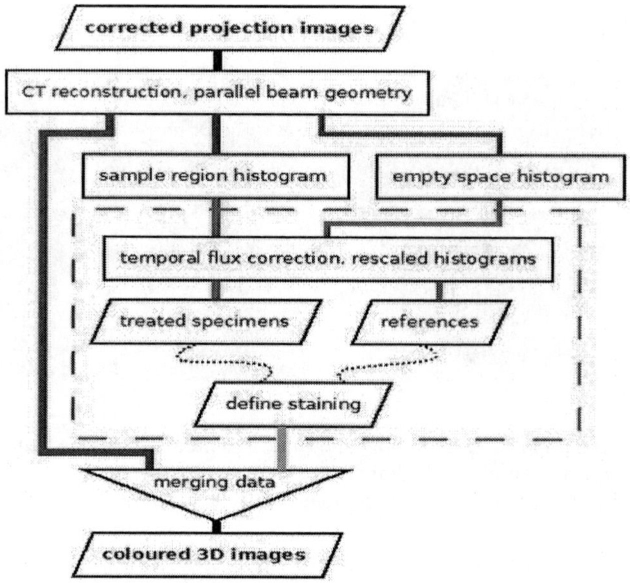

Fig. 4. Flow chart showing the data processing to show areas of increased density following a treatment. Coloured lined tracks: spatial 3D image data (blue), histogram data (red) and staining selection (green).

The raw images obtained from the neutron radiographic instruments were strewn with dotted interferences so they had to be filtered accordingly. This has been achieved with algorithms specially developed for this purpose [7, 8] before starting a 3D-reconstruction. Either standard software (Octopus or VGStudio MAX 2.2 and its supplement for tomography) or own software using the slice theorem have been used for the CT reconstructions. The overlay of sections from both, the X-ray-CT and the neutron-CT needed reorientations of the 3D-images as well as a brightness, contrast and γ adaption across the whole 3D-imaging spaces since they were obtained from different facilities with their own properties and individual positioning of the specimens in each instrument. This was achieved with own programs written on a standard Windows 7 PC using a Lazarus-Pascal-Compiler [9]. The resulting images consist of overlays of cuts through both, the neutron (blue or green) and the X-ray (red) data sets after after adjusting the neutron data to align with those from the X-ray reconstructions.

The steps of how to compare treated specimens with untreated controls in a mock-up study are listed in a flow chart shown in Fig. 4. Starting with the spatial reconstruction the image data are kept for subsequent staining of areas showing increased densities (blue line on the left). Partial volumes representing the body of the specimen and an empty space were selected for calculating histograms (red lines). The latter ones were used for correcting for intensity fluctuations from one CT scan to another one (90% limit as a "white" offset). By this way, all histograms have been adjusted to a common scale. Based on the histograms of the treated and the untreated specimens, the ranges of grey values representing the density gains due to the uptake of the consolidants as well as overlaps between both groups are assigned to colours ranging from blue to red. All processes in the dashed red box in the flow chart were completed within an ordinary spreadsheet program. The last step of merging the colour coding with the spatial image data was completed with another special own program [9]. As a tool for spatial presentation, the public domain program ImageJ has been used with the 3D viewer plug-in [10].

3. Results and Discussion

3.1 An example of small objects in a cold neutron beam

The specimen shown in Fig. 1 has successfully been interrogated both with cold neutrons and with X-rays. The combined result of both technologies can be seen in Fig. 5 as an overlay of cross sections at the same location where the neutron image is displayed in blue while the X-ray one is stained in red. It is rather obvious that more hydrogen material is located close to the surface indicating putative traces of previous interventions. However, any further interpretation in this direction requires chemical confirmation. At least, this result indicates the sites where to probe for samples in a future dedicated study.

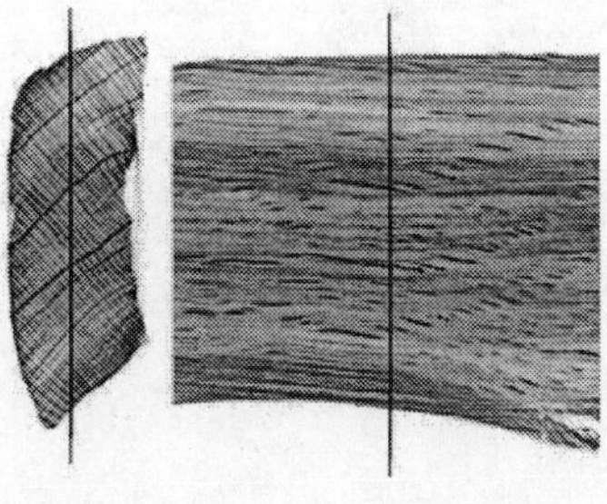

Fig. 5. Overlay of cross section images taken from neutron and X-ray CT of the specimen shown in Fig. 1. Blue colour: neutron-CT, red colour: X-ray-CT, left: cross section, right longitudinal section.

3.2 Interrogation with hight energy neutrons

The sculpture shown in Fig. 2 was interrogated with neutrons at the NECTAR instrument and with X-rays at BAM. The result is shown in Fig. 6 as overlays of cuts through the CT data of the respective technologies with the green colour for the neutron technology and the red colour for the X-ray absorption. The neck region and areas close to the surface with many worm holes contained more organic material than the core region. The horizontal nail on the left side was only seen in the X-ray image and penetrated by neutrons. Very obviously the head was replaced in one of the previous conservation campaigns that explains the flawless appearance of the face as it can be seen in Fig. 2. However, there is nothing mentioned in the conservation documents, only a suggestion in this direction.

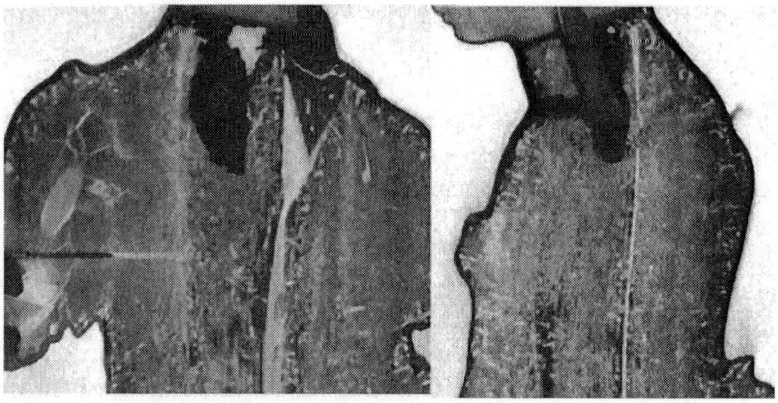

Fig. 6. Overlay of longitudinal cross sections of the sculpture shown in Fig. 2 on the left. The images were produced in the same manner as in Fig. 5, green: neutron-CT, red: X-ray-CT.

The model of the skull shown to the right in Fig. 2 was heavily soaked with carbolineum that could have been realised by the scent emanating. As a consequence, an attempt to interrogate this object at the NECTAR instrument remained without any success, neutrons of higher energy were requested. At that time, they were available at the PTB, and a neutron CT was successfully performed at that site. The result is shown in Fig. 7 in a spatial overlay of both CT results, red for the X-rays and green for the neutrons. The green of the hydrogen content is visible even through then red coating in the surface rendering presentation. While the coating obviously consisted of materials that absorb predominantly X-rays the whole inner space was prevailed by organic substances with minor variation in concentration. Obviously, the worm holes are filled with carbolineum.

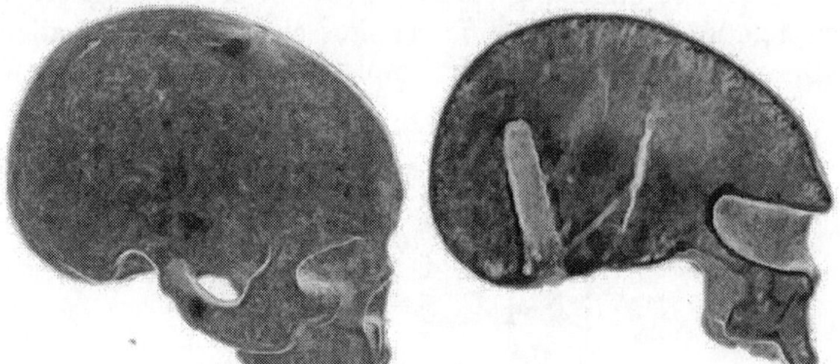

Fig. 7. X-ray CT (red) and neutron CT (green) overlay of the specimen to be seen in Fig. 2 on the right. The carbolineum impregnation is visible both, in the translucent surface rendering (left) and in the cut (right).

3.3 Correlating an uptake of consolidants with the spatial density distribution

It appears to be a recommendable practice to study preservation protocols with mock-ups rather than applying them directly to delicate original objects. The overall density of the interrogated specimens were easily determined by their weight and the volume calculated from the CT results. Generally, an increase in density was correlated with a shift towards higher values in the related histograms as expected, but not always quantitatively as demonstrated in Fig. 8 with the two consolidants linseed oil and PVB. In the first case, the density gain was 28% while it was 4% only in the latter one.

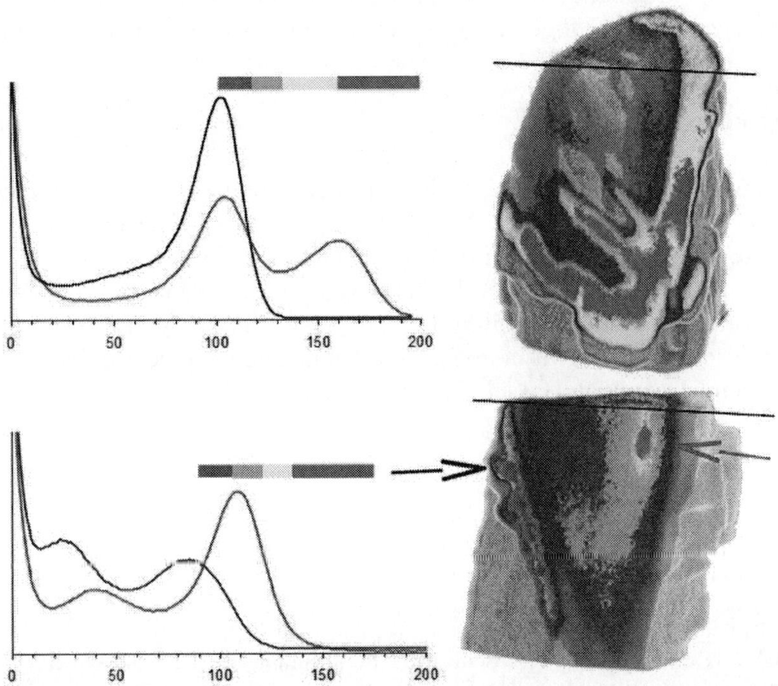

Fig. 8. Shifted density distributions in histograms and regions of increased densities in lateral cuts. Histograms with colour coding (left) and lateral cuts (right), treatment: linseed oil (top) and PVB (bottom), red line in histograms (frequency vs. grey values bright to dark): with treatment, black line: untreated control.

In case of the treatment with linseed oil, the density shift following a treatment seen in the related histogram is in good agreement with the expansion of areas with increased densities exceeding those of the control (yellow and red) with low areas of overlaps (blue and green). A different situation is encountered with the PVB where the density shift was less pronounced than in the previous example. However, it was larger than it could have been expected from the measured density increase of 4% only. A plausible explanation has been found in the distribution pattern shown in the lateral cut on the lower right within Fig. 8. Actually, two different sites of increased densities have been identified with distinctive characteristics related to their surrounding. One of them is located within the char close to the surface (black arrow), the other one within the bulk of the interior but with large adjacent areas of overlaps coded by the blue and green colour. As a consequence, only the increase detectable at the first site could be attributed to the treatment while the other one merely might be the result of structural differences between the treated specimen and the control not detectable at the beginning of the experiment. The red spot marked with the red arrow in Fig. 8 was attributable to a knot while nothing in this direction could be seen in the control. This example demonstrates the importance of an adequate interpretation of both together, numerical analytical results and images.

4. Conclusion

The combination of neutron and X-ray tomography revealed features of the interior of the objects interrogated that were not detectable from the surface. While the fine structures were resolved by X-rays clues on the material composition can be derived from the neutron CT. Finally, the overlay to the results from both technologies allow a more complex insight into the interior of the interrogated object that might provide valuable information to project future conservative measures. Analysing the density profiles of the neutron CT with subsequent spatial colour tagging the areas characterised by certain density ranges help to identify probing sites for further chemical analysis. Having both, numerical results and images together provides more information than each one by itself.

Acknowledgements

This project was supported by the Deutsche Bundesstiftung Umwelt (DBU) under the grant Az. 30165 – 45 and by the Berliner Antike-Kolleg (BAK-TT8-2015). The valuable help received from several colleagues is gratefully acknowledged, namely from S. Söllradl at the NECTAR instrument, B. Redmer, S. Hohendorf and C. Bellon for achieving the X-ray tomography results and from M. Freitag, N. Wrobel, S. Simon, M. Hilgert and U. Ewert for many fruitful discussions. The charred wood specimens were kindly provided by the Harzköhlerei Stemberghaus in Hasselfelde, Germany.

References

[1] K. Osterloh, D. Fratzscher, M. Jechow, T. Bücherl, B. Schillinger, A. Hasenstab, U. Zscherpel, U. Ewert (2011): Limited View Tomography of Wood with Fast and Thermal Neutrons, DIR 2011 International Symposium on Digital Industrial Radiology and Computed Tomography, 20-22 June 2011, Berlin, Germany, P14, https://www.ndt.net/article/dir2011/papers/p14.pdf,

[2] K. Osterloh, D. Fratzscher, A. Schwabe, B. Schillinger, U. Zscherpel, U. Ewert (2011): Radiography and partial tomography of wood with thermal neutrons, Nucl. Instr. & Meth. in Physics Res. (NIM) A 651, pp. 236-239,

[3] K. Osterloh, A. Nusser, S. Röhrs, S. Radujkovic, A. Schwabe, S. Söllradl, I. Reiche (2015): Verteilungsstudien von Festigungsmitteln in verkohlten Objekten mittels der Neutronen-Computertomographie, 8. N.i.Ke.-Workshop, 2015, N.i.Ke. Schriftenreihe des Netzwerks zur interdisziplinären Kulturguterhaltung in Deutschland, ISSN 2567-1251, 94-99,

[4] M. Schulz, B. Schillinger: Heinz Maier-Leibnitz Zentrum et al. (2015). ANTARES: Cold neutron radiography and tomography facility. Journal of large-scale research facilities, 1, A17. http://dx.doi.org/10.17815/jlsrf-1-42,

[5] T. Bücherl, S. Söllradl: Heinz Maier-Leibnitz Zentrum et al. (2015). NECTAR: Radiography and tomography station using fission neutrons. Journal of large-scale research facilities, 1, A19. http://dx.doi.org/10.17815/jlsrf-1-45,

[6] A. Nusser, M. Freitag, A. Fuhrmann, C. Herm, K. Osterloh, I. Reiche, A. Schwabe, S. Simon (2015): Analyse historischer Carbolineumimprägnierungen als Grundlage konservatorischer Dekontaminierung und modellhafte Erprobung neuer Sanierungstechnologien, Abschlussbericht zum Projekt, gefördert unter Az 30165von der Deutschen Bundesstiftung Umwelt, Berlin, 2015, Hornemann Institut, HAWK Hochschule für angewandte Wissenschaft und Kunst, Hildesheim/Holzminden/Göttingen, https://www.hornemann-institut.de/german/epubl_projekte149.php,

[7] K. Osterloh, T. Bücherl, Ch. Lierse von Gostomski, U. Zscherpel, U. Ewert, S. Bock (2011): Filtering algorithm for dotted interferences, Nucl. Instr. and Meth. in Physics Res. A 651, 171-174,

[8] K. Osterloh, T. Bücherl, U. Zscherpel and U. Ewert (2012): Image recovery by removing stochastic artefacts identified as local asymmetries, Journal of Instrumentation 7 C04018, http://iopscience.iop.org/1748-0221/7/04/C04018,

[9] Free Pascal Lazarus Project, version 1.6 (2016), http://www.lazarus-ide.org/

[10] K.U. Barthel (2014): Volume Viewer, https://imagej.nih.gov/ij/plugins/volume-viewer.html, W. Rasband: ImageJ 1.48v (2012), https://imagej.nih.gov/ij/.

New Concepts for the Measurement of Focal Spot Parameters of Nano- and Microfocus X-Ray Tubes

Uwe EWERT [1], Gerd-Rüdiger JAENISCH [2], Andreas DERESCH [3], Benjamin A. BIRCHER [4], Felix MELI [4]
[1] X-Ray-Net, DIN, NA 062-08-22 AA, CEN TC 138 WG1, Teltow, Germany
[2] Bundesanstalt für Materialforschung und -prüfung (BAM), Berlin, Germany
[3] YXLON International GmbH, Hamburg, Germany
[4] Federal Institute of Metrology METAS, Laboratory for Length, Nano- and Microtechnology, Bern-Wabern, Switzerland

Contact e-mail: uwe@ewert-net.de

Abstract

Accurate measurement of focal spot sizes of X-ray tubes, betatrons and linear accelerators is still an open problem. Several standards are currently active, which provide different results depending on the focal spot shape. CEN published the standard series EN 12543 (five parts) in 1999, which describes the measurement of focal spot sizes of X-ray tubes for NDT applications. The measurement procedure and its tolerances deviate from the measurement standards for medical X-ray tubes, e.g. EN 60336:2006 / IEC 60336:2005. EN 12543 describes different measurement methods for X-ray tubes with focal spots in several size ranges, defined as macrofocus, minifocus and microfocus. CEN TC 138 WG 1 und ASTM E 07.01 are revising (CEN) or have just revised (ASTM) the standards for focal spot size measurements, basing them on pinhole or edge measurements only. Slit cameras are exclusively applied for medical tube evaluation. Tomographic methods for focal spot size shape reconstruction were proposed and will be discussed, but have not yet been used for standardization. The currently available standards cover the range of spot sizes from five µm up to several millimetres. During the last years, more and more nanofocus tubes have been developed for high-resolution applications and microscopy. Manufacturers and users apply different methods for measurement of nanofocus spot sizes and image resolution, which differ by a factor of up to two. These well-known methods and new alternative methods for the measurement of nanofocus spots (size and shape) will be evaluated and improved in the EMPIR project "NanoXSpot" in order to develop a related measurement standard.

X-Ray-Net

02.07.2019

NEW CONCEPTS FOR THE MEASUREMENT OF FOCAL SPOT PARAMETERS OF NANO- AND MICROFOCUS X-RAY TUBES

U. Ewert[1] , G.-R. Jaenisch[2] , A. Deresch[3] , B.A. Bircher[4] , F. Meli[4]

[1]Radiology Committee of DGZfP, X-Ray-Net, Berlin, Germany,
[2]Bundesanstalt für Materialforschung und -prüfung (BAM), Berlin, Germany,
[3]YXLON International GmbH, Hamburg, Germany,
[4]Federal Institute of Metrology METAS, Laboratory for Length, Nano- and Microtechnology, Bern-Wabern, Switzerland

Standards for Measurement of Focal Spots

- **Standards and Committees:**
  - **CEN TC 138 WG 1** EN 12543 Part 1-5:2011
  - **ASTM International E07.01** E 1165-17, E2903-18
  - **IEC** (International Electrotechnical Commission) IEC 60336:2005, IEC 62976:2017
  - **NEMA** (National Electrical Manufacturers Association) XR 5:1992(R1999)
- **Standards are different and provide different results**.
  - This is critical, if measurement ranges overlap
- **CT algorithms are used** to substitute **pin hole camera measurements** with better sensitivity
- **A standard for measurement of spot sizes of nanofocus tubes will be developed in the** EMPIR-Projekt „**NanoXSpot**" as prEN 12543-6.

The Currently Unsatisfying Situation ...

- **EN 12543 parts 1 - 5 provide different measurement values for the same focal spot in the overlap range of the standards**
- **ASTM and IEC standards provide again different values**
- **IEC 60336 (Medicine) permits different tolerances in comparison to NDT**
- **User standards in NDT shall provide the "right" focal spot size for**
  - **Calculation of the minimum magnification** achieving the standard requirements of image unsharpness U_{im}
- **User unsharpness measurements are based on line pattern image quality indicators (IQI) such as**
  - Duplex wire IQI, ISO 19232-5, ASTM E 2002
  - Line pattern IQIs (e.g. Siemens star, JIMA-Target)

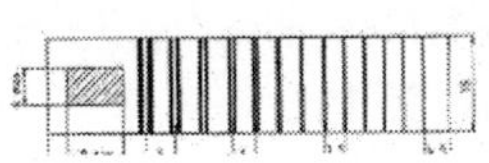

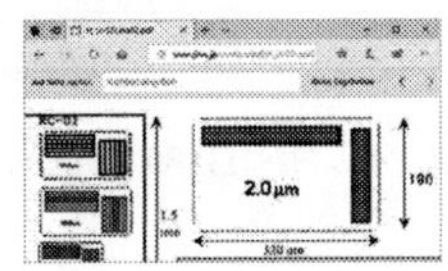

ISO 17636-2: Weld testing
EN 12681-2: Casting testing
ASTM E 2698: Standard practice for DDAs

$$u_{im} = \frac{1}{v}\sqrt{u_G^2 + \left(2 \times SR_b^{detector}\right)^2}$$

Cubic formula in ASTM E 2698

Requirement:

- Measurement of **image unsharpness U_{im}** based on detector unsharpness, geometric unsharpness and magnification
- Not exceeding maximum limits for U_{im} oder SR_b^{image}

Measurement of image unsharpness U_{im} und SR_b^{image} is based on duplex wire IQI measurements in most standard practices based on ISO 19232-5 und ASTM E 2002.

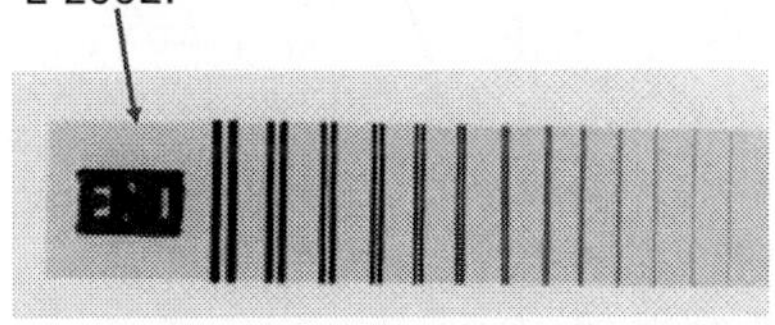

Only focal spot sizes measured with **Edge or Line Camera Methods**

→ conform to the results of measurements with duplex wire IQIs!

→ **Consequently, the results of all measurement methods shall be converted to these values.**

Calibrate the pixel size and calculate the integrated line profile of the pin hole camera image

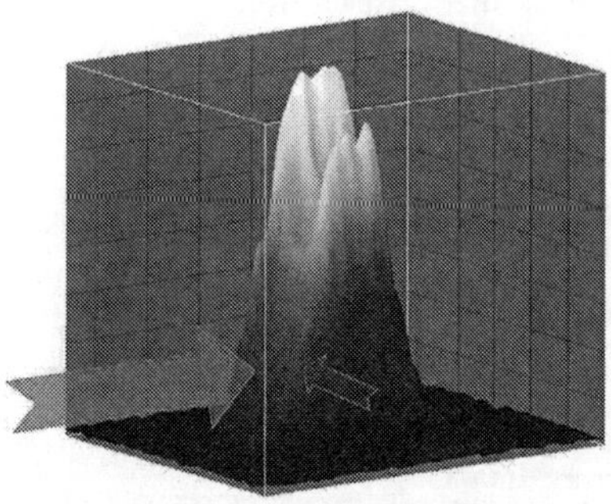

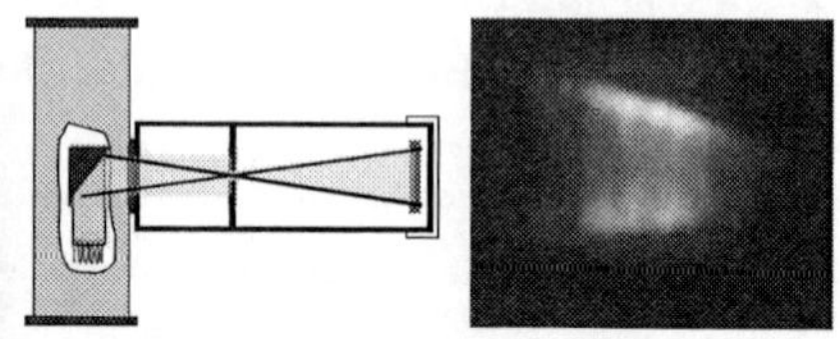

- Scheme of a pin hole camera
- 2D-image and pseudo 3D image of a focal spot radiograph (0,57 x 0,53 mm²)

Calculation of the focal spot size with profile integration corresponding to ASTM E 1165

Evaluation with „Integrated Line Profile Method“ (ILP)

Set cursors at 16% and 84% in the integrated line profile
(Klasens method, corresponds to one standard deviation of normal distribution: 68%)

Measure the distance between the cursors (388µm)

Multiply with 1.47
(68% → 100%)

→ **570 µm focal spot length**

→ **Free Software ISee_demo!**
http://www.uzscherpel.de/BAM/ic/index.html

See Klasens method for measurement of unsharpness in ASTM E 1000

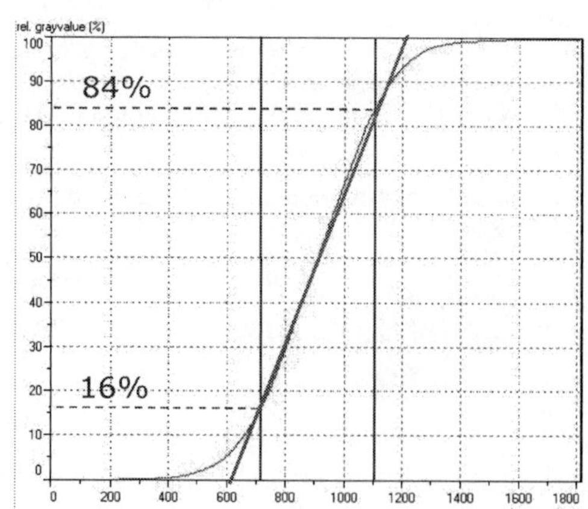

See Bavendiek et al., https://www.ndt.net/article/wcndt2012/papers/346_wcndtfinal00346.pdf

Focal Spot Length calculated with „Integrated Line Profile Method“

570 µm

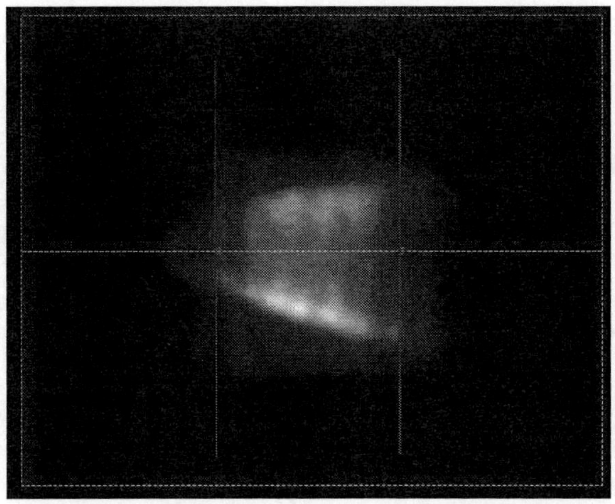

Focal Spot CT with Slit Camera

- **Scanner-Method is technically very complex and requires long measurement times**
- Slit cameras are significantly more sensitive
- With a rotating slit a spot CT can be performed with high SNR
  - Reduced measurement time, less expensive, stable results
  - Exact adjustment of rotational axis required
  - Special slit camera design required

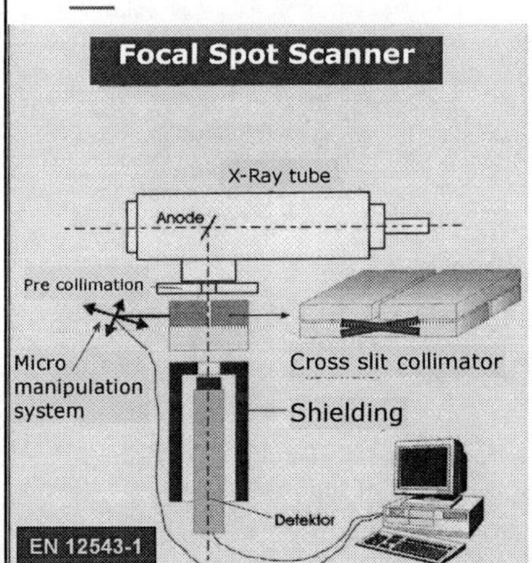

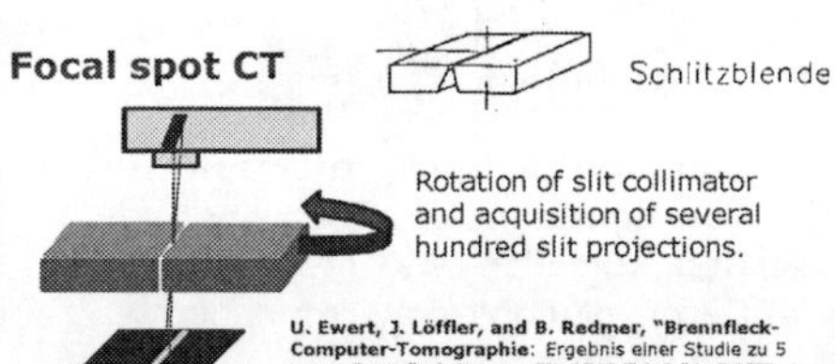

U. Ewert, J. Löffler, and B. Redmer, "Brennfleck-Computer-Tomographie: Ergebnis einer Studie zu 5 neuen Brennfleck-Normen EN12543 T.1-5," in DGfZP Jahrestagung, 1997, pp. 289–296

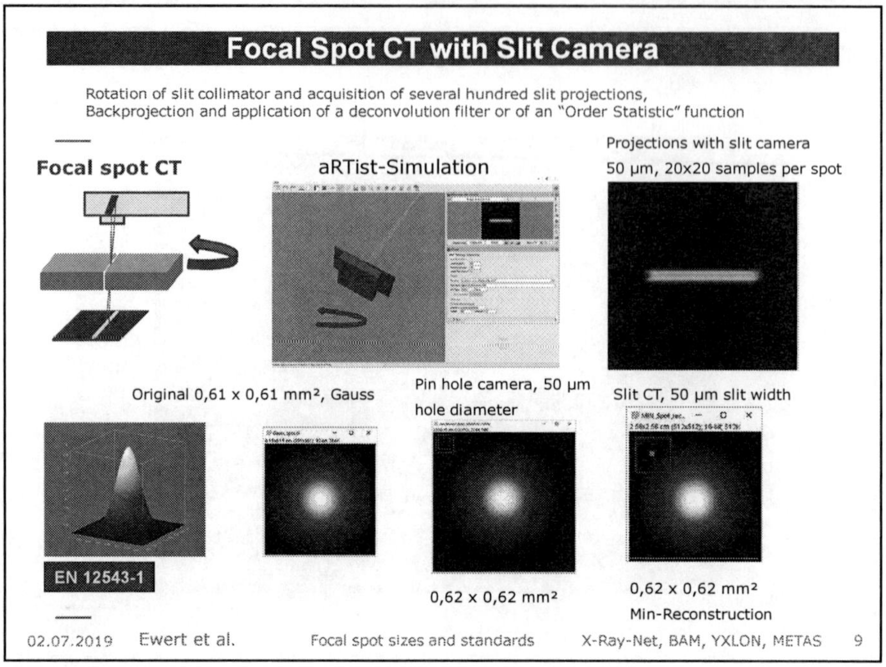
Focal Spot CT with Slit Camera
Rotation of slit collimator and acquisition of several hundred slit projections,
Backprojection and application of a deconvolution filter or of an "Order Statistic" function
Focal spot CT
aRTist-Simulation
Projections with slit camera
50 µm, 20x20 samples per spot
Original 0,61 x 0,61 mm², Gauss
Pin hole camera, 50 µm hole diameter
Slit CT, 50 µm slit width
EN 12543-1
0,62 x 0,62 mm²
0,62 x 0,62 mm²
Min-Reconstruction
02.07.2019
Ewert et al.
Focal spot sizes and standards
X-Ray-Net, BAM, YXLON, METAS
9

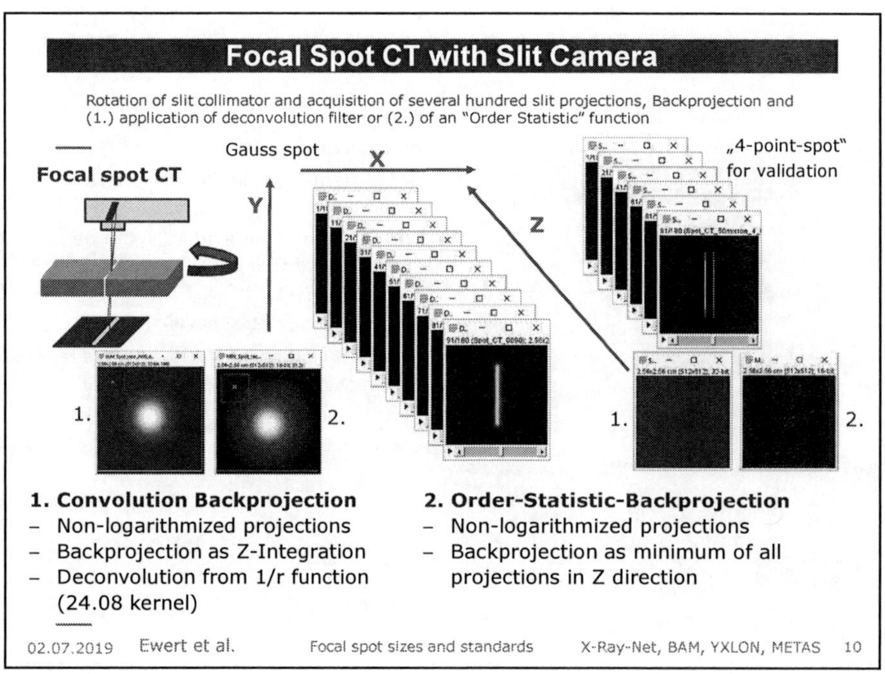
Focal Spot CT with Slit Camera
Rotation of slit collimator and acquisition of several hundred slit projections, Backprojection and
(1.) application of deconvolution filter or (2.) of an "Order Statistic" function
Focal spot CT
Gauss spot
X
Y
Z
„4-point-spot"
for validation
1.
2.
1.
2.
1. Convolution Backprojection
- Non-logarithmized projections
- Backprojection as Z-Integration
- Deconvolution from 1/r function (24.08 kernel)
2. Order-Statistic-Backprojection
- Non-logarithmized projections
- Backprojection as minimum of all projections in Z direction
02.07.2019
Ewert et al.
Focal spot sizes and standards
X-Ray-Net, BAM, YXLON, METAS
10

Direction Dependent Edge Method with large Hole IQIs

ASTM E 1165 ANNEX A

Focal spot CT with Rotating Edge

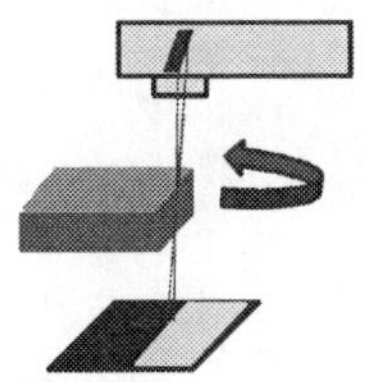

Focal Spot CT from a "large" hole exposure → „Single Shot CT"

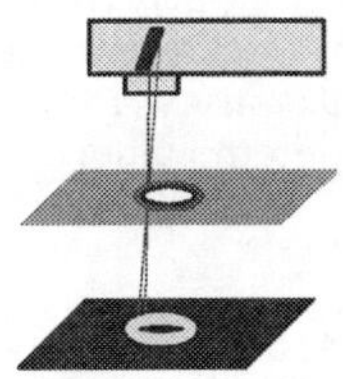

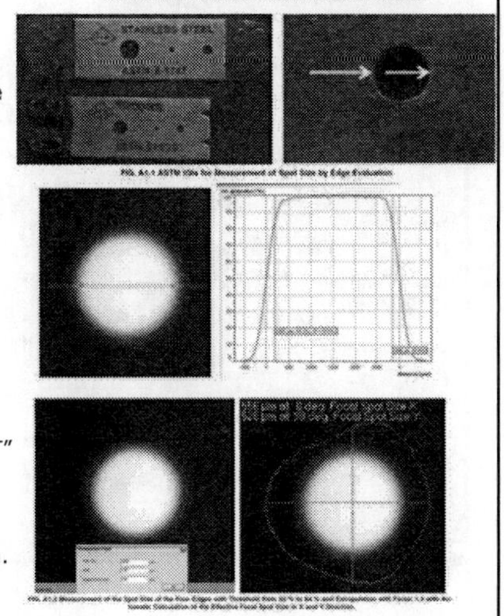

A "slit camera CT" can be substituted by an "edge CT" based on the derivation of the edge profiles.

This is equivalent to the reconstruction from edge unsharpeness profiles of hole exposures as sinogram.

„Single Shot" Focal Spot CT

U. Ewert , G.-R. Jaenisch , A. Deresch , B.A. Bircher , F. Meli, Neue Konzepte zur Messung von Parametern der Brennflecke von Nano- und Mikrofokus-Röntgenröhren, DGZfP-Jahrestagung, 27. - 29. Mai 2019 in Friedrichshafen

Focal spot CT from edge profiles of a hole exposure: „Single Shot CT"

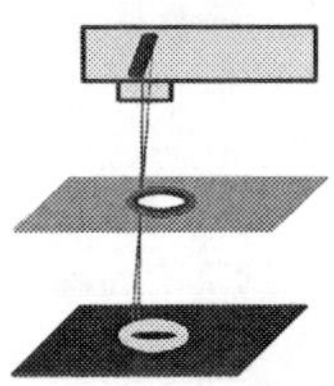

Derived edge profiles

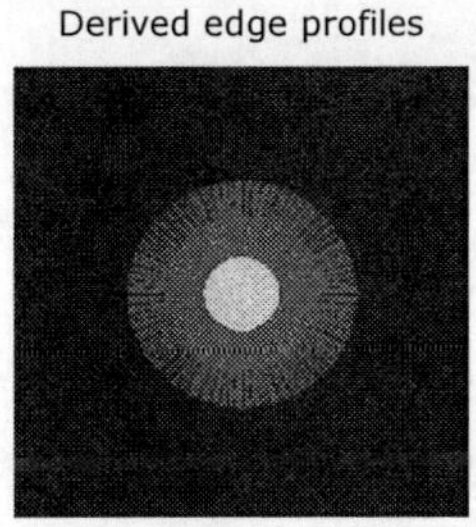

Focal spot distribution

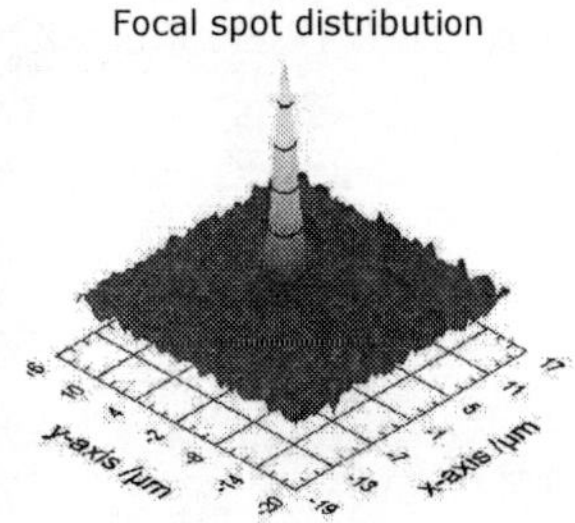

- CT of a focal spot distribution from derived edge profiles.
- Low measurement time, low hardware requirements, stable results.
- Restrictions arise from edge penetration, exposure geometry, kV and adjustment of hole IQI.

Nanofocus Spot Size Measurement with Line Pattern IQIs

Procedure also valid for spot sizes < 5 µm

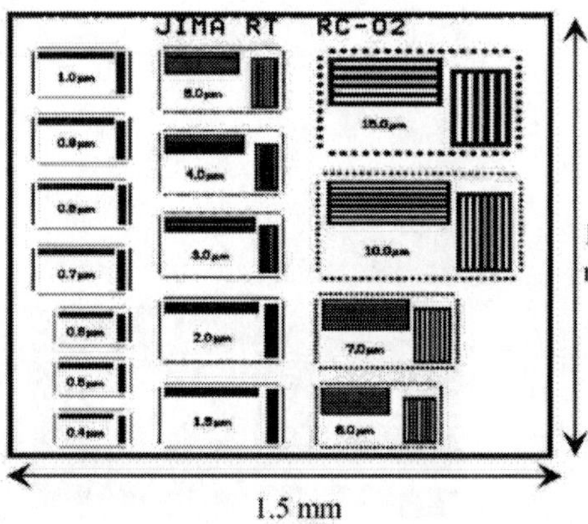

Line pattern targets are available with different line widths and distances. They can be used for determination of focal spot sizes.

Attention: The numbers are given as line widths in µm. Multiply all values by **2** to obtain a spot size value.

Nanofocus Spot Size Measurement with Line Pattern IQIs

Procedure also valid for spot sizes < 5 µm

Alternative line pattern targets with high Aspect ratio (YXLON)

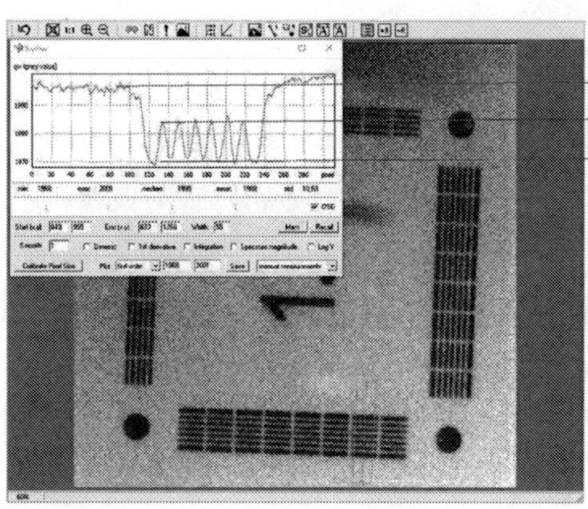

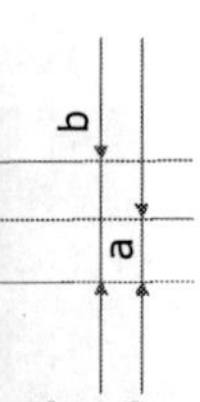

Simple method for users
(see also ASTM E 2445)

Approximation of modulation depth Md(%) = 100 * a/b,

If the result is 20% modulation depth, it is equivalent to MTF_{10}

Nanofocus Spot Size Measurement with Line Pattern IQIs

Procedure applies also for spot sizes < 5 µm

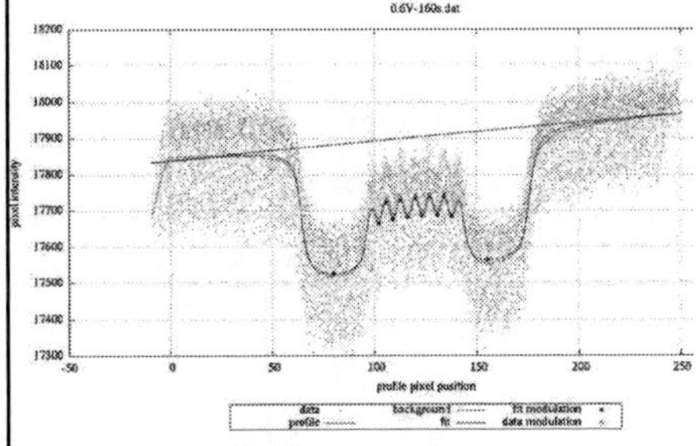

Complex procedure for manufacturers (EMPIR-Projekt „NanoXSpot")

Determination of modulation depth with a numeric fit procedure

- Fit with Pseudo Voigt Profile
  - Weighted sum of Gaussian and Lorentzian Profile ("long foothills")
  - Defined by the half width at half height and Lorentzian to Gaussian ratio
- Results are the 10% value of the modulation depth, MTF_{10}, and the line shape factor Lorentzian/Gaussian for better evaluation of the influence of the focal spot shape on the detail visibility.

Pseudo Voigt Profile

Profile functions of Voigt profiles with different Lorentzian weights

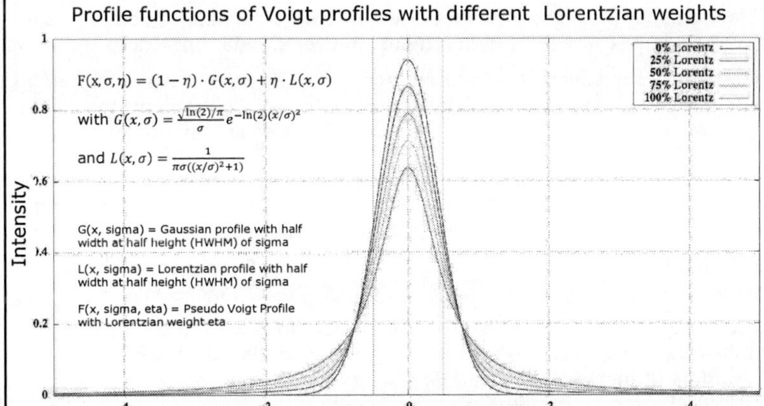

Focal Spot Sizes

Measurement of focal spot form factors and half value widths of X-Ray tubes of different manufacturers

Fit results of real measurements

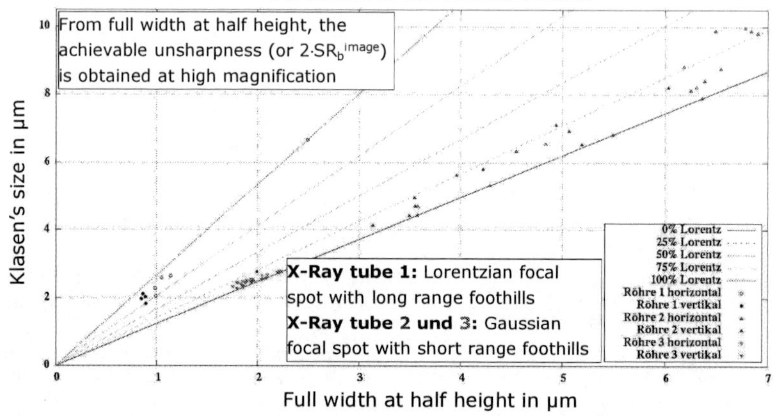

02.07.2019 Ewert et al. Focal spot sizes and standards X-Ray-Net, BAM, YXLON, METAS 17

Summary

- ASTM standards for focal spot size measurements were revised, to obtain values conforming to duplex wire IQI measurements in user standard practices.
- CEN standards of the series EN 12543 are under revision to be harmonized with the related ASTM measurement standards and the given requirements for maximum unsharpnesses of published standard practices in radiography.
- The scan method EN 12543-1 shall be withdrawn.
- CT algorithms will be developed and tested to calculate focal spot distributions from slit camera exposures and "unsharp" hole radiographs.
- A "Single Shot CT" shall be developed to reconstruct the focal spot distribution from exposures of very small hole targets (Project EMPIR NanoXSpot).
- A draft for a new standard on measurement of focal spot sizes and shapes of X-Ray tubes with spot sizes < 5 µm shall be developed, based on special line pattern targets (Project EMPIR NanoXSpot).
- CEN TC 138 WG 1 supports the EMPIR project NanoXSpot and plans to publish a new standard part EN 12543-6 based on the project results.

02.07.2019 Ewert et al. Focal spot sizes and standards X-Ray-Net, BAM, YXLON, METAS 18

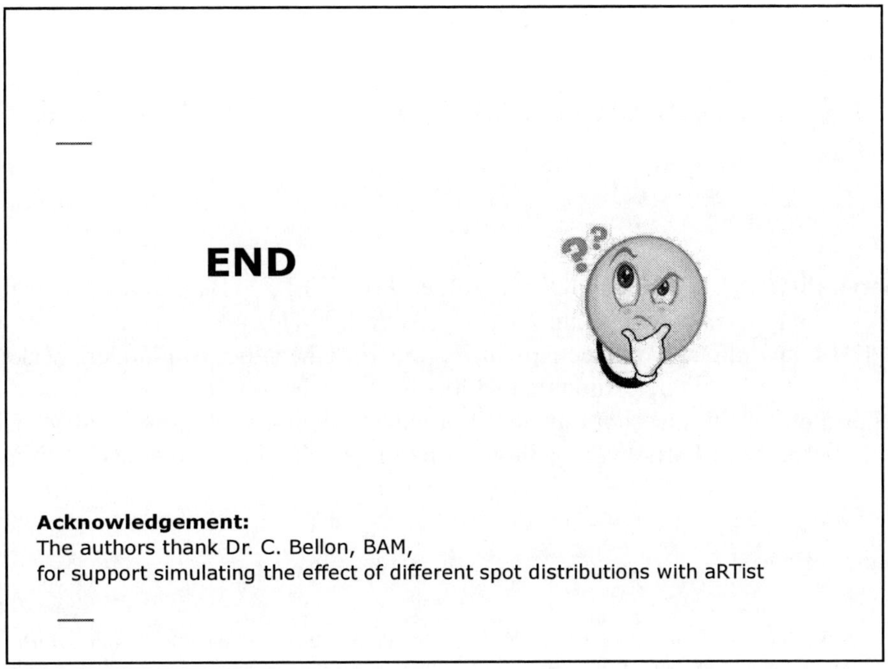
END
Acknowledgement:
The authors thank Dr. C. Bellon, BAM,
for support simulating the effect of different spot distributions with aRTist

Focal spot characterization of an industrial X-ray CT scanner

Gabriel PROBST [1], Qinhan HOU [2,3], Ruben PAUWELS [1], Bart BOECKMANS [1], Yongshun XIAO [2,3], Wim DEWULF [1]

[1] KU Leuven, Dept. of Mechanical Engineering, Member of Flanders Make, Celestijnenlaan 300, Heverlee, Belgium

[2] Department of Engineering Physics, Tsinghua University, Beijing, 100084, China

[3] Key Laboratory of Particle & Radiation Imaging (Tsinghua University), Ministry of Education, China

Contact e-mail: gabriel.probst@kuleuven.be

Abstract. The focal spot of an X-ray tube, combined with detector modulation transfer function, are the main contributors of image blurriness. Blurriness or unsharpness makes it difficult, or even impossible, to correctly segment material and background. Components with micro-features, due to excessive blurriness, might end up blending in with their surroundings, making their assessment difficult. This paper presents the results of a focal spot characterization, in terms of its size, shape, and intensity distribution, of an industrial X-ray CT scanner. A Siemens star pattern was used for focal spot size and shape assessment throughout a series of one-hour long scans. Validation of the proposed methodology was carried out through simulation. Initial results demonstrated focal spot stability throughout a scan, as well as the dependency of size and shape on the exposure settings (i.e. power). A study on the target material influence on both focal spot shape and size is also presented. Moreover, the influence of focal spot distortions on the measurement volume is investigated through simulation of real focal spot shapes. This case study illustrates how the focal spot might influence dimensional measurements with an X-ray CT system.

1. Introduction

X-ray Computer Tomography (XCT) has become a powerful inspection technique since its first introduction as a quality control tool in 2005 [1]. For the identification of small features such as pores and cracks, high measurement resolution is of the utmost importance. The two main factors limiting measurement resolution in XCT are: the focal spot and the detector modulation transfer function (MTF). International standards, such as the EN 12543-5 and ASTM E1165, describe methods to assess focal spot size [2][3]. These methods, however, do not fully characterize the focal spot and its influence on the measurement result.

In this research, a newly developed methodology to assess focal spot shape, size, and intensity distribution is presented. The method works by imaging a Siemens Star Pattern (SSP) attached to the X-ray tube, and looking at the edge profile transition between lines.

The circular profile provides enough information to recover and reconstruct the focal spot used to image thereof.

2. Method

The proposed methodology uses a sinogram, in combination with the core of the Filtered Back-Projection (FBP) algorithm, to reconstruct a focal spot. The sinogram is acquired by looking at the edge spread function (ESF) of a commercially available SSP containing 120 stripes [4]. A CAD model with identical features of the physical object was developed, and the aRTist [5] simulation software from the Bundesanstalt für Materialforschung und –prüfung (BAM), was used for the validation of the algorithms.

2.1. Center point determination

A squared focal spot having dimensions of (10 x 10) µm in size and a uniform intensity distribution was simulated. The developed method works by collecting the ESF along a radial path. All points should be equidistantly acquired in relation to the center of the object. Fig. 1 illustrates the SSP used and the imaging technique employed to filter the circle and identify the center of the test object. A canny edge segmentation, with subpixel interpolation, is used to highlight the edges of the object. Subsequently, two orthogonal vectors, u and v, are hovered in each of the quadrants of the image to filter out data points that do not belong to the edges of the circle. Once all edges have been identified, an ellipse is fitted and its center point coordinates extracted [6]. The current methodology achieves center point identification errors smaller than 0,5 pixels on simulated data for both directions (u and v).

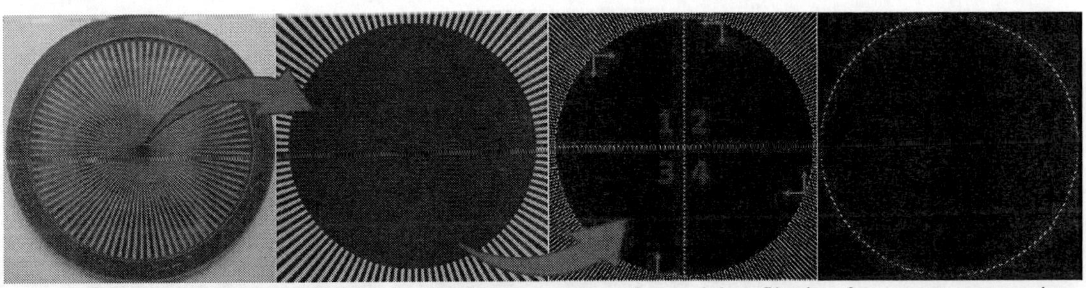

Fig. 1. Siemens Star Pattern used for the focal spot reconstruction and data filtering for correct center point identification.

2.2. Focal spot reconstruction

With the center point coordinates identified, a radial path is created and data is collected along it (blue and red circles in Fig. 2 A). A linear interpolation is used to compute measurement points between pixels that do not belong to either the same rows or columns. A magnification of 97 is used as this value is equivalent to what can be achieved with the industrial XCT scanner used to collect experimental data further in this paper.

Fig. 2 B illustrates the reconstruction of the simulated squared focal spot for the red radial path in Fig. 2 A. The presence of stripes in the reconstructed image can however be noticed. Those stripes emerge due to the fact that the SSP provides limited data, in intervals of 1,5 degrees, as it contains only 120 stripes. In order to overcome this issue, and have a higher fidelity reconstruction, all radial information from the blue to the red circles in Fig. 2 A are used. In the experiments conducted, approximately 350 radial paths were utilised. The outcome of this endeavour is shown in Fig. 2 C, with a squared focal spot presenting a uniform intensity distribution. The streak artifacts are no longer present.

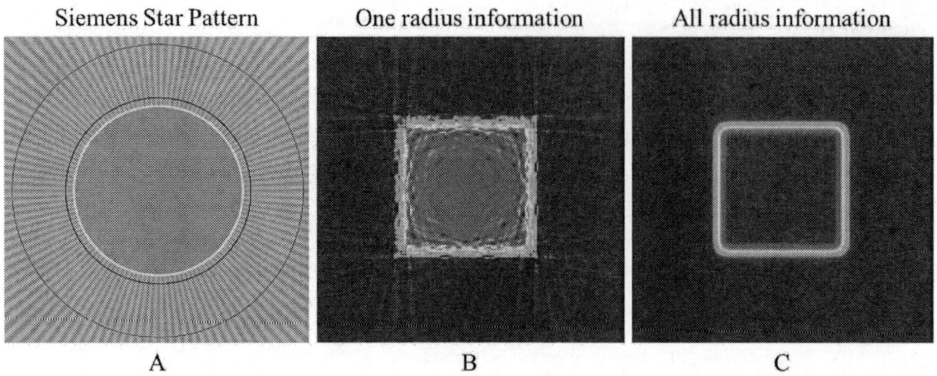

Fig. 2. A) Radial paths for the collection of data for the focal spot reconstruction. B) Focal spot reconstructed based on one radius path (red circle in A). C) Focal spot reconstructed with all radius paths (from the blue to red circle in A).

2.3. Methodology verification

To verify whether the developed methodology would work for focal spots with unknown sizes and shapes, an oddly shaped focal was created (Fig. 3 A) for the simulation of an SSP X-ray image. The focal spot of the simulated SSP was thereafter reconstructed (Fig. 3 B) and compared to the original (Fig. 3 A). Fig. 3 C shows the local difference between both reconstructions. A slight difference in intensity, about 6,8%, is observed in the center of the image. The fact that the reconstruction managed to capture the designed focal spot, in terms of its size and shape, gives confidence that the employed methodology for focal spot reconstruction works in adverse conditions.

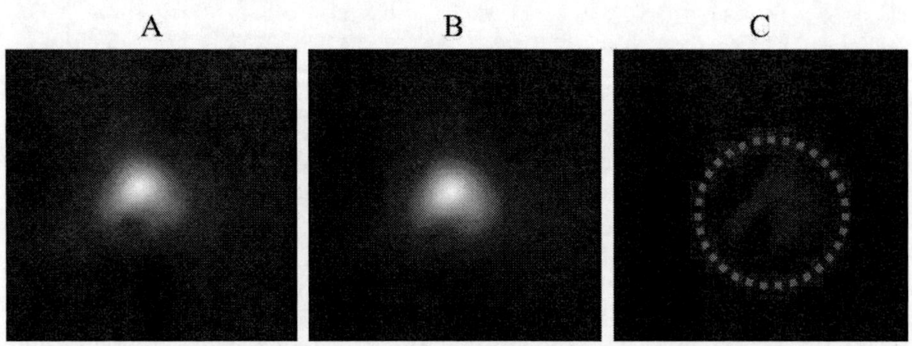

Fig. 3. A) Designed focal spot. B) Reconstructed focal spot based on an aRTist simulation of the designed focal spot in A). C) Difference between the designed focal spot and the reconstructed one.

2.4. Focal spot edge and size determination

The intensities change gradually across a focal spot and its edge is not crisp. Therefore, a threshold value has to be defined in order to define a focal spot size. For that, a focal spot with Gaussian intensity distribution [7][8][9][10] having 10 μm in diameter was simulated on aRTist, and a minimization method used to identify what threshold value yielded the simulated size. The minimization procedure was carried out by selecting a horizontal and vertical line crossing the center of the reconstructed focal spot, as shown in Fig. 4. The outcome of this minimization corresponds to the threshold value of 45,928% used to estimate focal spot size in the experimental section of this paper.

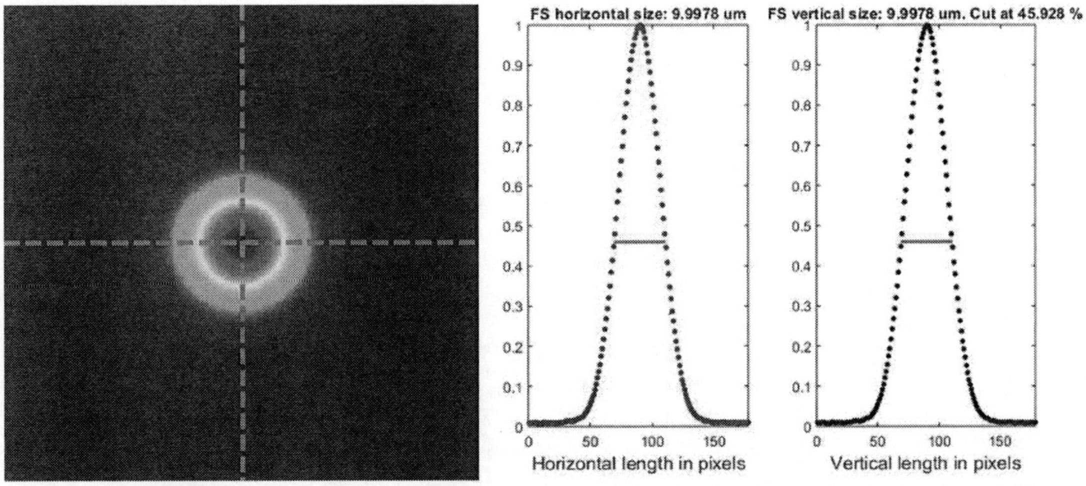

Fig. 4. Identification of threshold value for focal spot size estimation based on a circular focal spot with a Gaussian distribution. Threshold value defined at 45,928%.

3. Results

A set of experiments were carried out in order to understand how the focal spot behaves throughout a CT scan. For that, measurements of one-hour long were taken with the SSP attached to the X-ray gun, Fig. 5. Each of the radiographs is individually processed, resulting in a stack of reconstructed images of the focal spot. Each of the experiments acquired a total of 3600 projections.

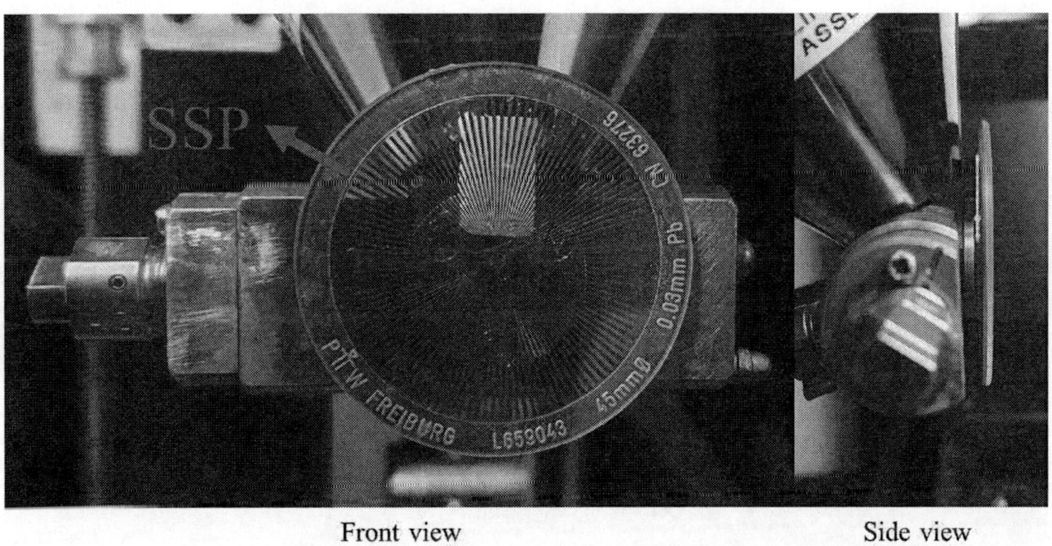

Fig. 5. Siemens Star Pattern placement at the X-ray gun window.

For a better visual understanding of how the focal spot shape varied during the one-hour long experiments, a dedicated color mapping has been designed. First, all focal spot images are computed and the focal spot edge is determined as described in Section 2.4. Subsequently the focal shape is approximated with an ellipse, the average between major and minor axis is defined as the spot size, and a conversion between pixels and micrometers is realized.

The first focal spot image is used as a reference point, and the maximum absolute difference is identified. In the example of Fig. 6, the maximum absolute difference is observed to be 0,5 μm. That value is then doubled originating a full color range, from red to blue, between 4,0 μm and 3,0 μm. This color mapping allows for a local observation of how the focal spot size and shape varied along the one-hour long scans. For creating the graphs displayed in the next sections, after segmenting the focal spot images, the edges of the focal spot are extracted, a color is assigned given its diameter and the focal spot contour is plotted in a Cartesian map. Afterwards, the subsequent focal spots are processed, following the same methodology, and plotted along the z-axis. This gives all graphs a three-dimensional form where the z-axis represents time (projections).

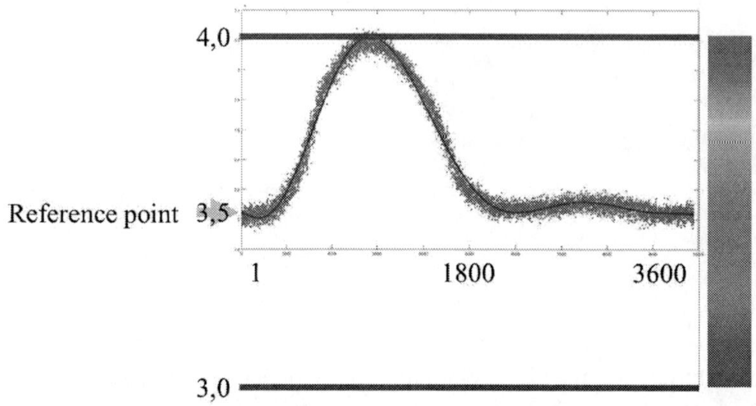

Fig. 6. Implementation of a dedicated color mapping. X-axis representing the individual reconstructed focal spots and, y-axis the corresponding radius in μm.

3.1. Focal spot behaviour during an XCT scan

The Nikon XT-H 225 ST is equipped with a target that can be manually rotated. This rotation allows for selecting different target materials for different X-ray spectra. The system is equipped with four materials: tungsten, copper, molybdenum and silver. Each of these targets is used for a specific type of application, e.g. tungsten for measuring metals and molybdenum plastic components. The target component is hollow, allowing cooling liquid to go through the target preventing the component from melting. Excessive use of high measurement power at the same target location can however induce pitting at the surface of the component. Fig. 7 depicts a used target, with the red dashed ellipse indicating several small craters created along the years due to overexposure due to excessive measurement power.

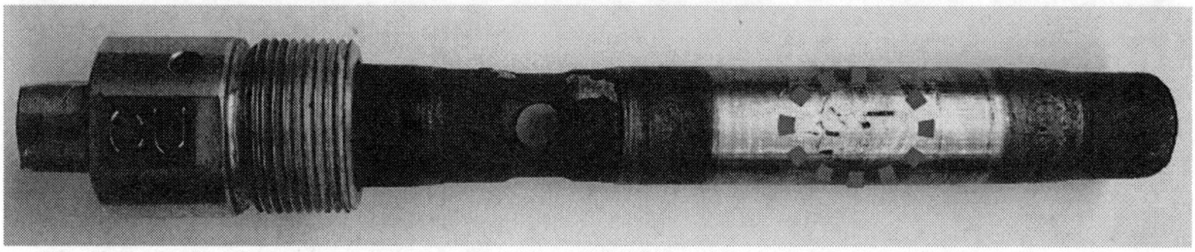

Fig. 7. Example of a target component of a Nikon XT-H 225 ST with several small craters (red ellipse)

This section therefore describes how target positioning can influence focal spot shape and size. Fig. 8 shows the results of four measurements where the target material was slightly rotated in between each one of them. X-ray tube voltage for these experiments was set at 80 kV and amperage at 80 μA. Color mapping followed the schematic described above. The maximum and minimum focal spot size diameter for all four scans were used to create the color range.

It can be noticed in Figs. 8 and 9 that the focal spot shape has been largely influenced by the target component. Interestingly, the second experiment seems to contain a second, less bright spot, which is not present in the graph as it is filtered out during the thresholding phase defining focal spot size. Experiment 4 presents a constant flat side that could have been caused due to the electron beam hitting the target material exactly where pitting is located. Experiments 1 and 3 seem to not have been affected by the target material, indicating that the location where they were being generated did not contain anomalies, such as pitting.

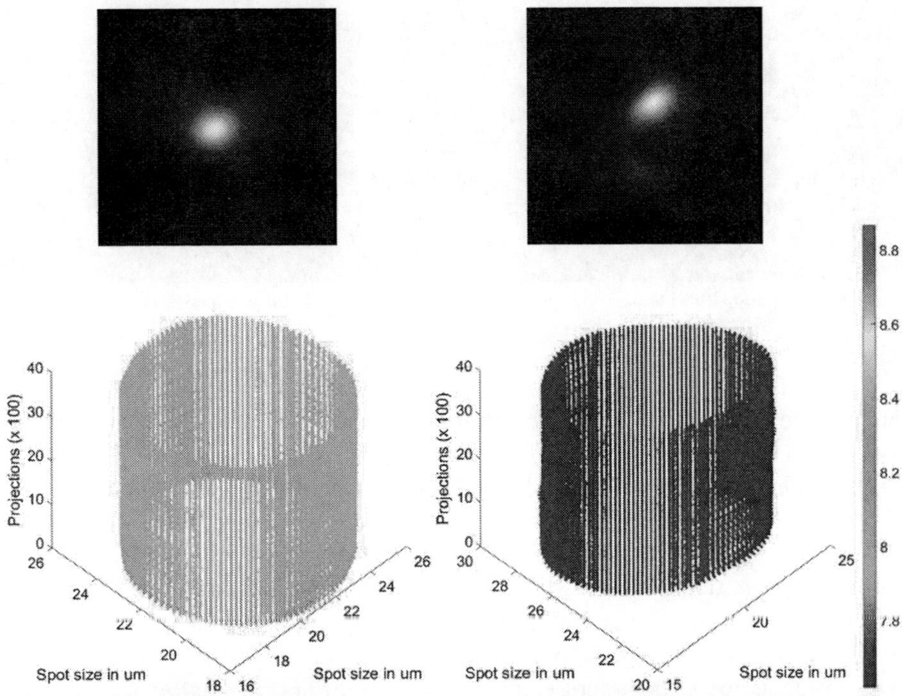

Fig. 8. Influence of the target material on the focal spot shape and size. Experiments 1 and 2.

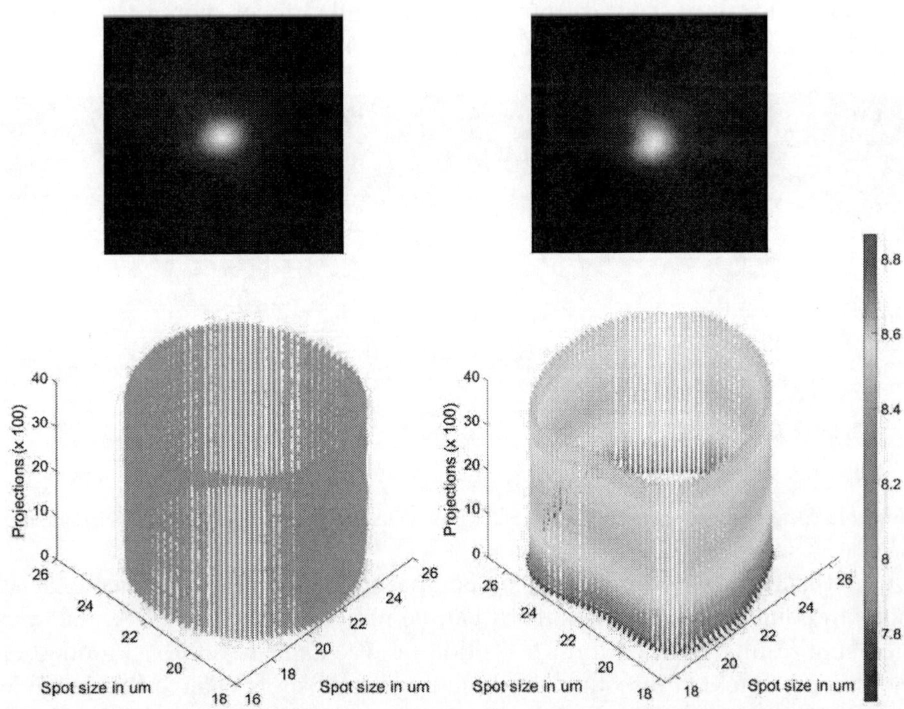

Fig. 9. Influence of the target material on the focal spot shape and size. Experiments 3 and 4.

In Fig. 10 several other occurrences of real focal spots, measured on used targets, are presented. Focal spot shapes are far from circular, imposing a new challenge in terms of defining size. In this research, after segmentation, the edges of the focal spot are extracted and an ellipse is fitted to the data. The major and minor axes were then used for an estimation of the observed focal spot size.

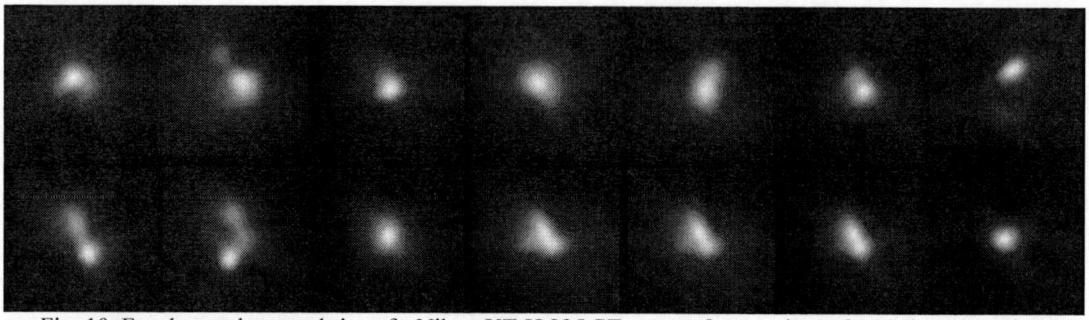

Fig. 10. Focal spot shape and size of a Nikon XT-H 225 ST system for a variety of scanning strategies.

3.2. Focal spot influence on a reconstructed volume

To understand how different focal spot shapes would influence the 3D volume, a single sphere was used as the object of study. The reason for choosing a sphere was twofold. Firstly, it allowed for an easy identification of local deviations induced by the focal spot shape, and secondly, it reduces simulation time. Single projection simulation, with realistic focal spot, takes on average 15 minutes in aRTist. Using a sphere as the measurand, investigation time was reduced, from an average of 37 days, to a few hours as only one projection needs to be simulated. To complete the dataset, and therefore allowing a volumetric reconstruction, the single projection was replicated 3600 times.

The influence of 3 focal spot shapes (Fig. 11), of approximately 25 μm in diameter, on the volumetric model were evaluated. A Gaussian shape spot was used as a reference as this shape can be ideally expected from a well-controlled focal spot at the target material. Other two shapes, namely U- and O-Shape are used for comparison.

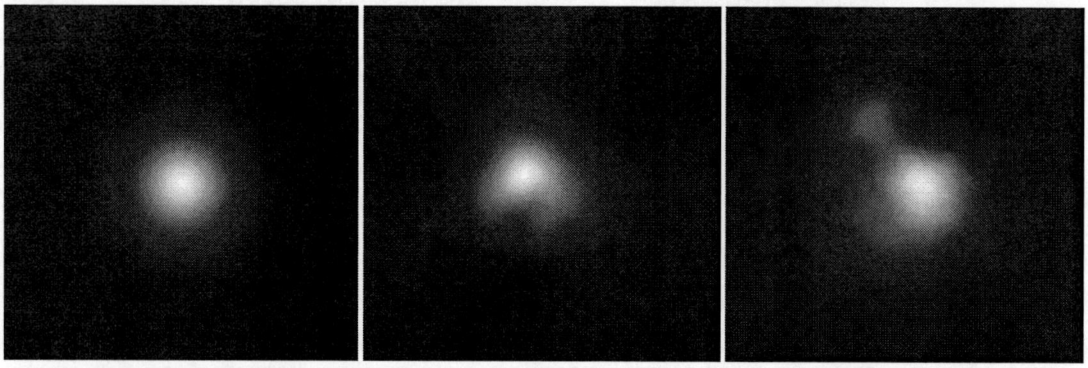

Gaussian spot U-Shape spot O-Shape spot

Fig. 11. Focal spot shapes used in aRTist for the evaluation thereof on the volumetric model.

A first analysis was carried out by comparing the simulated projections, with real focal spots, to point source simulations. It can be noticed from Fig. 12 A, that a Gaussian distributed spot results in an uniform transition between sphere and background along the entire measurand edge. On the other hand, oddly shaped spots change that behaviour. The edge transition becomes dependent on how the focal spot shape is defined, indicating a sharper or blurrier edge which is locally affected.

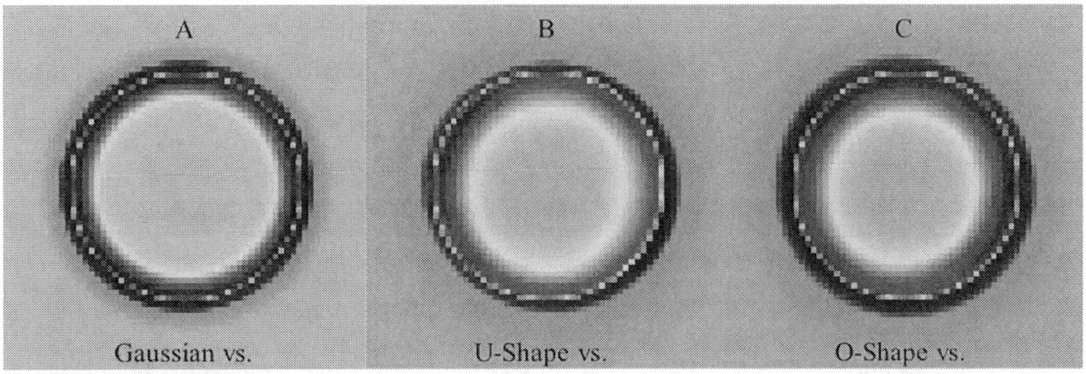

point source point source point source

Fig. 12. Edge comparison in the projection level between real focal spots and point source.

All three volumes were simulated at a magnification of 104 and resolution of 1,9 μm. These were ideal simulations with a monochromatic X-ray spectrum and no noise induced. Reconstruction happened with Nikon's proprietary software CT Pro 3D and subsequent data analysis with Volume Graphics. All volumes were compared to their STL model used in the simulation phase. Fig. 13 illustrate the obtained results.

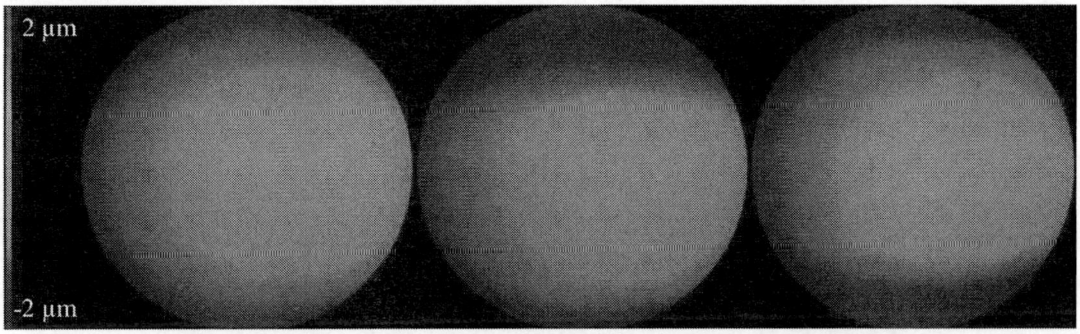

Gaussian spot U-Shape spot O-Shape spot

Fig. 13. Focal spot shape influence on the volumetric volume. Focal spot size estimated to be of 25 μm.

As can be seen in Fig. 13, every focal spot influences the volume differently. The Gaussian reference slightly deviates at top and bottom whereas its equator is in good agreement with the STL model. On the other hand, U- and O-Shape spots have their poles much more influenced by the abnormal focal spot shape. Although focal spot shape influences sphericity, its overall influence is limited. Having a maximum deviation that corresponds to the measurement resolution of 1,9 μm for a focal spot diameter of 25 μm. For smaller focal spots, shape influence is insignificant. Fig. 14 shows the results of both U- and O-Shape spots, for a size estimated to be of 10 μm in diameter, simulated at the same resolution of the experiments presented in Fig. 13.

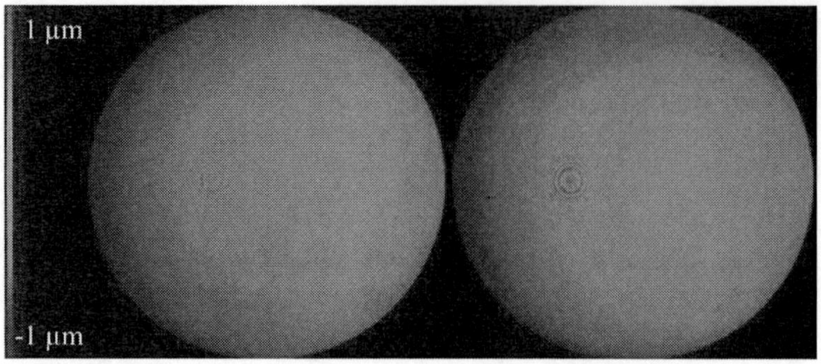

U-Shape spot O-Shape spot

Fig. 14. Focal spot shape influence on the volumetric volume. Focal spot size estimated to be of 10 µm.

Another important aspect regards voxel size. Focal spot shape seems to only influence the dataset at very high measurement resolution. In the experiments of Figs. 13 and 14, measurement resolution was set to 1,9 µm, which in practice is unrealistic as the magnification is set to a factor of 104. For X-ray system working with a cone-beam geometry, a trade-off between sample size and measurement resolution is present. Large sample sizes demand a low magnification (generally smaller than 5), whereas smaller sample sizes can reach a magnification factor of up to 25. Another difference between both approaches usually lies in the measurement power. Large samples (low magnification) frequently need high X-ray tube power (larger focal spots) due to larger penetration lengths. The opposite is also true for small samples at high magnifications, where measurement power is limited (small focal spots) due to the short measurement lengths. In Fig. 15 the U-shape spot was used for both simulations and the focal spot is again about 25 µm in size (high X-ray tube power). As illustrated, the focal spot shape influence diminishes as magnification becomes lower, to a point where it can be completely neglected.

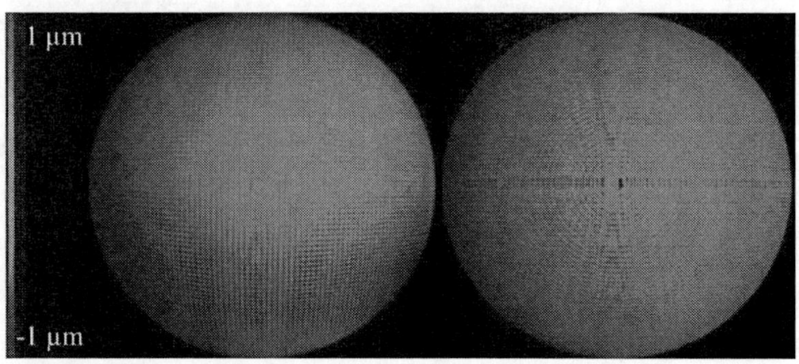

Mag of 24 Mag of 10

Fig. 15. Influence of U-Shape spot of estimated size of 25 µm at magnification factors of 24 and 10.

4. Conclusion

As demonstrated in this research, a Siemens Star Pattern can be used for a focal spot image reconstruction. Its edge information is used to create a sinogram, and through back projection an image representing both focal spot shape and size can be recovered.

The proposed methodology was used to investigate how the target material of an X-ray source influences both focal spot shape and size. It has been shown that pitting at the target material greatly influence thereof. Moreover, it has also been shown that for low measurement power, both features are stable throughout one-hour long scans replicating real measurement conditions.

Focal spot shape has shown to introduce measurement error when a high measurement resolution is being used. Its contribution in a real dataset could be nevertheless negligible compared to other influence factors, such as beam hardening and noise.

A rather important question to be answered falls within the definition of size, as focal spot size is one of the factors defining maximum detectability in XCT.

Acknowledgments

The authors would like thank FWO-SBO MetroFlex project for all the support.

5. References

[1] KRUTH, J.P., *et al.* Computed tomography for dimensional metrology, CIRP Annals, Volume 60, Issue 2, 2011, Pages 821-842, ISSN 0007-8506, https://doi.org/10.1016/j.cirp.2011.05.006.

[2] ANSI 1999 Characteristics of Focal Spots in Industrial X-ray Systems for Use in Non-destructive Testing—Part 5: Measurement of the Effective Focal Spot Size of Mini and Micro Focus X-ray Tubes Document DIN EN 12543-5 (Berlin: Beuth).

[3] ASTM E1165-12(2017), Standard Test Method for Measurement of Focal Spots of Industrial X-Ray Tubes by Pinhole Imaging, ASTM International, West Conshohocken, PA, 2017.

[4] PTW-Freiburg, X-ray test patterns. https://ptw.de/home_start.html?&no_cache=1, Access on Feb. 26, 2019.

[5] BAM, aRTist – Analytical RT Inspection Simulation Tool. http://www.artist.bam.de/, Access on Feb. 26, 2019.

[6] DENG, L. *et al.* A method to determine the detector locations of the cone-beam projection of the balls' centers. Phys. Med. Biol. 60 (2015) 9295–9311. https://doi.org/10.1088/0031-9155/60/24/9295

[7] HILLER, J., MAISL, M., REINDL, L. M. Physical characterization and performance evaluation of an x-ray micro-computed tomography system for dimensional metrology applications, Measurement Science and Technology, v. 23, 18 p., 2012. https://doi.org/10.1088/0957-0233/23/8/085404

[8] LÉTANG, J-M., PEIX, G. On-line X-ray focal spot assessment based on deconvolution using standard imaging devices. NDT&E International 36 (2003) 303–317. https://doi:10.1016/S0963-8695(03)00010-0

[9] OLIVEIRA, A. D., *et al.* The determination of the focal spot size of an X-ray tube from the radiation beam profile. Radiation Measurements (2015), https://doi.org/10.1016/j.radmeas.2015.09.010.

[10] GAMBACCINI, M., *et al.* Measurement of focal spot size in a 5.5 MeV linac. Nuclear Instruments and Methods in Physics Research B 269 (2011) 1157–1165. https://doi.org/10.1016/j.nimb.2011.02.089.

Focal Spot Intensity Distribution Estimation of X-ray tube via Machine Learning

Wenju PENG [1], Yongshun XIAO [1], Qinghan HOU [1]
[1] Department of Engineering Physics, Tsinghua University, Beijing, China
Contact e-mail: xiaoysh@mail.tsinghua.edu.cn

Abstract. X-ray tube is widely used on various occasions such as digital radiography and computed tomography for industrial defects detection that require a high imaging accuracy even up to micron. The focal spot distribution can blur the result of X-ray radiography, which is equivalent to superimposing a point spread function on the image. A high-precision focal spot intensity distribution estimation is needed for evaluating and improving image quality. There are many methods for determining the focal spot size of X-ray tubes. Such as the method EN 12543-5 described, FocView method, pinhole imaging method and so on. However, most methods require high-precision test workpiece and high magnification, but are hard to acquire the accurate distribution of focal spot intensity. In this paper, a novel focal spot intensity estimation method based on X-ray image feature extraction via machine learning was proposed. Some typical focal spot distributions and a large number of random focal spot distributions were used to simulate projection images by using multiple Gaussian distributions to simulate focal spot. Square mesh, radial mesh of equal radian division and radial mesh of equal area division were used to divide these images into regions of appropriate size to capture image features at different locations. The captured image features which are edge direction, edge gradient and image contrast in each region as well as parameters of projection simulation were normalized to train a BP machine-learning network to estimate the distributions of focal spot intensity. The results of computer simulation experiments showed that the method can effectively estimate the focal spot distribution of X-ray tube from the projection image.

Introduction

The quality of X-ray imaging is affected by many factors. One of the most important factors is focal spot size of the X-ray tube. Therefore, it is necessary to measure the size of the focal spot. Actually, this measurement is regard as primary importance in acceptance tests and image quality evaluations of clinical radiography systems

The most common method to estimate the focal spot size are pinhole method slit method and star resolution pattern method[1],[2]. In the past, these methods obtained the image of focal spot by using direct exposure film[3]-[4],[6]. With the development of detector technology, imaging with a digital planar detector array allows many computed methods to be used to measure the spot size. Therefore, the accuracy of above methods have improved a lot. However, the classical methods require specific test object, and the focal spot size is always changing during practical imaging process. If we can get the corresponding focal spot information from an actual image, this information can improve the image quality more accurately and even we can get the third dimension information from one image.

In this paper, a method based on machine learning is proposed. It aim to acquire focal spot intensity distribution from a projected image of an unknown object.

It is worth mentioning that the anisotropy of the energy spectrum of the X-rays emitted by the focal spot is not considered in the simulation. Normalization was done by default when processing projected images to eliminate the influence caused by anisotropy.

In order to verify the feasibility of the method, a confirmatory simulation experiment was performed. The experimental results show that the method is feasible and operative, and it helped us find a way to build a universal method.

1. Method and Confirmatory Simulation

In order to obtain numerous projection image which contain the information of focal spot, a numerical simulation was performed to generate data set. In order to simulate the position and intensity distribution of the source and to ensure the completeness of the data, the source intensity distribution is superimposed by multiple random circle distributions. In addition, to simplify the model, we assume that five random circles can simulate most of the intensity distribution. There were three parameters corresponding to each circle, which are the two coordinates of the center of the circle and the radius of the circle, so that, there were 15 parameters to generate the source intensity distribution corresponding to the five random circles. In the simulation, the simulated focal spot is a 300um-by-300um plane with a resolution of 10 micron, and the simulated phantom was Shepp-Logan phantom. The detector was a 200mm-by-200mm planar detector with 1024-by-1024 pixels, and the magnification is 20.

According to the above, 3,000 examples were calculated and 2723 examples were left after eliminating the non-converged results, of which 2,523 were used as training sets and 200 were used as test sets. The projection image is shown in Figurc 1.

1.1 CNN Method

At first, the method of directly fitting the source intensity distribution based on CNN (Convolutional Neural Network) was considered. Considering the size of the input image is large and the difference between the images is small, it was difficult to achieve the desired result by using the transfer learning method. So, a simple CNN was trained. The network architecture for focal spot distribution estimation is shown in Figure 1.

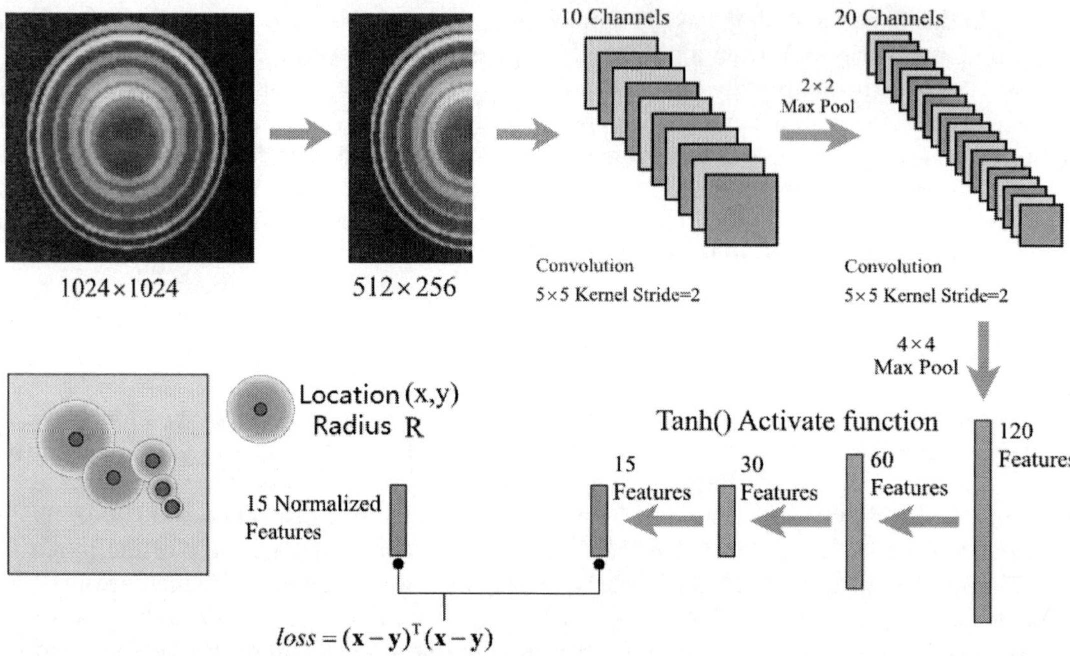

Figure 1 CNN model

All convolutional layers are pooled with a maximum value; the activation function between all fully connected layers selects a hyperbolic tangent (Tanh) function. The output layer consists of 15 values. The 15 output values are used to regress the normalized 15 values of the source intensity distribution. The loss function is defined as:

$$loss = (\mathbf{x}-\mathbf{y})^{\mathrm{T}}(\mathbf{x}-\mathbf{y})$$

1.2 Manual Feature Extraction Method

In fact, as described above, the key to this problem is to obtain subtle changes in the image, and the features extracted by CNN are likely to erase these subtle changes. So it make sense to manually extract features and then train a machine learning network from these features.

Based on prior knowledge and observation of the simulated dataset, we selected three aspects of image features: luminance (directly related to the intensity of the source strength), position of each projection feature (directly related to the position of the source strength), Edge definition (reflecting the degree of defocus of the source).

Each feature was quantified. The quantitative characteristics are listed below:

Luminance—the average and variance of the brightness of each elliptical ring region, counted as 18.

Position—the central coordinate (x, y), the major axis, and the minor axis length (a, b) of all elliptical boundaries, counted as 40.

Edge definition—the mean and variance of the gradient on each 1/8 boundary, counted as 80.

The reason for selecting the edge sharpness values in multiple directions is that the defocus behavior will show significant directionality on the projected image. The description of the 1/8 boundary is shown in Figure 2.

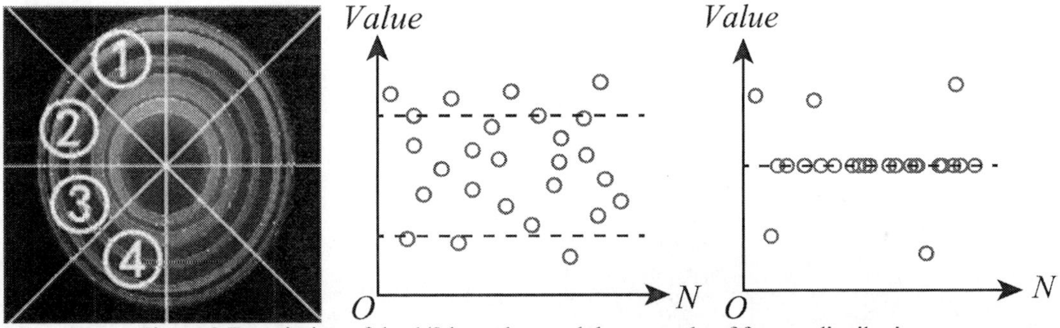

Figure 2 Description of the 1/8 boundary and the example of feature distribution

The main process of manually extracting features is as follows: the image edges were first extracted by the Canny operator, and then the redundant points were removed and boundary fractures were connected so that the boundaries of each ellipse were closed. Next, images were cut into a plurality of regions according to these elliptical boundaries, and the boundaries and internal pixels of each region were counted to extract features. After obtaining the quantified values of all features, the features needed to be filtered and normalized. As shown in figure, if the feature had a relatively uniform distribution on the training set, the feature was retained, and the 0.1and 0.9 quantiles are linearly mapped to [-1, 1]. If the feature's distribution on the training set was concentrated, then the feature is discarded.

After filtering, 115 normalized features were retained and used as inputs to a fully connected neural network. The fully connected neural network consists of three hidden layers with the size of 150, 75 and 35. The calculation of the output and loss functions of the neural network is consistent with the previous description. The network architecture for manual feature extraction is shown in Figure 3

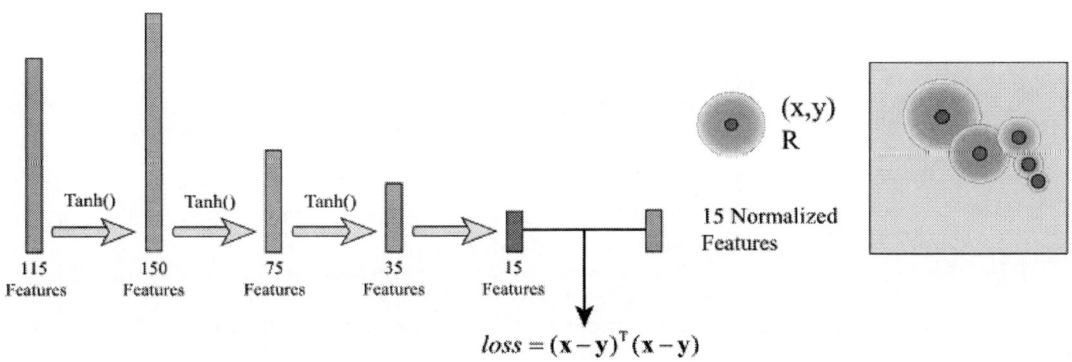

Figure 3 Fully Connected Network

1.3 Results and Discussion

Two types of networks were implemented and trained using the open source deep learning framework Pytorch. After the CNN and fully connected network training was over, the performance of the neural network was tested on 200 samples on the test set. The test criteria was: restore 15 parameters of the neural network output and 15 parameters of the test set label to the source intensity distribution on the 2D plane, and compare the two pictures with the structural similarity function SSIM[7]. The SSIM takes a value between 0 and 1, the closer its value is to 1, the more similar the two image structures are. The distribution histogram of the SSIM values of the two network results in a certain period of time is shown in the Figure 4.

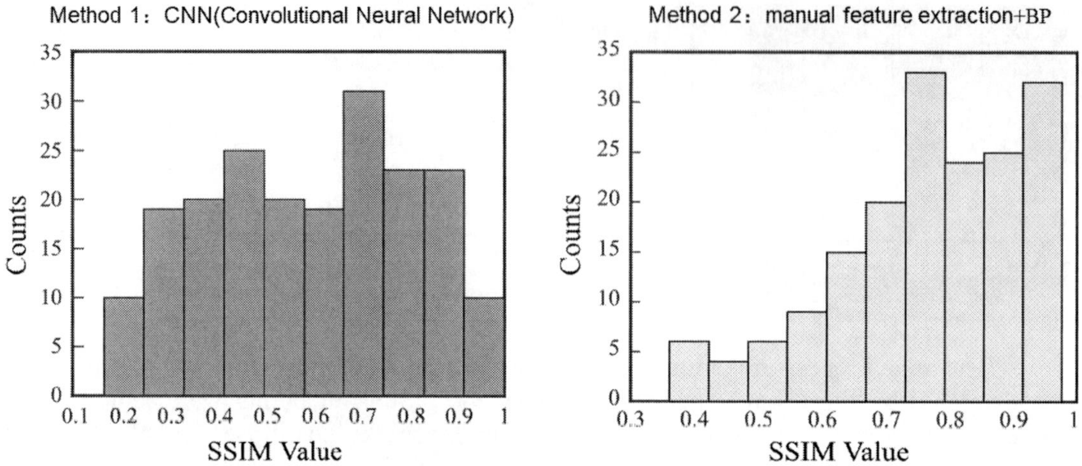

Figure 4 SSIM distribution of CNN and BP

Therefore, in the same training time, the results from method two based on manual feature extraction is better than CNN method. The error reduction process of a fully connected network is shown in the Figure 5. After training 200,000 Epoch, the loss function dropped to about 0.18. From the definition of the loss function, the loss function value meant that each parameter has an error of about 6%, which is acceptable. And the average SSIM on the test set was 0.76, which was improved relative to the correlation of 0.69 given by the standard test method of EN-12543-5.

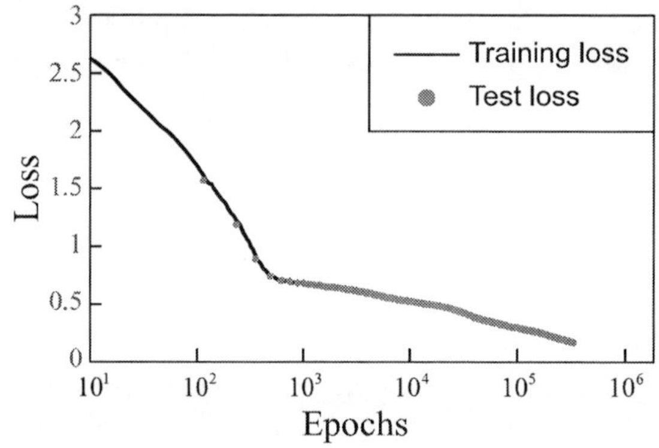

Figure 5 Convergence of training loss

The following Figure 6 is a sample of a fully connected neural network with good inference results (SSIM high) and a poorly inferred result (low SSIM). The source-intensity distribution of samples with good performance in the fully connected network is relatively concentrated, while the samples with poor inference performance tend to be very scattered in the source intensity distribution. The reason can be that for the case where the source intensity distribution is very concentrated (in an extreme case, the ideal point source), the manually extracted features are sufficient to reflect the position of the source (the manually extracted information includes the position of each ellipse). In the case where the distribution is very scattered, the number of conditions for estimating source intensity distribution from the image is large.

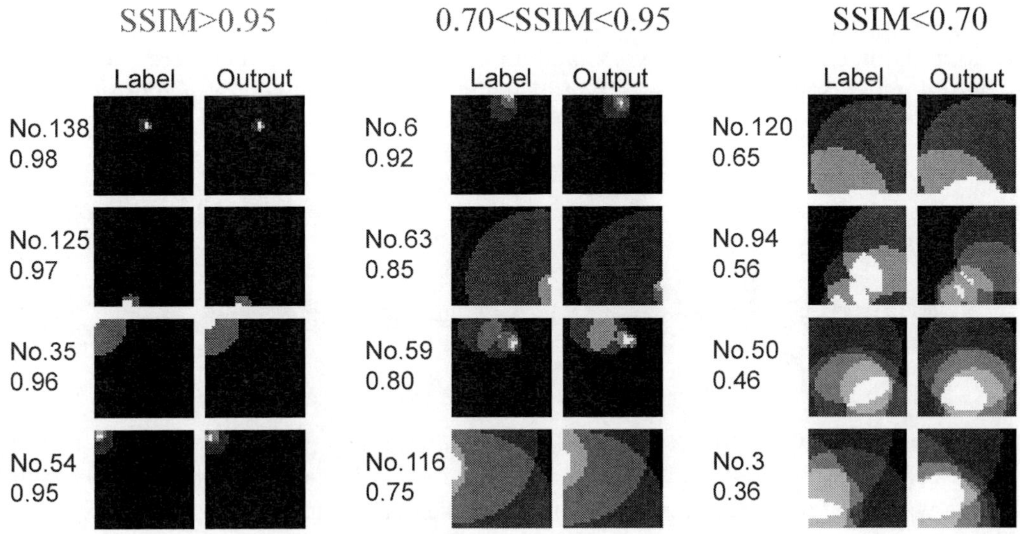

Figure 6 Contrast between labels and outputs

Obviously, the method based on manual feature extraction and fully connected networks can estimate source intensity distribution from one projection image efficiently. If we can extract more image feature and improve network architecture, maybe we can estimate nearly accurate source distribution from a projected image of an unknown test object.

2. Simulation

Different from confirmatory simulation, the virtual planar detector used here was a 40mm-by-40mm plane with 1024-by-1024 pixels. In addition, the phantom whose side length was 10mm consists of three cylinders. The central axes of the three cylinders are parallel to the coordinate axes and are randomly located. The magnification was three. In this case, the focal spot distribution was simulated by using 10 different Gaussian distributions. According to the parameters, 3200 examples were calculated of which 3000 examples were used as train data set to train the neural network. The rest 200 examples were used as test data set.

2.1 Method

To acquire the information about focal spot from the projected image, square mesh, radial mesh of equal radian division and radial mesh of equal area division were used to divide these images into regions of appropriate size to capture image features at different locations. The visualization of three mesh is shown is Figure 7. In order to ensure that the area in most of regions is equal, the outer radius of each ring from the outside to the inside was 724, 512, 362, 256, 181, and 128 respectively. The previous number was $\sqrt{2}$ times of the next number.

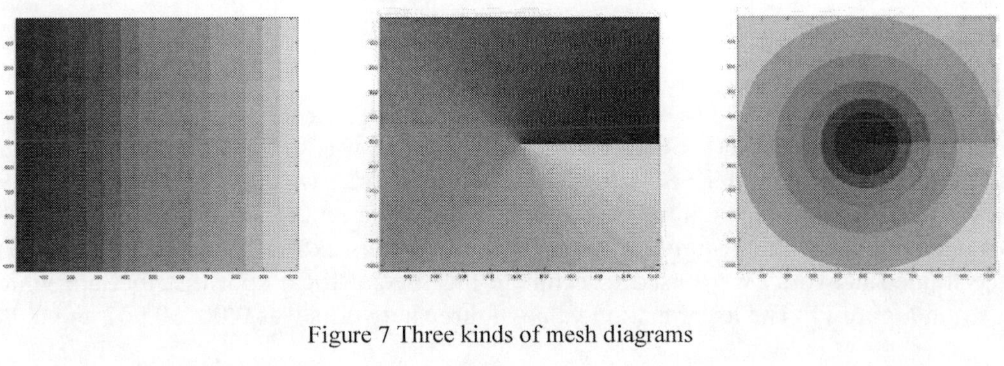

Figure 7 Three kinds of mesh diagrams

Calculated average gradient values and average gradient directions of each regions were used as inputs to neural network, and an array with 30 elements were used as labels. The neural network was a 6-layer fully connected network, whose architecture is shown in Figure 8.

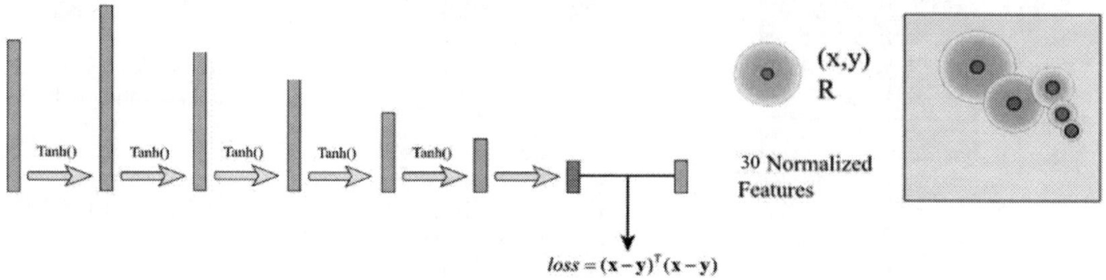

Figure 8 Six-layer fully connected network

Because of the mesh was different, the number of inputs was different. The number of inputs was 512 for square mesh, 480 for radial mesh of equal radian division and 720 for radial mesh of equal area division. The number of features of the 6 layers of square method is 512, 480, 240, 120, 60, 30 respectively, 480, 480, 240, 120, 60, 30 of radial mesh of equal radian division, 720, 480, 240, 120, 60, 30 of radial mesh of equal area division.

2.2 Results

After training 100,000 Epoch for the three networks, the loss function of the three networks dropped to around 0.002, and the third method was the lowest, which was lower than 0.001. With the validation of test data set, there was no overfitting phenomenon, which is shown in Figure 9.

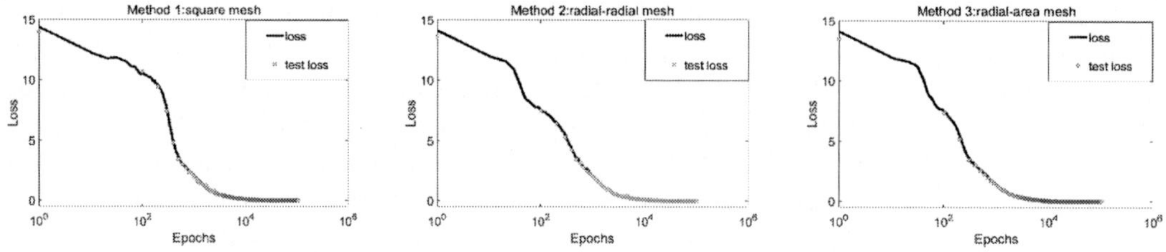

Figure 9 Convergence of training loss of three method

The SSIM have been used to evaluate the network performance on test data set. The distribution histogram of the SSIM values of the three network results is shown in the Figure 10.

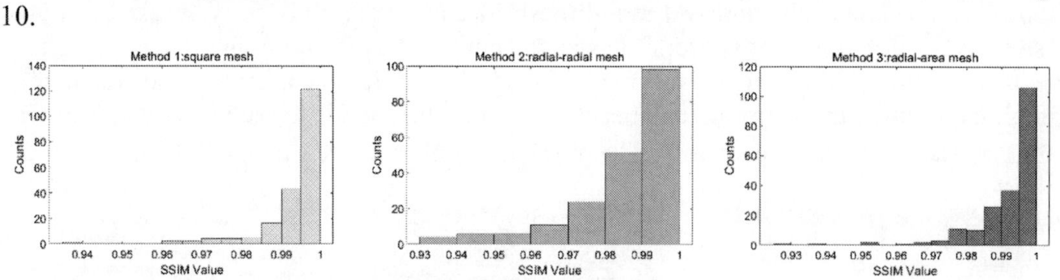

Figure 10 SSIM distribution of three methods

The average SSIM of these three networks was 0.9936, 0.9851 and 0.9917 respectively. The minimum SSIM for these three networks was 0.934, 0.9313, and 0.9281 respectively. Therefore, from the results of SSIM calculation, it is best to estimate the focal spot intensity distribution simply by using 512 parameters extracted by square mesh. These three trained networks were used to estimate the special focal spot distribution, which is shown in Figure 11. The loss function value of three networks was 0.002, 0.002 and 0.00097 respectively. And the SSIM value is 0.9790, 0.9953, 0.9953. The result from radial mesh of equal area division method was the best one. Because of the SSIM values were closer to 1, It is difficult for the human eye to find the difference between the two images. The output from method 3 is also shown in Figure 11.

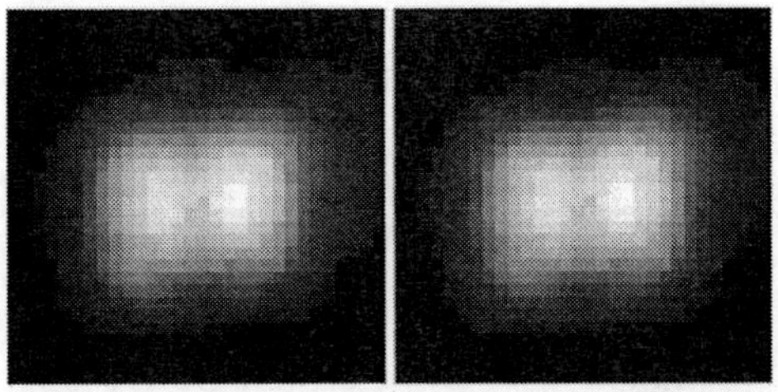

Figure 11 Focal spot distribution and estimation results
From left to right, simulated focal spot and estimated results

3. Discussion

In this paper, a method for focal spot distribution estimating was implemented and tested. This method has better accuracy for estimating the focal spot with a relatively concentrated intensity distribution. Limited by the amount of data and computational complexity, it was only able to acquire focal spot intensity distribution from a single projected image of an unknown object under specific geometric conditions.

If possible, larger data and computing resources could be used in the future research to give the model more complex capabilities, such as variable depth, variable magnification, variable detector size, etc.

It can be concluded that this method can now be used as a reference for the real-time estimation of focal spot size of a single device, if a large amount of train data can be obtained or simulated according to the parameters of the device.

This work was supported by The National Key Research and Development Program of China(2017YFF0104104).

References

[1] Everson, Jeffrey D., and J. E. Gray. "Focal-spot measurement: comparison of slit, pinhole, and star resolution pattern techniques." Radiology 165.1 (1987): 261-264.

[2] Non-destructive testing—Characteristics of focal spots in industrial X-ray systems for use in non-destructive testing—Part 5: Measurementof the effective focal spot size of mini and micro focus X-ray tubes, European Committee for Standardization, CEN, September 1999, EN 12543-5:1999 E

[3] Salamon, M., et al. "Comparison of different methods for determining the size of a focal spot of microfocus X-ray tubes." Nuclear Instruments and Methods in Physics Research Section A: Accelerators, Spectrometers, Detectors and Associated Equipment 591.1 (2008): 54-58.

[4] Di Domenico, Giovanni, et al. "X - ray focal spot reconstruction by circular penumbra analysis—Application to digital radiography systems." Medical physics 43.1 (2016): 294-302.

[5] Jain, Amit, et al. "Focal spot measurements using a digital flat panel detector." Medical Imaging 2014: Physics of Medical Imaging. Vol. 9033. International Society for Optics and Photonics, 2014.

[6] Nishiki, Masayuki. "Evaluation of the effective focal spot size of x-ray tubes by utilizing the edge response analysis." Medical Imaging 2015: Physics of Medical Imaging. Vol. 9412. International Society for Optics and Photonics, 2015.

[7] Hore, Alain, and Djemel Ziou. "Image quality metrics: PSNR vs. SSIM." 2010 20th International Conference on Pattern Recognition. IEEE, 2010.

Wind Turbine Rotor Blade Testing by Dual-Energy Laminography

David SCHUMACHER [1], Dongsheng OU [2], Carineh GHAFAFIAN [1], Uwe ZSCHERPEL [1], Volker TRAPPE[1]
[1] Bundesanstalt für Materialforschung und –prüfung (BAM), Berlin, Germany
[2] Department of Engineering Physics, Tsinghua University, Beijing, China
Contact e-mail: David.Schumacher@bam.de

Abstract. Modern wind turbine rotor blades consist of sandwich shell segments made from glass fiber reinforced polymers. During manufacturing, defects can arise which could lead to failure of the whole component under dynamic mechanical and thermal loads. Hence during operation defects can arise which, if detected, can be repaired locally and in-situ by applying repair patches instead of taking the whole rotor blade down and repair it remotely. This method is much more time and cost effective, since the shut-down time of the energy converter is limited to a minimum. These repair patches can, however, also lead to new defects if not applied optimally. Therefore, it is necessary to control the quality of the repair patches to ensure the best possible restoration of structural integrity of the component. As a rotor blade is an object with a large aspect ratio, X-ray laminography is predestined to provide 3D information of the objective volume. To enhance the amount of information gained from laminographic reconstruction, we use in this study a photon counting and energy discriminating X-ray detector and apply a material decomposition algorithm to the data. By inherently separating the incident spectra within the detection process into two distinct energy bins, the basis material decomposition can provide material resolved images. Choosing glass and epoxy resin as basis materials and numerically solving the inverse dual-energy equation system, the reconstructed laminographic datasets contain highly valuable information about the distribution of the basis materials within the structure. Furthermore, cross- artifacts arising from the limited angle of the projection data can be reduced by this method which allows to investigate structures that were hidden underneath the artefacts.

Introduction

Non-destructive testing of wind turbines is important in the life cycle maintenance of these systems, since they are subjected to high dynamic loads and cyclic fatigue. Offshore systems are exposed to especially extreme mechanical and thermal loads due to higher wind and temperature variations [1]. The segment of a wind turbine blade investigated in this study consists of a sandwich shell segment made from glass fiber reinforced polymer (GFRP), a standard material of wind turbine blades. Each sandwich segment contains a polyvinyl chloride (PVC) foam core with two layers of biaxial (BIAX) glass fiber plies on each side [2, 3]. The shells of turbine blades are manufactured per hand. The high amount of manual fabrication facilitates the formation of manufacturing flaws which have the potential to grow into harmful defects due to the cyclic loading, hence compromising the integrity of the turbine blade or the whole system. The most likely and severe damages are structural failure initiating cracks in the blade shell that occur long before the maximum service life time of 20 years is reached [1].

The in-service detection and validation of flaws in turbine blades is still being improved to-date. In case a flaw has been located and identified, special techniques exist to locally repair the defective volume and thus restore the component design lifespan. Due to time, environmental constraints and ease of availability, however, the quality of the repairs often remains unmonitored in the field. As the repairs are installed per hand on the ropes, variations such as gaps between repair and parent foam, for instance, could result in resin channels. Such imperfections in the repair patches could be further propagated during service and thus turn into potential sources for damage initiation.

To find out if and how sustainably a repair patch can restore the design lifespan of a damaged wind turbine blade, the BAM is conducting research project in which GFRP sandwich shell specimens are manufactured and instantly repaired with various predefined methods. The repaired parts are tested under cyclic load to simulate the fatigue behavior of these specimens [4].

In this study, an X-ray dual-energy laminography technique addresses this matter, enabling a detailed quality assurance imaging method of repair patches installed on blade shells. Like CT, laminography can deliver a three-dimensional volume of the investigate part, but due geometric constrains the depth information is limited [5]. To enhance the reliability of this NDT technique, the laminography is extended by a material resolution gained from spectral information of the detected X-ray photons. Therefore, a photon counting and energy discriminating X-ray detector (PCD) in combination with a basis material decomposition (BMD) algorithm is used. This BMD algorithm enables to separate between two predefined base materials (here: glass and epoxy resin) in the resulting data set.

1. Methods

1.1 Samples

The test specimens were produced as a curved sandwich structure with SAERTEX GFRP laminate face sheets and an AIREX C70.55 polyvinyl chloride (PVC) foam core, representing a portion of a wind turbine rotor blade shell. Two layers of ±45° 812 g/m^2 non-crimp glass fiber fabric were used as face sheets, with additional 1200 g/m^2 unidirectional reinforcement layers in the clamping region to maintain homogeneity in the specimen thickness, illustrated in **Fig. 1**. The vacuum-assisted resin infusion (VARI) process was utilized to produce the shell specimens with the RIMR135/RIMH137 epoxy resin system from Hexion. Specimens were cured first at room temperature for 48 hours, then for 15 hours at an elevated temperature of 80°C.

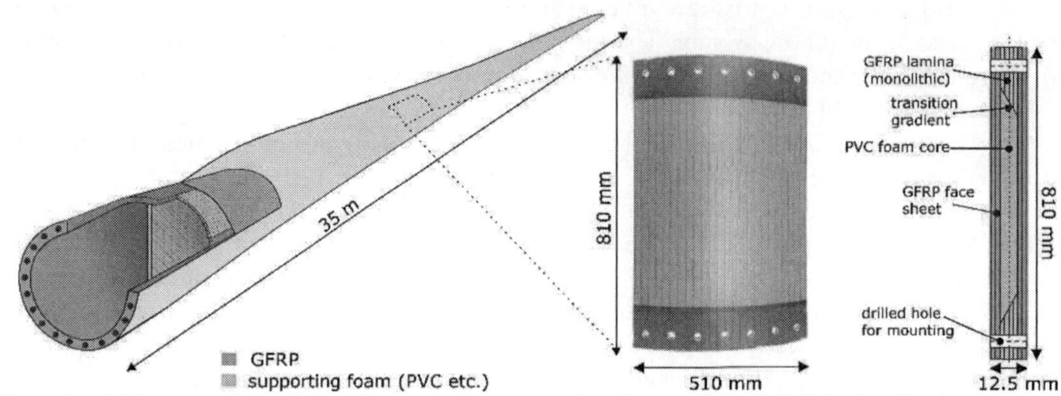

Fig. 1: Sandwich test specimens designed as a representative portion of a rotor blade shell. Side view (far right) shows the ±45° GFRP face sheets, as well as the monolithic clamping region at the top and bottom.

The specimens were then repaired with an adhesively bonded patch, a method often utilized on composite materials, especially in the aerospace industry, where the lost load path is replaced with a new material that is joined to the parent structure [6, 7]. The specimen examined in this study was repaired with an external bonded repair, a relatively quick and simple method often used to repair blade shells, illustrated in **Fig. 2**. This method involves removing the damaged region, then cleaning and applying a filler material and adhesive, upon which the external repair plies are then placed. The loads are thus transferred over and around the damage due to the repair patch, with a decrease in localized stress concentration at the damage boundaries [6]. The shell specimen was repaired entirely by hand, as is done with rotor blade repairs in the field.

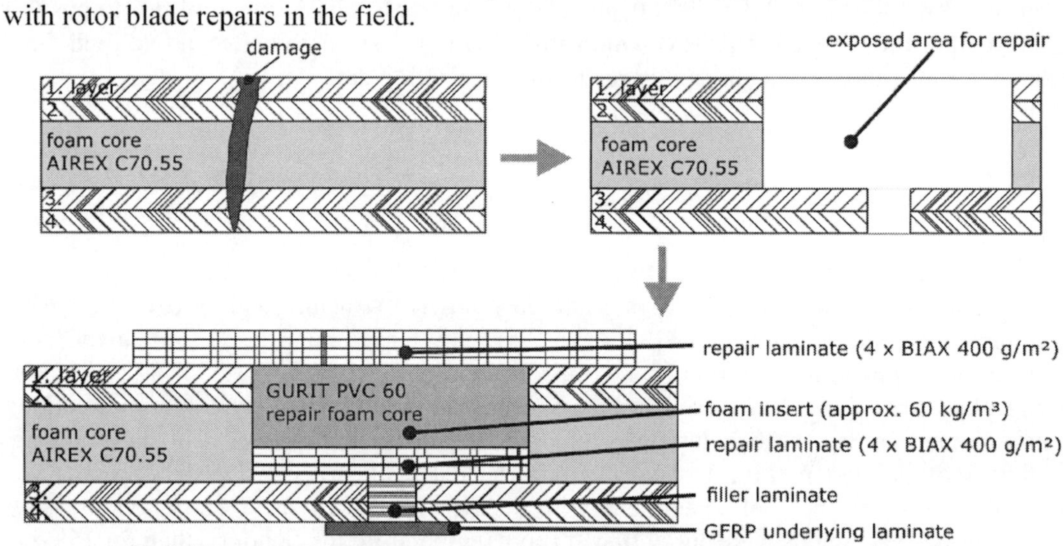

Fig. 2: External bonded repair patch schematic, as applied to sandwich specimen in this study.

The adhesively bonded specimen was then cyclically tested under uniaxial tensile-compressive load (R = -1) to study the effect of the repair patch on the damage and failure mechanisms of the structure.

1.2 Photon Counting Detectors

The direct detection principle of photon counting detectors (PCDs) differs from common indirect detector technology by renouncing scintillation layers (e.g. CsI or Gd_2O_2S) in which the incident X-rays are converted to visible light before the optical photons can be detected (**Fig. 3**). PCDs are based on semi-conductor materials (e.g. CdTe or Si) in which the incident

X-rays are directly converted into electrical signals. Since the amount of charge generated by an X-ray photon is directly proportional to its energy, PCDs are able to determine the energy of the incident X-ray photons and therefore generate spectral information. Additionally, the absence of optical scattering (from the scintillation layer) leads to a much lower inner unsharpness (basic spatial resolution equals usually pixel size) and therefore much sharper images. [8–10]

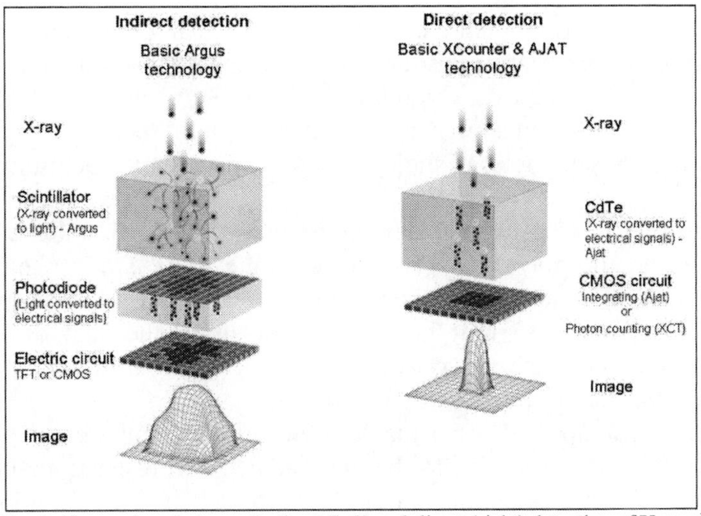

Fig. 3: Comparison between indirect (left) and direct (right) detection of X-rays[1]

The PCD used in this study is based on CdTe monocrystalline. The spectral information is generated by comparing the charge pulse height (i.e. amount of charge) of a detected event with a predefined threshold. Only if the detected pulse height is larger than the threshold set, the counter is incremented. The number of pulse height comparators per pixel gives the number of available energy bins. In the present detector, each pixel is equipped with two pulse height discriminators resulting in two energy bins per pixel (**Fig. 4**). The first threshold cuts off background noise from the detector electronics and the second threshold divides the detected spectrum into a low energy (*le*) and a high energy (*he*) part. This spectral information can be used to generate material specific information from the investigated object.

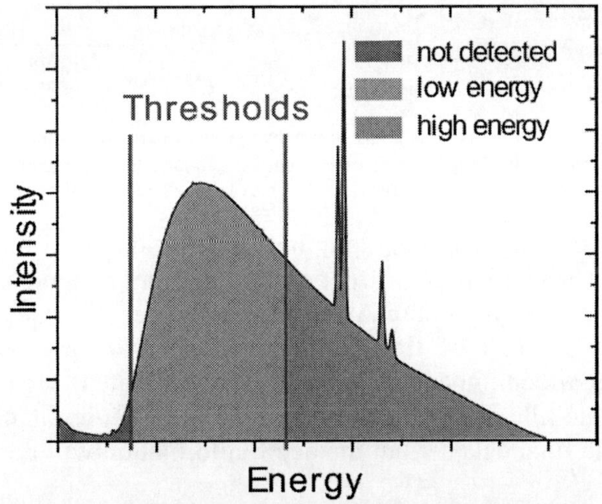

Fig. 4: Principle of spectrum splitting by internal energy thresholding of the PCD.

---

[1] from: http://directconversion.com/technology/

1.3 Laminography

The large aspect ratio (approximately 1:40) of the shell segments does not allow CT to be used to investigate the 3D structure of the volume. Instead, computed laminography is used to reveal the 3D information of the repair patches. Computed laminography is a technique which was developed in the 1930s by B. G. Z. des Plantes [11] for medical purposes. There exist many types of laminographic techniques which are mainly distinguished by the characteristic trajectory. "Rotational laminography" or "swing laminography" are based on an inclination of the sample relative to source and detector. Translational techniques like "co-planar translational laminography" which is used in this study are based on a relative linear movement between sample and source and/or detector. In all techniques, projection images are acquired under different angles of incidence (AOI) from which depth information can be reconstructed.

In the co-planar translational laminography, the X-ray tube executes a linear movement parallel to the horizontal pixel lines of the PCD. During this manipulation a sequence of projection images is acquired by the PCD where each projection image represents the object from a different AOI. After acquisition, the stack of projection images is passed to a *weighted filtered shift-average* reconstruction algorithm [12] which delivers a three-dimensional data set of the investigated object.

Fig. 5 depicts the principle of a co-planar translational laminography. The geometric constraints in **Fig. 5**b show, that due to the max. AOI α and restrictions in source-detector distance (SDD) and object-detector distance (ODD), the achievable resolution in the resulting data set is limited.

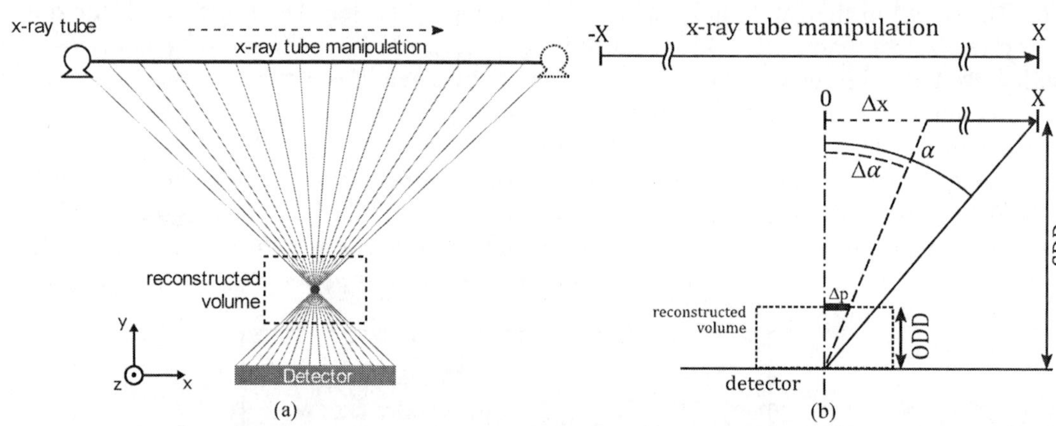

Fig. 5: (a) Principle of co-planar translational laminography to reconstruct a three-dimensional volume. (b) The geometric constraints, especially the limited angle of incidence α and the source-detector-distance (SDD) or object-detector distance (ODD), restrict the achievable resolution in the data set.

As the co-planar translational laminography intrinsically implies a limited view due to the limited AOI, artefacts arise after reconstruction. The amount and intensity of those artefacts is mainly depending on the maximum AOI and the number of projection images acquired during the scan. As a result of the maximum AOI, cross-artefacts may arise in the reconstructed data set which impede the analysis or (automated) segmentation of defective volume. By appropriate filtering of the projection data, the representation of cross-artefacts may be reduced in the final data set but the depth information will be further distorted. [5, 13]

Fig. 6 and **Tab. 1** show the setup and corresponding radiographic parameters for the dual-energy laminography on the repaired and cyclically loaded sandwich shell segments.

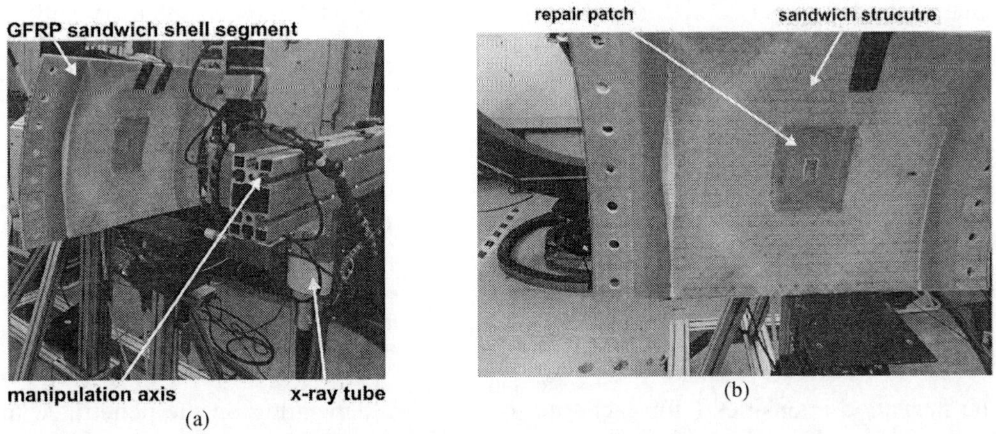

Fig. 6: (a) Setup for dual-energy laminography on a GFRP sandwich shell segment of a wind turbine rotor blade. (b) The repair patch in the middle of the sandwich structure.

Tab. 1: Parameters for dual-energy laminography on a GFRP sandwich shell segment.

tube voltage[kV]	100
tube power [W]	230
spectrum filtering [mm Cu]	1.0
detector type	photon counting
sensor material	CdTe
sensor thickness [mm]	0.75
pixel size [mm]	0,1
detector area [mm²]	102.4 x 51.2
frame rate [Hz]	10
frames per projection	40
exposure time per projection [s]	4
No. of projections	971
max. AOI	± 40.3°
threshold le	$\sim 21\ keV$
threshold he	$\sim 52\ keV$

1.4 Basis Material Decomposition

A method to process the spectral information resulting from the internal energy thresholding of the PCD, is the basis material decomposition (BMD) which was developed by Alvarez and Macovski in the 1970s for medical purposes [14]. The BMD is not specific for PCDs, but since numerous energy bins per projection can be acquired within a single shot and be used for a material reconstruction, PCDs are advantageous for CT or laminographic applications.

The BMD uses the spectral information from the projection data to gain material specific images. The aim is to reconstruct the material thicknesses of two predefined basis materials by measuring the le and he intensities in the X-ray projection images.

In the simplified monoenergetic case of a transmission measurement of two materials i=1, 2 with thicknesses t_1 and t_2, the log. intensity according to Lambert-Beers law is:

$$\ln\left(\frac{I}{I_0}\right) = -\mu_1 t_1 - \mu_2 t_2 \tag{1}$$

Where I is the measured attenuated intensity, I_0 is the unattenuated intensity and μ_i is the attenuation coefficient of the corresponding material i.
If now two energy bins $j = le, he$ are defined, the following equation system can be formulated:

$$\begin{cases} I^{le}(t_1, t_2) = \int\limits_{E_{min}^{le}}^{E_{max}^{le}} I_0(E)\, e^{-\mu_1(E)t_1 - \mu_2(E)t_2}\, dE \\ I^{he}(t_1, t_2) = \int\limits_{E_{min}^{he}}^{E_{max}^{he}} I_0(E)\, e^{-\mu_1(E)t_1 - \mu_2(E)t_2}\, dE \end{cases} \tag{2}$$

The measured intensities I for each energy bin j are depending on the penetrated material thicknesses t of the two basis materials and the energy E. For this polychromatic case, the log. intensity p^j for each energy bin j and measured in each projection image can be expressed by:

$$p^j(t_1, t_2) = -\ln\left(\frac{\int_{E_{min}^j}^{E_{max}^j} I_0(E)\, e^{-\mu_1(E)t_1 - \mu_2(E)t_2}\, dE}{\int_{E_{min}^j}^{E_{max}^j} I_0(E)\, dE}\right) \tag{3}$$

This non-linear relation between the measured intensities p and the penetrated thicknesses t_i can now be approximated for each energy bin j by an appropriate function. This yields an equation system with a matrix of material thicknesses $\mathbb{T}$ and the measured log. intensities as the solution vector $\vec{p}$:

$$\vec{p} = \mathbb{T} \cdot \vec{c} \tag{4}$$

The coefficient vector $\vec{c}$ for this calibration system needs to be determined by known sampling points (i.e. material thicknesses). Depending on the application and positions of the sampling points, different functions may be chosen. In literature, commonly polynomials of degree n or conic functions are proposed [14–16], e.g.:

$$p^j = c_0 + c_1 t_G + c_2 t_E \tag{5}$$

$$p^j = c_0 + c_1 t_G + c_2 t_E + c_3 t_G^2 + c_4 t_G t_E + c_5 t_E^2 \tag{6}$$

$$p^j = \frac{c_0 + c_1 t_G + c_2 t_E + c_3 t_G^2 + c_4 t_G t_E + c_5 t_E^2}{1 + c_6 t_G + c_7 t_E} \tag{7}$$

with

$$\vec{c} = \begin{cases} \vec{a} & \text{für} \quad p^{le} \\ \vec{b} & \text{für} \quad p^{he} \end{cases} \tag{8}$$

Since in the projection images the intensities p^j are given (i.e. measured) and the combination of penetrated material thicknesses t_1, t_2 is searched, the inverse of the calibration system needs to be determined:

$$\vec{t} = \mathbb{P} \cdot \vec{c}^* \tag{9}$$

A (numerical) solution of this measurement system reveals the combination of basis material thicknesses t_1, t_2 for each measured combination of p^{le}, p^{he}. The relation between calibration and measurement system is shown in **Fig. 7**.

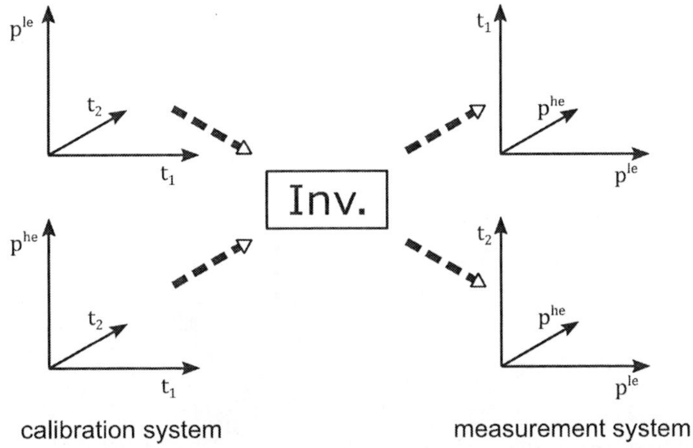

Fig. 7: Relation between calibration and measurement system of the BMD.

This algorithm is used to reconstruct the material thicknesses of glass (representing glass fibers) and epoxy resin based on the energy specific projection images of the laminographic scan in advance of the three-dimensional reconstruction.

2. Results and Discussion

To create a calibration system (described by Eq. 4), the *le* and *he* log. intensities are measured at different penetrated thickness combinations of glass and pure epoxy resin. The resulting sampling points are plotted in **Fig. 8**.

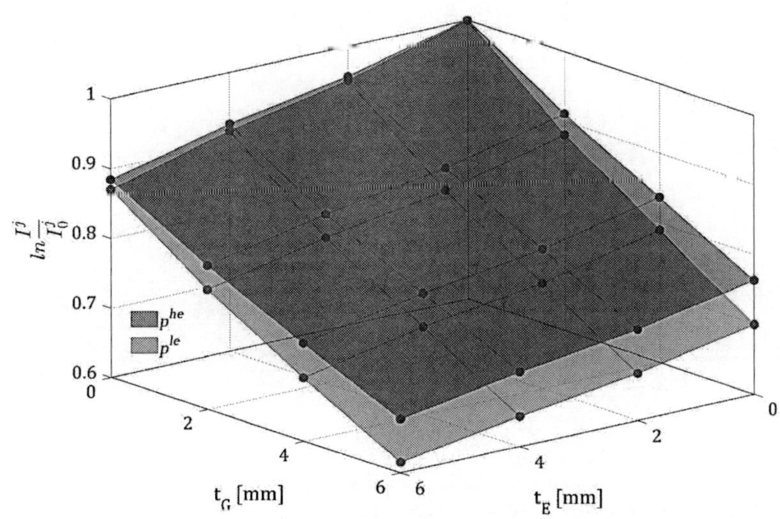

Fig. 8: Calibration system of an example pixel for the BMD of glass (G) and epoxy resin (E) and two distinct energy bins (*le* and *he*).

For reasons of simplicity and stability of the equation system, each of these planes (*le* and *he*) is fitted by a polynomial of degree 1 (Eq. (). The inverse functions of these fits yield the measurement systems which can be used to calculate the present material thicknesses of glass or epoxy resin for a given combination of measured *le* and *he* intensities (**Fig. 9**).

(a)

(b)

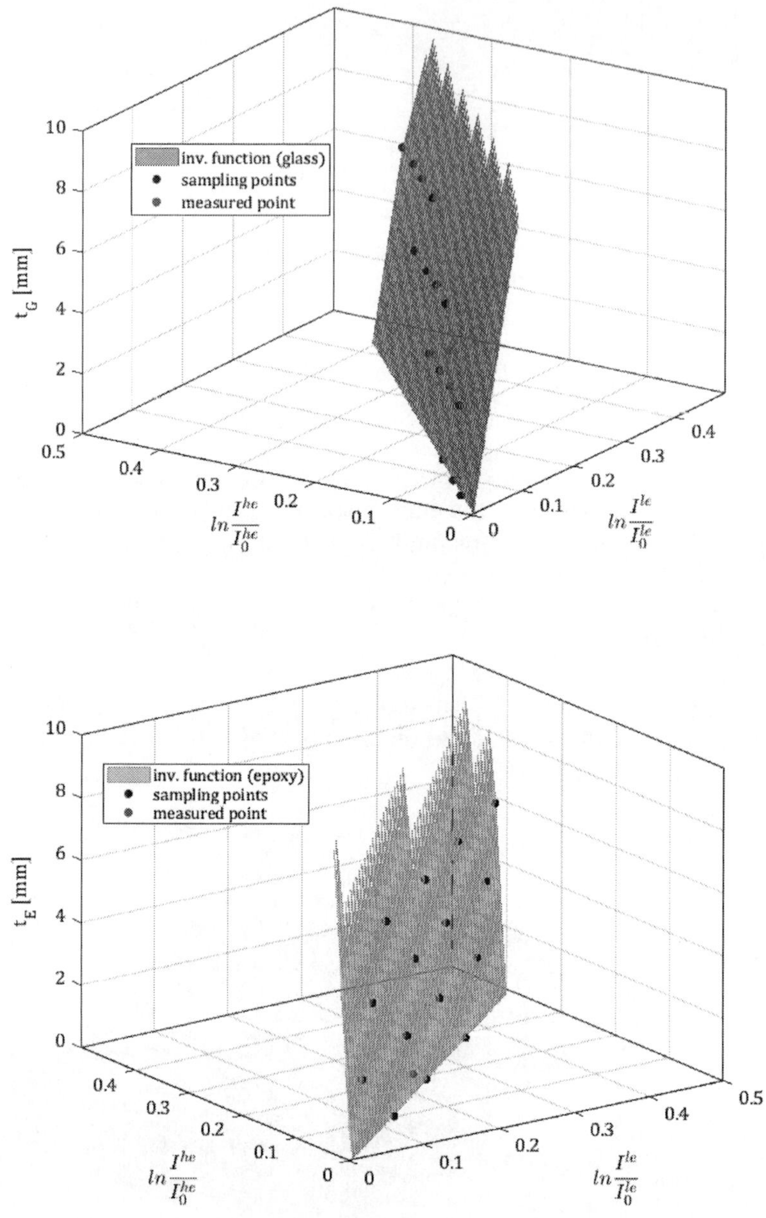

Fig. 9: Inverse fitting functions of an example pixel to calculate the material thicknesses of (a) glass and (b) epoxy resin for a given combination of measured intensities in the low and high energy bin. The block dots represent the sampling points and the red dot represents the measured le and he combination and hence the resulting material thickness in this pixel.

The results of the dual-energy laminography using a BMD to reconstruct the material thicknesses of glass and epoxy resin are shown in Figures 10 – 13.
Each of the Figures 10 – 12 represents another layer at the same lateral position in the reconstructed 3D volume. The subfigures contain the conventional laminography result (without BMD) next to the laminographies with the reconstructed basis materials glass and epoxy resin. The higher the intensity in the BMD data sets, the higher the amount of the corresponding basis material in this position.

Fig. **10** represents a slice close to the front surface of the sandwich shell segment. The intensive artefacts in case of the conventional laminography (Fig. **10**a) impede an analysis of these and the surrounding locations. These artefacts are suppressed by the BMD due to material selective reconstruction. The fine bright lines in Fig. **10**a could be misinterpreted as glass fibers, but Fig. **10**b reveals that only bundles of fibers and the gaps in-between can be resolved because the spatial resolution of the detector is too high to resolve single fibers. The epoxy resin data set in **Fig. 10**c shows dark spots which correspond to voids in the epoxy resin. The large number of voids in this slice cannot be recognized in **Fig. 10**a due to the overlapping material information.

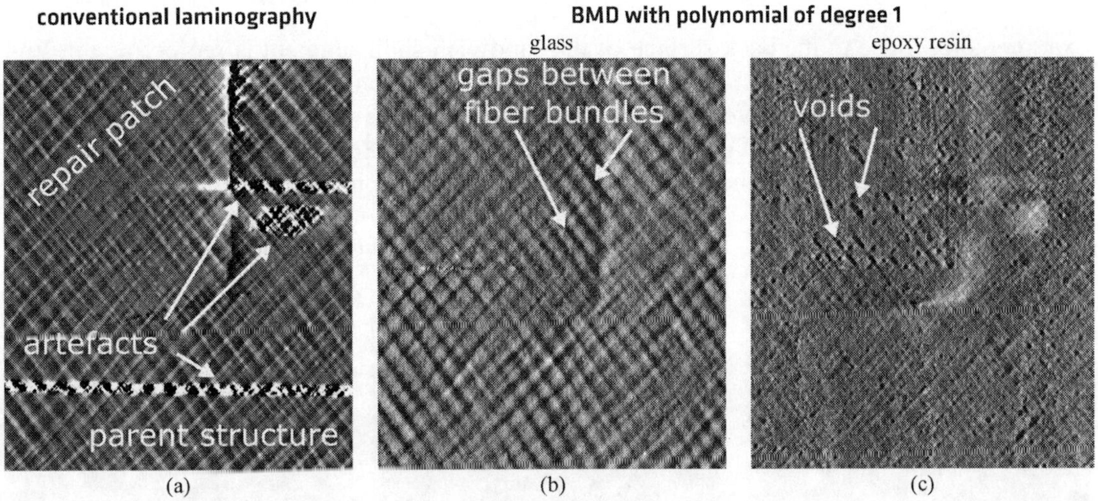

Fig. 10: Reconstructed slice close to the front surface of the sandwich shell segment. (a) Conventional laminography and the result of the BMD for (b) glass and (c) epoxy resin. Bright areas indicate a high amount of the present basis material.

Fig. 11 represents a slice in the middle of the sandwich shell segment where no glass fibers and rather only foam are present. Due to the artefacts in case of the conventional laminography (**Fig. 11**a) it is not clear, whether the edge of the repair patch and the drill hole are completely filled with resin or not. The epoxy resin reconstruction reveals, that the hole is homogeneously filled with resin, but the outer edge of the repair patch is not. The boundary of the repair patch consists partially of foam where an evenly filled resin channel was supposed to be. This indicates an insufficient connection between patch and parent structure which can be the origin of future failure initiation.

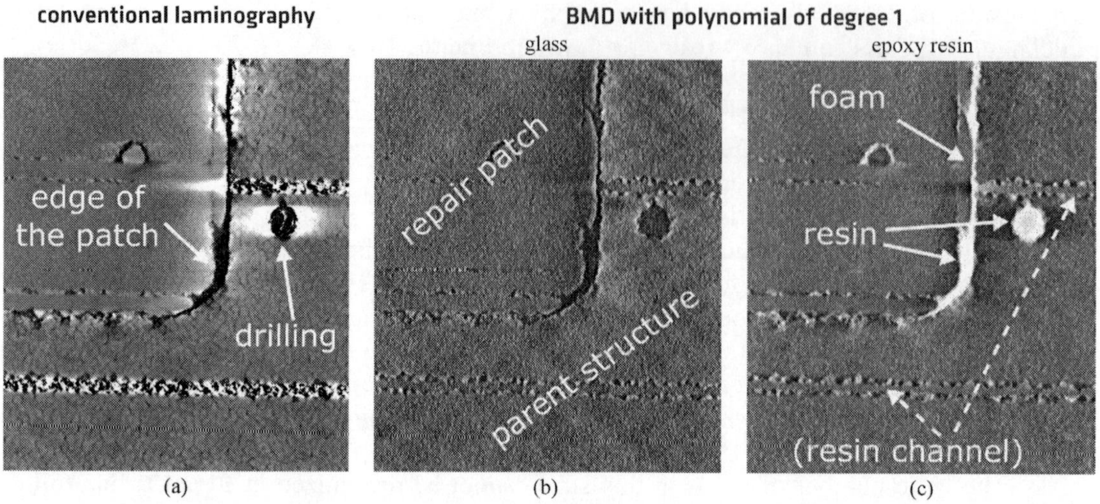

Fig. 11: Reconstructed slice in the middle of the sandwich shell segment. (a) Conventional laminography and the result of the BMD for (b) glass and (c) epoxy resin. Bright areas indicate a high amount of the present basis material.

Another slice close to the back surface of the sandwich shell segment is represented in **Fig. 12**. It shows that an additional layer of glass fiber fabric which was used for the repair patch, since in Figures 12a and 12b only the area of the repair patch indicates glass fiber bundles. The parent structure is free of glass fibers in this slice. **Fig. 12**c reveals that both, the edge of the repair patch and the drill hole are evenly filled with resin in this slice.

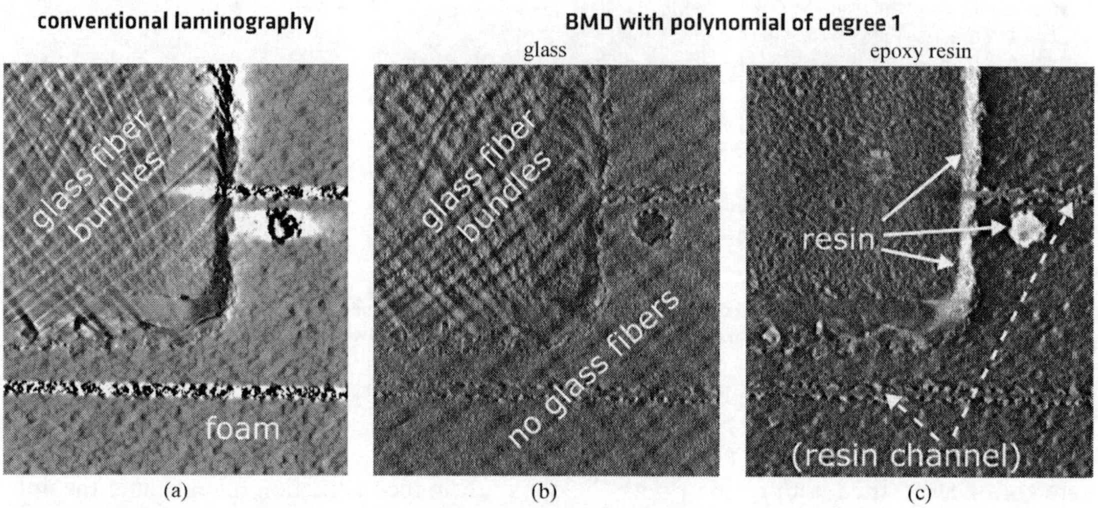

Fig. 12: Reconstructed slice close to the back surface of the sandwich shell segment. (a) Conventional laminography and the result of the BMD for (b) glass and (c) epoxy resin. Bright areas indicate a high amount of the present basis material.

The representation of the reconstructed volume in orthogonal slices allows to investigate the structure in more detail. Digital cross-sections of the conventional laminography and the epoxy resin reconstruction in **Fig. 13** show how the cross-artefacts are being suppressed by the BMD. In case of the conventional laminography (**Fig. 13**a), the artefacts overlay parts of the structure. Especially in the XY-plane, the artefacts of the drilled hole overlay a large part of the cross-section which impedes a distinct evaluation of the integrity of the component. In the epoxy resin reconstruction of the BMD (**Fig. 13**b), the XY-plane reveals a homogeneously resin filled drilled hole. The XY- and XZ-plane likewise reveal that the edge of the repair patch is not filled with resin across the entire depth of the component. Close to the front surface (in y-direction), parts of the resin channel contain a piece of foam, signifying that the repair patch is not homogeneously surrounded by resin.

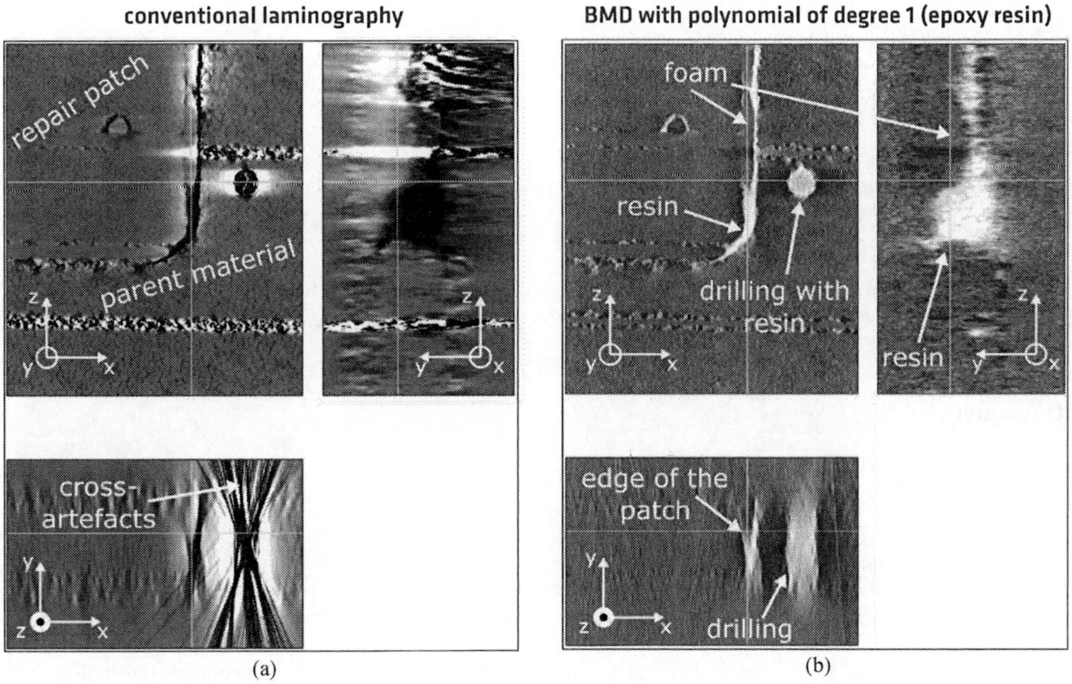

Fig. 13: Digital cross-sections through the reconstructed 3D volume from the laminographic scan. (a) Conventional laminography and (b) epoxy resin. Bright areas indicate a high amount of the present basis material.

The results also show an inherent drawback of the laminography as a 3D-reconstruction technique: No structures parallel to the scanning (x-)direction can be reconstructed due to direction selective high-pass filtering in advance of the 3D-reconstruction. This can particularly be observed in Figures 11c and 12c at the horizontal resin channels which seem not to be filled with resin but which actually are.

The use of different fitting functions (as in Eq. (6) or (7)) on the one hand yields a better approximation of the function with the sampling points but on the other hand results in a less stable equation system [17]. This in turn results in a high amount of salt-and-pepper noise after the BMD, especially in case of the conic function (Eq. (7)), and makes the results less accurate and reliable.

3. Conclusion

By using the spectral information of a photon counting detector and a suitable material reconstruction algorithm (*here:* basis material decomposition) the significance of a laminographic investigation on a wind turbine rotor blade shell could be increased. The material specific laminographic data sets revealed small voids, gaps between glass fiber bundles and lack of resin which were not detectable or not distinctly recognizable in the conventional laminographic data set. Heavy cross-artifacts in the conventional laminographic data set impeded a detailed investigation of the GFRP rotor blade shell segment since they overlaid structural information. These artefacts were suppressed by the material specific reconstruction using the BMD. The epoxy resin specific reconstruction shows unevenly distributed resin around the repair patch which results in an inhomogeneous connection between the repair patch and the parent structure. This method allows to detect damage and evaluate repaired structures on wind turbine blades more reliably and therefore ensure the design lifespan of wind turbines to be reached. By using a mobile miniature X-ray tube and an appropriate manipulation system, this method might have the potential to be applied on-site offshore turbines.

Acknowledgments

The authors would like to acknowledge the financial support provided by National Natural Science Foundation of China (No. 51727813) and Tsinghua University Initiative Scientific Research Program.

We would also like to acknowledge Dustin Nielow for the preliminary work which became the basis of the rotor blade focus study within the LeBeWIND project at the BAM.

References

[1] A. Kupsch *et al.*, "X-ray laminographic inspection of sandwich shell segments for wind turbine rotor blades," in *Proceedings 12th European conference on Non-Destructive Testing*, 2018, pp. 1–8.

[2] F. Grasse, "Beitrag zur Untersuchung des Betriebsfestigkeitsverhaltens von Rotorblättern für Windenergieanlagen im verkleinerten Maßstab," Technische Universität Berlin, 2013.

[3] M. Gaal, F. Schadow, D. Nielow, and V. Trappe, "Air-coupled ultrasonic ferroelectret transducers with additional bias voltage for testing of composite structures," in *10th International Symposium on NDT in Aerospace*, 2018.

[4] V. Trappe and D. Nielow, "Fatigue Loading of Sandwich Shell Test Specimens with Simulated Production Imperfec-tions and In-situ NDT," in *Proceedings of 7th International Conference on Fatigue of Composites*, 2018.

[5] D. Schumacher, U. Ewert, and U. Zscherpel, "Influencing parameters on image quality using photon counting detectors for laminography," in *7th European-American Workshop on Reliability of NDE*, 2017, pp. 1–11.

[6] M. A. Caminero, S. Pavlopoulou, M. Lopez-Pedrosa, B. G. Nicolaisson, C. Pinna, and C. Soutis, "Analysis of adhesively bonded repairs in composites: Damage detection and prognosis," *Compos. Struct.*, vol. 95, pp. 500–517, Jan. 2013.

[7] D. J. Lekou and P. Vionis, "Report on Repair Techniques for composite parts of Wind Turbine blades," 2002.

[8] C. Ullberg, M. Urech, N. Weber, A. Engman, A. Redz, and F. Henckel, "Measurements of a Dual-Energy Fast Photon Counting CdTe Detector with Integrated Charge Sharing Correction," in *Medical Imaging 2013: Physics of Medical Imaging*, 2013, vol. Vol. 8668.

[9] K. Taguchi and J. S. Iwanczyk, "Vision 20/20: Single photon counting x-ray detectors in medical imaging," *Med. Phys.*, vol. 40, no. 10, 2013.

[10] D. Schumacher, U. Zscherpel, and U. Ewert, "Recent Developments of Photon Counting and Energy Discriminating Detectors for Radiographic Imaging," in *2017 Far East NDT New Technology & Application Forum (FENDT)*, 2018, pp. 281–285.

[11] B. G. Z. des Plantes, "Eine neue Methode zur Differenzierung in der Röntgenographie (Planigraphie)," *Acta radiol.*, vol. 13, no. 2, pp. 182–192, Mar. 1932.

[12] Andreas Deresch, K.-U. Thiessenhusen, U. Ewert, and C. Bellon, "Schneller Shift-Rekonstruktions-Algorithmus für die koplanare Translationslaminographie," in *DGZfP Jahrestagung*, 2014.

[13] K. Thiessenhusen, U. Ewert, B. Redmer, and K. Bavendiek, "Geometric Corrections in Coplanar Translational Laminography," in *International Symposium on Digital Industrial Radiology and Computed Tomography*, 2011, pp. 1–8.

[14] R. E. Alvarez and A. Macovski, "Energy-selective Reconstructions in X-ray Computerized Tomography," *Phys. Med. Biol.*, vol. 21, pp. 733–744, 1976.

[15] L. A. Lehmann, R. E. Alvarez, and W. R. Brody, "Generalized image combinations in dual KVP digital radiography," *Med. Phys.*, vol. 8 (5), pp. 659–667, 1981.

[16] H. N. Cardinal and A. Fenster, "An accurate method for direct dual-energy calibration and decomposition," *Med. Phys.*, vol. 17 (3), pp. 327–341, 1990.

[17] D. Ou, D. Schumacher, U. Zscherpel, and Y. Xiao, "Dual-energy materials characterization methods for laminography image enhancement based on photon counting detector," *J. Instrum.*, vol. 14, no. 02, p. P02018, Feb. 2019.

Multimodal Transfer Functions for Talbot-Lau Grating Interferometry Data

Lucas DA CUNHA MELO[1,2], Bernhard FRÖHLER[2], Johannes WEISSENBÖCK[2], Johann KASTNER[2], Christoph HEINZL[2]
[1]TU Wien, Vienna, Austria
[2]University of Applied Sciences Upper Austria, Wels, Austria
Contact e-mail: Lucas.daCunhaMelo@fh-wels.at

Abstract. Talbot-Lau grating interferometry X-ray computed tomography (TLGI-XCT) generates three modalities depicting different features in the scanned specimen. The combination of two or even all three modalities is often required for an in-depth analysis of the material. We present a multimodal transfer function widget for this purpose. It enables users to edit the combined transfer functions of three modalities, resulting in a customized fused image. We implement methods to arrange the individual transfer functions in two different intuitive layouts. The weighting for each modality can be easily modified and the resulting fusion image is immediately updated. We provide a combined histogram over all modalities within a triangular visualization which we call trimodal heatmap. Through this interactive analysis technique, the user can quickly gain insights into the analysed material. We show the effectiveness of our technique with the help of real-world TLGI-XCT datasets.

1. Introduction

Due to advances in X-ray computed tomography (XCT) hardware [1, 2], Talbot-Lau grating interferometry (TLGI) has recently become available in laboratory XCT devices. Such TLGI-XCT devices deliver three imaging modalities:

1. The attenuation contrast (AC), resulting from the attenuation of X-rays in the specimen, as also provided in conventional XCT devices.
2. The differential phase contrast (DPC), which contains information about the refraction of X-rays. It is highly sensitive to changes in the material, such as interfaces, delamination or fractures.
3. The dark-field contrast (DFC), which provides information about X-ray scattering DFC shows high responses for X-ray scattering, such as those from micro-pores.

Specimens are scanned with a TLGI-XCT device in particular when the attenuation contrast from a conventional XCT device provides insufficient information. With TLGI-XCT, it is possible, for example, to obtain additional clues on structures smaller than the voxel size (provided by DFC) or on boundaries between materials with very similar attenuation coefficients (provided by DPC). Therefore, it is necessary to combine the three modalities for a comprehensive analysis.

For most investigation scenarios, it is also vital to explore the transfer function (TF) design space and the blending of the modalities. The user needs to be able to interactively modify the TFs and immediately see the resulting blended image. For a single modality, the design of TFs is a well-addressed topic, but there are no fully established solutions for multimodal data and proposed techniques offer limited interactivity [9, 10]. Therefore, we propose a technique to interactively define multimodal TFs and blend them based on user-defined weights, specifically for the case where two or three modalities are analysed.

2. Related Work

As TFs for volume rendering are a central topic in the field of scientific visualization, being aware of the relationship with the field of materials science is essential. Heinzl and Stappen [11] identify the intersection of visual computing and materials science in a state-of-the-art report where they show how the two fields profit from each other. One of the acknowledged issues in the analysis of materials is that passive visualizations have reached their limits, emphasizing the need for interactivity.

With the lack of established solutions for interactive multimodal material analysis, it is often the case that the user's only option is side-by-side comparison. Extensions could be considered, such as checkerboard patterns or other multi-image views (e.g., as presented by Malik et al. [12]). However, these techniques applied to materials exploration would not provide as much spatial context as a fully fused view, where different modalities can intersect. Elaborate TFs can enable such views.

Several methods have been proposed to define multidimensional and multimodal TFs. In the next paragraphs, a short introduction to the methods most closely related to our approach is given. For a more comprehensive overview, referring to literature such as the state-of-the-art report on research surrounding TFs for direct volume rendering by Ljung et al. [4] is advised. They classify TFs into six categories, two of which are of particular importance for this work: dimensionality and user interfaces. These two aspects are also closely tied together, as an increased dimensionality poses a challenge in the design of a user interface: dealing with more dimensions means that there are more attributes to manipulate, which is a challenge, considering the 2D limitation of displays (e.g., computer monitors). To help illustrate the problem of dimensionality, consider the simplest type of TF (the 1D TF), which already requires two dimensions: intensity values on the horizontal axis, opacity on the vertical axis.

Kniss et al. [5] propose 2D TFs where the first and second derivatives of the scalar points are used additionally to enable a more precise voxel classification. It allows the user to manipulate regions (such as boxes and pyramids) and assign them a colour. Zhou et al. [6] combine this 2D TF technique with the 1D TF to enable up to three attributes in a 3D TF. Two attributes are selected using the 2D TF by defining regions with a lasso tool, which can be further selected using the third attribute. TFs using parallel coordinates plots for n-dimensional data such as those proposed by Zhao and Kaufman [7] and Guo et al. [8] are attractive due to the increased scalability (limited only by screen size), but at the cost that most relationships between attributes are lost, as each axis in this plot can connect to at most only two others.

Although these methods enable high-dimensional TFs to be interactively manipulated, they do not necessarily lead to satisfactory results for multimodal data, such as the data from TLGI-XCT. The additional attributes used are either extracted from the scalar volume (e.g., gradient magnitude, variance, skewness) or they use inherently multivariate data (e.g., pressure, temperature and speed from a hurricane dataset).

Lawonn et al. [9] offer an overview of multimodal visualization for medical data, where a large volume of publications is cited. However, none of them offer the level of interactivity this work aims at. Based on information theory, Bramon et al. [10] present a

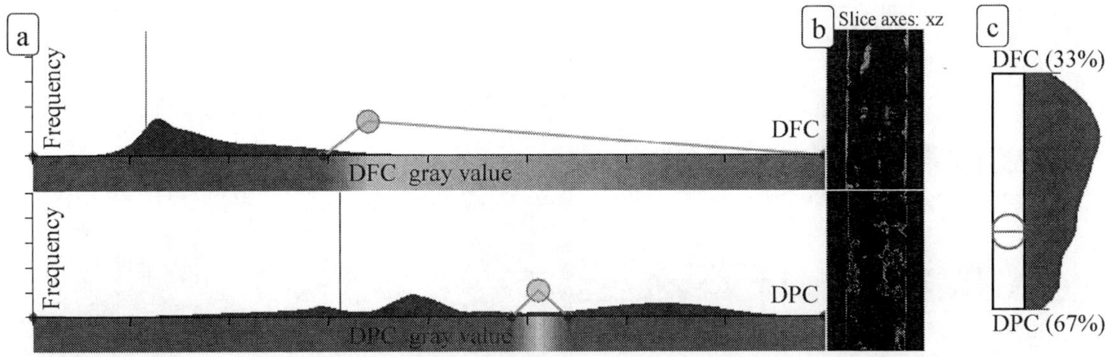

Fig. 1. Bimodal transfer function widget with two modality TFs (a), their respective modality slicers (b) and the weighting widget with co-occurrence histogram (c), which is a visualization of the dominance of grey values in the modalities. For instance: if a voxel from modality *A* has a higher value than the voxel of modality *B* with the same coordinates, this pair of voxels contributes to the histogram in a position closer to modality *A*'s end of the slider.

multimodal TF design as well, but their techniques also defy our interactivity goal, as their method can function entirely without user intervention, optionally allowing the user to make only two choices (choosing a reference image for the colour fusion and proposing a target distribution to decide which features to enhance).

A concept study for the widget proposed in this work has been presented at the 9th international conference on industrial computed tomography, where the developed tool applies the proposed TF in 2D slices only [3]. In this work, we expand that concept by applying the TF to a 3D visualization of the volume, as well as proposing a method to provide the user with visual context of the data.

3. Multimodal Transfer Function Design

Our current multimodal transfer function designs support the analysis of two or three modalities. For both scenarios, the proposed TF definition is divided into two aspects: the first is the individual TF definition for each input modality, described in Section 3.1; the second is the weighting of each modality (and thus their corresponding TF) in the final visualization, as described in Section 3.2. Currently, all modalities are required to have the same dimensions and resolution.

3.1 Per Modality Transfer Function Definition

For the definition of each modality's individual TF, we employ a 1D TF widget, as shown in Fig. 1(a) and Fig. 2(a). It offers a histogram of binned scalar attribute data to help the user gain context of the modality's intensity distribution. The colour and opacity are set by manipulating control points to create linear piecewise functions. The colour of the TF for each intensity is shown underneath the axes of the chart; the shown colour is however made half-transparent such that the axes are always visible. Where grey is shown in Fig. 1(a), the image therefore shows black. We chose this design due to its popularity, especially outside the visualization research community [4]. In theory, however, any TF design can be used here, e.g., the 2D TF [5] referenced in Section 2.

Next to each of these TF widgets, a volume slicer is displayed, as can be seen in Fig. **1**(b) and Fig. **2**(b). Any change in the individual TFs triggers an instant update of the slicer, showing the result of the defined TF on its corresponding modality. The slicer was chosen due to its relatively cheap rendering, in contrast to a 3D visualization. Displaying either or both is, however, possible. The slice number can be changed, as well as the slice axis (XY,

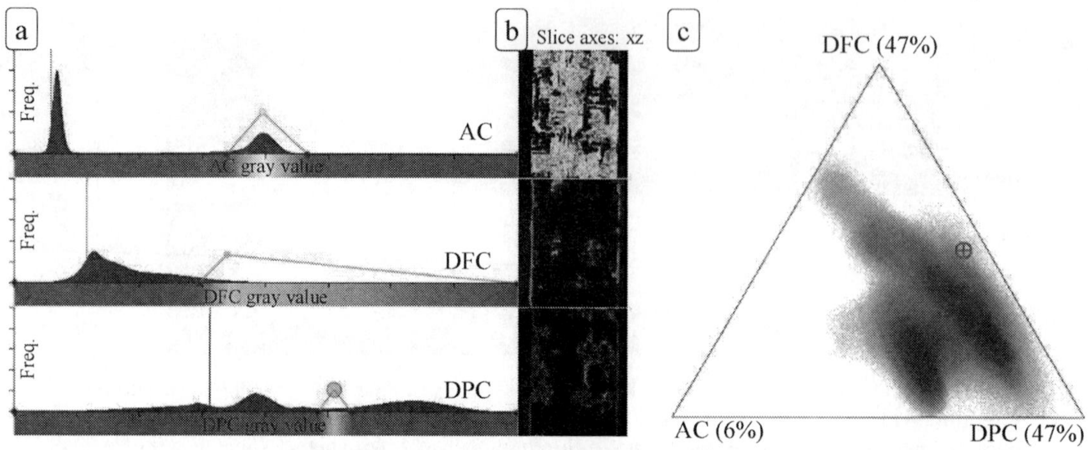

Fig. 2. Trimodal transfer function widget with three modality TFs in the stacked layout (a), the respective modality slicers (b) and the weighting widget with trimodal heatmap (c).

XZ, YZ). Panning and zooming interactions are also provided. When appropriate TFs are selected, the user can switch to selecting weights such that the blended visualization fits their needs. This process is not strictly sequential; the user can switch back to refining the TFs at any time.

3.2 Weighting

The bi- and trimodal variants of our weighting widgets differ, although both are based on the same principles. For the bimodal case, a slider is provided to set the weights of the modalities, as can be seen in Fig. 1(c). On each end of the slider, a label of one modality is displayed alongside its influence. For a pair (A, B) of modalities, the slider is positioned initially halfway between A and B, resulting in the weights $(0.5, 0.5)$. Moving the slider all the way towards modality A, for instance, results in an increasingly stronger influence of modality A and finally in the weights $(1, 0)$. The slider thus controls an interpolation parameter that is used in the final renderings.

Next to the slider, we propose a histogram that provides information on the co-occurrence of the values of each modality. This histogram is calculated by first normalizing each modality (i.e., linearly mapping its intensity values to the range [0,1]) and then performing a voxel-wise normalization across the modalities. In other words, for two modalities A and B and each pair of voxels $a(\boldsymbol{x}) \in A$ and $b(\boldsymbol{x}) \in B$ with equal coordinates $\boldsymbol{x} = (x, y, z)^T$, the resulting voxels are $a'(\boldsymbol{x}) = \frac{a(x)}{a(x)+b(x)}$ and $b'(\boldsymbol{x}) = \frac{b(x)}{a(x)+b(x)}$. Note that $a'(\boldsymbol{x}) = 1 - b'(\boldsymbol{x})$, which means only the value for a' needs to be computed and stored in a new volume, the value of b' is computed dynamically when required.

The resulting volume is then used to draw the histogram next to the slider. A high frequency towards modality A would mean that there is a large number of voxels where $a'(\boldsymbol{x}) > b'(\boldsymbol{x})$, whereas a high frequency halfway between A and B means there is a large number of voxels where $a'(\boldsymbol{x}) \approx b'(\boldsymbol{x})$. In Fig. 1(c), for instance, the higher values towards the DFC modality indicate that a larger weight for the DPC modality may be necessary, hence the chosen weights of 33% for DFC and 67% for DPC.

For the trimodal case, a triangular weighting widget is provided instead of a slider. The previously mentioned slider can be interpreted as a means of manipulating one interpolation parameter t based on the handle position on the slider axis, which is used to weight two modalities. Analogously, a handle inside the triangle's area is used to manipulate two interpolation parameters α and β, defined as the barycentric coordinates of the handle (with $\gamma = 1 - \alpha - \beta$).

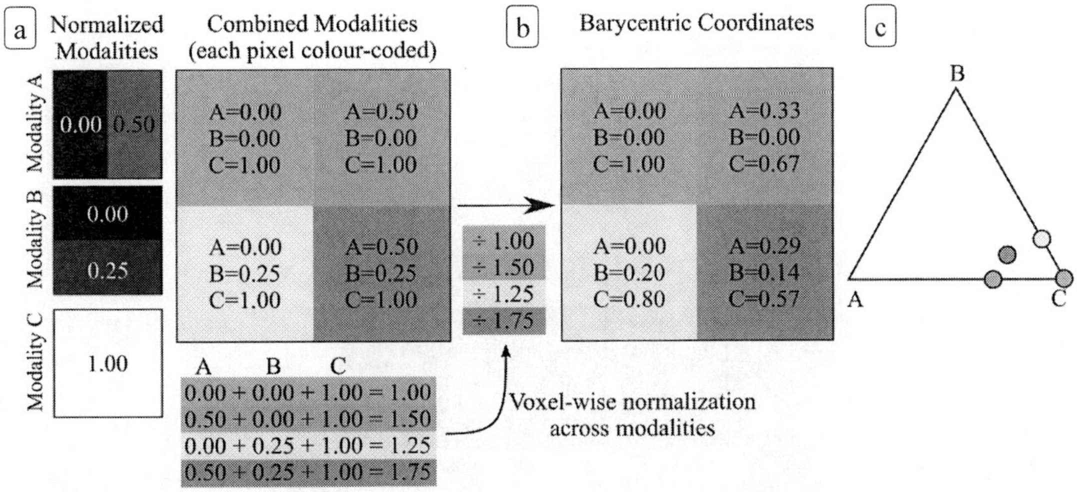

Fig. 3. The trimodal heatmap computation on the example of three 2x2 modality images. First, each modality image is normalized separately to values in the range [0,1] (a). Then for each pixel/voxel a normalization across the modalities takes place (b). The resulting triple is interpreted as barycentric coordinates, the frequency at the corresponding location in the triangle is increased (c). For the trimodal heatmap, each location in the triangle is colour-coded by its frequency, a value of zero is coloured white, while the highest occurring frequency is coloured black; frequencies in between are coloured in an interpolated grey tone.

In the triangle's area, we propose a visualization of the correlation of the data analogous to the one provided for bimodal case, which we call trimodal heatmap, as shown in Fig. 2(c). A graphical representation of how this visualization is computed is shown in Fig. 3. Analogously to the slider's histogram, the three modalities are first normalized, then fused, as shown in Fig. 3(a). Let A, B and C be the modalities. For each voxel triple $a(\boldsymbol{x}) \in A, b(\boldsymbol{x}) \in B and\ c(\boldsymbol{x}) \in C$ with equal coordinates $\boldsymbol{x} = (x, y, z)^T$, the resulting voxels are $a'(\boldsymbol{x}) = \frac{a(\boldsymbol{x})}{a(\boldsymbol{x})+b(\boldsymbol{x})+c(\boldsymbol{x})}$, $b'(\boldsymbol{x}) = \frac{b(\boldsymbol{x})}{a(\boldsymbol{x})+b(\boldsymbol{x})+c(\boldsymbol{x})}$ and $c'(\boldsymbol{x}) = \frac{c(\boldsymbol{x})}{a(\boldsymbol{x})+b(\boldsymbol{x})+c(\boldsymbol{x})}$, i.e. the values are normalized across modalities such that their sum equals 1, as shown in Fig. 3(b). Again, note that $c(\boldsymbol{x}) = 1 - a(\boldsymbol{x}) - b(\boldsymbol{x})$, meaning that only two of the volumes need to be calculated and stored, as the third volume can be calculated dynamically based on them.

The resulting values are then interpreted as barycentric coordinates and visualized in a trimodal heatmap on the triangular widget, as Fig. 3(c) illustrates. High frequencies towards modality A_1 would mean that there is a large number of voxels where $a(\boldsymbol{x}) > b(\boldsymbol{x})$ and $a(\boldsymbol{x}) > c(\boldsymbol{x})$, whereas high frequencies near the triangle's barycentre means there is a large number of voxels with $a(\boldsymbol{x}) \approx b(\boldsymbol{x}) \approx c(\boldsymbol{x})$. High frequencies near the edge between, for example, modalities A_1 and A_2 would mean that there is a large number of voxels with $c(\boldsymbol{x}) < a(\boldsymbol{x})$ and $c(\boldsymbol{x}) < b(\boldsymbol{x})$.

3.3 Trimodal Transfer Function Layouts

The layout presented in Fig. 2, which we call the "stacked" layout (as TFs are positioned on top of each other), occupies an ample horizontal space on the screen. As a more compact alternative, we propose the "triangle" layout, shown in Fig. 4. In this layout, the TF widgets are positioned on the edges of a triangle, and the weighting triangle is rotated so that its corners (each corresponding to a modality) touch their respective TF widget. That results in an intuitive manipulation of the handle on the weighting triangle: the closer the handle is placed to a TF widget (i.e., a vertex of the triangle), the more influence its modality will have in the final visualizations.

This layout leaves available space on the corners of the triangle created by the TF widgets. We use this space to display the modality slicers, previously positioned to the right

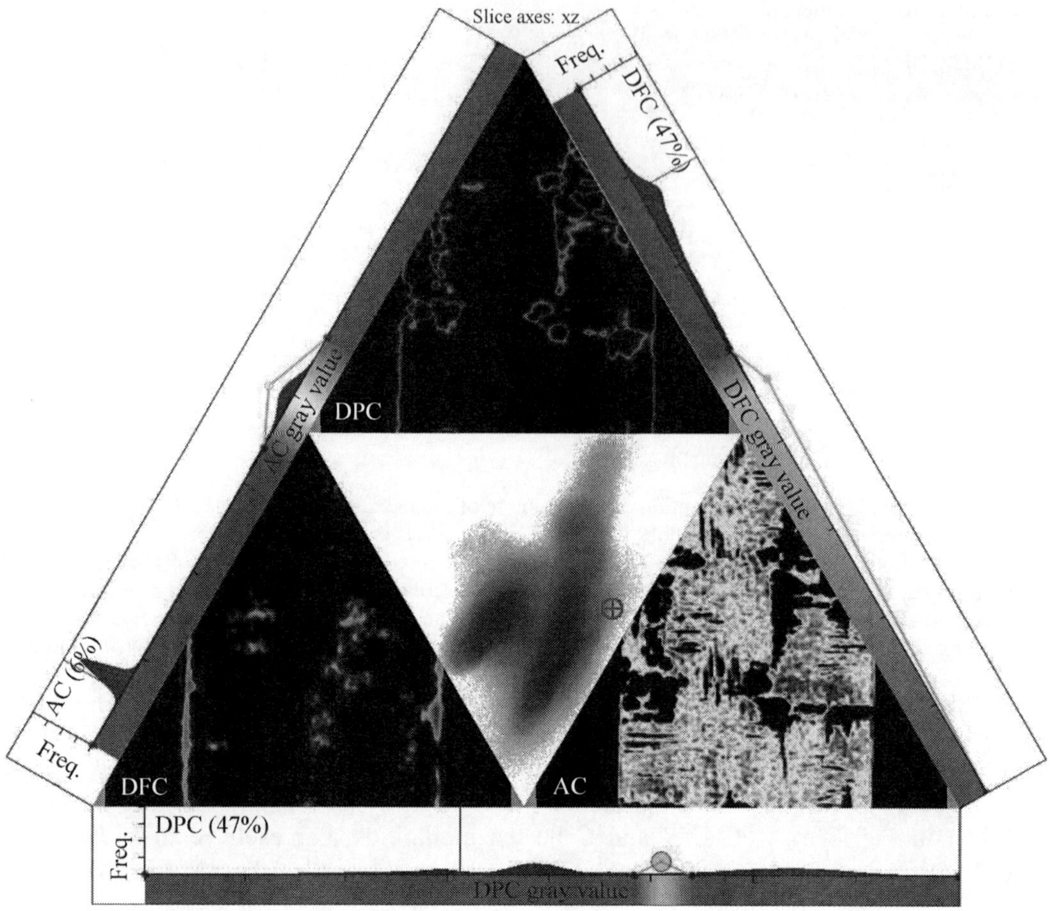

Fig. 4. Trimodal transfer function widget in its triangular layout.

of each TF widget. The slicers are placed on the triangle region on the opposite side of their corresponding TF widget.

4. Evaluation

The material scientists we worked with expressed a preference over the stacked layout, as, with the triangle layout, the interpretation of histograms becomes more difficult due to their differing rotations. In the implemented tool, the user can dynamically switch between the two layouts.

We evaluate our technique on two datasets; both scanned with a Bruker SkyScan 1294. The device delivers pre-registered datasets of the three imaging modalities AC, DPC and DFC. In Section 4.1, an open cell polyurethane foam is explored, followed by a carbon fibre reinforced polymer in Section 4.2.

4.1 Case Study: Foam data

We used our technique to explore a 6 mm x 6 mm x 6 mm open-cell polyurethane (PU) foam. This specimen was measured in our TLGI-XCT device at a tube voltage of 35kV with 1mA current, acquiring 1800 projection images using an Aluminium filter of thickness 0.25 mm with 4 phase steps. The measurement time was approximately 19 hours. On the detector side,

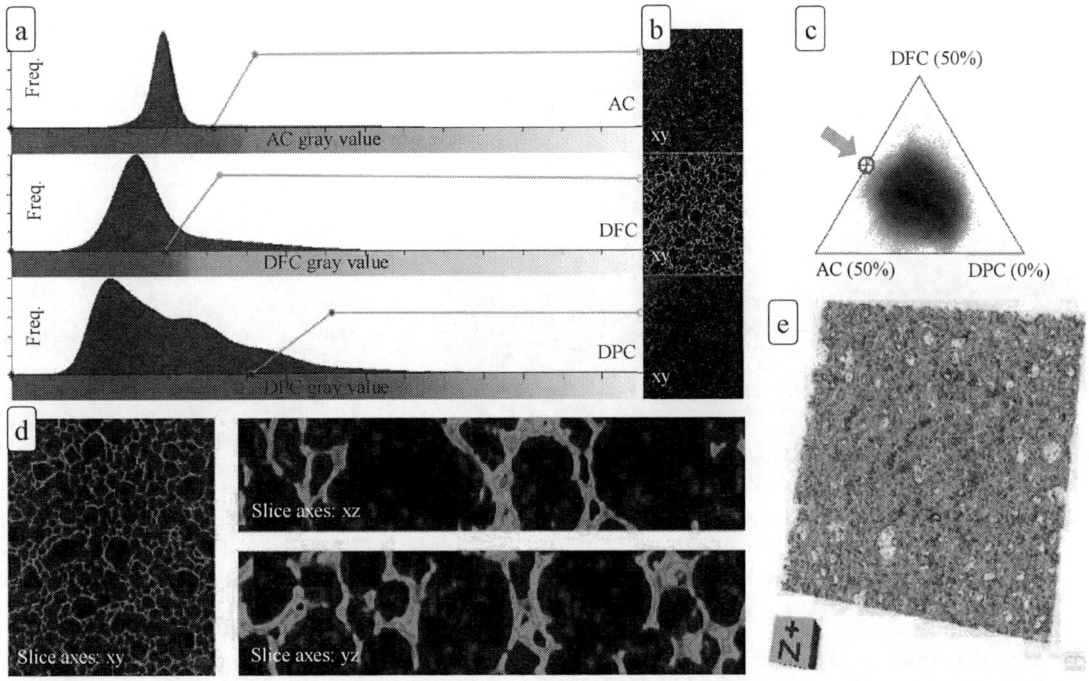

Fig. 5. A polyurethane foam dataset is analysed with the proposed technique. The individual TFs for each modality (a) are shown next to a slice image of each modality (b). Blending is controlled by the control point (marked with an orange arrow) in a triangular widget (c) which also contains the trimodal heatmap. Blended images are shown in axis-aligned slicer images (d) and in a 3D view (e). The slices with axes *xz* and *yz* were zoomed-in to make details better visible.

five-fold averaging and 2x2 binning at an integration time of 1440 ms was used, resulting in an effective voxel size of 11.4 µm.

The task, in this case, was to detect the foam cell walls. To some degree, the cell walls are visible in all modalities. In the AC image, however, visualized in red in Fig. 5, they are visible only partially. As can be seen in the enlarged slicer images showing the blended images Fig. 5(e), the DFC image, shown in green, shows much more signal on the cell wall structures. The DPC has an inhomogeneous distribution of grey values, as can be noticed through the visible wall structures in blue in the upper left corner of the DPC image in Fig. 5 (b). Therefore, we chose to assign a weight of 50% to both AC and DFC for the blended visualization. We see that the DFC modality can resolve the cell wall structures much better in comparison to the AC modality. This comparison through superposition helped the material experts gain insight into how well the foam structures can be recognized in each modality for this scan. The fusion by weighting made the early detection of the low relevance of the DPC modality possible, and its influence was decreased accordingly.

4.2 Case Study: CFRP data

The second analysed specimen is a carbon fibre reinforced polymer (CFRP) laminate. In this case, the specimen was approximately 1.4 cm x 1.8 cm x 0.3 cm. It was measured in our TLGI-XCT device at a tube voltage of 35kV with 1mA current. 1499 projection images were taken using an Aluminium filter of thickness 0.25 mm with 4 phase steps. Here, we employed six-fold averaging and 4x4 binning at an integration time of 650 ms, resulting in an effective voxel size of 22.8 µm. A typical task in the analysis of CFRP materials is identifying fibres and pores, which is set as the goal of this case study. For that, we use the proposed bimodal widget, as the DPC and DFC modalities contain the sought information. The DFC image contains high responses for fibre structures and the DPC image, for the hulls of pores.

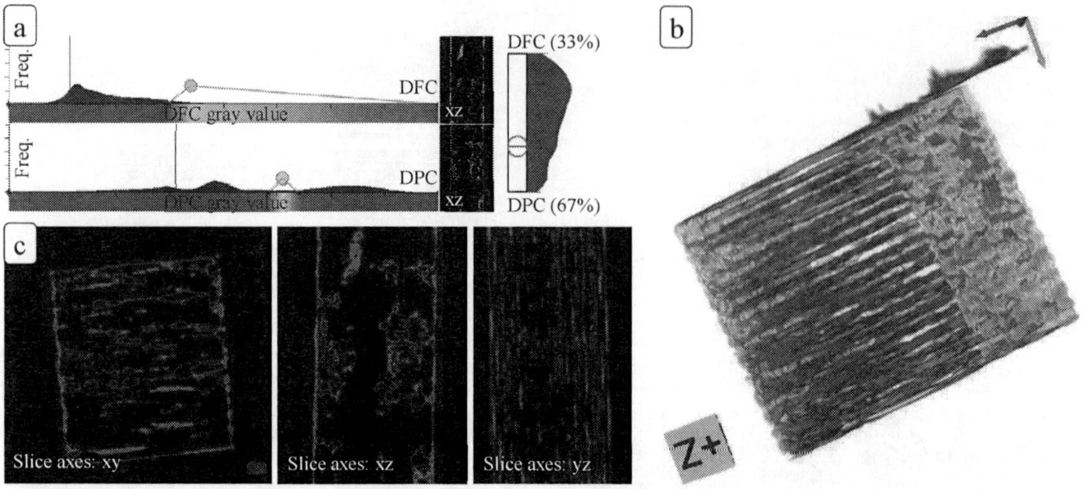

Fig. 6. A carbon fibre reinforced polymer is analysed with the proposed tool. The individual TFs of each modality are positioned next to their respective slicer and the interpolation slider (a). These components are configured exactly as in Fig. 1. The blended results are shown in two types of visualizations: a 3D view (b) and axis-aligned slicer images (c).

The colours are defined in the TF in Fig. 6(a) and each of the resulting visualizations in Fig. 6(b-c) displays the two features in a combined view. These results accelerate and facilitate the comparison of structures from different modalities, otherwise only possible with a side-by-side comparison.

5. Discussion and Future Work

Material experts working with our technique voiced very positively about the interactivity of our approach. Especially the separate 2D slicers of each modality together with the simultaneous blended visualization can tremendously improve the analysis through their quick feedback.

However, the experts remarked that determining a suitable TF for single modalities remains a difficult task. Therefore, in future work, we plan to incorporate additional guidance for defining these TFs, as well as methods of providing context to the multimodal data, additionally to the trimodal heatmap.

The first developed widget [3] was also limited to exactly three modalities. The experts expressed a wish to extend this to other numbers of modalities, which we have begun to address with the bimodal widget. We plan to address the exploration of a larger number of modalities in future work.

6. Conclusion

We propose a technique for the interactive analysis of TFs for multimodal datasets. For each modality, an individual TF can be specified, and its result is visualized immediately. The modalities can be weighted individually to arrive at a blended image that merges the required information from all modalities.

The two presented cases demonstrate that the introduction of a weighting component (in the presented widgets: a slider and triangle) is sufficient to accelerate and facilitate the exploration of the multiple modalities of Talbot-Lau grating interferometry X-ray computed tomography datasets. Overall, the experts noted that our tool is much easier to use for the task of defining and manipulating multiple transfer functions and the blending of multimodal datasets in comparison to more generalized visualization applications.

Acknowledgements

This project partly received funding by the project *"Multimodal and in-situ characterization of inhomogeneous materials"* (MiCi) by the federal government of Upper Austria and the European Regional Development Fund (EFRE) in the framework of the EU-program IWB2020. The research leading to these results has also received funding from the Research Foundation Flanders (FWO) and the Austrian Science Fund (FWF) under grant numbers G0F9117N and I3261-N36, *"Quantitative X-ray tomography of advanced polymer composites"*.

References

[1] F. Pfeiffer, T. Weitkamp, O. Bunk, C. David, Phase retrieval and differential phase-contrast imaging with low-brilliance X-ray sources, Nature Physics, 2 (2006) 258–261.

[2] F. Pfeiffer, O. Bunk, C. David, M. Bech, G. L. Duc, A. Bravin, P. Cloetens, High-resolution brain tumor visualization using three-dimensional X-ray phase contrast tomography, Physics in Medicine and Biology, 52 (23) (2007) 6923–6930.

[3] B. Fröhler, L. da Cunha Melo, J. Weissenböck, J. Kastner, T. Möller, H.-C. Hege, E. Gröller, J. Sanctorum, J. De Beenhouwer, J. Sijbers, C. Heinzl, Tools for the Analysis of Datasets from X-Ray Computed Tomography based on Talbot-Lau Grating Interferometry, Proceedings Industrial Computed Tomography Conference (2019), Padua, Italy.

[4] P. Ljung, J. Krüger, M. E. Gröller, M. Hadwiger, C. Hansen, A. Ynnerman, State of the Art in Transfer Functions for Direct Volume Rendering, Computer Graphics Forum, 35 (3) (2016) 669–691.

[5] J. M. Kniss, G. L. Kindlmann, C. D. Hansen, Multi-Dimensional Transfer Functions for Interactive Volume Rendering, IEEE Transactions on Visualization and Computer Graphics, 8 (2002) 270–285.

[6] L. Zhou, M. Schott, C. D. Hansen, Transfer function combinations, Computers & Graphics, 36 (2012) 596–606.

[7] X. Zhao, A. E. Kaufman, Multi-dimensional Reduction and Transfer Function Design using Parallel Coordinates, Volume Graphics, (2010) 69–76.

[8] H. Guo, H. Xiao, X. Yuan, Scalable Multivariate Volume Visualization and Analysis Based on Dimension Projection and Parallel Coordinates, IEEE Transactions on Visualization and Computer Graphics, 18 (9) (2012) 1397–1410.

[9] K. Lawonn, N. N. Smit, K. Bühler, B. Preim, A Survey on Multimodal Medical Data Visualization, Computer Graphics Forum, 37 (1) (2018) 413–438.

[10] R. Bramon, M. Ruiz, A. Bardera, I. Boada, M. Feixas, M. Sbert, Information Theory-Based Automatic Multimodal Transfer Function Design, IEEE Journal of Biomedical and Health Informatics, 17 (2013) 870–880.

[11] C. Heinzl, S. Stappen, STAR: Visual Computing in Materials Science, Computer Graphics Forum, 36 (3) (2017) 647-666.

[12] M. M. Malik, C. Heinzl, M. E. Gröller, Comparative Visualization for Parameter Studies of Dataset Series, IEEE Transactions on Visualization and Computer Graphics, 16 (5) (2010), 829-840.

Analysis of Drill Cores by Means of RoboCT 3-D X-ray Imaging – Evaluation of High Throughput Approaches

Julija LUCIC [1], Wolfgang HOLUB [1], Alexander ENNEN [1], Rebecca WAGNER [1], Markus FIRSCHING [1]
[1] Fraunhofer Development Center X-ray Technology EZRT, Fürth, Germany
Contact e-mail: markus.firsching@iis.fraunhofer.de

Abstract. X-ray Computed Tomography (CT) is one established method for sophisticated 3-D analysis of drill cores. It allows the three-dimensional evaluation of different features such as porosity, grain size and shape or rock quality in addition to structural information. However, application of lab CT is difficult as multiple fractured drill cores often reside within one necessarily horizontal tray and therefore not eligible for CT measurements without laborious preparation and time consuming scans of every single piece of core. Scanning of the whole drill core tray could be an advantageous procedure to get valuable information about the drill cores in a much shorter time and without handling every single drill core. Nonetheless, geometry and orientation of those trays pose some challenges to conventional X-ray CT approaches.

To cope with this task we propose the application of RoboCT – i.e. robot based computed tomography – for the analysis of whole drill core trays. Limited angle and coplanar laminographic CT measurements are being realized through a system of cooperating industrial robots as manipulation system for the X-ray equipment which thereby offers unusual object accessibility of the horizontal trays. This approach promises to enable the fast and automated examination of whole batches of drill core trays in reasonable time.

Within this work, we compare the results of RoboCT measurements of an entire drill core tray to regular 3-D Micro-CT scans of the same isolated drill cores. While the approach proposed is expected to provide results in inferior quality the gain on measurement speed and the ease of handling may compensate therefor and lead to much broader application of 3-D analysis of drill cores than today's sporadic use of laboratory X-ray CT in this field. Those data may allow geologists to gain more objective and significant data in shorter time.

1. Introduction

X-ray computed tomography (CT) has proven to be a valuable technique for analysis of drill cores. In contrast to other sensor-based techniques like X-ray fluorescence (XRF) or optical inspection, which can only provide information about the surface of the drill cores, CT measurements can provide information on the entire core – including internal structures. It allows thereby the three dimensional evaluation of different features of the drill core, such as grain size and shape or rock quality and also provides structural information.

This information can be valuable for geologists, as the need for automated inspection and computer aided analysis rises constantly with the huge amounts of cores coming from drilling activities. Especially for exploration purposes, in-depth information of the drill cores can be very helpful.

2. Motivation

The application of lab CT for three-dimensional analysis of drill cores is a time consuming procedure, as the drill cores do often reside as many fragments within their tray. Taking every single piece of core out from the tray is inconvenient. It is also sensitive to errors like mixing-up samples or putting them back into the tray at wrong positions and orientations.

Scanning of the whole core tray would be a huge advantage over the current approach for scanning of drill cores, avoiding the need to handle every single drill core piece and it also provides results in a much shorter time than conventional CT approaches.

However, conventional lab CT of the tray would require the rotation of the tray between X-ray source and detector (see **Fig. 1**) which is hardly possible in regular CT system setups where the upright trays would need to be placed on a conventional CT turn table. Furthermore, for certain directions the penetration length would than equal the width of the tray with multiple cores in a row. Thus, the measurement time and / or the energy of the X-ray source needs to be very high to achieve a good reconstruction result. Instead, we propose the application of robot based CT (RoboCT). Different trajectories for e.g. limited angle and coplanar laminographic CT measurements are being realized through a system of cooperating industrial robots as manipulation system for the X-ray equipment, which thereby offers unusual object accessibility of the horizontal trays. This approach promises to enable the fast and automated examination of whole batches of drill core trays in reasonable time.

3. Background on X-ray Tomography

3.1 Conventional 3-D Micro-CT

During the last years, 3-D micro-CT has been applied for the non-destructive visualization in many applications. Formerly, this technology was only used for medical imaging. Nowadays, CT is used as a standard tool in industrial applications for material analysis. X-rays allow to penetrate the specimen and to see not only its surface but also the structure inside. With X-ray CT, the 3-D volume information of objects can be reconstructed using X-ray projections of the object from a series of perspectives, for example from a circular trajectory around the object (see **Fig. 1**). The reconstructed volume data set consists of grey levels, which represent information about the X-ray attenuation depending on the mass attenuation coefficient, the density distribution of the material and the spectral distribution of the X-ray source.

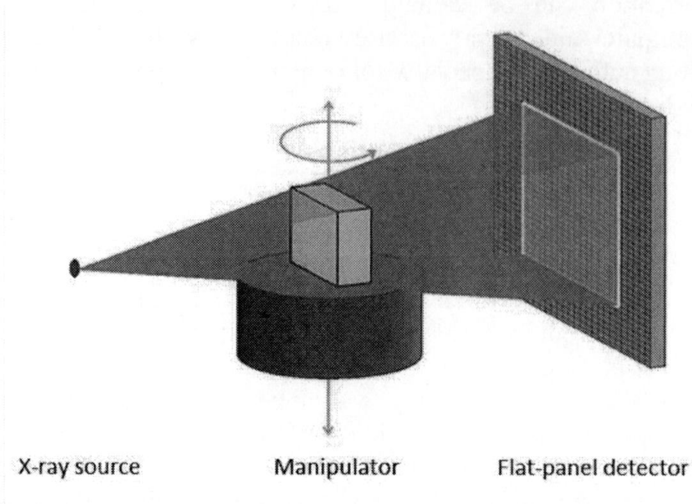

Fig. 1: General working principle of 3-D CT [Fraunhofer EZRT]

3.2 RoboCT – Robot Based CT

During the past years, EZRT has developed the technology of RoboCT, i.e. the application of at least two cooperating industrial robots as manipulation system for the X-ray equipment (that is at least one of each, source and detector) (see **Fig. 2**). Such systems allow to access regions-of-interest (ROIs) in very large workspaces where they are capable to realize different, almost arbitrary forms of X-ray trajectories, which cannot be realized with conventional CT systems.

EZRT has solved the challenges to control and parametrize this hardware in its very high dimensional parameter space and to perform CT measurements with such inaccurate manipulators that are imprecise for several orders of magnitude compared to laboratory CT systems built on large basis of granite.

As such, these RoboCT systems also allow performing CT measurements to study applications of yet unbuilt hardware, like here a possible portal CT setup.

Fig. 2: One implemented RoboCT system with four cooperating robots for two X-ray sources and two detectors

3.2.1 Limited Angle CT

Whilst conventional CT requires rotation of the object about 360° (or respective rotation of the X-ray equipment around the object), it is also possible to perform less than 360° rotation, of course. Once, one does only acquire perspective information from less than 180° (plus opening angle of the acquisition cone), the resulting three-dimensional information available for CT reconstruction is mathematically incomplete and does no longer allow correct CT reconstruction. These limited angle CT scans – that are being applied whenever either object geometry or penetration lengths do not allow to perform regular computed tomography – represent the most straightforward means of laminographic imaging, i.e. performing CT with limited information in contrast to regular circular CT. The resulting laminography reconstructions typically result in CT images with good image quality in the slices of one orientation whilst, due to the missing information, quality decreases in orthogonal slices.

3.2.2 Coplanar Laminographic CT

One more sophisticated method on laminographic imaging is the class of coplanar laminographies, where only source and detector are being moved on trajectories parallel to each other – and to the desired ROI in the object (**Fig. 3**). The advantage is that in the case of the drill cores, there is no need to take them out of their tray, but still good structural information on the drill cores can be achieved.

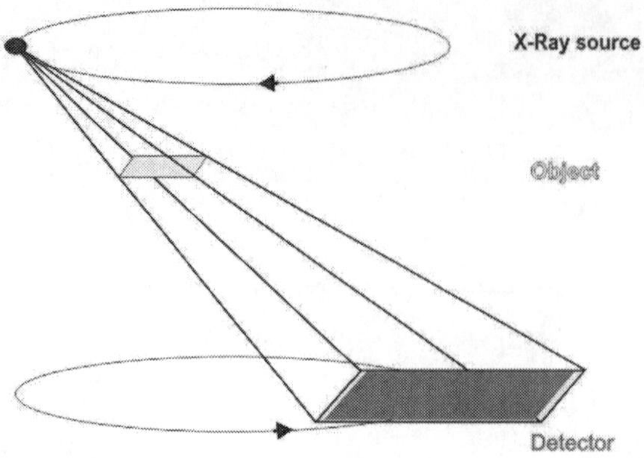

Fig. 3: Shows the general working principle of X-ray coplanar laminographic CT: 2D Translation – Circular scan, where the source and the detector are moving on parallel trajectories. [Fraunhofer EZRT]

4. Measurement Setup

4.1 Conventional 3-D Micro-CT

The used scanning geometry was the axial 3-D CT. The measurement parameters used for the axial 3-D CT measurement were 220 kV with 0.5 mm copper filter. The spatial resolution was approx. 68 µm. The drill cores were mounted in a three-jaw chuck and measured separately one by one (see **Fig. 4**). The exposure time was 250 ms/projection and 1600 projections were acquired. The used measurement protocol was a stop and go procedure so that the total scan time was approx. 23 minutes for approx. 10 cm of drill core.

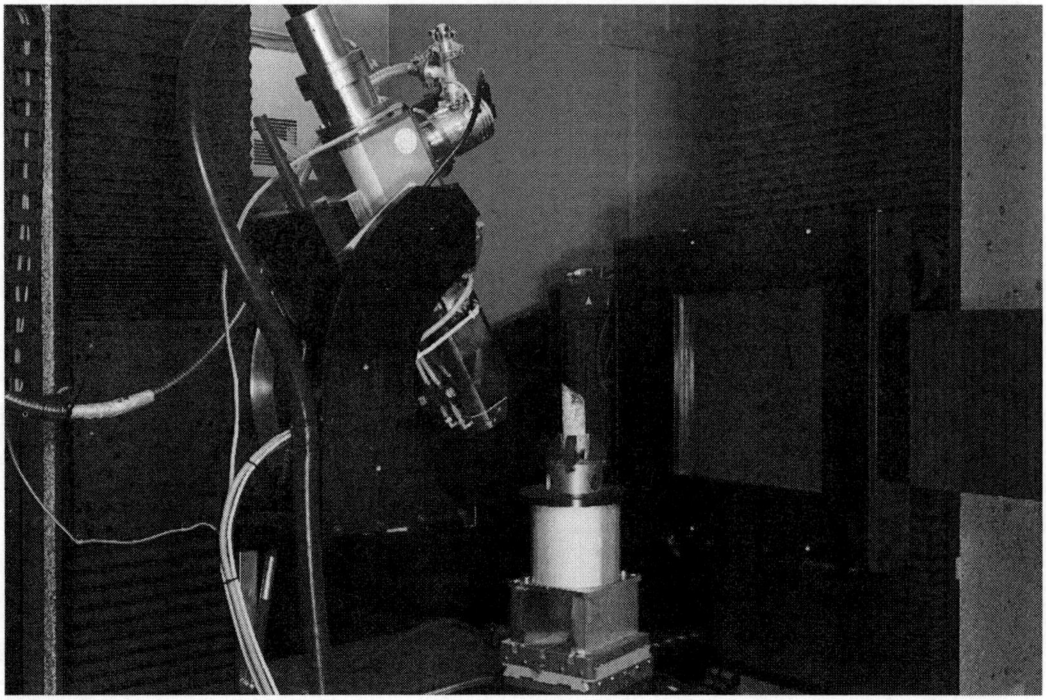

Fig. 4: Measurement setup used for 3-D measurement of the single drill cores. The trajectory of the axial 3-D CT is a 360° circle.

4.2 RoboCT

The measurement setup used for the limited angle CT and for the coplanar laminographic CT and the three times measuring field extension (3 x MFE) coplanar laminographic CT can be seen in **Fig. 5**.

Fig. 5: Measurement setup used for the limited angle CT and the coplanar laminographic CT of the drill core tray.

4.2.1 Limited Angle CT

The trajectory used for the limited angle CT was a circle with 187°, which is nearly 180° plus the opening angle of the cone beam. The measurement parameters used for the measurement were 225 kV with 1 mm copper filter. The spatial resolution was approx. 63 µm. The exposure time was 2 s/projection and 750 projections were acquired. The total scan time was approx. 57 minutes for 42 cm of drill core. The measurement covered three drill cores with a ROI of 14 cm each.

4.2.2 Coplanar Laminographic CT

The trajectory used for the coplanar laminographic CT was a combination of three concentric circles to reach laminographic angles of 10°, 20° and 30°. The measurement parameters used for the measurement were 225 kV without a filter. The spatial resolution was approx. 63 µm. The exposure time was 250 ms/projection with eight times averaged image acquisition and 120 projections were acquired. The total scan time was approx. 17 minutes for 42 cm of drill core. The measurement covered three drill cores with a ROI of 14 cm each.

Additionally, a 3 x MFE coplanar laminographic CT was used as a trajectory to cover the entire area of the tray. This results in total in nine concentric circles to reach laminographic angles of respectively 10°, 20° and 30° per measuring field. The total scan time for this procedure was 49 minutes for 84 cm of drill core. The measurement covered six drill cores with a ROI of 14 cm each.

5. Comparison of Results

The comparison of the measurement setups in terms of exposure time, measurement time and scanned length of core is shown in **Table 1**.

Table 1: Comparison of the different used measurement setups in terms of exposure time, measurement time and scanned length of core.

	Axial 3-D CT	Limited Angle CT	Coplanar Laminographic CT	3 x MFE Coplanar Laminographic CT
Exposure time	6.6 min	25 min	4 min	12 min
Measurement time	25 min	57 min	17 min	49 min
Scanned length of core	0.24 m/h	0.44 m/h	1.6 m/h	1 m/h

5.1 Conventional 3-D Micro CT

The result of the filtered back projection reconstruction is full 3-D information of the measured object. The inner structures of the drill cores are clearly visible and shown exemplarily in **Fig. 6**. It would be possible to extract information about grain size and shape and its distribution along the whole drill core.

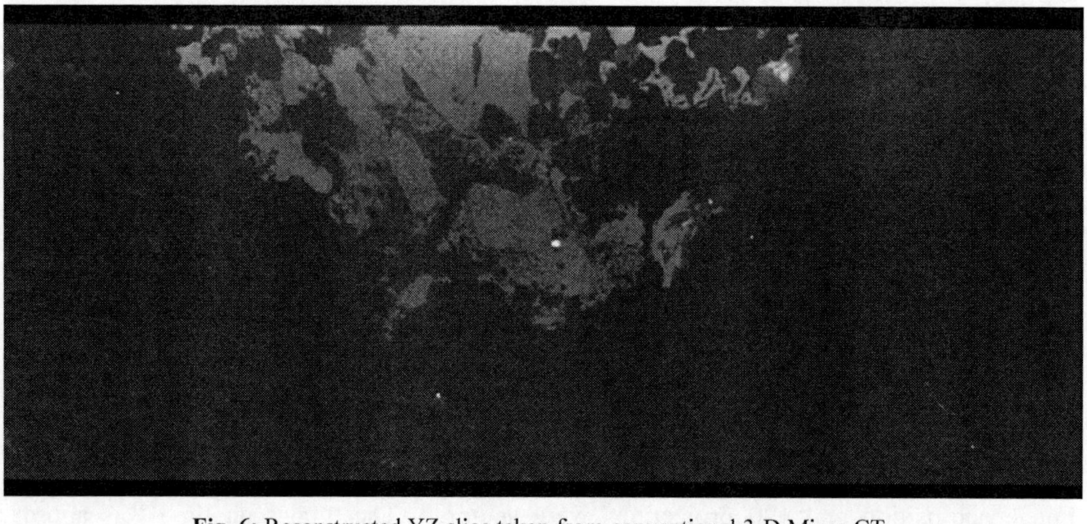

Fig. 6: Reconstructed YZ slice taken from conventional 3-D Micro CT.

5.2 RoboCT

5.2.1 Limited Angle CT

Fig. 7 shows a slice taken from the reconstructed limited angle measurement using a RoboCT setup. While the overall structure within the core is visible, typical limited angle artefacts reduce the recognisability of details and thus complicate automated image processing. However, the image quality could be sufficient for certain use cases.

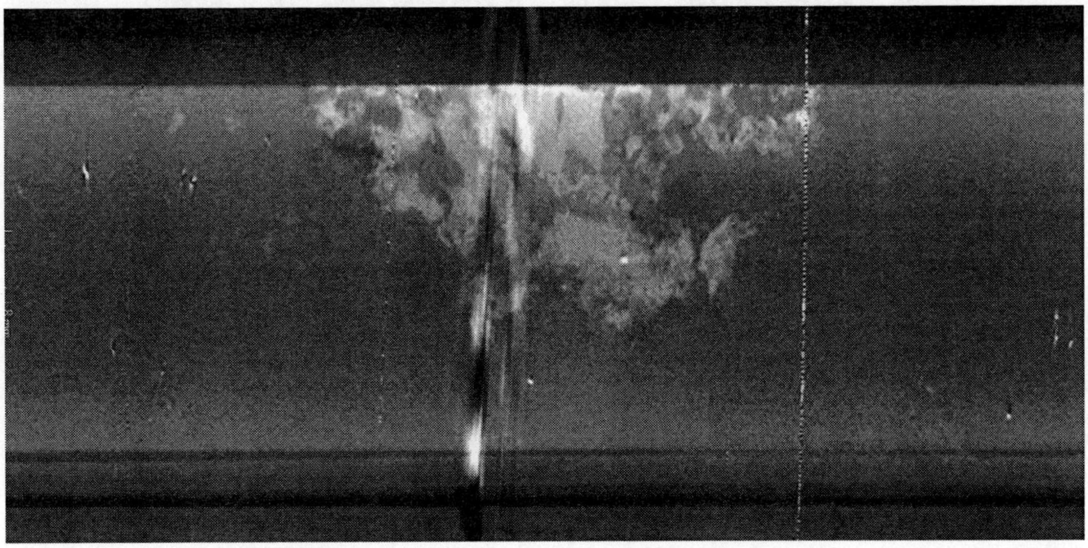

Fig. 7: Reconstructed YZ slice taken from limited angle CT.

5.2.2 Coplanar Laminographic CT

The coplanar laminographic CT has a limited depth information and therefore not the full structural information is given. Still there is a major part of the structure visible and evaluation of the information can be made. The result for one drill core is shown in **Fig. 8**. The increased field of view allowed for scanning of three respectively six cores at the same time within the tray.

Fig. 8: Reconstructed XY slice taken from coplanar laminographic CT

As can be seen in **Fig. 9** there are artifacts resulting from the limited depth information. While the information within the focus plane is displayed in good quality, moving out of the focus plane decreases image quality and geometric information available. Image processing for e.g. structural tracking would only be possible within regions not effected by cone artefacts and therefore providing sufficient information. Structures orthogonal to the focus plane are effected by the same artefacts and are thereby hard to analyze.

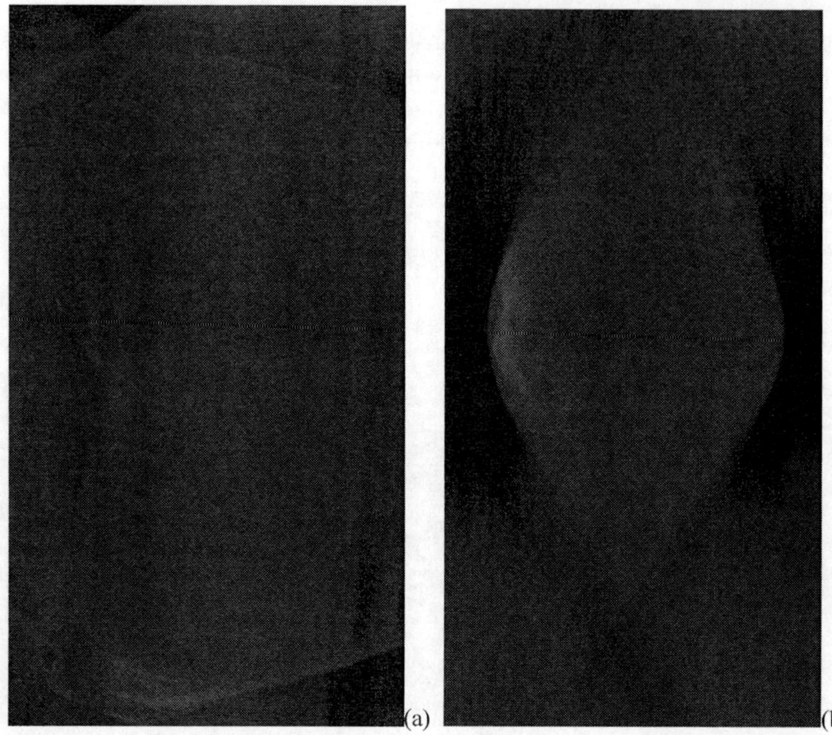

Fig. 9: **(a)** XZ view taken from coplanar laminographic CT, **(b)** YZ view illustrating laminographic artefacts.

6. Conclusion and Outlook

RoboCT is a promising attempt for the measurement of entire trays of drill cores and enables multiple trajectories (e.g. limited angle CT or Coplanar Laminographic CT) for image acquisition given the high degree of freedom from axis movement. While the image quality of the laminographic setup proposed is lower in regards to the limited 3-D-information, handling of the drill cores and measurement speed is greatly increased. The proposed system enables the creation of 3-D information of entire trays of drill cores without the need to remove the drill cores from it and hereby avoid the tedious and error prone handling of single cores.

With the used system, a measurement speed of approximately 1.6 m/h could be achieved. Within the performed measurements, the exposure time aggregates only to 4 minutes while the rest of the measurement time was needed for the movement of the manipulation system, hence taking up most of the time needed to complete the scan. Using a more specific setup, a scanning speed of approximately 16 m/h would be achievable, utilizing optimized axis movement and measurement setup.

This would enable the creation of depth information from drill cores in a fast and reliable way.

References

[1] C. Schorr, "Optimierung iterativer Rekonstruktionsverfahren bei unvollständigen Daten zur Anwendung in der Computerlaminographie," ZfP-Zeitung 141, pp. 51-54, October 2014.

High definition X-ray Source based on Carbon Nanotube Emitter for Micro CT application

Xiaobing LI [1], Menglong LIU [1], Juntao ZHOU [1], Xuanyi WU [1], Qingyang WU [2], Liang REN [2], Jiewen ZHENG [2], Zexiang CHEN [2]
[1] University of Electronic Science and Technology of China, Chengdu, China
[2] Guangzhou Haozhi Imaging Technology, Guangzhou, China
Contact e-mail: zxchen@uestc.edu.cn, zexiang_chen@haozhigp.com,

Abstract. A micro-focus CNT-based X-ray tube was developed successfully, in which CNT circular rings grown by microwave plasma CVD employed field emission cathode to obtain large and stable emission current. The size of the CNT ring is 50 micrometer in diameter and the wall thickness is 2 micro meter. A high convergence ratio electron gun with grid gate and focus electrode was designed for this X-ray tube to achieve a micron level X-ray focal spot. The full vacuum-sealed micro-focus CNT-based X-ray tube exhibited a very small focal spot with a diameter of about 5 micro meters and stable current of more than 50 μA, the corresponding average field emission current density of the CNT emitter is more than 2.55A/cm^2. The X-ray images of IC indicated this CNT x-ray source is competent for high definition DR and micro CT application.
Keywords: Carbon Nanotube, Field Emission, X-ray Tube, Micro-focus spot..

Introduction

Carbon nanotubes (CNTs), with a small tip radius and a high aspect ratio combined with chemical inertness and mechanical strength, was considered as a very attractive potential field emission material. In the past years, emission current density of CNTs was improved sharply, CNT emitter has been demonstrated owning stably current density of more than 1 A/cm^2 [1-4]. The high stable emission current of CNT are promoting greatly for the application of cold cathode in electronic vacuum devices, such as field emission display, microwave amplifier, X-ray tube and so on.

It is well known that field emission based x-ray tubes possess massive potential applications in various important fields due to its fast switch, digitally addressing, and rapid snapping. Zhou's group developed a CT system using CNT-based electron source for X-ray tube to achieve high quality CT image of pulsatile cardiac of small mouse. A fully vacuum-sealed miniaturized X-ray tubes using CNT-emitter as electron source without an active pump had been reported by several research groups already [5-10].

Generally, there are two typical methods to fabricate CNT emitter. An economic and efficient method is a two-step method, in which CNTs were grown firstly and then mounted on a substrate by screen printing or electrophoresis. However, this method is hard to get uniform vertical aligned evenly distributed CNTs. Another method is chemical vapor deposition (CVD), including thermal CVD, PECVD and microwave plasma CVD. However, the electric field screening effects in CNT arrays is difficult to be overcome. On the whole, the emission current density of CNT cathode still cannot meet the need of high power vacuum electron devices.

Here, we reported the fabrication of the CNT bundles by microwave plasma CVD to obtain well-aligned CNT array with stable large current density, and the design of a CNT-based X-ray tube to obtain micro focus X-ray spot to achieve micrometer level X-ray image..

Experimental

The silicon substrate was lithographed into two small ring structure, the outer ring is 50 μm in diameter and the inner ring is 20 μm, Then a catalyst stack were deposited by vacuum evaporation deposition method. Catalyst structure was Fe/Al/Cu/TiN with thickness of 2 nm, 2nm, 200 nm and 200nm respectively. The CNT growth was performed in a microwave plasma CVD reactor operating at a frequency of 2.45 GHz [11]. Typical growth conditions were: microwave power level of 1 kW, gas flow rate of 200 sccm with a CH4 + H2 mixture at a ratio of 9: 1, the reactor pressure of 25 mbar, the deposition temperature of approx. 650°C. After keeping the sample at a constant temperature of 650°C for 5 minutes, then methane was injected into the hydrogen plasma to induce CNT-growth for 3 minutes.

The CNT samples were thoroughly analyzed using a high resolution scanning electron microscope (SEM, Philips XL-30 FEG). The field emission measurement was carried out in a vacuum chamber with a base pressure of about 10-7 mbar.

Results and discussion

The SEM image of as-grown CNT-rings was shown in Fig. 1a. It can be seen from the SEM picture that the CNT has a regular length of about 10 μm, and the thickness of rings is about 2 μm. All CNTs were oriented vertically to the substrate and formed an excellent emitter pattern. Uniformity in length and diameter of CNTs enable every single CNT contributes evenly to the field emission current, avoiding the phenomenon of "hot spots" and burning of individual CNTs that stick out from the cathode plane.

The field emission specification of CNT emitter was measured using a classical diode structure before the CNT emitter was amounted into X-ray tube. The distance between the anode and the cathode was about 250 μm. Fig. 1b is the I-V curve of the as-prepared CNT emitter. A total emission current of 240 μA was obtained from the two ring of CNTs at an anode voltage of 2.8 kV. The average emission current density can reached to higher than 12.2 A/cm^2. If we only calculated the effective CNT emitter, the current density of the tip of CNT was over 54.5 A/cm^2.

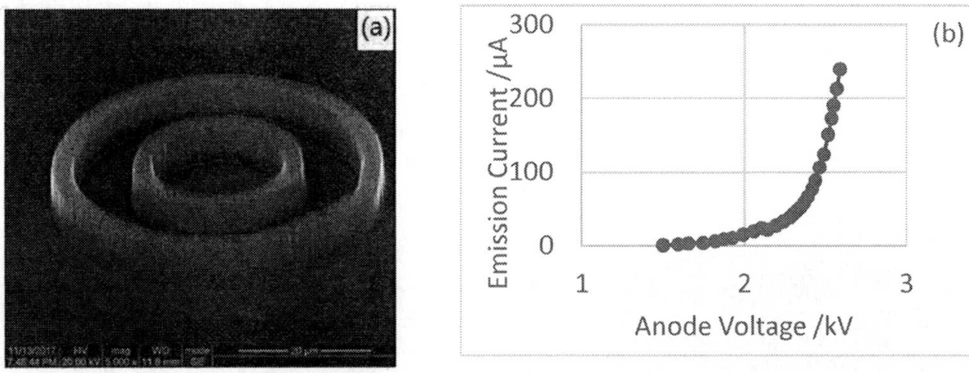

Fig. 1. The SEM picture of as-grown CNT array (a) and the corresponding I-V curve (b)

The electron gun for X-ray tube was optimized to acquire a very fine convergence of electron beam. The Fig. 2 is the electron gun structure and the electron beam track and electron beam section on the target. The diameter of the electron beam on the target was less than 4.5 µm, which indicated the X-ray spot was less than 5 µm certainly.

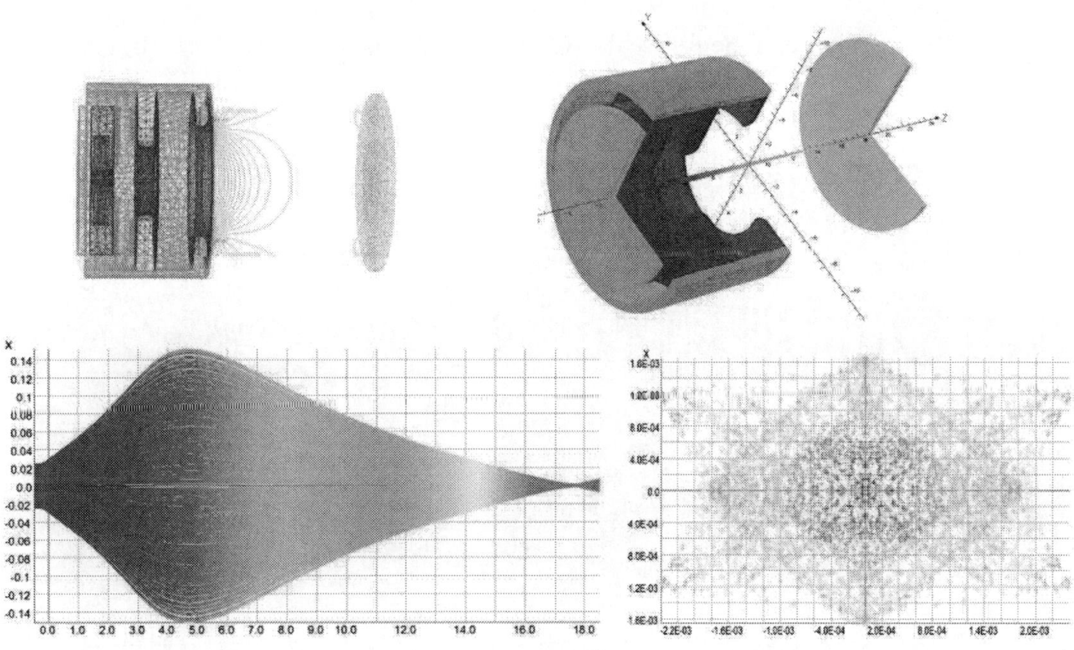

Fig. 2. The simulation results of the electrostatic gun. The electron density plot of the cross section on the target shows that the electron spot diameter is about 4.5µm.

Fig. 3a shows the object of the prepared X-ray tube. The tube was sealed at the base pressure of about 5×10^{-8} mbar after being baked and exhausted. The highest anode voltage was more than 110 kV. The I-V curve of the fully vacuum-sealed X-ray tube was also measured as shown in Fig. 3b. In this case, a voltage of 6 kV was applied to the anode to ensure the emission electrons to be collected by the anode target. The low anode voltage could avoid the X-ray generation effectively when doing the I-V curve measurement. The I-V curve of the fully sealed X-ray tube displayed the similar field emission feature to that of the CNT array operated at diode mode. An anode current of higer than 50 µA was achieved at the gate voltage of 3 kV. And we can find out that the gate voltage in the fully x-ray tube is higher than that of diode structure, the reason is that the distance between the cathode and the gate is about 300 μm and the gate is a hole structure to make the emitted electrons penetrate the gate to anode. In most case, 50 μA anode current is enough for micro focus imaging.

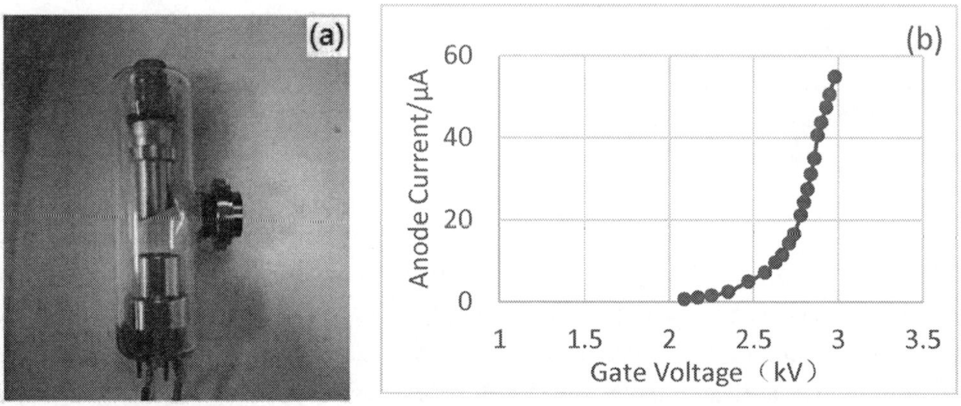

Fig. 3. (a) is the picture of the full-sealed CNT-base micro-focus x-ray tube, (b) is the I-V curve of the X-ray tube. The anode voltage was set to 6 kV, the gate voltage was adjusted from 1 kV to 3.5 kV.

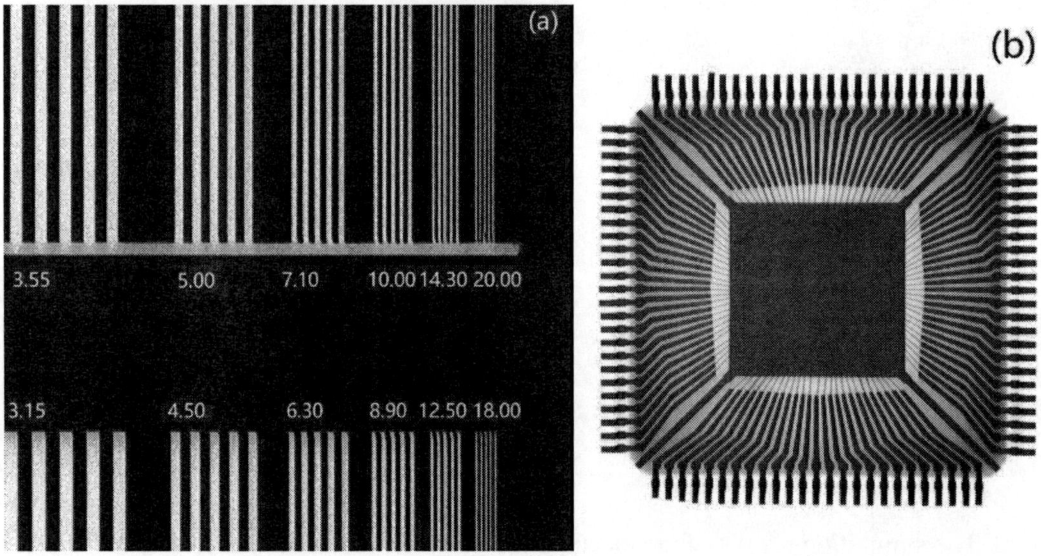

Fig. 4. X-ray Images taken by the as-prepared X-ray Tuber, image of X-ray line alignment card (a) and image of ULSI chip (b)

The actual definition of CNT-base micro-focus tube was firstly measured by the high definition resolution X-ray line alignment card. The image was shown in Fig. 4a, the anode voltage and the anode current were set to about 90 kV and 20 μA respectively when doing the X-ray imaging. And a CMOS X-ray detector with pixel size of 49.5 μm (Shad-o-Box 1K

HS, Teledyne Dalsa) was used. The test image showed that it is easily distinguished the lines at 20 lp/mm. In order to have a directly evaluation of the X-ray image taken by the CNT-base x-ray tube, a ultra large scale integration (ULSI) was sampled to X-ray imaging, as shown in Fig. 4b. In this case the anode voltage was set to 95 kV and the anode current was 20 μA. The lead distributions with finest gold wires inside the ULSI chip was clearly displayed, in which the gold wire in the chip was generally about 20μm. The above X-ray images indicated that the as-prepared micro-focus X-ray tube could get high spatial resolution.

Conclusion

A macro-focus X-ray tube based on CNT array was successfully developed, in which a total anode current and anode voltage of 50 μA and 110 kV can be reached respectively. A very small x-ray focal spot size with less than 5 μm has been obtained. The full vacuum-sealed CNT based X-ray tube could potentially produce very high special resolution X-ray image for micro CT applications.

References

[1] A. V. Eletskii, "Carbon nanotube-based electron field emitters," Physics-Uspekhi, vol. 53, p. 863, 2010.

[2] X. Calderon-Colon, H. Geng, B. Gao, L. An, G. Cao, and O. Zhou, "A carbon nanotube field emission cathode with high current density and long-term stability," Nanotechnology, vol. 20, p. 325707, Aug 12 2009.

[3] C. Li, Y. Zhang, M. Mann, D. Hasko, W. Lei, B. Wang, D. Chu, D. Pribat, G. A. J. Amaratunga, and W. I. Milne, "High emission current density, vertically aligned carbon nanotube mesh, field emitter array," Applied Physics Letters, vol. 97, p. 113107, 2010.

[4] Y. Z. Lee, L. Burk, K. H. Wang, G. Cao, J. Lu, and O. Zhou, "Carbon Nanotube based X-ray Sources: Applications in Pre-Clinical and Medical Imaging," Nucl Instrum Methods Phys Res A, vol. 648 Supplement 1, pp. S281-S283, Aug 21 2011.

[5] E. Samei, S. Sultana, X. Calderón-Colón, G. Cao, O. Zhou, J. Lu, and N. J. Pelc, "Design and characterization of a carbon-nanotube-based micro-focus x-ray tube for small animal imaging," Proc. of SPIE, vol. 7622, pp. 76225G-76225G-9, 2010.

[6] X. Qian, A. Tucker, E. Gidcumb, J. Shan, G. Yang, X. Calderon-Colon, S. Sultana, J. Lu, O. Zhou, D. Spronk, F. Sprenger, Y. Zhang, D. Kennedy, T. Farbizio, and Z. Jing, "High resolution stationary digital breast tomosynthesis using distributed carbon nanotube x-ray source array," Med Phys, vol. 39, pp. 2090-9, Apr 2012.

[7] H. J. Kim, J. M. Ha, S. H. Heo, and S. O. Cho, "Small-Sized Flat-Tip CNT Emitters for Miniaturized X-Ray Tubes," Journal of Nanomaterials, vol. 2012, pp. 1-6, 2012.

[8] S. H. Heo, H. J. Kim, J. M. Ha, and S. O. Cho, "A vacuum-sealed miniature X-ray tube based on carbon nanotube field emitters," Nanoscale Res Lett, vol. 7, p. 258, 2012.

[9] J. W. Jeong, J. W. Kim, J. T. Kang, S. Choi, S. Ahn, and Y. H. Song, "A vacuum-sealed compact x-ray tube based on focused carbon nanotube field-emission electrons," Nanotechnology, vol. 24, p. 085201, Mar 1 2013.

[10] J.-W. Jeong, J.-T. Kang, S. Choi, J.-W. Kim, S. Ahn, and Y.-H. Song, "A digital miniature x-ray tube with a high-density triode carbon nanotube field emitter," Applied Physics Letters, vol. 102, p. 023504, 2013.

[11] Z. Chen, Q. Zhang, P. Lan, B. Zhu, T. Yu, G. Cao, and D. Engelsen, "Ultrahigh-current field emission from sandwich-grown well-aligned uniform multi-walled carbon nanotube arrays with high adherence strength," Nanotechnology, vol. 18, p. 265702, Jul 4 2007

Multi-energy X-ray image fusion

Valerie KAFTANDJIAN [1], Fatima Zahra OUJEBBOUR [2]
[1] INSA-Lyon Laboratoire Vibrations et Acoustique, Villeurbanne, France
[2] Division Applications industrielles, Unité Contrôle Non-Destructif, Centre National de l'Energie des Sciences et des Techniques Nucléaires, Rabat, Morocco
Contact e-mail: valerie.kaftandjian@insa-lyon.fr

Abstract. Different X-ray conditions are often necessary to obtain a good contrast for parts having different thickness ranges. In this context, several images are obtained and our aim is to combine them in a single one, keeping as much as possible the original contrast information, or even improving it. An example case study is conducted in order to show the potential of different combinations.

Introduction

For parts having very different thicknesses or different attenuation coefficients, only one X-ray acquisition is sometimes not sufficient, or not optimal for all the regions of the part. Thus, several X-ray images should be acquired. Moreover, new detectors are now coming with the ability to acquire spectrometric images, typically with four to six energy bins. It is thus interesting to have a procedure to show a single image to the operator, where the "best of" each acquisition is combined in a global image. This single image can also be an input for a tomographic reconstruction if necessary.
However, fusion is different from superimposition of images. The difficulty is to make use of redundancy between images, without creating artificial structures into the new image. The Dempster Shafer (DS) evidence theory already proved successful in the field of NDT, for high level fusion. For example, a DS-based classifier was developed during a collaboration between INSA and Fraunhofer EZRT and IZFP for defect classification. This classifier was used for 2D images of castings, 3D CT images of castings, and 3D ultrasonic images of composites [1-6]. DS theory was also used after a fuzzy C-means segmentation for X-ray images combination by INSA, but the result of the combination was a segmented image [7]. Here our challenge is to obtain a final image where the grey-levels are obtained by fusing the original grey-levels.
In the following we show the results obtained with DS fusion, and compare it with the thresholding & logical based fusion, where a grey-level range is selected in each image and associated with the others, without use of redundancy. Color-based image combination will also be shown, where each original image is assigned a specific color range. The results are still not perfect but we will show our work in progress.

1.Case study

In this application case, a medicine box is considered as an exemplary case, because it contains several elements of interest having different attenuation, from a metal spring (high attenuation) to medicine powder (low to medium attenuation depending on total thickness) and plastic elements (gear and box, low attenuation). Figure 1 shows the three acquisition conditions and corresponding images obtained to correctly identify all the elements of interest.

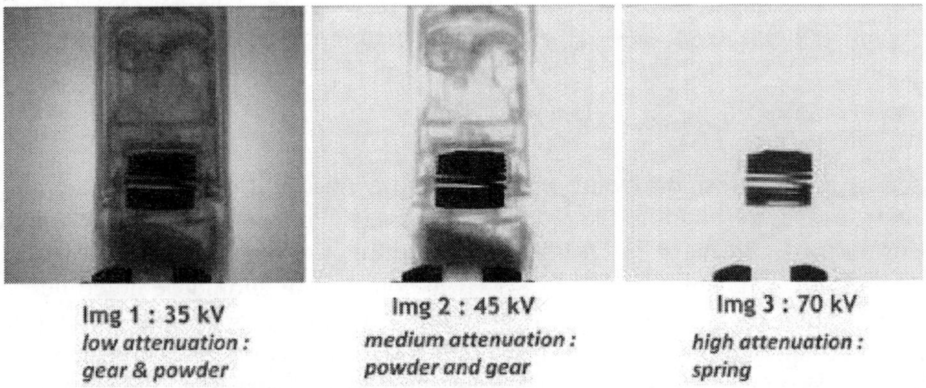

Fig. 1 : Case study : Three images of the same object (medicine box) are acquired : for each kV range, only a part of the information is present.

The aim of this multi-energy image fusion case is to build a unique image gathering "all the best" of the three images. Before showing some results, we explain some basic points of the Dempster-Shafer fusion method.

2. Data fusion method

The Dempster-Shafer (DS) [7-8] theory also known as evidence theory, is an extension of the classical Bayesian probability theory. Let's consider a set of assumptions which is called the frame of discernment Ω : it is the set of mutually exclusive and exhaustive hypotheses of the application considered. In our case it corresponds to the different classes of materials in the images: class H_1 = high attenuation range (e.g. spring), H_2 = medium attenuation (e.g. powder), H3 = low attenuation (e.g. plastic), H4 = no attenuation (air background). In a probabilistic framework, any piece of information would be distributed between these singleton hypotheses. However the DS theory supports the consideration of union hypotheses. For example, if a source of information cannot decide if an object is H_1 or H_2, which can be for example the real case of thick powder, there is an hesitation, which is modeled in the theory by introducing a further hypothesis, equal to the union of both hypotheses H_1 or H_2 to represent hesitation or ignorance ($H_{12} = H_1 \cup H_2$). This difference between DS theory and probability theory is fundamental as we will see afterwards.

Once the frame of discernment Ω is defined, any information from a measurement source must be represented by a value called "mass" m, "piece of evidence" or "basic belief assignment" that can be assigned to a single hypothesis or union of hypotheses. Thus, the workspace is no longer Ω, but rather the 2^{Ω} which is composed of all single hypotheses and all their possible unions. The set 2^{Ω} includes all proposals for which the source can bring a piece of evidence. Any mass different from zero is said to be "focal", while a null mass is called "vacuous".

Obtaining the distribution of masses values m(Ai), $0 \leq m\,(Ai) \leq 1$ with Ai $\in 2^{\Omega}$, is an important step because it represents all the knowledge, the accuracy as well as the uncertainty associated with the measurement.

Once the masses are assigned for two sources of information 1 and 2, the masses can be combined using the so called Dempster rule of equation (1). If more than two sources of information are available, the formula remains the same but more products are present in the sum.

$$m(H_i) = m_1(H_i) \oplus m_2(H_i) = \frac{1}{1-K} \sum_{H_a \cap H_b = H_i} m_1(H_a) m_2(H_b) \quad (1)$$

Where $K = \sum_{H_a \cap H_b = \emptyset} m_1(H_a) m_2(H_b)$

K is used as a normalization factor, and it quantifies the conflict between the sources of information, as it corresponds to the product of masses having no hypothesis in common. On the opposite, the fused mass on Hi is built by multiplying together the masses of the two information sources having the hypothesis Hi in common. It should be noted here that the higher the K value, the less sense means the combination of information, and in practice, it is necessary to apply a threshold on it in order to prevent combining conflicting sources. Some authors recommend to use the Dempster rule without normalization (in this case it is called the conjunctive rule), because the higher the K, the higher is the fused mass which is obviously not a good result. However in this case, the final mass is not between 0 and 1.

An underlying assumption of the theory is that the sources must be independent. If not, another rule has been introduced by Denoeux, called the cautious rule.

In our case, we will use the equation (1) in order to get normalized fused images and we keep the conflict image as a result to be considered in the final decision.

Thus, the main part of the process is the mass assignment step. In our case, a segmentation by multi-thresholding of each image allows first to separate pixels belonging to the same "family". Depending on the acquisition settings, we decide which hypotheses are to be considered. If a doubt exists, a union of hypotheses is used. Then, the relevant piece of knowledge about attenuation is contained in the grey-level. Thus, the grey-levels are normalized, and assigned to masses in the different hypotheses depending on the image under consideration (cf results section). Dempster rule is then applied for each pixel. The final image is then obtained by the sum of all the fused masses, and the conflict is kept as a separate image. Other mass assignment procedures are detailed in [1,6] for other applications where the aim is to classify defects/false alarms using feature extraction. In this case, the features are sued as sources of information.

3. Results

The result of multi-thresholding is shown as an example for image 1. The hypotheses corresponding to the different grey-level ranges, and the number of ranges have been chosen by the expert. This is for the moment not an automatic process. Joint hypotheses are used when the grey-level range is such that it can belong to one or the other family (low or medium attenuation, medium to high, etc…).

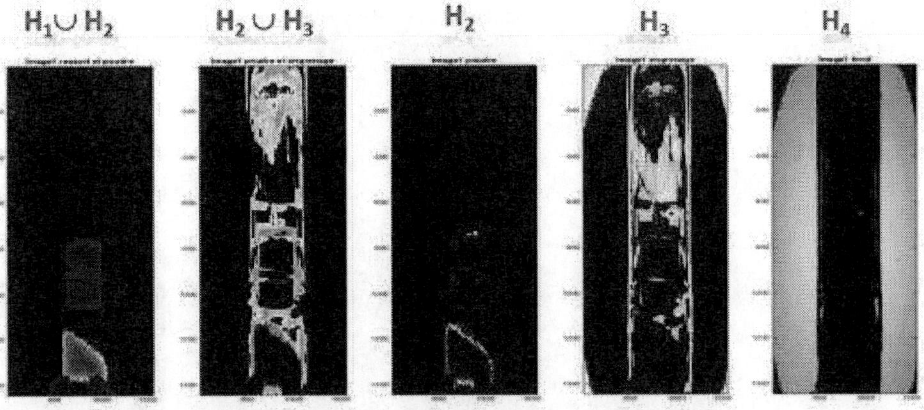

Fig. 2: Example result of multi-thresholding (a) on image 1 of figure 1. Note that the scale is not respected in order to show all the images on the same figure. The hypotheses assigned to each grey-level family of the histogram are chosen by the expert considering the acquisition conditions.

Fig. 3 ;Left : image obtained after DS fusion after summing all the masses on the 4 hypothesis for each pixel; right : conflict image.

The result obtained on figure 3 is not perfect if we look to the spring. Indeed, the initial image #3 is better than the fused image for this high attenuation part. The reason is that the two other images #1 and #2 are not informative, and this lack of information is modeled as an ignorance. However, the ignorance is part of the fusion process, and this decreases the final confidence on these pixels. On another hand, for the low and medium attenuation parts, the final image after fusion is relevant, and better than each of the single acquisition.
The result can be improved by considering a max rule combination for the hypothesis "high attenuation": in this case, the maximum mass among the three information sources is kept, which correspond in our case to the image #3. This is a "contextual" fusion in the sense that different combination rules are applied depending on the hypothesis. However an a priori knowledge is necessary to conduct such a data fusion process.

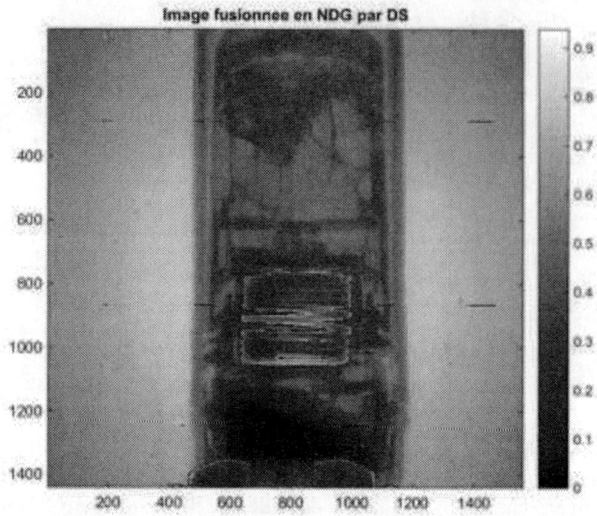

Fig. 4 : Image after contextual fusion : DS fusion for the hypothesis H1, H2, and H4, and max rule for H3.

In order to compare the results obtained with data fusion, we also tried simpler procedures, like logical operations and color imaging.
Indeed, as we have three information sources, a composite RVG image can be built as shown in figure 5. Here, image #1 was kept as the red scale, the green one is the negative of image #2, and the blue one is the negative of image #3. Other combinations could also be tried to get different visual aspects. However, it is not a fusion in the sense of a mathematical operation between the information sources, here it is more a superimposition of information.

Fig. 5 : RGB image resulting of the superimposition of R = im#1, G=negim#2, B = negim#3.

Finally, thresholding and logical operations were also used to realize an image fusion where only complementary information is used. The idea is that the low kV image will get information about low attenuation parts : thus thresholding operations are applied to hide medium and high attenuation pixels (see figure 6), the same is done for the medium attenuation/kV image, and the last image is kept as it is. Then all the images are summed together. Since low attenuation pixels are still present in image # after thresholding, their value is summed with that of image #1, which gives a new information in the final image.

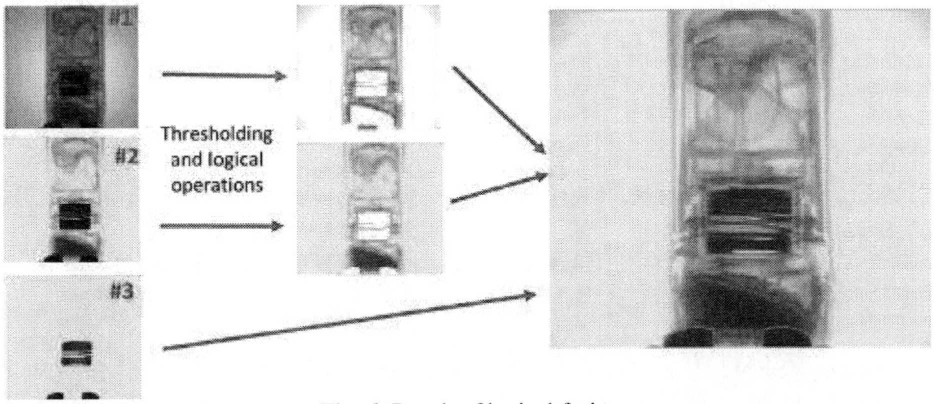

Fig. 6: Result of logical fusion

4. Conclusion

With respect to our previous works where the final fused image was a segmented image [7] also using different acquisitions as inputs, here the final result is a grey-level image containing as much as possible the original contrast information, or a better one. We have shown different ways to combine the different X-rays acquisitions. Results are not perfect, other combination rules have also been tried with rather similar performances. A weak point is the multi-thresholding which is for the moment supervised by the expert in order to assign the right hypotheses to each family range identified in the image. Our future works will focus on the automatization of this part. On another hand, an image quality assessment is necessary to really quantify the interest of fusion with respect to a simple superimposition. Again, future works are necessary to investigate more deeply the potential of such techniques. However, the interest is clear at least to make use of future detectors able to acquire simultaneously several energy channels, and this fusion step can be the input to a single tomographic reconstruction once the projections have been combined. For parts having different attenuation ranges, this can be a solution. It is worth noting that the challenge here is high, because we want to build a final grey-level image, i.e. where the information about best contrast is still present (as best as the best acquisition for this attenuation range), while all attenuation ranges are fused together.

References

[1]. A. Osman, V. Kaftandjian, and U. Hassler. "Improvement of X-ray castings inspection reliability by using Dempster-Shafer data fusion theory". Pattern Recognition Letters, Vol. 2, Issue 32, pp: 160-180, 2011.

[2]. A. Osman, V. Kaftandjian, and U. Hassler. "Automatic classification of 3D segmented CT data using data fusion and support vector machine". Journal of Electronic Imaging of Spie, Vol. 21, May 2012.

[3]. A. Osman, U. Hassler, V. Kaftandjian, J.Hornegger, Automated segmentation of ultrasonic volumetric data of composite materials, Insight - Non-Destructive Testing and Condition Monitoring 2015; 57(3), p. 1-8.

[4]. A. Osman, V. Kaftandjian, U. Hassler, Application of data fusion theory and support vector machine to X-ray castings inspection. Proceedings of 10th European Conference on Non-Destructive Testing (ECNDT), 7-11 June 2010, Moscow Russia.

[5]. A. Osman, U. Hassler, V. Kaftandjian, Interest of data fusion for improvement of classification in X-ray inspection, Digital Industrial Radiology and Computerized Tomography DIR2011, Berlin, June 2011

[6]. A. Osman, Automated evaluation of three dimensional ultrasonic datasets. Phd thesis, university of Erlangen-Nuremberg and INSA-Lyon, June 2013.

[7]. Y.M. ZHU, L. BENTABET, O. DUPUIS, V. KAFTANDJIAN, D. BABOT, M. ROMBAUD, Automatic determination of mass functions in Dempster-Shafer theory using fuzzy-C-means and spatial neighborhood information for image segmentation, Optical Engineering, 2002, vol.41 (04), p. 760-770.

[8]. A. Dempster, "Upper and lower probabilities induced by multivalued mapping", Annals of Mathematical Statistics, 38, pp. 325-339, 1967.

[9]. G. Shafer, "A mathematical theory of evidence", Princeton University Press, Princeton, pp. 297, 1976.

Sensor fusion and correlation of X-ray tomography and XRF data for drill core analysis

Jonathan HANSEN [1], Virginia VOLAND-SALAMON [1], Tobias C. KAMPMANN [2], Markus FIRSCHING [1], Oleksiy SAMODVIGA [1], Alexander ENNEN [1], Norman UHLMANN [1]

[1] Fraunhofer Development Centre X-ray Technology EZRT, Fürth, Germany

[2] Department of Civil, Environmental and Natural Resources Engineering, Division of Geosciences, Luleå University of Technology, Luleå, Sweden

Contact e-mail: jonathan.hansen@iis.fraunhofer.de

Abstract. State of the art analysis techniques on drill cores for exploration purposes, including X-ray fluorescence (XRF), laser-induced breakdown spectroscopy (LIBS) or Raman spectroscopy are used to derive mineralogical information. Since this sensor data corresponds to materials that occur on the surface of the core, inclusions (e.g. diamonds) cannot be detected. In addition, information outside of the measurement position is not taken into account and may lead to misinterpretation or the miss of certain elements. X-ray computed tomography (CT) and radioscopy provide data about the entire sample as well as inlying structures based on X-ray absorption. As a drawback, CT is time-consuming and the material information is not explicit.

For the enhancement of geological interpretation, we propose to apply sensor data fusion techniques in order to unite both depth information as well as reliable material data information from surface measurement techniques. This leads to more substantial information of the drill core.

For further insights in the feasibility we investigate the correlation of XRF data at varying abstraction levels with CT data, i.e. grey value information.

The applied XRF technique involves the fact that the data is not acquired continuously but discrete point by point. This is accompanied by the circumstance that the spatial resolution of the acquired data has a different magnitude than the CT-data. Both facts result in the challenge to register the XRF data coming from a one-dimensional scan line with a micro-CT volume. The experiments must be planned in a way that location and orientation of the scan data are well-known and reproducible.

In the experiments, the acquired and registered data of a defined drill core is analysed with respect to correlation and fusion capability. The experimental setup will be presented and results will be discussed.

1 Introduction

Sensor based analysis of drill cores has been well established in mining and exploration activities in order to support geologists extracting valuable information from drill cores and samples. With huge amounts of cores coming from drilling activities, the need for automated inspection and computer aided analysis rises constantly. Different analysis techniques like LIBS, XRF or optical inspection are able to derive detailed and reliable information from the core but are limited to analyse only the surface. In addition providing measurement data covering most of the surface is unpractical due to handling of the drill cores or samples. While most of the measurement techniques provide information based on a line or spot scan, the results from the surface must be extrapolated to the entire core.

In contrast, computed tomography (CT) measurements are able to provide information on the entire core but lack the possibility to deduce chemical or geological information in the same detail and precision.

In order to obtain precise mineralogical information as well as volumetric data considering the entire core, we propose a way of combining both measurement techniques with the aim of improving the overall available data of the core.

2 Experimental Method

2.1 Drill core samples and data generation

Two drill core samples of 12 cm and 16 cm length respectively, have been chosen in close collaboration with the Geological Survey of Finland (GTK) and its drill core archive. The samples have been characterized petrographically using standard optical methods, with special emphasis on the chosen locations for the analytical lines (see section 3). To follow the attempt of combining multiple measurement techniques we performed XRF and CT scans on the selected drill cores. The measured data was then superimposed and combined.

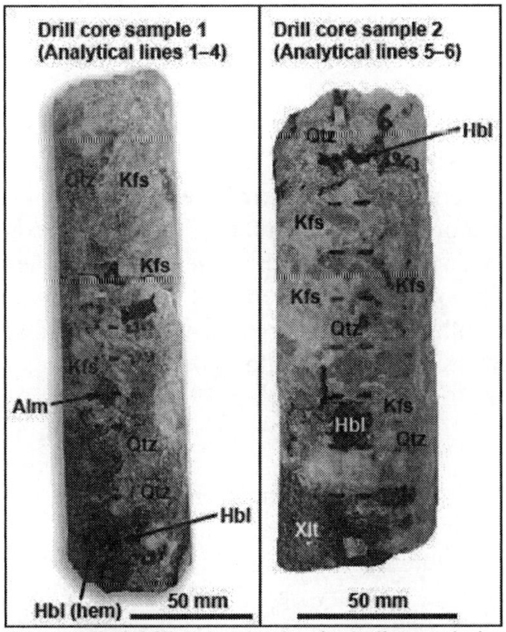

Fig. 1. Optical photographs of the studied drill core samples, including analytical lines 1 (drill core sample 1) and 6 (drill core sample 2). Macroscopically present minerals are marked according to the petrographic analysis. See Table 2 and text for more information. Abbreviations: Qtz = Quartz; Kfs = K-feldspar; Alm = Almandine garnet; Hbl = Hornblende; Hbl (hem) = Hornblende with hematite staining; Xlt = Xenolith fragment consisting of fine-grained biotite, hornblende and plagioclase.

2.1.1 XRF

XRF-data were generated manually using a *Bruker S1 TITAN* series handheld XRF analyser. The locations on the surface of the core where chosen in a way that we defined six lines along

the core samples that looked promising in measuring all present minerals. This fostered differentiating between them in CT-analysis. Then, measurement sites were marked with a pen. The spots were placed in regular intervals of 10 mm (scans 1 and 2) or 20 mm (scans 3 and 4) respectively. The *S1 TITAN* has an internal camera that was used to measure as close as possible around the predetermined spots.
Mineral distribution was ascertained using the *GeoChem* feature provided by *Bruker*.

2.1.2 CT

3D data was generated using a stop-and-go CT scan procedure that is common in diamond drill core scanning of samples similar to the ones used in this study. Scan parameters are listed in Table 1.

Table 1. Scan parameters as used for the CT-scans

Source-detector-distance (mm)	663.384	Tube voltage (kV)	220
Source-object distance (mm)	326.482	Tube current (mA)	0.18
Number of projections	1200	Exposure time (ms)	300
Voxel size (μm)	94.1765	Number of frames averaged	1

Apart from the raw data, the most important information to be used is the edge length of a single voxel within the 3D dataset. Based on this we can calculate the volume of a voxel cluster we estimate to be of a specific material.

2.1.3 Matching

Matching the data of different sources was possible by marking the XRF measurement segments with zinc plates, which are easily identifiable in the CT data due to their known shape and high absorption. Therefore, we were able to calculate the individual measurement locations in terms of 3D coordinates. Automation of this procedure is possible, due to the difference in signal of the zinc plates from the other present minerals in the samples. It may also be possible to superimpose the measurements utilising contained materials or minerals with significant signal in both techniques or simply by a known alignment of the sensors to each other within the measurement coordinate system.

2.2 Image Processing

The aim with the segmentation of the aforementioned areas is to determine the amount of Fe in the core samples. The segmentation process creates a mask that marks each voxel as either Fe or not. In future work, it should be investigated if a weighted mask is more suitable. Such mask would not only represent a binary decision (Fe: yes or no) but an approach closer to fuzzy logic.
There are two issues regarding the raw CT scan data:

1. Cupping: cupping artefacts are caused by beam hardening and lead to radial, non-equal grey value distributions in the core volume (Fig. 2). We managed to minimize the cupping effect by first calculating a median over the z-index of the whole core and then calculating a radial mean value over 100 angles.

Fig. 2. Median of z-projection showing higher grey values on the border (lighter) than on the inside (darker).

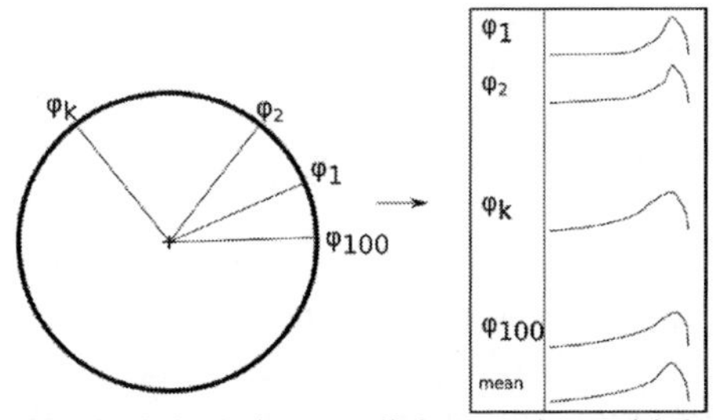

Fig. 3. Calculating a radial mean for multiple angles from the result of Fig. 2.

The averaged data was fitted via polynomial curve fitting to get a smooth curve (see Fig. 3). Based on the curve, the whole core volume data was corrected to achieve the same grey values for equal materials and densities in the core, independent of their distance to the core surface, which is crucial for approximating the total concentration of a certain material. In addition, the grey value outside the radius of the core was set to zero.

2. Noise: This is partly because of the acquisition method and partly due to the fact that minerals can be of a specific structure that may mimic a noise signal in 3D data. To eliminate most of the noise, we applied a specific type of median to the core volume data that would ignore values that are zero. This was necessary to ensure that the air around the core would not affect the calculated median value.

After these two problem solving steps, we noticed a better data quality (i.e. a more distinct grey value profile) by examining single slices of the 3D volume and individual plots. The denoising step was important to increase the overall contrast between different areas and lower the contrast within them, which was the starting point to begin segmentation.

2.3 Segmentation and Binarisation

For segmentation a k-means algorithm was applied. Alternative methods would include watershed segmentation and manual separation of histogram values. We settled with the k-means approach to be able to automate the process in the future. Some specific values for the algorithm were chosen manually: we set k=3 to segment the data into background/air (low values), Si (mid values) and Fe (highest values). Manual starting points for the clustering were gathered by considering the volume histogram. We performed this step in order to ensure the reproducibility of the results.

We segmented solely for the grey-value of the CT scan and did not track any other features such as the size of specific connected clusters. After the segmentation we only kept the label representing Fe, the other voxels were set to zero. One main intention was to ensure the interpretability of the analysis results. The labelled volume contained clusters of different sizes, especially many small voxel clusters or even single voxels. The petrographic analysis suggested that the small clusters and single voxels do not correspond to Fe in reality. We hence deleted all clusters with sizes smaller than $5mm^3$. This threshold was selected arbitrarily according to the petrographic analysis and had minor impact on the analysis result itself.

3 Results

3.1 Petrographic analysis

Both samples are gneisses with granitic to granodioritic composition and coarse grain sizes which in part resemble a pegmatitic texture. They belong to a suite of 1.8 – 1,9 Ga metamorphic granite gneisses representing a portion of the bedrock around Helsinki, Finland (Alan Butcher, personal communication 2019). Identified minerals in the samples include K-feldspar, quartz, plagioclase, hornblende (often hematite-stained) and almandine (Fe-rich) garnet. In core sample 2 (analytical lines 5 and 6), a xenolithic fragment occurs which consists mainly of fine-grained hornblende, biotite, quartz and plagioclase. This drill core sample generally contains lower amounts of quartz and higher portions of hornblende, probably representing a more intermediate composition compared with sample 1 (analytical lines 1–4). The encountered main minerals in the drill core samples, including their ideal chemical formulas, are summarized in Table 2.

Table 2. Main mineralogy of the drill core samples utilized in this study as identified in the petrographic analysis, including ideal chemical formulas. Chemical substitutions and trace elements can occur.

Mineral	Chemical formula
K-feldspar (orthoclase)	$KAlSi_3O_8$
Quartz	SiO_2
Plagioclase	$(Na,Ca)(Si,Al)_4O_8$
Hornblende	$(K,Na)_{0-1}(Ca,Na,Fe,Mg)_2(Mg,Fe,Al)_5(Al,Si)_8O_{22}(OH)_2$
Hematite (present as staining on hornblende)	Fe_2O_3
Almandine garnet	$Fe_3Al_2(SiO_4)_3$
Biotite	$K(Mg,Fe)_3AlSi_3O_{10}(OH,F)_2$

3.2 Matching

In general, grey values in a reconstructed CT-volume correspond to attenuation coefficients of the respective material: the higher the grey value, the higher the attenuation coefficient. In our case, the grey values are represented by arbitrary units. Correlation of the XRF- and CT-data sets provided the insight that high grey values (corresponding to the bright spots in the cross sections) show a correlation with the fraction of Fe compounds measured by the XRF scans. To provide this data, we listed the Al and Fe compounds, which represented the major share of minerals besides Si and plotted them onto an unwrapped view of the drill core surface provided by CT data (see Fig. 4). In order to analyse correlation, we created a line plot containing the Al- and Fe- fractions on the one hand and grey values on the other hand (Fig. 5). The impact of Fe on the grey values is clearly visible, while Al has minor impact.

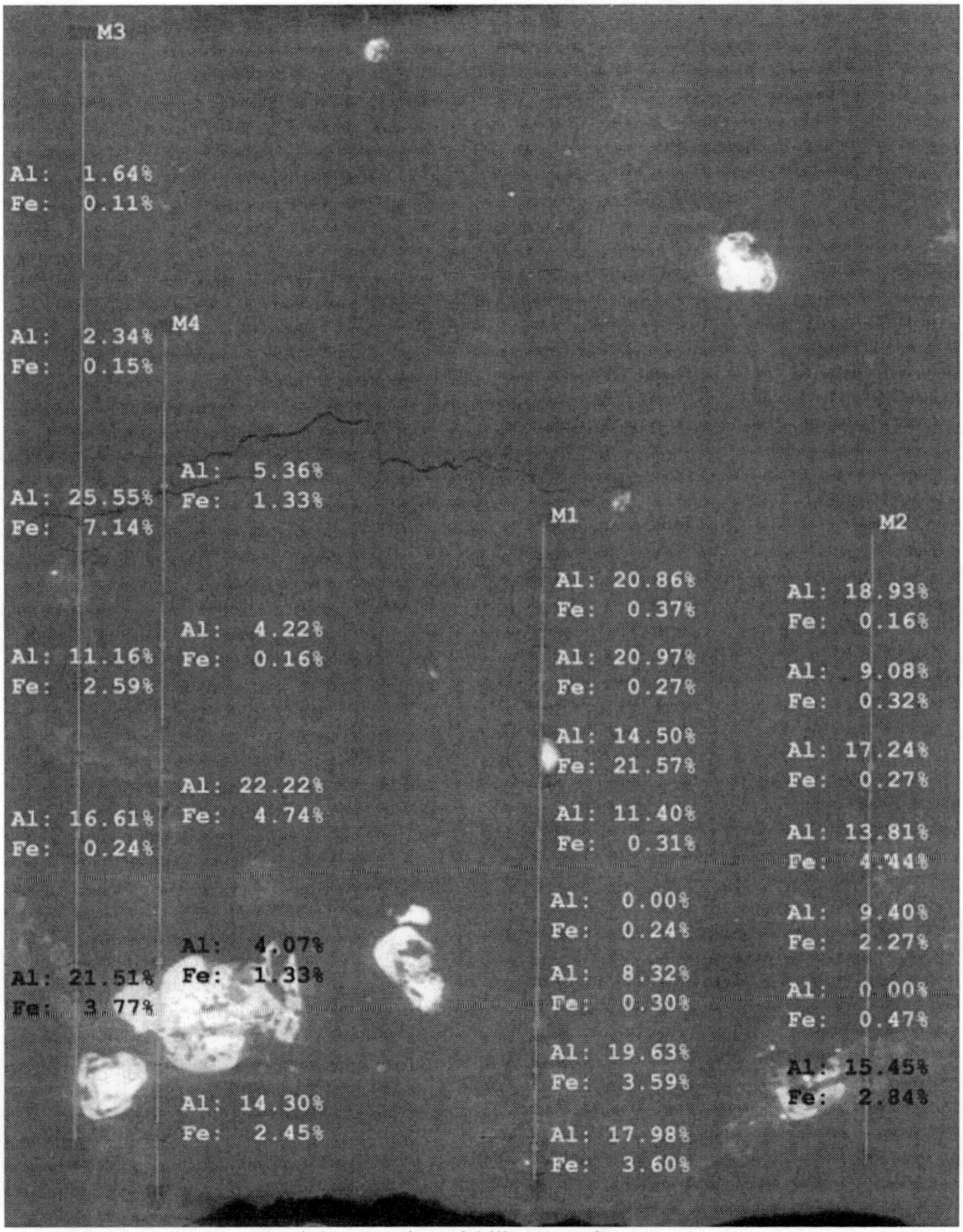

Fig. 4. Unwrapping of drill core surface.
XRF Measurements marked in green, Fe marked by red circles, Al findings in blue.

Other minerals such as Al compounds could not be identified with good correlation in the CT data. This should be further investigated in using different approaches based on what features to correlate (not the grey values itself but the mean grey value in a defined area combined with the standard deviation in the region of interest for example). In order to be able to correlate more minerals in XRF and CT data, we need more drill cores and methods how to correlate the data.

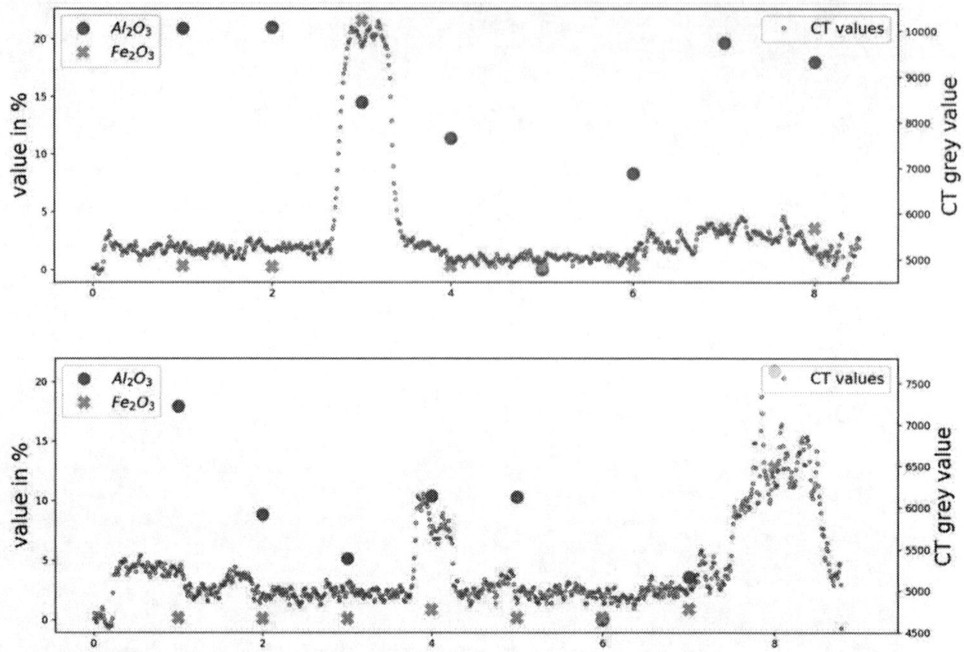

Fig. 5. Correlation between XRF data (blue o, orange x) and grey values from CT (green dots). It is clearly visible, that the occurrence of Fe leads to an increased grey value, while the occurance of Al has comparably less impact.

3.3 Volume data and image processing

Following the workflow outlined above (see 2.2), we have a volume that contains binary information on whether the individual voxel in the original volume is to be taken as Fe or not. Calculating the number of voxels that were labelled to be Fe, we obtained the following data (Table 3):

Table 3. Combined results of the core analysis. The method allowed a determined fraction of Fe of 1.5 %, which correlates with the petrographic analysis.

	volume (cm^3)	number of voxels	fractional part
drill core	204.48	244,806,465	100 %
Fe	3.07	3,673,824	1.50 %

The following images illustrate the steps applied to the volume. Every row shows one specific 2D slice from the same 3D volume in different stages of the processing.

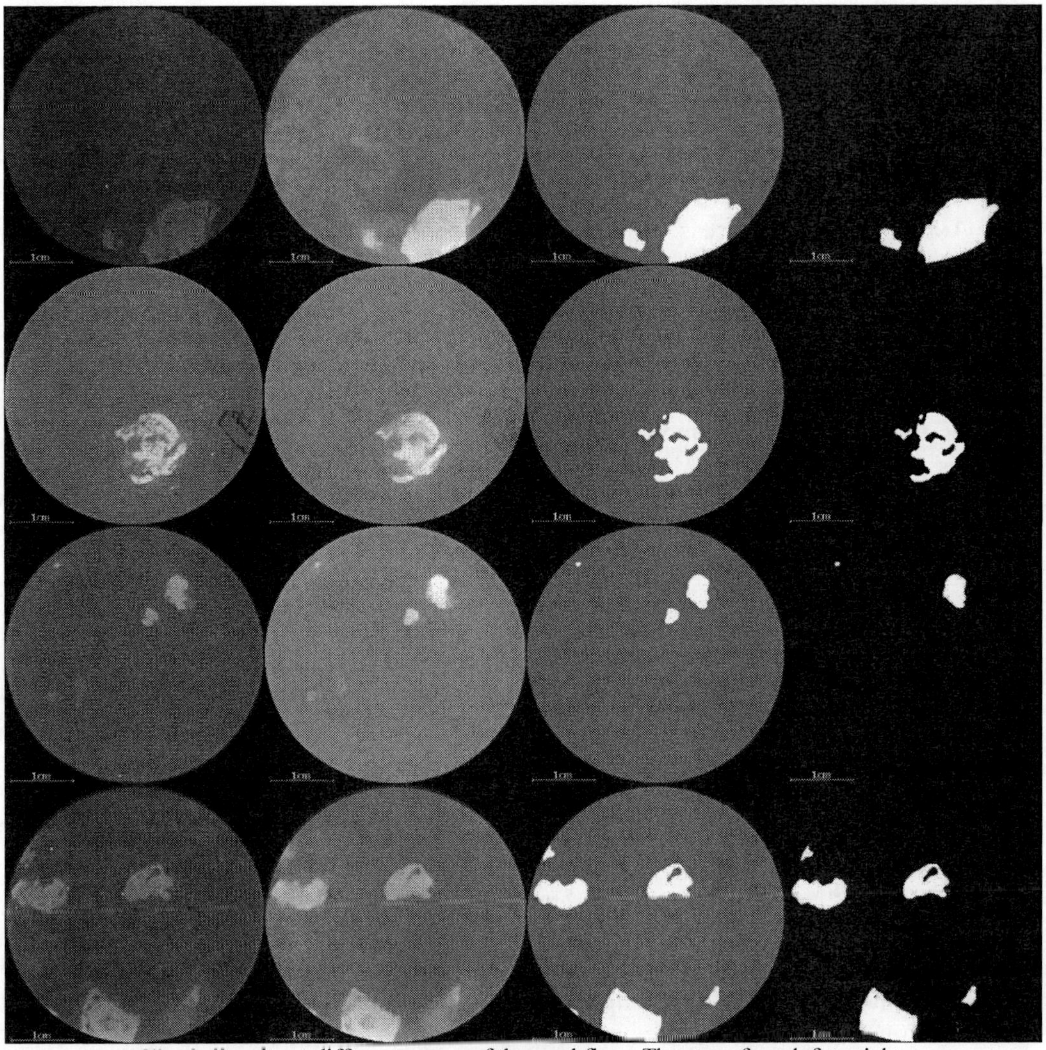

Fig. 6. Results at different stages of the workflow. The steps from left to right:
1: after having corrected the cupping artifact
2: after having applied denoising
3: after segmentation
4: after having eiminated small grains

The images show 2D cross-sections of a three-dimensional data set. All in step 4 eliminated areas belong to clusters of size smaller than $5mm^3$. All remaining single white pixels are directly connected to clusters larger than $5mm^3$. These clusters in general extend over several slices and may therefore appear smaller than they are.
All calculations were done using python with packages from *numpy*[2] and *scipy*[3] (*scipy.ndimage* in particular, [1]-[4]).

4 Conclusion and outlook

4.1 Performance of the developed methodology and possibilities for improvement.

The overall goal of extending the surface data into the 3D volume could be partially achieved. When working with the data we observed that handling the segmentation between Si and Fe was relatively stable while trying to separate Al or K from other materials did not work with the same reliability.

There is, of course, opportunity for improvements, especially when we consider investigating more samples. A few examples of improvements that will be worked on are:

- Extending the range of detectable materials by sharper CT-segmentation and gathering combinations of XRF-measurements and responding CT-information.
- Using the graphics card as a way to improve calculation speed.
- Automating the process of matching the coordinates of XRF- and CT-data by segmenting the CT data for markers.
- Finding an automatic way to provide parameters for the k-means algorithm.
- Identifying the radius and overall extent of the core in the 3D data automatically.

These steps should help to develop an algorithm that receives XRF- and CT-data as input, producing an estimate for Fe as an output. In regards to the precision of the proposed method, comparison to established surface technique based scanners as well as a direct chemical analysis is part of the ongoing work to validate the generated.

4.2 The significance of precise measurement of iron content in the context of geology and exploration

The developed methodology has been demonstrated to accurately determine the presence and three-dimensional distribution of Fe in the processed drill core sample. Iron is present in the crystal lattice of nearly all minerals present in the studied core samples (Table 2), with the notable exception of quartz (pure SiO_2). The minerals with the highest Fe content in these samples are expected to be almandine garnet (up to 33.66 wt% Fe), as well as hematite staining on hornblende (69.94 wt% Fe). The comparison between the petrographic analysis and the CT images of the cores leads to the interpretation that the brightest areas in the CT images represent almandine garnet. Hematite staining is also identifiable, but not as prominent which is probably due to its subordinate character (staining on hornblende).
The distribution of Fe in geological samples has important implications for exploration for metallic ore deposits. Alteration haloes of hydrothermal ore deposits, such as volcanogenic massive sulphide deposits, for example, typically experienced Fe enrichment to varying degrees during fluid flow and ore formation. This will lead to the formation of Fe-rich minerals such as chlorite and almandine garnet, the latter especially in metamorphosed deposits. Recent studies demonstrate that the three-dimensional distribution of Fe content in rocks surrounding hydrothermal ore deposits, based on both whole-rock geochemical data and mineral-chemical analysis, can be utilized to identify vectors towards mineralization ([5]). For this reason, methods that are able to accurately determine the distribution of Fe-rich minerals and general Fe content in rocks will aid geologists in the efficient targeting of ore deposits. In this context, we evaluate the herein developed method as a valuable contribution towards more objective and automatized processes in exploration, worthy of further investigation and development.

5 References

[1] https://www.python.org
[2] http://www.numpy.org
[3] https://www.scipy.org
[4] https://docs.scipy.org/doc/scipy/reference/tutorial/ndimage.html
[5] Kampmann TC, Jansson NJ, Stephens MB, Majka J, Lasskogen J (2017) Systematics of Hydrothermal Alteration at the Falun Base Metal Sulfide Deposit and Implications for Ore Genesis and Exploration, Bergslagen Ore District, Fennoscandian Shield, Sweden. Economic Geology 112, 1111–1152.

Multiscale and Multimodal Approaches for Three-dimensional Materials Characterisation of Fibre Reinforced Polymers by Means of X-ray based NDT Methods

Bernhard PLANK [1,2], Marcel SCHIWARTH [1], Sascha SENCK [1], Johanna HERR [1], Santhosh AYALUR-KARUNAKARAN [3], Johann KASTNER [1]
[1] University of Applied Sciences Upper Austria, Wels, Austria
[2] University of Augsburg, D-86135, Augsburg, Germany
[3] FACC Operations GmbH, Ried im Innkreis, Austria
Contact e-mail: bernhard.plank@fh-wels.at

Abstract. Non-destructive testing (NDT) and three-dimensional materials characterisation of fibre reinforced polymers using X-ray based methods can be carried out at different length scales and by using different modalities. This work gives an overview of different X-ray based NDT methods and their characteristics.
Multiscale X-ray computed tomography (XCT) usually includes scanning an entire part at lower resolution – governed primarily by specimen diameter. Subsequently, a smaller sample is cut out of the respective specimen and scanned at a higher resolution. Accordingly, in this work typical XCT resolutions ranging from (135 µm)³ voxel size down to (250 nm)³ are presented.
Using different XCT modes such as region of interest scans or laminography (XCL) modes this multiscale approach is also possible without destroying or cutting the sample in smaller pieces. However, some limitations in image quality and sample geometry have to be considered. We show that cracks with a width between 122 and 56 µm can be clearly seen at a relatively low resolution of (135 µm)³ voxel size in one example of a larger carbon fibre reinforced polymer (CFRP) sample from the aeronautic industry. With XCL voxel sizes down to (0.75 µm)³ can be reached, showing clear structures in the range of 16 µm. Main disadvantage of XCL is that only a view layers and not the full 3D-microstructure can be represented.
Using an XCT resolution in the range of (2 µm)³ voxel size for CFRPs may lead to misinterpretation in relation to porosity because of propagation-based phase contrast effects. High-resolution region of interest XCT scans at (250 nm)³ voxel size show that epoxy-rich areas between individual C-fibres smaller than 6 µm are leading to relatively dark grey values, easily misinterpreted as voids.
Multimodal XCT data was generated using a Talbot-Lau Grating Interferometer (TLGI) XCT to obtain modalities such as dark-field contrast and differential phase contrast in addition to standard attenuation contrast. In one example it is shown that metal artefacts in CFRP issued by a Cu-mesh can be significantly reduced by TLGI-XCT. This provides improved image quality and the possibility to segment voids close to metallic components.
For easier interpretation and a better understanding of material features the open source software open_iA was used with new implemented visualisation approaches for multimodal and multiscale data-visualisation.

1. Motivation and Introduction

Due to high complexity and rising demand for new materials such as fibre reinforced polymers (FRP), in many cases a three-dimensional materials characterization is essential to get better knowledge of the material behaviour. Therefore non-destructive testing (NDT) methods based on X-ray attenuation contrast (AC), such as X-ray computed tomography (XCT) are becoming more and more the state of the art for research and development. However, different XCT methods need to combine in a unified approach to extract information about different length scales. [1]

Three-dimensional materials characterisation of a specimen using X-rays is usually done by scanning an entire part at a relatively low resolution – governed primarily by the specimen´s diameter. To achieve higher resolution which is limited by the size, a multiscale approach is adopted by scanning a smaller part (cut-out) of the specimen. Accordingly, typical lab based XCT resolutions range from approximately $(150\ \mu m)^3$ to around $(0.3\ \mu m)^3$ voxel size [1].

Using different XCT-modes such as region of interest scans (ROI-XCT) or laminography modes (XCL), a multiscale approach is also possible without destroying or cutting the sample into smaller pieces. [2, 3]

Using additional modalities, such as differential phase contrast (DPC) or dark-field contrast (DFC) are creating new possibilities for materials characterization. For instance, using a lab based equipment such as a Talbot-Lau grating interferometer XCT (TLGI-XCT) one can gain three different modalities (AC, DPC, DFC) in one scan using a phase stepping approach [4-6].

However, besides all advantages of the mentioned XCT-modes and XCT-modalities, several limitations in measurement time, artefacts, image quality and sample geometry have to be considered and will be further discussed in this paper for different FRP samples.

2. Experimental Setup

2.1 Investigated Samples

In this work, carbon fibre reinforces polymer (CFRP) samples from research projects in the field of aeronautic and automotive industry are used to investigate using different XCT systems, modes, and modalities.

Sample #1 is a CFRP plate with dimensions of 100x100x1 mm^3 made out of Prepregs in plain weave style (PRG09229 - CC 120 ER450 43%). In this plate, mainly porosity is expected.

Sample #2, is a sub-component of an aeroplane spoiler assembly, a so-called *center hinge fitting* (CHF), with dimensions of approximately 450x640x150 mm^3. A complete scan of the sample was not possible due to sample dimensions on the available devices in the project consortium, thus only a region of interest (ROI) could be scanned. To discuss data quality, additional reference samples from the same material were carried on the specimen, which were scanned separately in advance to achieve a higher resolution.

Sample #3 is a piece from a CFRP sheet moulding compound (C-SMC) with recycled carbon fibres and automotive grade epoxy resin (HexPly® M77). Images were obtained from a sample cross section of approximately 2x2 mm^2. Main interest lies in the present of voids and resin rich areas.

Sample #4 is characterized by a CFRP sample with a copper mesh near the surface used as lightening protection. It consists of two layups, multiple prepreg layers made of woven carbon fibre bundles embedded in aerospace grade epoxy resin, which are bonded by an adhesive film. This adhesive film has a higher absorption contrast compared to the epoxy matrix. Important material characteristics of the CFRP sample are the distribution and orientation of the fibre bundles (HexPly® F584) and the distribution of pores. The sample cross section was 9x4 mm².

2.2 X-ray based Measurement Systems and Scan Parameters

All scans were performed at the XCT-laboratories of University of Applied Sciences Upper Austria which is equipped with four different XCT-devices. More details and specifications are shown in Table 1. The devices used in this work were chosen in dependence of sample size, applicable modality and necessary scanning-mode.

Table 1. Used X-ray based measurement systems and main specifications

XCT device	RayScan RayScan 250E	GE phoenix Nanotom 180 NF	Bruker SkyScan 1294	RX Solutions EasyTom 160
X-ray source	225 kV µ-focus & 450 kV mini-focus (fix ~0.4 mm)	180 kV sub-µ-focus	60 kV µ-focus (fix ~ 30 µm)	160 kV nano-focus
Detector system(s)	2048*2048 pixels (flat panel)	2304*2304 pixels (flat panel)	4008*2672 pixels (CCD camera)	1920*1536 px (flat panel) & 4008*2672 px (CCD camera)
Min. voxel size	~ 5 µm	~ 0.5 µm	~ 5.7 µm	~ 50 nm
Max. sample diameter	< 300 mm	< 68 mm	< 20 mm	< 200 mm
Sample height	< 2 m	< 150 mm	< 60 mm	< 700 mm

All used devices and scan parameters for the individual samples are listed in Table 2.

Table 2. Applied scan parameters for the individual samples

Sample	XCT system	Image modalities/ XCT modes	Scanning parameters (tube voltage; Tint; images; Target-material)	Voxel size [µm]	Scanning - time
#1	RayScan 250E	XCT	160 kV; 999 ms; 1610; W	(70 µm)³	51 min
		ROI-XCT	160 kV; 1999 ms; 1440; W	(15 µm)³	99 min
	Nanotom 180 NF	XCT (1 mm² cut out)	50 kV; 1000 ms; 1700; Mo	(2 µm)³	258 min
	EasyTom 160	XCL	60 kV; 500 ms; 230; W	4*4*4.1 µm³	32 min
		XCL	60 kV; 500 ms; 230; W	1*1*0.94 µm³	137 min
		XCL	60 kV; 500 ms; 228; W	0.75*0.75*0.94 µm³	60 min
	(1 mm² cut out):	ROI-XCT	60 kV, 800 ms; 1568; W	(0.25 µm)³	220 min
#2	RayScan 250E	ROI-XCT	190 kV; 3750 ms; 1440; W	(135 µm)³	138 min
ref. #2.1	Nanotom 180 NF	XCT	60 kV; 500 ms; 1500; Mo	(17 µm)³	89 min
ref. #2.2		XCT	60 kV; 500 ms; 1500; Mo	(13 µm)³	89 min
#3	Nanotom 180 NF	XCT	60 kV; 600 ms; 1800; Mo	(6.5 µm)³	150 min
		XCT	60 kV; 600 ms; 1800; Mo	(2 µm)³	150 min
		ROI-XCT	60 kV; 1000 ms; 1800; Mo	(0.9 µm)³	215 min
#4	SkyScan 1294	AC, DPC, DFC	35 kV, 650 ms; 720; W	(22.8 µm)³	960 min

2.3 Used Software Tools

For reconstruction, the tools of the XCT-manufacturer, mainly based on filtered back projection algorithms, were applied. Standard software for voxel dataset handling and manual registration of each individual scan was VGStudio MAX 3.1 (Volume Graphics GmbH., Heidelberg, Germany).

Open_iA [7] was applied for new approaches regarding multimodal data-visualisations [8, 9]. For multiscale data visualisations, XCT-datasets have to be adapted in a way that they have exactly the same amount of voxels. That means that the voxel size of all registered datasets has to be recalculated to the highest resolution gained in one scan series. For example, for a scan series done with (70 µm)³, (15 µm)³ and (0.75 µm)³ voxel size the (70 µm)³ and (15 µm)³ were recalculated and extracted with (0.75 µm)³. To optimise visualisation performance the file size was kept to a minimum. Therefore, only small regions of individual voids or inclusions were extracted out of the entire data set for multi-scale visualisations in open_iA.

3. Results and Visualisation Approaches

Figure 1 shows the results of a 100x100x1 mm³ CFRP plate (Sample #1) scanned by conventional XCT with a resolution of (70 µm)³ voxel size to get an overview of the entire plate. By applying an ROI-XCT scan in the centre of the plate, a resolution of (15 µm³) was achieved. The resolution of ROI-XCT scans are mainly limited due to geometrical limitations of the used systems, the achievable geometric magnification and in most cases the necessity to fulfil a full 360° rotation of the entire specimen without hitting the X-ray tube or detector during the acquisition process. To achieve a higher resolution, using XCL was chosen as NDT-solution to get further details from this specimen. Using the XCL mode, voxel sizes between (4 µm)³ and (0.75 µm)³ were achieved. Applying XCL, it would theoretically be possible to investigate the entire plate in high resolutions, keeping in mind measurement time and the huge amount of data. In addition it has to be noted, that the voxel size for XCL is not isotropic, thus for easier comparison to other methods, we are only referring of the in-plane voxel size in this study. Exact voxel size is denoted in Table 2. For better comparison of individual material features scanned at different resolutions, one smaller void (Fig. 1, 2nd row) and one higher dense particle (Fig. 1, 3rd row) are shown separately.

The void is clearly seen at all resolutions but is only represented by a view voxels in XCT mode at (70 µm)³ voxel size. A quantification of void dimensions is not useful without any prior knowledge of the actual size. Using an ROI-XCT scan at (15 µm)³ voxel size, the shape of the void is already well represented and applying the correct threshold a segmentation is possible. By applying XCL, this void is represented very well. In addition, propagation based phase contrast effect [10] occurs between air and the material, which helps to clearly identify the border of the void structure. Using a higher resolution than (4 µm)³ for this relatively large 300 µm void does not lead to any additional benefits. On the contrary, higher resolution for bigger voids reduces the edge enhancement by the phase contrast effect.

XCL scans at (1 µm)³ and (0.75 µm)³ voxel size clearly shows the shape and dimensions (18x16 µm²) of the higher dense inclusion, which is not visible at lower resolutions. Using ROI-XCT this higher dense inclusion is clearly visible, even if the particle dimensions are almost equal to the voxel size of (15 µm)³. A quantification of the size would be impossible. In the normal XCT scan at (70 µm)³ voxel size this inclusion could not be resolved anymore.

It has to be noted that XCL was not able to resolve any fibre structures of the carbon fibres in this material. This is a bit surprising, because the resolution should be high enough to resolve individual carbon fibres with a typical diameter of ca. 6 µm. Conventional XCT scans at voxel sizes < (2.75 µm)³ on smaller samples sizes usually clearly resolve individual carbon fibres [10].

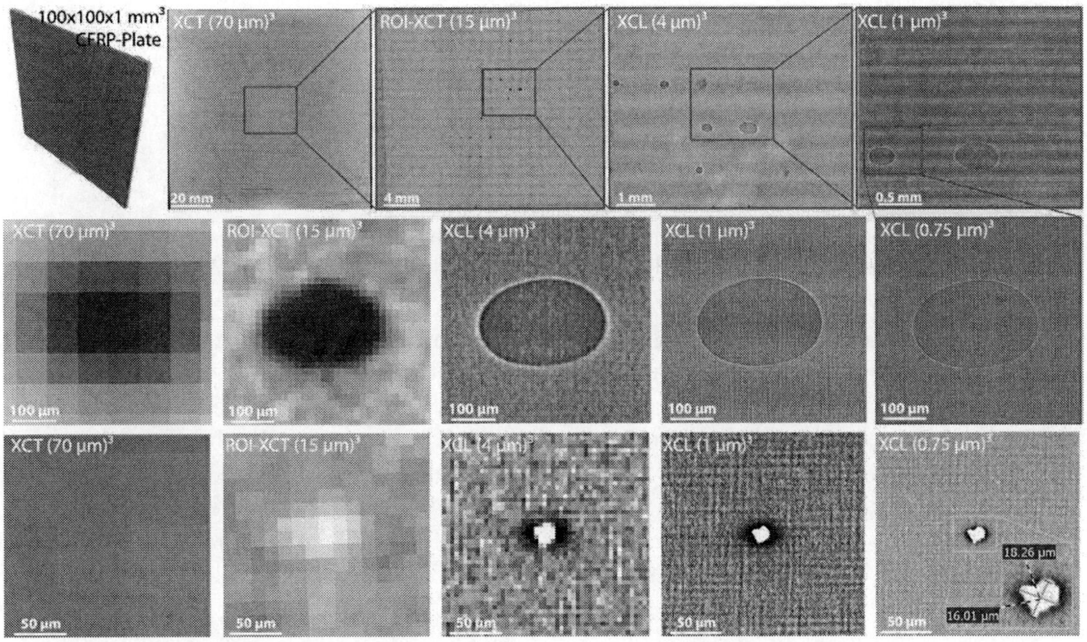

Fig. 1. XCT, ROI-XCT, and XCL images of Sample #1 showing different scan regions of the plate and corresponding voxel size (first row). In the second row one small void (diameter ~300 µm) and in the third row a higher dense inclusion (diameter ~16 to 18 µm) is shown at different resolutions.

Figure 2 shows visualisation approaches implemented in open_iA for visualizing three different datasets showing the void from sample #1. By varying the threshold (TH) of the ROI-XCT scan and using the edge enhancement effect due to phase contrast from XCL (4 µm)³ as reference, the segmentation threshold for the entire ROI-XCT scan can be approximated, by using TH = 12,366. For each individual dataset a transfer function can be adjusted, to show the necessary features (left). After setting up the transfer function (e.g. threshold), with an weighting widget with trimodal heatmap (center) a weighting fraction of the datasets (A, B and C) can be chosen, mainly adjusting the transparency of the individual files shown in 2D or as 3D rendering (right).

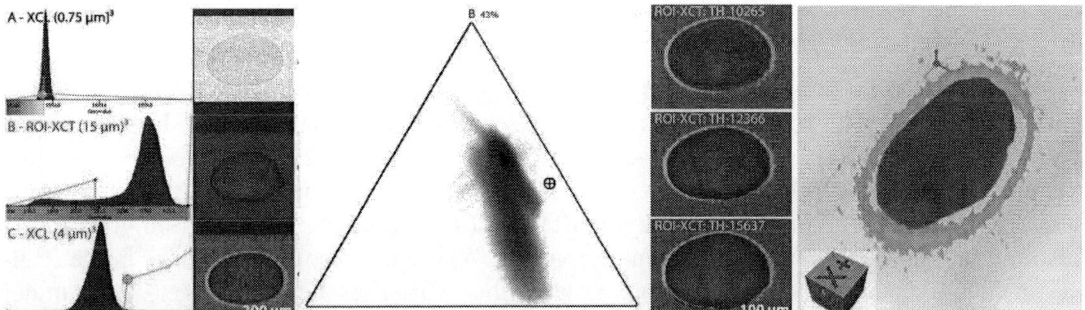

Fig. 2. Visualisation approaches implemented in open_iA for visualizing three different datasets showing the void from sample #1. By varying the threshold (TH) of the ROI-XCT scan and using the edge enhancement effect due to phase contrast from XCL (4 µm)³ as reference, the segmentation threshold for the entire ROI-XCT scan can be approximated, by using TH12,366.

In Figure 3 the higher dense inclusion from sample #1 is shown. This picture clearly shows that it is not possible to properly reconstruct the geometry seen by XCL (0.75 μm)³ (red) by varying the threshold of the ROI-XCT scan (yellow).

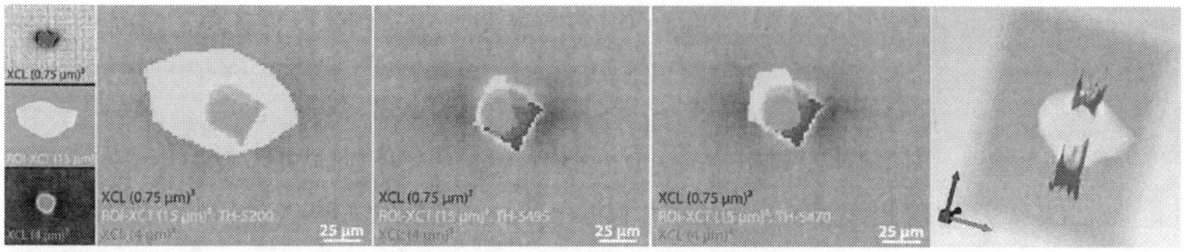

Fig.3. Visualisation approaches implemented in open_iA for visualizing three different datasets showing the higher dense inclusion from sample #1.

In Figure 4 (left) photographs and the defined region of interest for an ROI-XCT scan of an aeronautic sub-component (Sample #2) with a cross section of ~450x150 mm² are shown. Close to the critical regions of the component, some reference samples (ref #2.1 & #2.2) were glued, which were scanned in advance at higher resolution by means of XCT.

Focusing on the ref. sample #2.1 scanned at (17 μm)³ voxel size, a clear crack and some smaller voids are visible. The measured crack width is between 122 μm and 54 μm, which means, it is clearly below the voxel size of the ROI-XCT scan done with (135 μm)³. Nevertheless, this crack can also be very clearly represented in the ROI-XCT scan.

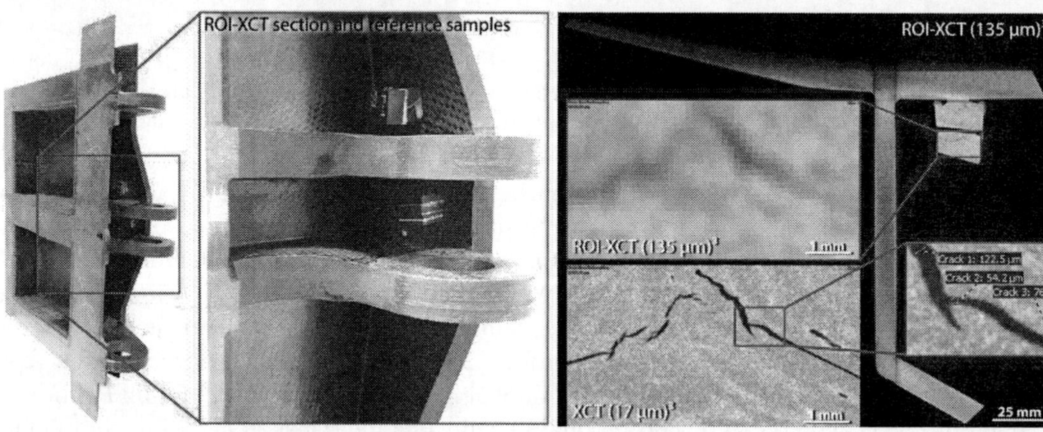

Fig.4. Photographs (left) and XCT slice images (right) of an aeronautic sub-component (sample #2) and reference samples #2.1 & #2.2 scanned with (135 μm)³ and (17 μm)³ respectively. A clear crack with a width between 122 μm and 54 μm is visible.

Figure 5 is showing an overlay of both datasets using open_iA. The crack and smaller voids are well resolved in the (17 μm)³ scan and are depicted in red. In blue are the results of the ROI-XCT scan performed with (135 μm)³ voxel size. It is clearly visible that the crack width is significantly overrepresented at the low resolution scan. In addition some areas are highlighted in blue, were no defects (#1) could be observed in the higher resolution scan. On the other hand, smaller voids (#2) are not resolved anymore in the ROI-XCT image data. In the 3D image (right) this variation between over- and under segmentation is visible more clearly.

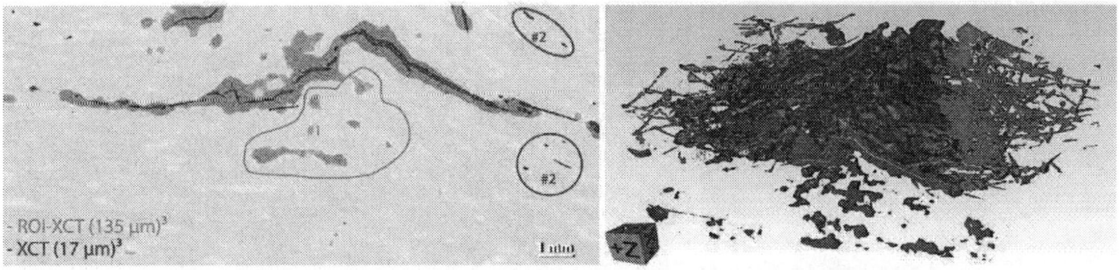

Fig.5. Visualisation approaches implemented in open_iA for visualizing ref. sample #2.1 at (17 µm)³ and (135 µm)³ voxel size respectively, showing crack structures and voids.

In Figure 6 the focus is on the voids existing in the ref. sample #2.2 scanned in advance at (13 µm)³ voxel size. By decreasing the threshold, it can be clearly seen that the small voids are mainly visible in some "junction areas" and the size of them is strongly overestimated at lower resolution (135 µm)³. By further reduction of the threshold, areas without voids are going to be segmented, which are mainly representing measurement artefacts.

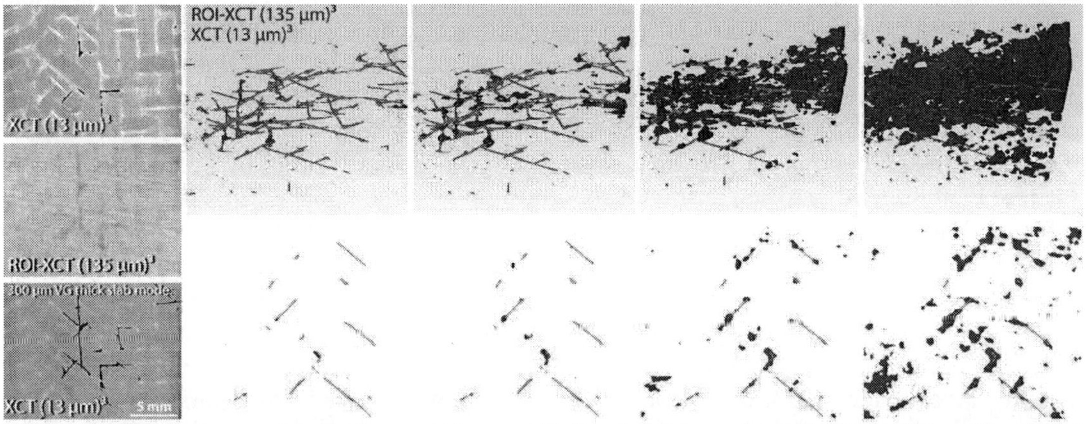

Fig.6. Visualisation approaches implemented in open_iA for visualizing ref. sample #2.2 at (13 µm)³ and (135 µm)³ voxel size respectively. For this images, the transfer function (threshold) was decreased step wise, to show how the void structures will be represented at low resolution (blue).

Several studies in the recent years have shown, that for porosity determination in woven CFRP samples [11-13] a voxel size in the range of (10 µm)³ is sufficient. Therefore, for sample #3 an initial voxel size of (6.5 µm)³ was chosen for XCT investigations shown in figure 7. At this resolution, only a few voids were present. To determine an proper threshold value, a higher resolution scans with (2 µm)³ was performed, showing that there are much more smaller voids in the resin rich areas. In addition, the (2 µm)³ scan clearly shows individual carbon fibres in the resin rich areas. These chopped fibres were gained by a recycling process. Many of these small voids are not clearly visible, thus an additional ROI-XCT scan with (0.9 µm)³ has to be performed. At this resolution, it is visible that no smaller voids are present. Smallest represented voids in this material type are in the range of 3.8 µm and therefore much smaller as individual carbon fibres with a diameter of around 6 µm. Usually for CFRP samples, it is expected that micro-voids in the range of the carbon fibre-diameter has a quite low impact of the overall porosity of material. In this case, these small voids have quite a high impact on total volume. An initial porosity evaluation with ISO50 threshold [12] delivers a porosity of 0.2 vol.% for the XCT scan at (6.5 µm)³ compared to a porosity of 0.8 vol.% evaluated in the same region of the (0.9 µm)³ scan.

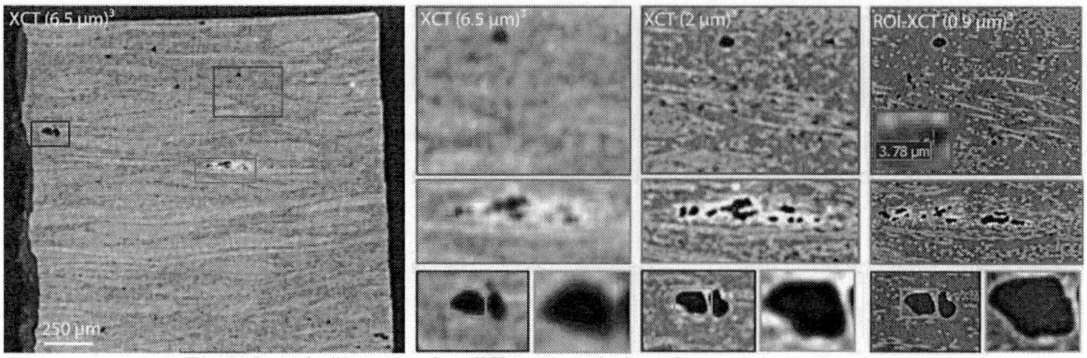

Fig.7. Sample #3 scanned at different resolutions for porosity estimation.

Usually for each multi-scale approach several scans have to be performed and afterwards a registration of the individual datasets has to be done. In many cases also different systems have to be used to cover the full necessary scale. To generate multimodal data using a TLGI-XCT device, three different datasets can be reconstructed out of one single scan, showing absorption contrast (AC), phase contrast (DPC) and dark-field contrast (DFC). Using open_iA, all three modalities can be visualized. A time consuming fusion of datasets as shown in Gusenbauer et al. [14] can be skipped, if the visualizations of material features are sufficient enough. In Figure 8 (row 1 and 2), all three modalities of the same slice image of sample #4 are shown. Keeping the focus on the containing copper mesh near the surface, in the AC images strong black areas resulting by metal artefacts are visible. In addition due to scattering noise, voids close to the copper wires have a very poor contrast. Looking at the DFC image, bright structures in the matrix are visible, representing individual carbon fibre bundles mainly oriented perpendicular to the rotation axis of the measurement setup respective the gratings [4]. In the DPC images, the voids close to the metal structures are represented very well with less metal artefacts. In the bottom row of Figure 8, the grey value histograms and chosen transfer function (left) as well the weighting with trimodal heatmap is shown, resulting in the final multimodal visualisation (right).

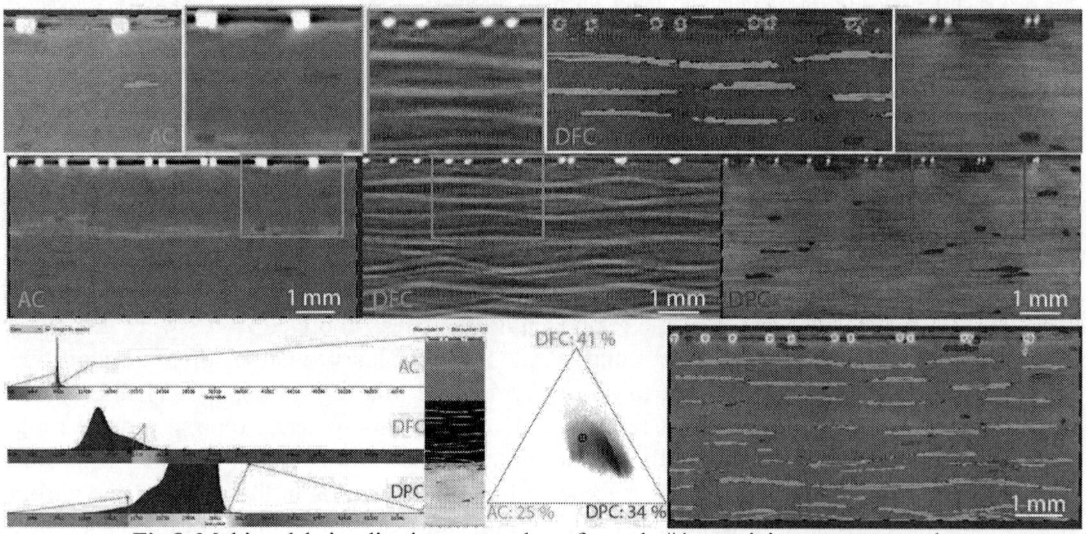

Fig.8. Multimodal visualization approaches of sample #4 containing a copper mesh.

4. Discussion and Conclusion

The main purpose of this work was the comparison of different scale levels on CFRP specimens achieved by a variety of different X-ray based imaging modes. If it is possible to prepare and destroy the samples as small as necessary, a conventional XCT scan would lead to the best results. If the entire part or sample should not be destroyed by additional sample preparation, some limitations have to be taken into account. In Table 3, a comparison of the applied modes and modality's are listed. For other laboratory or special devices like XXL-XCT [15] or Robot-XCT/ XCL [16] different limitations are presented, especially for features like maximum sample dimensions or file size.

Table 3. Comparison of the applied modes and modality's

Pros/ Cons	XCT	ROI-XCT	XCL	TLGI-XCT
Voxel size [µm³]	Depended on sample diameter > 150 … < 0.5 (isotropic)	Geometrical limited by tube-sample or sample-detector distance > 150 … < 0.2 (isotropic)	Limited by sample thickness > 150 … < 0.5 (non-isotropic)	Limited by design energy and grating geometry > 5.7 (isotropic)
Max sample Geometry	(D0.3*2) m³	~(D0.6*2) m³ at (150 µm)³ voxel size	Limited by movement of tube and detector: < 100*200 mm²	(D20*50) mm³
Modality's	Mainly absorption contrast. In addition at high resolution propagation based phase contrast can be strongly present.			AC, DPC and DFC
3D-Micro-structure	Yes	Yes	No, structures are only sharp at certain 2D layers and C-fibre structure was not visible. (see Fig. 9).	Yes, with limitations and dependencies of material features to grating orientation.
File size	Dependent on detector pixel count, ~16 GB for 16 bit file type	Dependent on detector pixel count, ~16 GB for 16 bit file type	Strongly depends on chosen field of view and resolution	One separate dataset for each modality dependent on detector pixel count.
Artefacts	Less for pure CFRP, Strong for metal structures inside	Average for pure CFRP, Very strong for metal structures inside	Strong Artefacts from sample surface or other material features out of focus layer	Less metal artefacts in DPC images. Interpretation of DFC can be difficult. Micro cracks can also lead to strong scattering contrast [17].

According to DIN ISO 15708-4:2017, if ROI-XCT does not lead to clear interpretation results and XCL is not possible due to sample geometry, small reference samples of the same material type, including relevant structures or natural defects to be detected, can be measured in advance at highest possible XCT resolution. In a second step, this reference samples can be scanned at the same time as the relevant object. If there is no possibility to scan the measurement object and reference sample at once, at least a scan with the same resolution and measurement parameters can be performed on the reference sample.

In Figure 8, the main disadvantages of XCL are shown, depicting that no real 3D microstructure can be extracted. The void and the particle shown in Figure 1 are only represented well in very few layers and getting blurred or inducing artefacts out of the focal layer. In this case, only the maximum size of the features is represented well, depending on the feature orientation in the specimen, exact sample positioning and exact movement trajectories of X-ray tube and detector during data acquisition.

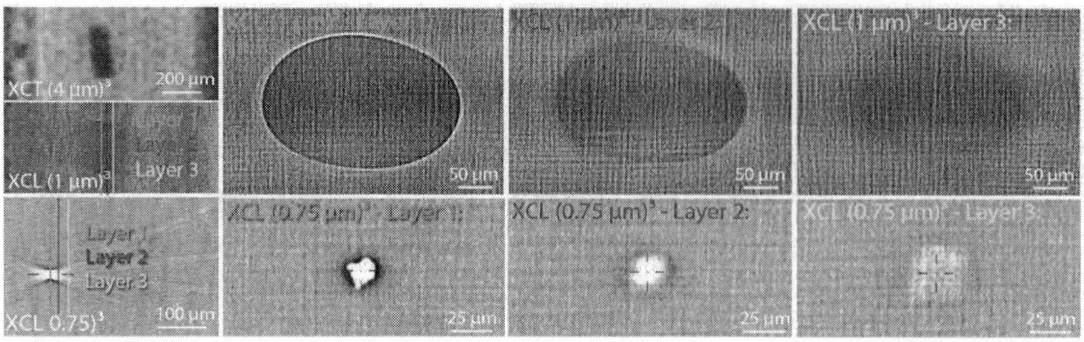

Fig.8. Sagittal slice image (left) and corresponding layers 1-3 of a void (top row) and a higher dense particle (bottom row) generated by XCL at (1 µm)³ and (0.75 µm)³ voxel size respectively.

As already mentioned, in the high resolution XCL measurements no fibre structures were visible in sample #1. Thus, on a 1x1 mm² cut-out additional reference scans at (2 µm)³ and (0.25 µm)³ voxel size were done (Figure 9). In the XCT (2 µm)³ scan (left), individual fibres in the fibre bundles can be clearly seen, but in addition darker areas (indicated by red arrows) occur, which can easily interpreted as micro-voids. Those potential micro-voids were not seen in the XCL scans and therefore the initiator for further high-resolution ROI-XCT scans at (0.25 µm)³ voxel size. Looking at these results (right) at the same regions where micro-voids are expected (green arrows), it can be clearly seen that in this areas only epoxy rich areas are present between the individual C-fibres. Potentially, these darker regions (red arrows) are mainly induced by propagation based phase contrast in this (2 µm)³ scan and can be easily misinterpreted.

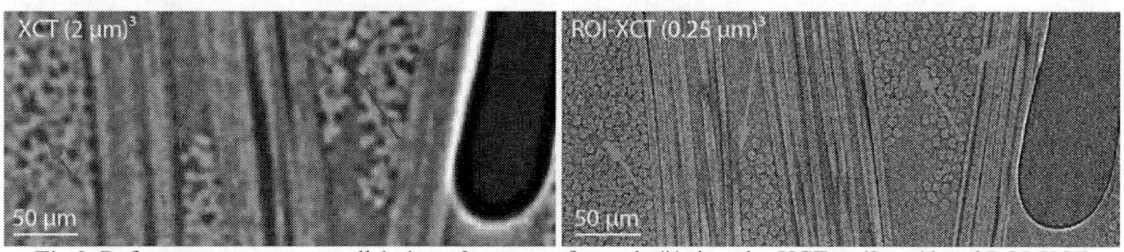

Fig.9. Reference scans on a small 1x1 mm² cut-out of sample #1 done by XCT at (2 µm)³ and ROI-XCT mode at (0.25 µm)³ respectively.

Acknowledgment

The work was financed by the project "Interpretation and evaluation of defects in complex CFK structures based on 3D-CT data and structural simulation" (DigiCT-Sim) funded by the federal government of Upper Austria and Austrian Research Promotion Agency (FFG). In addition this work was supported by the project "Multimodal and in-situ characterisation of inhomogeneous materials" (MiCi) by the federal government of Upper Austria and the European Regional Development Fund (EFRE) in the framework of the EU-program IWB2020. Sample #3 was provided by the Johannes Kepler University Linz and the 0-WASTE project under the lead of P. S. Stelzer and Z. Major.

References

[1] Kastner J, Heinzl C, Plank B, Salaberger D, Gusenbauer C, Senck S (2017), "New X-ray computed tomography methods for research and industry", In 7th conference on industrial computed Tomography (iCT2017), Leuven.

[2] Maisl M, Porsch F, Schorr C (2010) "Computed Laminography for X-ray Inspection of Lightweight Constructions", In 2nd International Symposium on NDT in Aerospace 2010, pp. 7.

[3] Kastner J and Heinzl C (2018), "X-Ray Tomography", In Handbook of Advanced Non-Destructive Evaluation. , pp. 1-72, Springer. DOI: 10.1007/978-3-319-30050-4_5-1.

[4] Revol V, Plank B, Kaufmann R, Kastner J, Kottler C and Neels A (2013), "Laminate fibre structure characterisation of carbon fibre-reinforced polymers by X-ray scatter dark field imaging with a grating interferometer", NDT & E INTERNATIONAL., May, 2013. Vol. 58, pp. 64-71, DOI: 10.1016/j.ndteint.2013.04.012.

[5] Plank B, Hannesschläger C, Revol V and Kastner J (2015), "Characterisation of anisotropic fibre orientation in composites by means of X-ray grating interferometry computed tomography", MATERIALS SCIENCE FORUM., May, 2015. Vol. 826(1), pp. 868-875. DOI: 10.4028/www.scientific.net/MSF.825-826.868.

[6] Gusenbauer C, Kastner J, Reiter M, Plank B, Salaberger D and Senck S (2018), "POROSITY DETERMINATION OF CARBON AND GLASS FIBRE REINFORCED POLYMERS USING PHASE-CONTRAST IMAGING", JOURNAL OF NONDESTRUCTIVE EVALUATION., December, 2018. Vol. 10921(4), DOI: 10.1007/s10921-018-0529-6.

[7] Fröhler B, Weissenböck J, Schiwarth M, Kastner J, Heinzl C (2019, "open_iA: A tool for processing and visual analysis of industrial computed tomography datasets", Journal of Open Source Software, 4 (35), 2019, 1185, DOI: 10.21105/joss.01185.

[8] Fröhler B, da Cunha Melo L, Weissenböck J, Kastner J, Möller T, Hege H.-C, Gröller E, Sanctorum J, De Beenhouwer J, Sijbers J, Heinzl C (2019), "Tools for the Analysis of Datasets from X-Ray Computed Tomography based on Talbot-Lau Grating Interferometry" In proceedings of the Conference on Industrial Computed Tomography 2019, Padua, Italy, pp. 8.

[9] da Cunha Melo L, Fröhler B, Weissenböck J, Kastner J, Heinzl C (2019), "Multi-Modal Transfer Functions for Talbot-Lau Grating Interferometry Data", In proceedings of the International Symposium on Digital Industrial Radiology and Computed Tomography, Fürth, Germany, pp. 9.

[10] Kastner J, Plank B and Requena G (2012), "Non-destructive characterisation of polymers and Al-alloys by polychromatic cone-beam phase contrast tomography", MATERIALS CHARACTERIZATION., February, 2012. Vol. 64(2), pp. 79-87, DOI: 10.1016/j.matchar.2011.12.004.

[11] Kastner J, Plank B, Salaberger D and Sekelja J (2010), "Defect and porosity determination of fibre reinforced polymers by x-ray Computed Tomography", In 2nd Int. Symposium on NDT in Aerospace. Hamburg, D, Germany, November, pp. 12.

[12] Plank B, Rao G and Kastner J (2015), "Evaluation of CFRP-Reference Samples for Porosity made by Drilling and Comparison with Industrial Porosity Samples by Means of Quantitative X-ray Computed Tomography", In Proceedings 7th International Symposium for NDT in Aerospace. Bremen, Germany, November, pp. 10.

[13] Plank B, Mayr G, Reh A, Kiefel D, Stössel R and Kastner J (2014), "Evaluation and Visualisation of Shape Factors in Dependence of the Void Content within CFRP by Means of X-ray Computed Tomography", In Proceedings of 11th European Conference on Non-Destructive Testing (ECNDT 2014). Prague, Czech Republic, October, pp. 9.

[14] Gusenbauer C, Reiter M, Plank B, Senck S, Hannesschläger C, Renner S, Kaufmann R and Kastner J (2017), "Multi-modal Talbot-Lau grating interferometer XCT data for the characterization of carbon fibre reinforced polymers with metal components", In Proceedings Industrial Computed Tomography (iCT2017). Leuven, Belgium, February, pp. 9.

[15] Holub W, Haßler U (2013), "XXL X-ray Computed Tomography for Wind Turbines in the Lab and On Site" In NDT in Canada 2013 Conference, https://www.ndt.net/article/ndt-canada2013/presentations/67_Holub.pdf (last access: June 5th, 2019).

[16] De Chiffre L, Carmignato S, Kruth JP, Schmitt R, Weckenmann A (2014), "Industrial applications of computed tomography". CIRP Annals - Manufacturing Technology Vol. 63, pp. 655–677, DOI: doi.org/10.1016/j.cirp.2014.05.011.

[17] Senck S, Scheerer M, Revol V, Plank B, Hannesschläger C, Gusenbauer C and Kastner J (2018), "Microcrack characterization in loaded CFRP laminates using quantitative two- and three-dimensional X-ray dark-field imaging", COMPOSITES PART A-APPLIED SCIENCE AND MANUFACTURING., October, 2018. Vol. 115, pp. 206-2014, DOI: 10.1016/j.compositesa.2018.09.023.

Self-supervised Learning for Pore Detection in CT-Scans of Cast Aluminum Parts

Patrick FUCHS[1,2], Thorben KRÖGER[1], Christoph S. GARBE[2]
[1]Volume Graphics GmbH, Heidelberg, Germany
[2]Interdisciplinary Center for Scientific Computing (IWR),
Heidelberg University, Heidelberg, Germany

Contact e-mail: patrick.fuchs@iwr.uni-heidelberg.de

Abstract. Automatically detecting pores in computed tomography (CT) scans of cast aluminum parts is a difficult task. This is especially the case when the data quality suffers from image artifacts. These can arise due to constraints from the imaging process and the short scan times that are common in industry. In previous work we have shown that using the synthetic ground truth generated from a realistically simulated CT data set is sufficient to train modern machine learning algorithms to detect pores on real data sets—even when heavy image artifacts like noise, beam hardening, or ring artifacts are present.

Fully convolutional neural networks can yield promising results when moving from the simulated data set to real world examples. So, instead of first removing artifacts from the image and then applying defect detection algorithms, we propose to tackle the defect detection task directly without altering the original data.

In this paper we evaluate three models: (1) a fast slice-wise approach which does the prediction without the full 3D context, (2) an encoder-decoder-pair with an additional refinement step, which acquires context by reducing the spatial resolution, and (3) a plain convolutional neural network using dilated convolutions for context aggregation.

Further, we provide promising qualitative results on difficult real world cases, among others CT-scans from in-line scenarios and additively manufactured parts.

1. Introduction

The detection of pores, blowholes, cavities, and other types of defects in CT-scans of cast aluminum parts is known to be a challenging and time-consuming problem. Nevertheless, this has to be done extremely diligently to ensure the reliability of the inspected parts. The difficulties not only originate from having to cope with three-dimensional images, but especially arise from the limited image quality of real world CT-scans. The underlying physics of the scan process makes the CT-data very artifact-prone. Besides sampling artifacts from the digitization, notably noise, beam hardening, and ring artifacts affect CT-scans and impede an automated inspection. Recently, the first machine learning algorithms based on deep learning were reported to surpass human-level perception in common image classification [1, 2, 3] and semantic segmentation tasks [4, 5]. These modern machine-learning algorithms manage to detect and correctly classify their targets even in cluttered scenes. This accuracy is achieved by statistically evaluating huge data sets containing a broad variety of images with hand-labeled annotations. Due to the high accuracy and the invariance against disturbances like image clutter, deep learning seem to be well suited to leverage the automated inspection of cast aluminum parts. We show that properly trained convolutional neural networks (CNNs) are able to sustain a stable detection rate when the scan-quality drops, i.e. the scan time is decreased. Having a scan of given quality, CNNs reach higher detection rates of small defects than simple filter-based methods. Fast, highly parallelized implementations utilizing modern general purpose graphic processing units (GPGPUs) allow for a high data throughput. Therefore, these methods hold the potential to bring production processes a step further towards an automated inspection of all parts.

However, as neural networks only learn a statistical representation of what they have seen during the training phase, we have to be careful when dealing with unseen data. In the end, this means if the target structures and target modalities are not covered by the training set, the deep learning algorithm cannot provide a reliable prediction. This is of special importance when the neural network not only localizes and labels the defects but also makes the decision if a defect has a critical impact on the structural integrity of a part. (The classification of defects is not part of this work.) Another downside of deep learning-based algorithms is the vast amount of labeled data which is necessary for proper training. Usually, this data is obtained by manually annotating thousands and thousands of samples. As this takes quite a while and can become very expensive depending on how qualified the labeling experts have to be, the trend goes towards training sets for which the ground truth can be derived automatically. These self-supervision approaches have been explored in several applications, e.g. by using simulated data [6, 7], or by exploiting other modalities like time-series [8]. Therefore, we propose a self-supervised method for pore detection in CT-scans of cast aluminum parts solely based on simulated data. We evaluate three different approaches to leverage the methods that have proven to be promising for two-dimensional image processing tasks to three-dimensional data. To validate the models, we evaluate their prediction performance on three hand-labeled CT scans of cast aluminum parts. Further, we evaluate the models on hard cases like composed parts of materials with very heterogeneous absorption and additively manufactured parts with smaller defects.

2. Related Work – Image Processing In Medical Science And Industry

The crucial point for a well trained machine learning system is a sufficient amount of labeled data. For single radiographs the $\mathbb{GDX}$ray data set [9] contains radiographs and annotations for many different applications. The part relevant for pore detection comprises a few radiographs of rims, brakes, and welds with bounding boxes of defects as ground truth. So far, there is no publicly available data set of CT-scans with annotated pores, which especially inhibits training deep learning algorithms for this task. However, the precise detection of pores in three-dimensional CT-scans can be extremely useful, e.g. for further consideration in pursuing analyses like wall-thickness analyses and stress test simulations. A first learning-based approach operating on three-diminesional data is specialized in the detection of air and gas inclusions [10]. This method is based on a traditional machine learning workflow which involves candidate selection, the time-consuming extraction of hand-crafted features, and the subsequent classification in a machine-learning algorithm. To unleash the full potential of CT-scans, we need to process a huge amount of data in short time, though. That means the detection methods have to be sufficiently fast. Here, deep learning is more suited as it combines candidate selection, feature extraction, and classification in a single module, using highly parallelizable operations.

Another field of application of CT is medical imaging. Because of the limitation in power and exposure, these CT-scans only have the quality that is really necessary for a diagnosis.

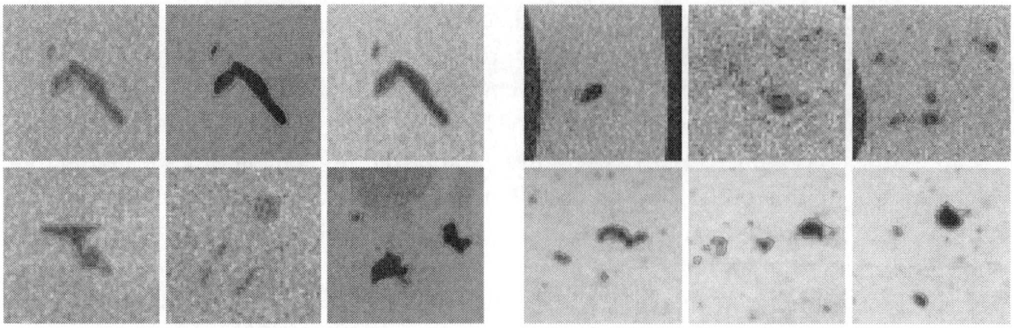

(a) Simulated data set for training (b) Real data set for validation

Figure 1: The data set for pore detection in CT-scans of cast aluminum parts. The orange overlay marks the outlines of the defect annotations. For the simulated data it is generated from the 3D models which were used for the simulation (see [24]), for the real data it is derived from the hand-labeled annotations.

Nevertheless, deep learning is successfully used to segment different parts of the human body like bones [11, 12], organs which consist of similar tissue [13, 14, 15, 16], or even individual tracts and neurons of the human brain [17, 18]. Moreover, CNNs are trained to detect defects in the human body, e.g. brain tumors [19] or lesions in vital organs like the liver [20]. In the industrial sector, first approaches use CNNs, for example, to segment individual short glass fibers in CT-scans of small samples of fiber composite materials [21]. Apart from this, generative deep learning methods are utilized to prepare CT-data for a subsequent inspection by a human expert. These methods are designed to remove common artifacts from the data, e.g. Compton scatter [22] or metal artifacts [23]. However, such pre-processing steps alter the original data and potentially induce unwanted side-effects like removing tiny structures along with the artifacts. Thus, we propose to head right for the actual task and learn to deal with image artifacts instead.

3. The Data Set – About Simulations And The Real World

When annotating casting defects in industrial CT-data we are confronted with several peculiarities complicating the task: (1) We need to handle three-dimensional data on a two-dimensional device, which makes it hard to grasp three-dimensional structures. Furthermore, in contrast to medical data, we do not have a specific model of the object under examination. Neither have the defects similar shapes like a liver always looks more or less the same, nor are they always in the same region. (2) CT yields artifact-prone data. In other domains, such as road scene segmentation, we only have to deal with slightly cluttered images. When searching for defects in CT-scans, however, we need to find tiny structures in images with high noise levels and other artifacts arising from the physics of the imaging process. (3) The structures we are looking for are very tiny. In common image processing data sets, small objects still are about 10 pixels large. Instances which are considered as too small belong to the background or an "undecided" class. However, we are exactly looking for anomalies which contain only very few image elements. (4) In different industrial branches different structures are considered as defective. We, however, want to find all defects, which are physically present in the part. Smaller defects are especially hard to spot and require scans of high image-quality.

Therefore, we use simulated data with precise per-voxel ground truth [24] for training and the small amount of hand-labeled real data we have for evaluation only. The training set consists of 675 simulated CT-scans of cast aluminum parts with defects, like pores or shrinkage, of varying shape and size. For each of the 25 parts there are 27 CT-scans varying the artifact strength of noise, beam hardening, and ring artifacts. This spans an "artifact subspace" within the space of possible scan parameters, which has three gradations along each axis. Further, the data set comes with 54 additional CT-scans for validation. Figure 1a shows an example of the simulated data set with different artifacts and the according ground truth.

However, simulations are only an approximation of the real world and even though the data is considered to be realistic enough to serve as training data, we need to validate our models on real data to ensure that we do not overfit to the simulations. Therefore, we have three real CT-scans which were hand-labeled by eight experts. To ease the annotation process we make a high-quality scan of a defective cast aluminum part; then we scan the same part again under "normal" conditions to obtain a scan for evaluation holding the challenges for the trained defect detection models. Figure 1b shows an example of the real data and the corresponding hand-labeled ground truth.

4. Methods – A Comparison of Neural Network Architectures

The success of deep learning for classification and segmentation has its origin in the processing of two-dimensional images and time-series, i.e. videos. We, however, are dealing with three-dimensional volumetric data. When moving from two dimensions to three, there are several ways to do so: we can (1) process the volume slice by slice along a principal direction [18], (2) evaluate for each voxel the three slices which are aligned along the three axes [12], (3) process the volume slice-wise along each dimension and learn how to combine the results [17], or (4) make the smallest operation—the convolutions—three-dimensional [13]. In this section, we compare three model architectures: a slice-wise model following approach (3), a similar model but with three-dimensional convolutions following approach (4), and a flat model with a different kind of context aggregation, following approach (4), too.

As we have way less "defective" voxels than "non-defective" ones, we have to deal with a highly imbalanced training set. Therefore, we train all the models using dice loss [25] which does not reward correct "non-defective" predictions and thus puts the focus on the defects. Further, we distinguish between defects which penetrate the surface of the scanned part and defects which completely lie within the part. With this separation, we avoid false positive predictions on the surface of the part, for example in screw threads. This diminishes the need to segment the part from its background beforehand.

In the wild, different CT-scanners work with a huge variety of scan parameters which not only affect the amount of artifacts in the image, but global image properties like the contrast as well. To be able to deal with the different scan modalities, we heavily use data augmentation in addition to the artifact variation of the training set. We specifically focus on augmenting the brightness, contrast and noise of our data.

4.1 Slice-wise Computation

First, we examine a slightly modified U-Net architecture [18]. It consists of a down-sampling part culminating in the bottleneck layers and a symmetric up-sampling part producing an output of the same size as the input. Between these two parts, which form the eponymous U-shape, skip-connections are added to provide information from earlier layers for a detailed up-sampling. We use strided convolutions instead of maximum pooling layers to reduce the spatial resolution, which has a positive effect on the computational effort. During training, we randomly pick axis-aligned slices from the volumes in the training set. For prediction, we slide a window along each axis through the test volume and later combine the individual results using a simple maximum operation. Alternatively, it is possible to train a separate

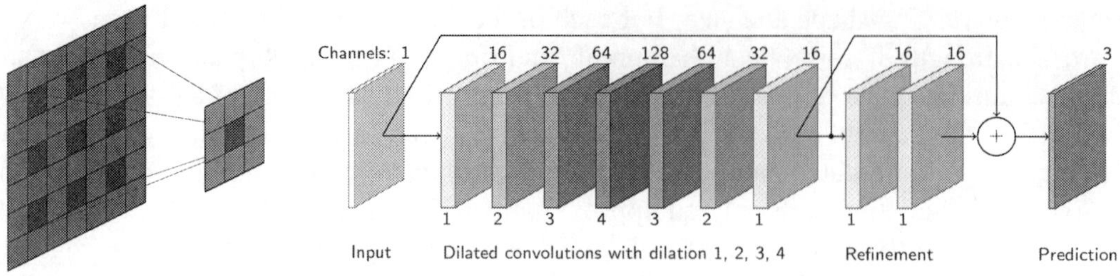

(a) Dilated kernel (b) Linear network architecture using only convolution layers

Figure 2: Dilated convolutions and the linear network architecture using a dilation pyramid.

model to combine the individual results, e.g. with the help of the original gray-values [17]. However, a separate classifier means more computational cost. Thus, we go for the maximum operation which already yields satisfying results in our two-class scenario. This method will further be referred to as "slice-wise detector".

4.2 Three-dimensional Convolutions

The second architecture basically is a three-dimensional version of the U-Net, as it is used in [13, 15]. Additionally, we add a refinement step which is introduced in [26] to segment thin and semi-transparent structures in images. We utilize it to refine the boundaries of the detected defects. The complete architecture is introduced in [24]. During training, we randomly pick small blocks from the training volumes, as they do not fit on the GPU in one piece. For faster convergence it is helpful to pick only blocks containing defects or parts of the surface, because they contain the most information. Fully convolutional networks (FCNs) do not rely on a fixed size input. So, we can vary the block size depending on the capability of the hardware. We only need to take care that the blocks overlap during prediction by half of the receptive field of the model so that the model has the full information for the inner part of each block. This method will further be referred to as "deep defect detector".

4.3 Dilated Convolutions

The two models above gather contextual information by reducing the spatial resolution of the data and subsequently upscale the results. These down-sampling steps might drop information necessary to detect tiny defects. A different way of context aggregation form dilated convolutions [27] which use dilated convolution kernels (see Figure 2a). A similar approach is used to detect small structures in remote sensing images [28]. The challenge when applying three-dimensional convolutions without any sub-sampling is the massive memory consumption. Therefore, we have to be content with less layers and less channels per layer. Thus, we only have nine convolutional layers. The full architecture is shown in Figure 2b. This method will further be referred to as "dilated defect detector".

4.4 Evaluation

We evaluate the prediction performance of the models on the separate validation set and the three hand-labeled real CT-scans of [24]. For a holistic analysis, two measures are taken into account: On the one hand, we evaluate the quality of the segmentation mask using the Jaccard-index [29], or intersection over union (IoU) measure, (see Figure 3a) and, on the other hand, the probability of detection (POD) [30] to determine how many of the actual defects we find (see Figure 3b).

The differences in the precision of the segmentation—shown by the IoU—are not significant. However, in the probability of detection a gap between the models can be observed:

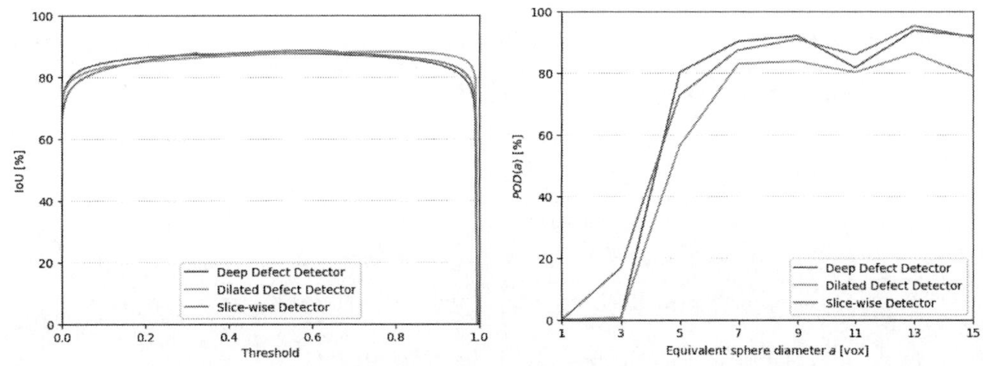

(a) Intersection over union (IoU) (b) Probability of detection (POD)

Figure 3: Results of the different neural network architectures.

The "dilated defect detector" suffers from insufficient depth and channels and, therefore, still fails to detect small defects. For small defects, the "slice-wise detector" and the "deep defect detector" yield comparable results. For larger defects, the "deep defect detector" slightly benefits from the three-dimensional context, until the defects become big enough so that a combination of all three swipes of the "slice-wise detector" yield comparable results again.

5. Discussion – Advantages and Risks

Finally, we provide some promising results of the "deep defect detector" for three especially challenging scenarios to demonstrate the capability of deep learning methods as well as the risks involved. (1) We show that the results of our "deep defect detector" is less sensitive to changes in the image quality compared to traditional filter-based approaches; (2) we show that, having a scan of given image quality, we can find smaller defects compared to traditional filter-based approaches—without introducing disturbing false positive predictions; and (3) we show that a pre-trained model can be adapted easily to more difficult scenarios. The task of defect detection is not limited to aluminum. What happens if we need to detect defects in a less absorbing material of a multi-material part? To provide an example, we have a CT-scan of a part made of three materials and want to detect defects in the material that has a way smaller attenuation coefficient than the enclosing aluminum.

Trade image quality for speed. As mentioned before, CNNs are able to yield consistent results despite the presence of image clutter. Moreover, in [24] it is shown that the impact of noise on the results of the deep learning method is not significant compared to the other methods. So, it should be possible to save time by reducing the scan time and accepting the resulting increase in image noise. To validate this, we scan a cast aluminum part making 3300 projections with an exposure of 1 s each. Then, we reconstruct the volume several times using a decreasing amount of projections down to only 33 projections. This comes up to a scan time of half a minute compared to the original 55 minutes. The scanned cast aluminum part has a diameter of about 12 cm and a height of about 4 cm. The scan was done at 175 kV with 30 µA. Despite noise, we introduce artifacts due to under-sampling to the image, as we do not consider enough projection angles. The top row of Figure 4 shows a slice of the resulting data in varying quality.

Then, we use the filter-based method from [24] and the "deep defect detector" to find pores in the data. The results are shown in the second and third row of Figure 4, respectively. The filter results get brittle when the quality drops. Moreover, tiny false positive responses

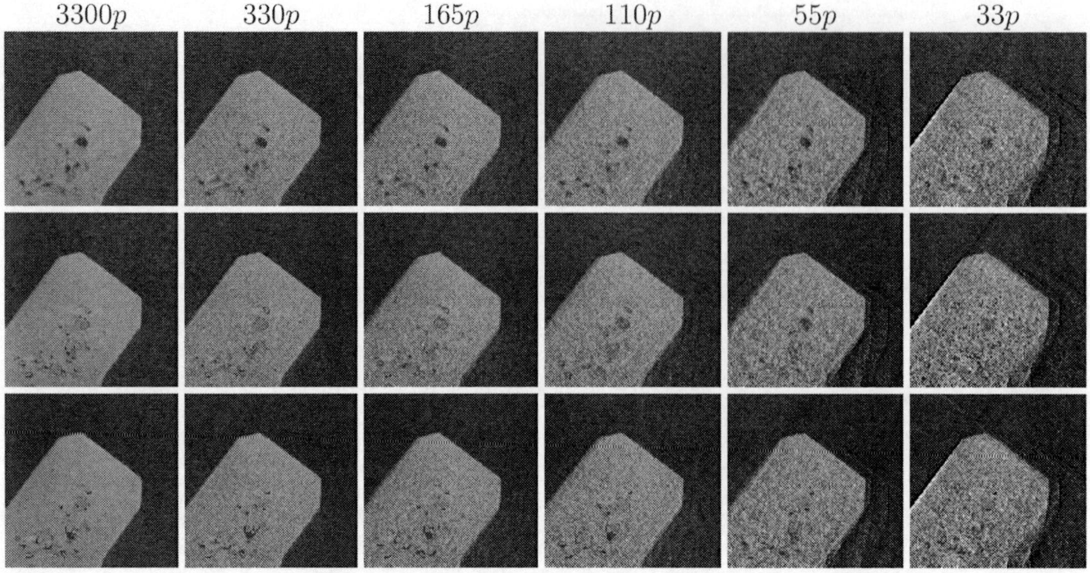

Figure 4: We reduce the image quality from left to right by leaving out projections. The top row shows a slice from the extracted data set to demonstrate the image quality. The mid row shows the results of the filter-based method, with the correctly identified defect voxels in green, the false negative predictions in red and the false positive predictions in blue. The bottom row contains the results of the CNN.

emerge. In contrast, until about 110 projections, the results of the "deep defect detector" remain consistent without introducing further false positive responses. The detected defects are recognizable as individual instances even in the image of the worst quality. However, some of the smaller defects become missed out. After that, the neural network starts to hallucinate false positives in the streaking artifacts. The benefits become clearer when evaluating the IoU measure (Figure 5): The results of the filter-based method drop rapidly as the number of projection decreases, the results of the "deep defect detector" remain almost constant.

That hand labeling pores in CT-scans of cast aluminum parts is a hard task can be seen in Figure 6. In the upper row, in red, we see a structural loosening which is annotated as defect but was not recognized as such. However, in the bottom row a similar structural loosening—which this time was recognized as defect—was not annotated as defect (colored in blue indicating false positive predictions). These inconsistencies lessen the overall numbers of the evaluation. Thus, using hand-labeled data for training can be problematic.

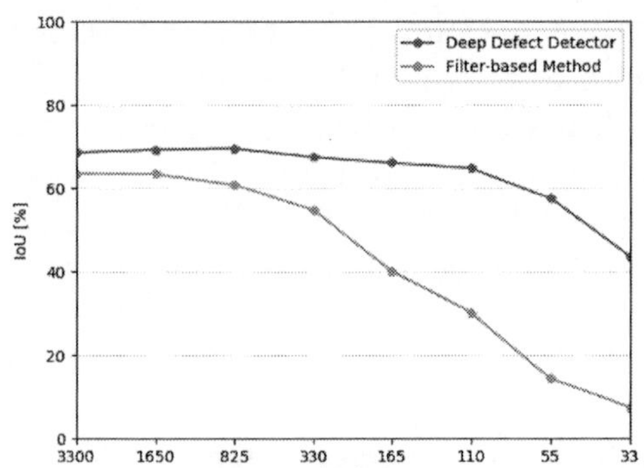

Figure 5: Prediction performance (evaluated as IoU) over varying image qualities.

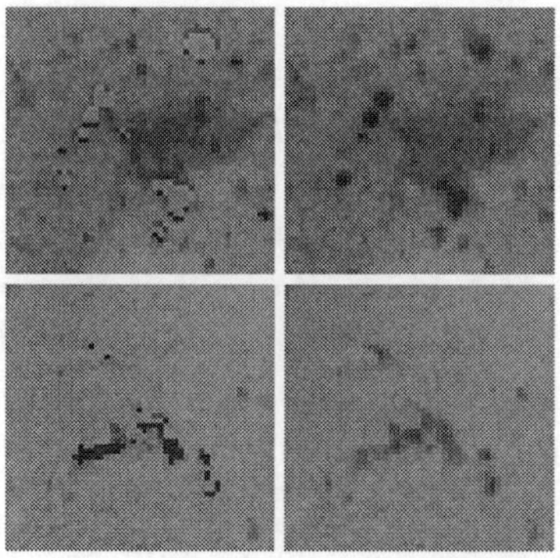

Figure 6: Inconsistency in the hand-labeled annotations.

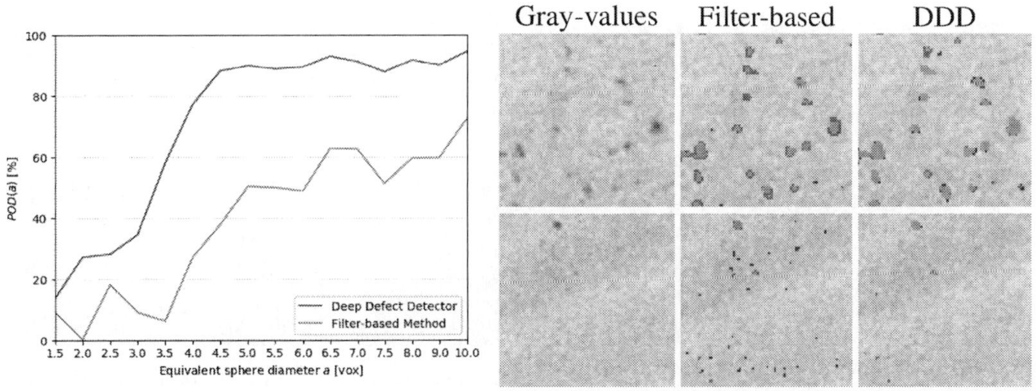

(a) Probability of detection (POD) (b) Qualitative evaluation

Figure 7: Detecting tiny defects in additively manufactured parts.

Find tiny defects. In the additively manufactured parts of the aerospace industry much smaller pores have to be detected than in the cast aluminum parts of the automotive industry. So, we evaluate the "deep defect detector" on an additively manufactured part looking for tiny pores and again compare it to the traditional filter-based method. Figure 7a shows the quantitative results by means of the POD at a confidence level of $90\,\%$ per defect. We see that the "deep defect detector" starts to yield reliable predictions (of more than $50\,\%$) for defects of an equivalent sphere diameter of 3 voxels, while the filter based method needs about 5 voxels. This difference can be explained by looking at the qualitative results in Figure 7b. In the top row we see that the filter-based method does not catch all the defects. If we configure it to be more sensitive more false positive responses emerge. The "deep defect detector", in contrast, detects more of the small defects and separates them more precisely from the material—without introducing more false positive responses. The bottom row shows that the filter-based method introduces many tiny false positive responses which are especially irritating when looking for tiny pores. Another advantage of the "deep defect detector" is that we do not necessarily need to segment material and background to restrict the analysis area.

Fine-tune for different data. Lastly, we consider some data which widely differs from anything we have in the training set. To see what happens when we encounter CT-scans of parts made not solely of aluminum, we have a look at a CT-scan of a thermal sensor which is composed of three materials with very different absorption properties: A hollow aluminum cylinder filled with wax, some kind of plastics, and another aluminum cylinder. To draw conclusions about the reliability of the part we need to detect all pores and cracks in the wax part, which has the smallest attenuation coefficient. The leftmost image of Figure 8 shows a slice of the CT-scan and a detailed view with a few highlighted defects.

When training CNNs for different image classification tasks, especially the early layers of a CNN tend to learn similar convolution kernels. These resemble Gabor-like filters or look like color-blob detectors. Sharing these weights across different tasks not only reduces the convergence time but can increase the accuracy of the models [31]. Therefore, we use the pre-trained "deep defect detector" as basis for fine-tuning and adapt it to the new problem by re-training the refinement step of the model. With the simulation pipeline from [24] we create a single CT-scan similar to the thermal sensor with about 500 defects. Adapting the pipeline to the new scenario requires only little manual effort. As the pipeline provides us with precise labels for the defects, too, we can use it as training input for the fine-tuning process without hesitation. The results of the fine-tuned model are shown in the rightmost image of Figure 8 and they significantly improved to the original "deep defect detector" (middle image).

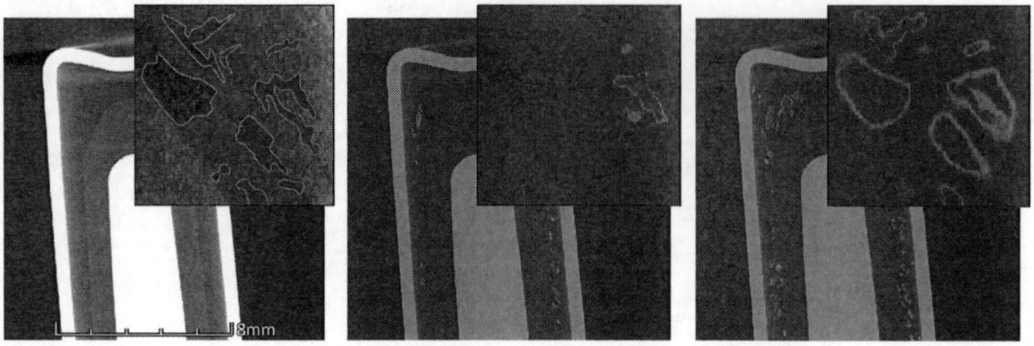

Figure 8: If we encounter data which widely differs from the training data, like the multi-material part on the left, the original "deep defect detector" fails to yield any reliable results (center image). However, when fine-tuning on a single simulation similar to the target, we can obtain better results again (right image). The orange outlines on the left indicate what we consider to be a defect.

As significant the results are, as dangerous the process of fine-tuning can be. We dramatically change the behavior of the model with the input of a single data set. We know that human annotators are not always consistent in their decision. Thus, we assume, if we fine-tune the model based on the decisions of a human inspector, e.g. in an in-line scenario, a bad day can have a tremendous impact on the prediction performance. It might even corrupt the whole system and all the effort would be in vain. We, however, use simulated data for training and fine-tuning. So, we can be sure to have precise and correct labels. Furthermore, as it is generated by a fully automated pipeline we only need to input a new model and the properties of the materials involved. This reduces the necessary manual effort to a minimum.

6. Conclusion

Deep learning methods are able to deal with noisy and artifact-polluted data and to detect smaller defects than filter-based approaches—if trained properly. For other tasks, fine-tuning might be necessary, which has to be done carefully. Already a small amount of data has a huge impact on the prediction performance. Of most importance for reliably trained machine learning algorithms is a precisely labeled ground truth. This might not be given if untrained human annotators produce new data in-line, but can be generated on the fly for simulated data with little effort. We demonstrate that deep learning methods are suitable for non-destructive testing by showing their capability to detect defects not only in CT-scans of cast aluminum parts but in those of additively manufactured parts, too. Further, we demonstrate that the good results are transferable to more complex scenarios, i.e. the testing of composed parts.

References

[1] Y. Taigman, M. Yang, M. Ranzato, and L. Wolf, "DeepFace: Closing the Gap to Human-Level Performance in Face Verification," *Proceedings of the 2014 IEEE Conference on Computer Vision and Pattern Recognition (CVPR)*, pp. 1701–1708, 2014.

[2] K. He, X. Zhang, S. Ren, and J. Sun, "Delving Deep into Rectifiers: Surpassing Human-Level Performance on ImageNet Classification," *Proceedings of the 2015 IEEE International Conference on Computer Vision (ICCV)*, pp. 1026–1034, 2015.

[3] L. Chen, S. Wang, W. Fan, J. Sun, and S. Naoi, "Beyond Human Recognition: A CNN-based Framework for Handwritten Character Recognition," *Proceedings of the 2015 Asian Conference on Pattern Recognition (ACPR)*, pp. 695–699, 2015.

[4] I. Kokkinos, "Pushing the Boundaries of Boundary Detection Using Deep Learning," *Proceedings of the 2016 International Conference on Learning Representations (ICLR)*, 2016.

[5] K. Lee, J. Zung, P. Li, and H. S. Seung, "Superhuman Accuracy on the SNEMI3D Connectomics Challenge," *arXiv preprint arXiv:1706.00120*, 2017.

[6] A. Shrivastava, T. Pfister, O. Tuzel, J. Susskind, W. Wang, and Webb, "Learning from Simulated and Unsupervised Images through Adversarial Training," *Proceedings of the 2016 IEEE Conference on Computer Vision and Pattern Recognition (CVPR)*, pp. 2107–2116, 2016.

[7] S. Richter, V. Vineet, S. Roth, and V. Koltun, "Playing for Data: Ground Truth from Computer Games," *Proceedings of the 2016 IEEE European Conference on Computer Vision (ECCV)*, pp. 102–118, 2016.

[8] M. Bautista, P. Fuchs, and B. Ommer, "Learning Where to Drive by Watching Others," *Proceedings of the 39th German Conference on Pattern Recognition (GCPR)*, pp. 29–40, 2017.

[9] D. Mery, V. Riff, U. Zscherpel, G. Modragon, I. Lillo, I. Zuccar, H. Lobel, and M. Carrasco, "GDXray: The database of X-ray images for nondestructive testing," *Journal of Nondestructive Evaluation*, pp. 34.4:1–12, 2015.

[10] F. Zhao, P. Mendonca, J. Yu, and R. Kaucic, "Learning-based Automatic Defect Recognition With Computed Tomographic Imaging," *IEEE International Conference on Image Processing (ICIP)*, pp. 2762–2766, 2013.

[11] A. Klein, J. Warszawski, J. Hillengaß, and K. H. Maier-Hein, "Automatic Bone Segmentation in Whole-body CT Images," *International Journal of Computer Assisted Radiology and Surgery*, 2018.

[12] A. Prasoon, K. Petersen, C. Igel, F. Laze, E. Dam, and M. Nielsen, "Deep Feature Learning for Knee Cartilage Segmentation Using a Triplanar Convolutional Neural Network," *Medical Image Computing and Computer-Assisted Intervention (MICCAI)*, pp. 246–253, 2013.

[13] Q. Dou, H. Chen, Y. Jin, L. Yu, J. Qin, and P. A. Heng, "3D Deeply Supervised Network for Automatic Liver Segmentation from CT Volumes," *Lecture Notes in Computer Science (including subseries Lecture Notes in Artificial Intelligence and Lecture Notes in Bioinformatics)*, pp. 149–157, 2016.

[14] F. Milletari, N. Navab, and S. A. Ahmadi, "V-Net: Fully Convolutional Neural Networks for Nolumetric Medical Image Segmentation," *Proceedings - 2016 4th International Conference on 3D Vision, 3DV 2016*, pp. 565–571, 2016.

[15] E. Gibson, Y. Hu, S. Bandula, D. C. Barratt, B. Davidson, S. P. Pereira, E. Bonmati, F. Giganti, K. Gurusamy, and M. J. Clarkson, "Automatic Multi-Organ Segmentation on Abdominal CT With Dense V-Networks," *IEEE Transactions on Medical Imaging*, pp. 1822–1834, 2018.

[16] H. Chen, Q. Dou, L. Yu, and P.-A. Heng, "VoxResNet: Deep Voxelwise Residual Networks for Volumetric Brain Segmentation," *NeuroImage*, pp. 446–455, 2016.

[17] J. Wasserthal, P. Neher, and K. H. Maier-Hein, "TractSeg – Fast and Accurate White Matter Tract Segmentation," *NeuroImage*, pp. 239–253, 2018.

[18] O. Ronneberger, P. Fischer, and T. Brox, "U-net: Convolutional Networks for Biomedical Image Segmentation," *Lecture Notes in Computer Science (including subseries Lecture Notes in Artificial Intelligence and Lecture Notes in Bioinformatics)*, pp. 234–241, 2015.

[19] M. Havaei, A. Davy, D. Warde-Farley, A. Biard, A. Courville, Y. Bengio, C. Pal, P.-M. Jodoin, and H. Larochelle, "Brain Tumor Segmentation with Deep Neural Networks," *Medical Image Analysis*, pp. 18–31, 2017.

[20] P. F. Christ, M. E. A. Elshaer, F. Ettlinger, S. Tatavarty, M. Bickel, P. Bilic, M. Rempfler, M. Armbruster, F. Hofmann, M. D'Anastasi, W. H. Sommer, S.-A. Ahmadi, and B. H. Menze, "Automatic Liver and Lesion Segmentation in CT Using Cascaded Fully Convolutional Neural Networks and 3D Conditional Random Fields," *Medical Image Computing and Computer-Assisted Intervention (MICCAI)*, pp. 415–423, 2016.

[21] T. Konopczyński, D. Rathore, J. Rathore, T. Kröger, L. Zheng, C. S. Garbe, S. Carmignato, and J. Hesser, "Fully Convolutional Deep Network Architectures for Automatic Short Glass Fiber Semantic Segmentation from CT scans," *8th Conference on Industrial Computed Tomography (iCT)*, 2018.

[22] J. Maier, S. Sawall, M. Knaup, and M. Kachelrieß, "Deep Scatter Estimation (DSE): Accurate Real-Time Scatter Estimation for X-Ray CT Using a Deep Convolutional Neural Network," *Journal of Nondestructive Evaluation*, pp. 1–9, 2018.

[23] Y. Zhang and H. Yu, "Convolutional Neural Network Based Metal Artifact Reduction in X-Ray Computed Tomography," *IEEE Transactions on Medical Imaging*, pp. 1370–1381, 2018.

[24] P. Fuchs, T. Kröger, T. Dierig, and C. S. Garbe, "Generating Meaningful Synthetic Ground Truth for Pore Detection in Cast Aluminum Parts," *9th Conference on Industrial Computed Tomography (iCT)*, 2019.

[25] C. H. Sudre, W. Li, T. Vercauteren, S. Ourselin, and M. Jorge Cardoso, "Generalised dice overlap as a deep learning loss function for highly unbalanced segmentations," *Lecture Notes in Computer Science (including subseries Lecture Notes in Artificial Intelligence and Lecture Notes in Bioinformatics)*, pp. 240–248, 2017.

[26] N. Xu, B. Price, S. Cohen, and T. Huang, "Deep Image Matting," *Proceedings of the 2017 IEEE Conference on Computer Vision and Pattern Recognition (CVPR)*, pp. 2970–2979, 2017.

[27] F. Yu and V. Koltun, "Multi-Scale Context Aggregation by Dilated Convolutions," *Proceedings of the 2016 International Conference on Learning Representations (ICLR)*, 2016.

[28] R. Hamaguchi, A. Fujita, K. Nemoto, T. Imaizumi, and S. Hikosaka, "Effective Use of Dilated Convolutions for Segmenting Small Object Instances in Remote Sensing Imagery," *IEEE Winter Conference on Applications of Computer Vision (WACV)*, pp. 1442–1450, 2018.

[29] M. Everingham, A. S. M. Eslami, L. Van Gool, C. K. Williams, J. Winn, and A. Zisserman, "The Pascal Visual Object Classes Challenge: A Retrospective," *International Journal of Computer Vision (IJCV)*, pp. 98–136, 2015.

[30] D. Kiefel, M. Scius-Bertrand, and R. Stößel, "Computed Tomography of Additive Manufactured Components in Aeronautic Industry," *8th Conference on Industrial Computed Tomography (iCT)*, 2018.

[31] J. Yosinski, J. Clune, Y. Bengio, and H. Lipson, "How Transferable are Features in Deep Neural Networks," *Proceedings of the 28th International Conferenceon Neural Information Processing Systems (NIPS)*, pp. 3320–3328, 2014.

Automatic Defect Detection for X-Ray inspection: Semantic segmentation with deep convolutional network

Roger BOOTO TOKIME [1], Xavier MALDAGUE [1], Luc PERRON [21] Université Laval, Québec, Canada[2] Lynx Inspection, Québec, Canada
Contact e-mail: rogerbooto@gmail.com

Abstract. Thoroughly testing every single part produced by a production chain is tedious and expensive work. To avoid excessive inspection costs, parts manufacturers often rely on random sampling for quality control. Random sampling may be sufficient in some cases, but demands for 100% parts inspection is on the rise in the industry, especially for safety-critical parts. To fulfill this requirement, we propose a deep learning method based on an end to end full Convolution Network (FCN) that performs semantic segmentation of a defect in an X-ray image. Semantic segmentation is a technique used in computer vision to detect objects in an image by using a binary mask that separates the foreground (defect) and the background (no defect). This work illustrates the capability of defect recognition for an FCN model that can be used in production. The proposed method is discussed in the paper and supported with preliminary results.

Introduction

X-rays were discovered by Wilhelm Conrad Röntgen in 1895[1,2,3] and during the second industrial revolution, many industries used radiographic images (film) to inspect newly manufactured parts. The third industrial revolution saw the birth of the digitalization of X-ray images. This generally speeds up inspection time, which makes the inspection less expensive, and paves the way for process automation. Despite considerable efforts made by the non-destructive testing community, the demand for parts inspection from the manufacturing industries continues to increase. Parts manufacturers use statistical techniques that allow the analysis of a large number of parts by inferring an acceptable quality level value [4, 5]. The acceptance quality level value is calculated by analyzing representative samples from several batches. Although effective, this method does not allow the inspection of 100% of the newly manufactured parts. The method proposed in this article uses a deep learning approach to allow for the analysis and inspection of 100% of the parts manufactured in a factory. To do this, we implemented a fully convolutional deep learning network model. The following sections of the paper will describe in detail what a convolutional network (1&2) is, the composition of the database (3), the composition and analytical metrics of the network model and the results obtained followed by a brief conclusion.

1. Convolutional network

Convolutional networks can be expressed as a mixture of the principles of discrete convolution in mathematics and simplistic modeling of the human neuron, called the artificial neuron. The basic operation of this type of network is the generation of a filter bank, also called feature map, in order to extract elementary visual patterns in the first layers. The combination of the latter then allows the extraction of higher-level attributes. Figure 1 shows that the first layers detect basic visual characteristics such as corners, endpoints, oriented edges or similar characteristics depending on the type of input signal [6]. Thus, an object detection system using an image as an input signal can be automated by applying techniques such as propagation and back-propagation of filter vector weights and errors across the network. This then allows the system to adjust the values of the feature maps to minimize detection and prediction errors. Each complete convolutional layer is composed of several feature maps with different weight vectors, allowing several features to be extracted at each location. In figure 2, we can see an example of a diagram representing the architecture of a typical neural network model. The first part represents a diagram of fully connected neurons and the second part represents a diagram of the different components of each artificial neuron. As mentioned earlier, convolutional networks inherit the properties of artificial neural networks, including the ability to have an activation function, which is explained in the following section.

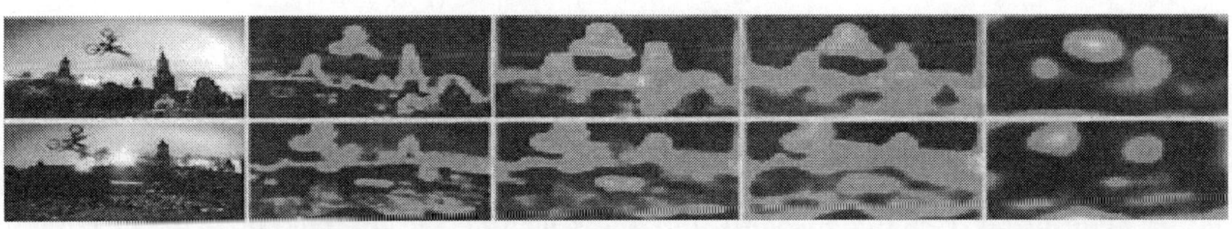

Figure 1: Example of convolution networks feature maps visualization [7]

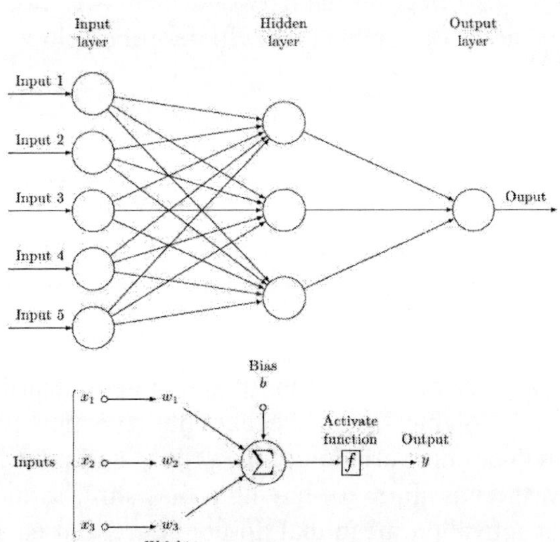

Figure 2: Neural network model diagram

2. Activation functions

In biology, a neuron is considered activated if it lets a signal received by the dendrites pass when the electrical voltage of this signal is high enough [8]. With artificial neurons, the activation is done through an activation function. This function takes the weighted parameter vectors as input and gives an output value. The output value activates the corresponding neuron if it is greater than or equal to a threshold. As we can see in figure 2 the activation of a neuron in one layer influences the output values of the neurons in the next layer. As a result, a neuron is considered activated if and only if the output value of the activation function is non-zero. In general, we want the distribution of the output values of a neuron to be normalized and centered around zero with a variance of one because this facilitates learning[9]. In the next subsections, we will explain the different activation functions that we have used for learning and prediction. In addition, we will detail the reasons why we did not use the ReLU activation function.

2.1 Rectified Linear Unit

The rectified linear unit (ReLU)[10] activates a neuron only in the presence of non-zero positive values. This activation function allows for a fast training of a network. As we can see in equation 1, where x is the sum of the input values, the propagated output value is simply the sum of the input values.

$$ReLU(x) = \begin{cases} 0 & \text{if } x \leq 0 \\ x & \text{if } x > 0 \end{cases}$$

However, when minimizing the prediction error, it may happen that almost all the patterns searched for by the network have values less than zero and these values will be reduced to zero. As a result, the network will begin to disable some neurons and prevent the propagation of the feature map learned by these neurons during training. This reaction of the network can interfere with learning. This phenomenon is called "dying ReLU" [11], and in order to minimize its effect on the network, there are several alternative activation functions including ELU and SELU which we will describe below.

2.2 Exponential Linear Unit

The linear exponential unit (ELU)[12] solves the problem of "dying RELU", as the function outputs non-zero values for all values in the domain when the latter is negative.

$$ELU(x) = \begin{cases} \alpha(e^x - 1) & \text{if } x \leq 0 \\ x & \text{if } x > 0 \end{cases}$$

Where the parameter α controls the mean of the output distribution. ELU does not introduce a bias to the following layers, because the function reduces the mean of the activation to zero and reduces the variation of the propagated information. This property of the function minimizes the possibility of having a bias shift, as the bias shift is caused by neurons with non-zeros activation mean that do not cancel out each other in the following layers when summing.

2.3 Scaled Exponential Linear Unit

The scaled exponential linear unit [9] is an activation function similar to ELU, but that allows the creation of a neural network that maintains the Gaussian distribution, centered on zero with a variance of one, of the output values of each neuron whose input values are normalized. We can see that the difference between equations 2 and 3 is the addition of the control parameter λ.

$$SELU(x) = \lambda \begin{cases} \alpha(e^x - 1) & \text{if } x \leq 0 \\ x & \text{if } x > 0 \end{cases}$$

Where λ and α are the parameters that control the variance and average of the output distribution. Using the SELU activation function in our network allows us to have a self-normalized neural network. In order for our self-standardized system to have a zero-centered mean and a variance of one, we used the values of 1.67326 for$\alpha$and 1.0507 forλ[9]. This ensured that we had a normalized distribution of our input data by performing a batch normalization operation before performing a SELU activation.

3. Database

The database used for this article is a public database called GDXray [13]. As shown in figure 3a, the database contains weld images captured by a radiographic system. The original images are about 4000x1500 pixels in width and height respectively. In order to make the training on our 8GB Nvidia GeForce GTX 1070 graphics card more efficient, we decided to select areas of interest, as shown in figure 3b. This increases the amount of data available for training. Then, we further increase the amount of data by performing simple transformations on the images. Among the transformations available were rotation, translation, intensity negation and the addition of artificial noise. When selecting the regions of interest, we were able to extract 360 images with a dimension of 256x256 pixels. By generating the segmented images associated with each original image (figure 3c), we produce a database of 720 images that we can use as a basis for our different transformations. It is important to note that a noisy or inverted image is associated with the same segmented image. Thus, after transformation, our database has grown from 720 images to 23742.

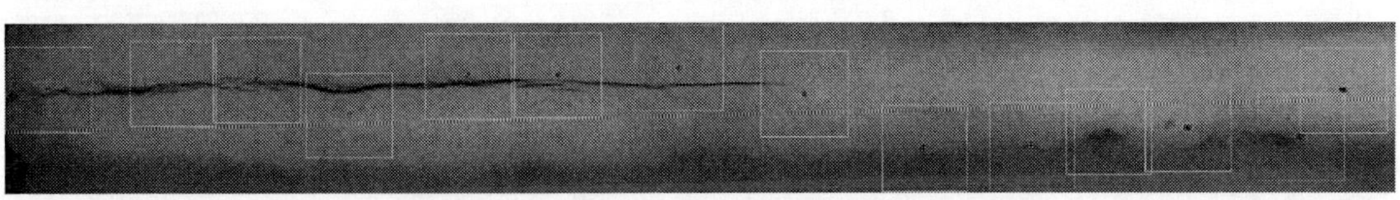

(a)

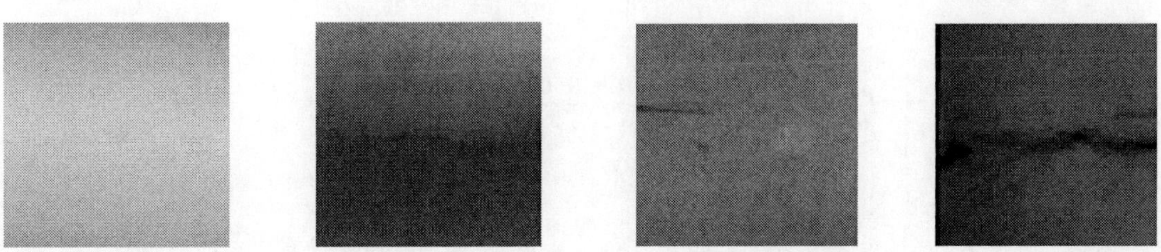

(b)

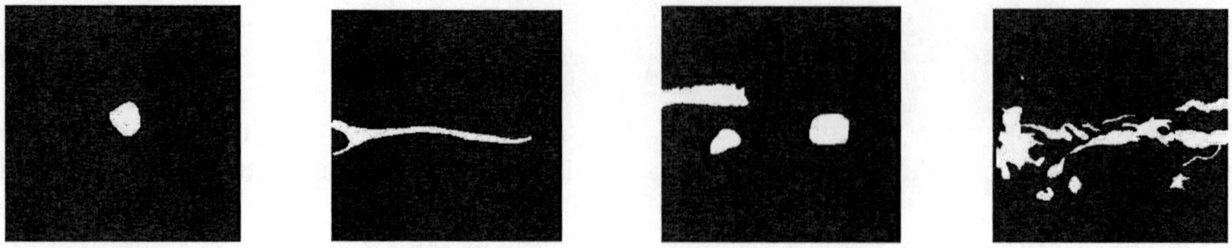

(c)

Figure 3: Example of how we created our database

4. Network Description

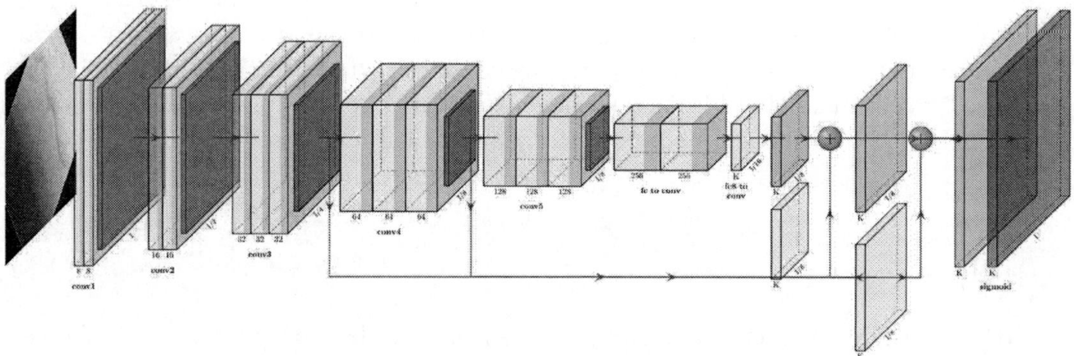

Figure 4: Our FCN architecture diagram

5. Model efficiency metric

The evaluation of our system was performed with the metric called F1 score. This metric provides a good estimate of the effectiveness of a prediction model because it takes into account both precision and recall factors. Precision is a measure of whether our model is able to distinguish between the two types of pixels we are looking for (defect, no defect).Recall is a measure of whether our model is able to classify a defect pixel without making mistakes. Therefore, the F1 score gives us a good measure of the effectiveness of the proposed model. During our experiments, we were able to achieve an F1 score of **70%** by comparing our predictions pixel by pixel on images of size 256x256, as shown in figure 6.

$$Precision = \frac{truePositive}{truePositive + falsePositive}$$

$$Recall = \frac{truePositive}{truePositive + falseNegative}$$

$$F1 = 2 * \frac{precision * recall}{precision + recall}$$

6. Results

In this section, we will present results obtained by applying our deep neural network model to images of porosities present in welds and on specimens unknown to the network. We recall that our network has been trained on welding images and we wanted to qualitatively evaluate the behavior of the system on images whose acquisition context is different from what the system has had the opportunity to observe. In this section, we will describe the results (1)of visualizing filter banks in different layers, (2) predictions on porosity images on welds, (3)predictions on porosity images on non-welded parts. The verification of the efficiency of our model was performed with 256x256 test images and to perform the prediction on large images, we used a method called a sliding window.

6.1 Visualization of feature maps

In figure 5 we present an overview of the feature map that activates the network on different layers. On the first row, we have placed 5 input images from 5 different welds. Each following row contains a visual representation of the areas (in red and yellow) that the network considers relevant for classification purposes. We can see that in the first two layers, the network focuses on the areas of the image where the contrasts are generally distinct, this represents the basic properties of our problem. In layers 3 and 4 the model seems to want to detect shapes and the last two layers seem to refine the detection of objects of interest, in our case, porosity defects. We would like to specify that this interpretation does not reflect in any way the method used by this type of network to learn. Figure 5 illustrates some examples of the solution found by the model in order to achieve our goal, which is the semantic segmentation of porosity defects found in images of welded parts.

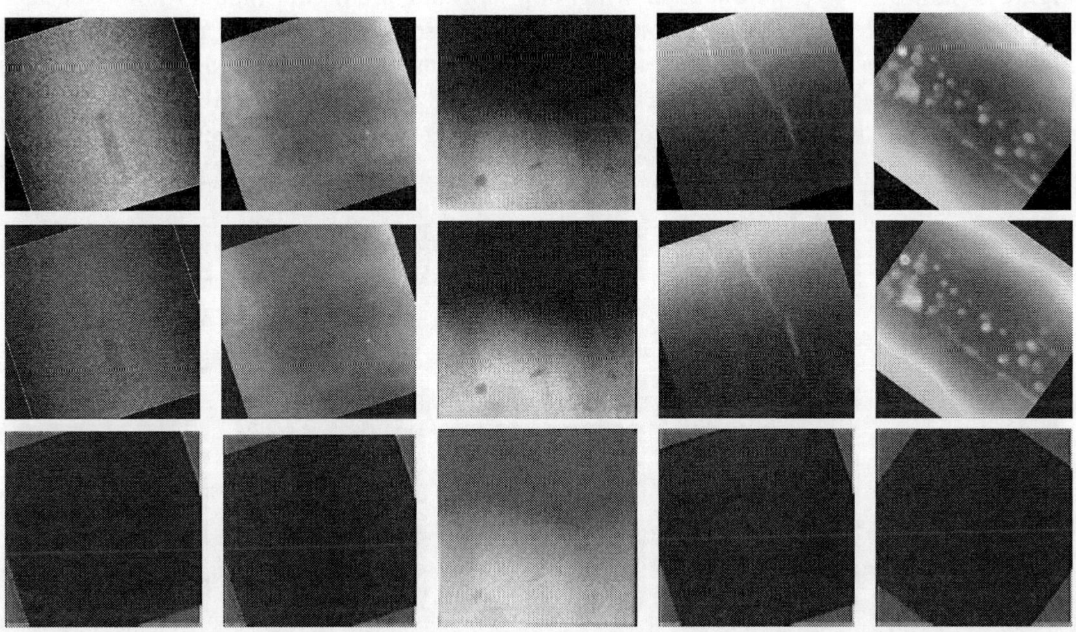

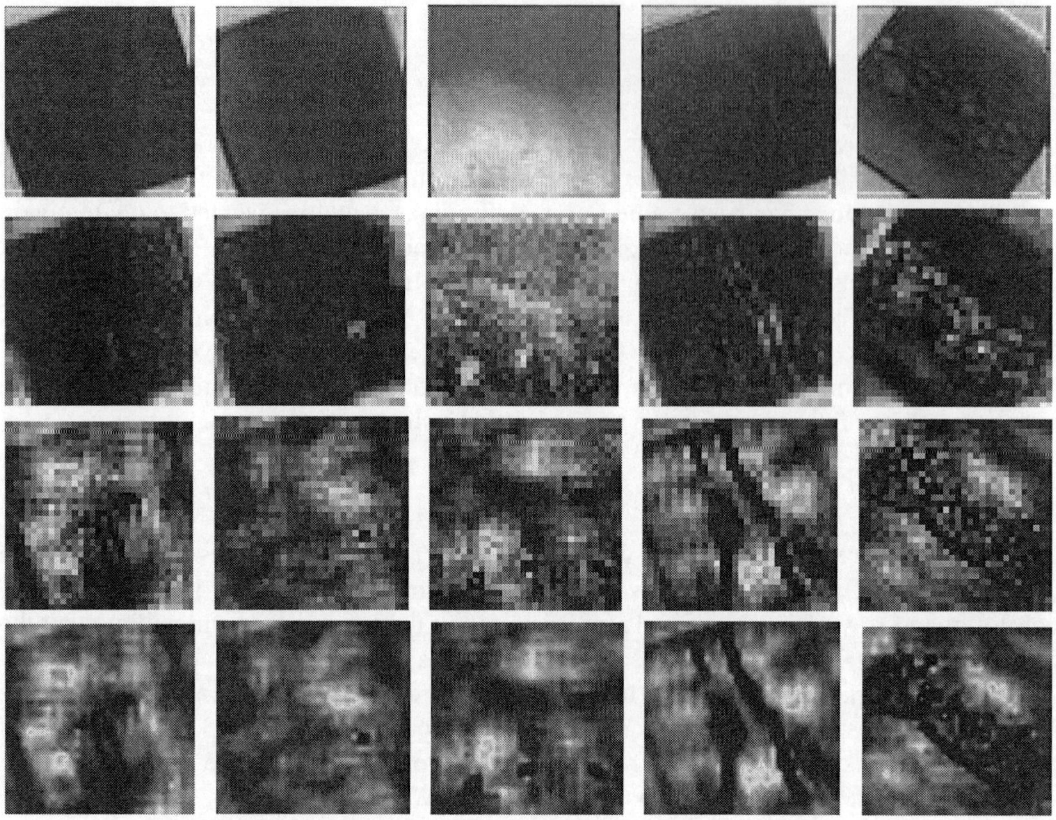

Figure 5: Visualization of some feature map in our network

6.2 Prediction on weld images

Figure 6 shows the results obtained when applying our prediction model to our test data. The first row represents the input images, the second row represents the predictions of our model and the last row represents the ground truth images. Our prediction images are heatmaps of the probability that a pixel represents a defect. In red, we have a high probability that the target pixel represents a defect while blue corresponds to a low probability.

Figure 6: Comparison between predicted defects and ground truth on cropped weld images

6.3 Prediction on non-weld images

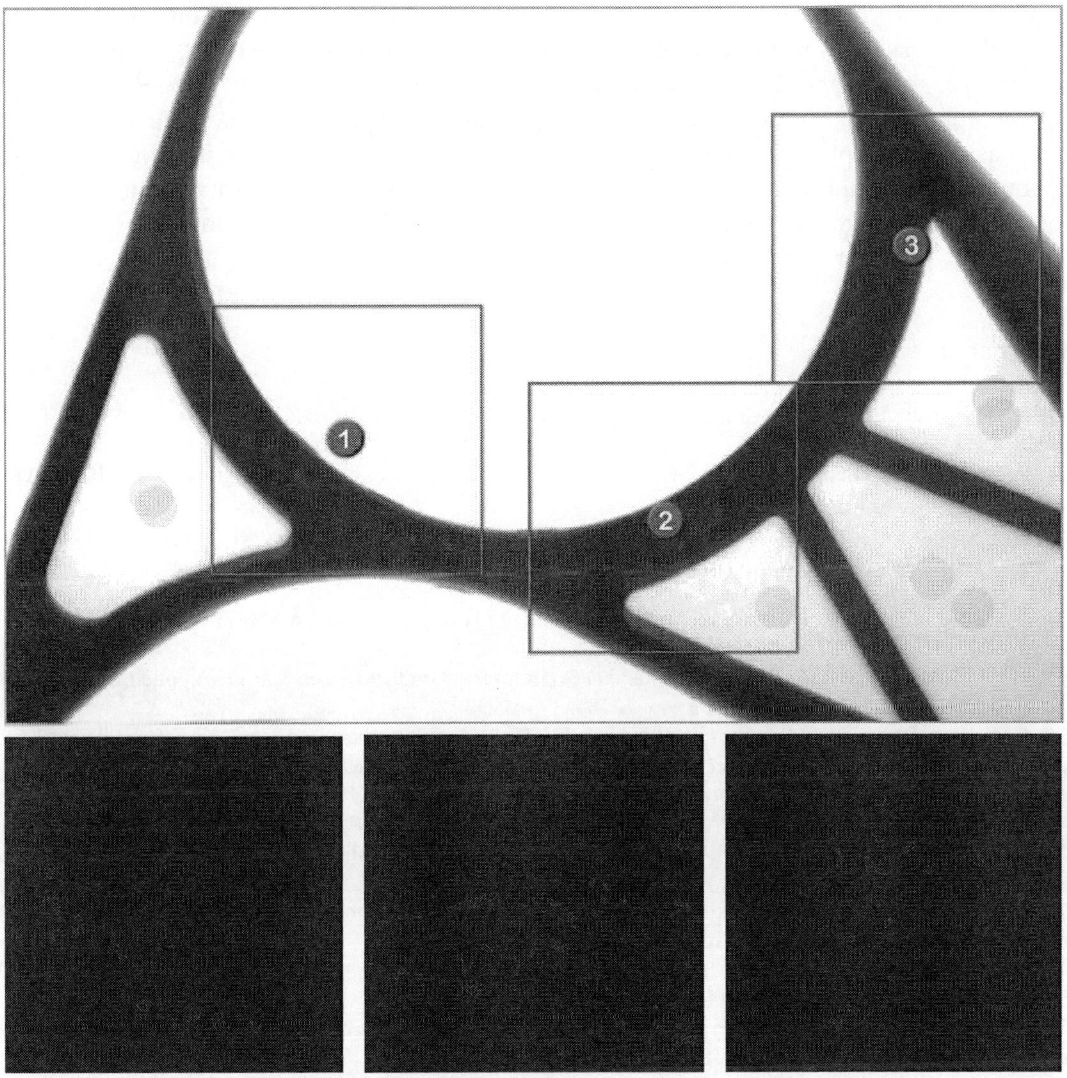

Figure 7: Prediction on non-weld images after sliding window.

7. Conclusion

In summary, we have demonstrated that a neural network model based on deep learning iscapable of assisting in the task of detecting defects in manufactured parts. Our network isin its primary stage but still yields promising results given the 70% value of the F1 scoreobtained. Furthermore, the proposed model appears to be one of the first that has beencompletely trained with radiographic images of porosities in welding. In our future work, wewant to build a more diverse data set so that we can train our model on porosity detectionin any context. In addition, we would like to improve the F1 score obtained by considering afull convolutional network of the auto-encoder type.

References

[1] A. B. Reed, "The history of radiation use in medicine," Journal of Vascular Surgery, vol. 53, no. 1, pp. 3S–5S, 2011.

[2] R. Halmshaw, "The discovery of x-rays and the early history of industrial radiography," INSIGHT, vol. 37, no. 9, pp. 669–671, 1995.

[3] R. I. Frankel, "Centennial of r ¨ontgen's discovery of x-rays," West J Med, vol. 164, pp. 497–501, Jun 1996. PMC1303625[pmcid].

[4] E. Corporation, "Quality control procedures qcp 15.1."

[5] J. A. Melchore, "Sound practices for consistent human visual inspection," AAPS Pharm-SciTech, vol. 12, pp. 215–221, Jan 2011. 21203872[pmid].

[6] Y. Lecun, L. Bottou, Y. Bengio, and P. Haffner, "Gradient-based learning applied to document recognition," inProceedings of the IEEE, pp. 2278–2324, 1998.

[7] C.-Y. Lee, S. Xie, P. Gallagher, Z. Zhang, and Z. Tu, "Deeply-Supervised Nets," in Proceedings of the Eighteenth International Conference on Artificial Intelligence and Statistics(G. Lebanon and S. V. N. Vishwanathan, eds.), vol. 38 of Proceedings of MachineLearning Research, (San Diego, California, USA), pp. 562–570, PMLR, 09–12 May 2015.

[8] Z. S. e. a. Lodish H, Berk A, Molecular Cell Biology. 4th edition. New York: W. H.Freeman. 2000. https://www.ncbi.nlm.nih.gov/books/NBK21535/.

[9] G. Klambauer, T. Unterthiner, A. Mayr, and S. Hochreiter, "Self-normalizing neural networks,"CoRR, vol. abs/1706.02515, 2017.

[10] R. Arora, A. Basu, P. Mianjy, and A. Mukherjee, "Understanding deep neural networks with rectified linear units,"CoRR, vol. abs/1611.01491, 2016.

[11] S. C. Douglas and J. Yu, "Why relu units sometimes die: Analysis of single-unit error backpropagation in neural networks,"CoRR, vol. abs/1812.05981, 2018.

[12] A. Shah, E. Kadam, H. Shah, and S. Shinde, "Deep residual networks with exponential linear unit,"CoRR, vol. abs/1604.04112, 2016.

[13] V. Z. U. M. G. L. I. Z. I. L. H. C. M. "Mery, D.; Riffo, "Gdxray: The database of x-ray images for nondestructive testing.," 2015. 34.4:1-12.

Octree CT Volume Based Reconstruction and Shape Extraction for Processing Massive X-ray Projection Data

Yutaka OHTAKE [1], Yukie NAGAI [2], Hiromasa SUZUKI [1]
[1] The University of Tokyo, Tokyo, Japan
[2] Tokyo Metropolitan University, Tokyo, Japan

Contact e-mail: ohtake@race.t.u-tokyo.ac.jp

Abstract

Demands on high-resolution CT scanning is rapidly increasing, which are intended for precise shape measurement and internal tiny-structure analysis in industrial applications. In order to satisfy such demands, cone-beam type CT scanners equipped with high-resolution (e.g. 4K by 4K pixels) flat-panel detectors are recently available. Since the projection data obtained from high-resolution CT scanners is massive (e.g. several hundred Giga bytes), processing the data is computationally intensive. Aimed to reduce the computational costs, several researchers have proposed to use adaptively sampled grid data structures, i.e. octree CT volumes, rather than the conventional uniformly sampled grid. However, these researches are mainly focused on only CT reconstruction.

In this research, we propose an efficient method combining reconstruction and shape extraction on octree CT volumes. The proposed method generates a precise mesh representing a single material object scanned by a high-resolution CT scanner. In the CT reconstruction stage, a pyramidal projection data set is created and used for FDK reconstruction of a coarse-to-fine octree volume. We utilize a simple octree-subdivision rule with analyzing a local variation of CT values according to changes of the subdivision levels. In the shape extraction stage, the dual marching cube algorithm, in which the standard marching cubes are applied to the grid cells constructed from the dual graph of an octree, is conducted to generate an initial mesh as an isosurface. We take into account the volume fractions of the material in the cells for interpolating the CT values inside the cells. The accuracy of the initial mesh is further improved by moving mesh vertices to the maxima of CT gradients. For estimating the gradients, we used the analytical differentials of FDK formula in order to avoid the difficulty of applying finite difference methods to octree structures.

Octree CT Volume Based Reconstruction and Shape Extraction for Processing Massive X-ray Projection Data

Yutaka OHTAKE [1], Yukie NAGAI [2], Hiromasa SUZUKI [1]

[1] The University of Tokyo, Tokyo, Japan
[2] Tokyo Metropolitan University, Tokyo, Japan

Contents

1. Reconstruction of octree CT volume
2. Isosurface extraction on octree CT volume
3. Gradient-maxima surface extraction

2

Algorithm Overview:
Reconstruction of Octree CT Volume

1. Construct pyramidal projection data with recursive down-sampling
   - Shepp-Logan filter is firstly applied to the finest images
2. Reconstruct a CT volume on an octree coarse-to-fine

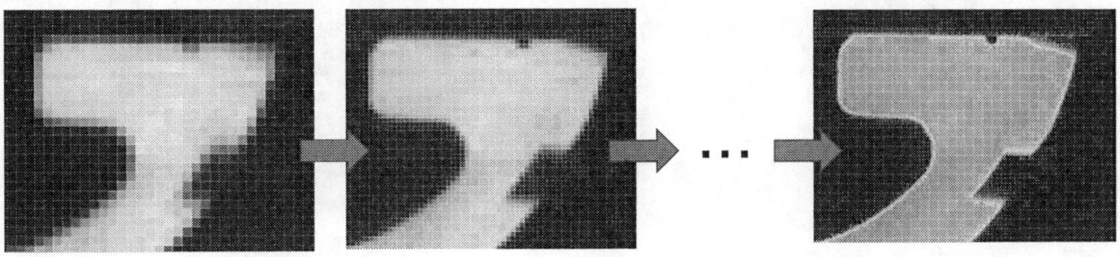

128^3 voxels corresponding to level 0 — Level 1 — Level 5

3

Construction of Pyramidal Projection Data

- Recursive down-sampling procedure
  - 2 by 2 binning for each projection image
  - Skip even-numbered projection images

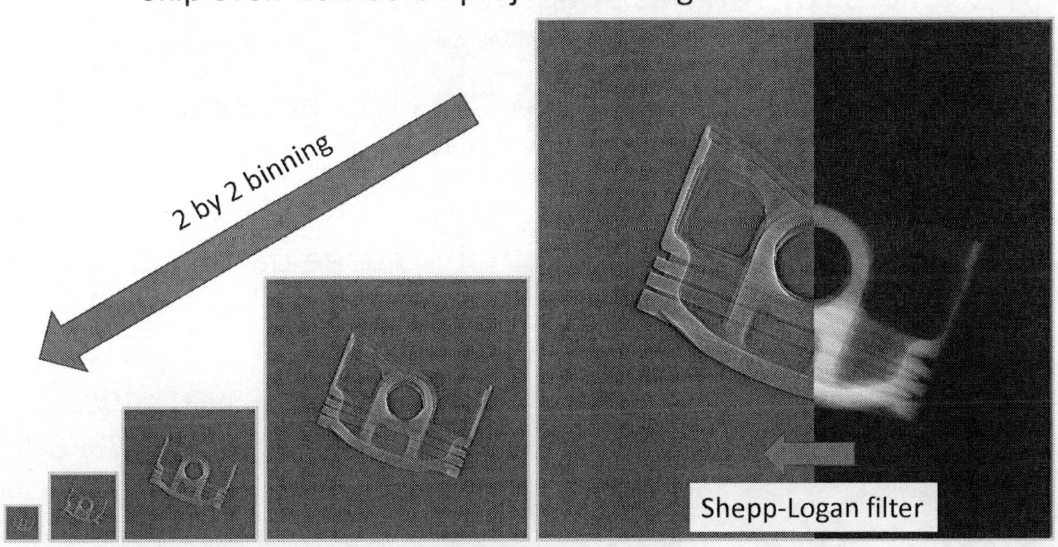

500 images — 1000 images — 2000 images

4

Division Rule of Octree Cells (Octants)

- If the change of CT values are small via octant division, then the division-level is kept.
  - CT values are evaluated at the octant center by the integration of backprojection values
  - Change of CT value: $\varepsilon \equiv \max\left(\max(v_i^{l+1} - v^l), \max(v^l - v_i^{l+1})\right)$

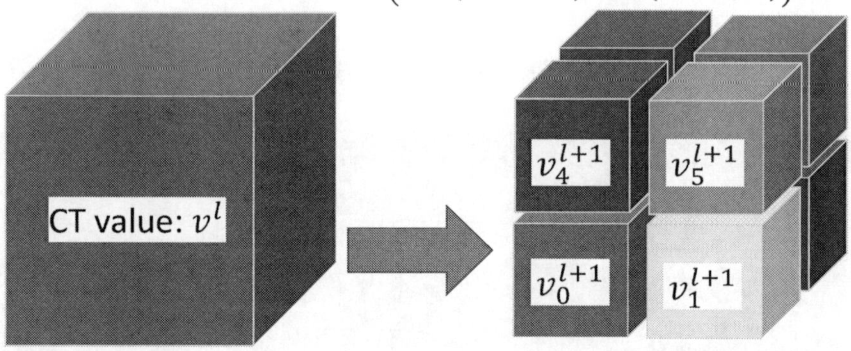

If $\varepsilon <$ threshold, the division is NOT applied

If $\varepsilon \geq$ threshold, the division is applied

5

Result of CT Reconstruction: Aluminum Piston Head

- Projection Data: 2000 images with 2048 × 2048 pixels
- Level 0: 128^3, 4 times adaptive divisions.

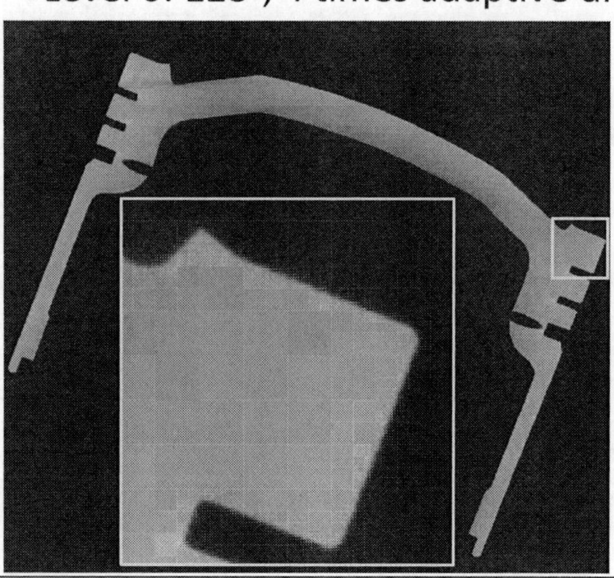

Filtering: 220 sec.
Backprojection: 41 sec. (including file IO)
Octree data size: 208 MB

by using
1× GeForce GTX 1080
1× Xeon E5-1650 3.6 GHz

6

Result of CT Reconstruction: Aluminum Step Cylinder

- Projection Data: 4000 images with 4096 × 4096 pixels
- Level 0: 128^3, 5 times adaptive divisions.

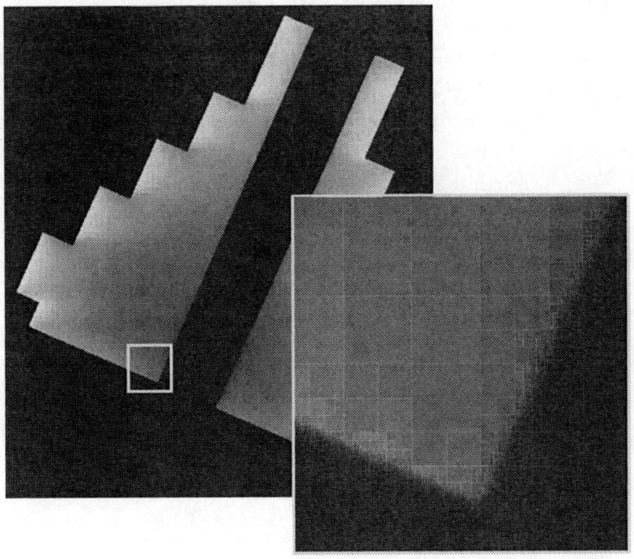

Filtering: 3295 sec.
Backprojection: 772 sec.
(including file IO)
Octree data size: 806 MB

by using
1 × GeForce GTX 1080
1 × Xeon E5-1650 3.6 GHz

7

Contents

1. Reconstruction of octree CT volume
2. Isosurface extraction on octree CT volume
3. Gradient-maxima surface extraction

8

Algorithm Overview: Isosurface Extraction on Octree CT Volume

Dual Marching Cubes for CT data [1]

1. Create dual-grid of octree
   - Conversion octants into grid-points, and grid-points into cells.
2. Apply marching cube algorithm to dual-cells
   - Points on isosurface are computed with taking into account octant-sizes.

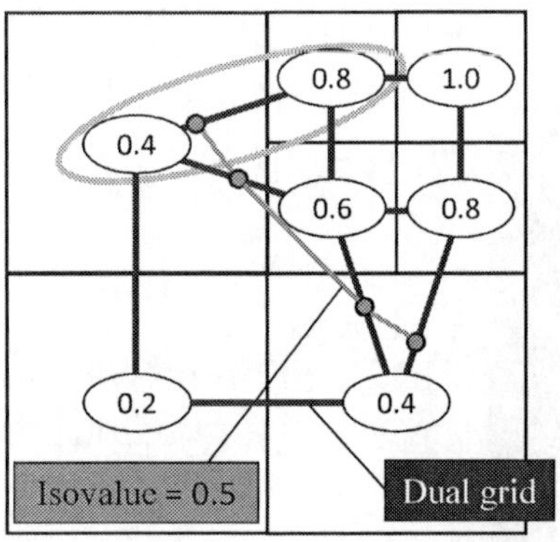

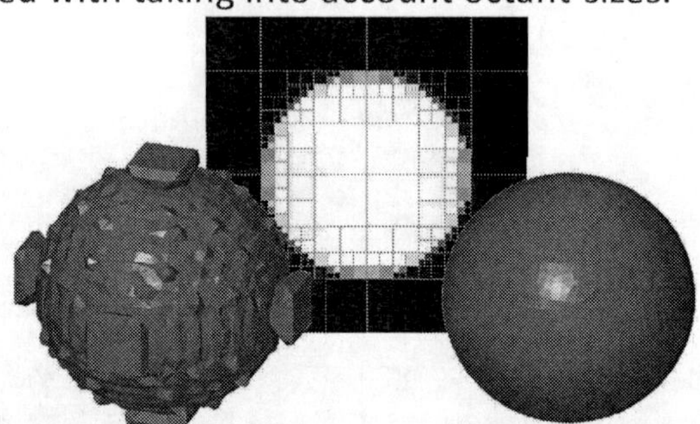

Without using octant-sizes

With using octant-sizes

[1] Kim, Ohtake, Nagai, Suzuki, "A Novel Interpolation Scheme for Dual Marching Cubes on Octree Volume Fraction Data", Computers & Graphics, 2017.

9

Result of Isosurface Extraction: Aluminum Piston Head

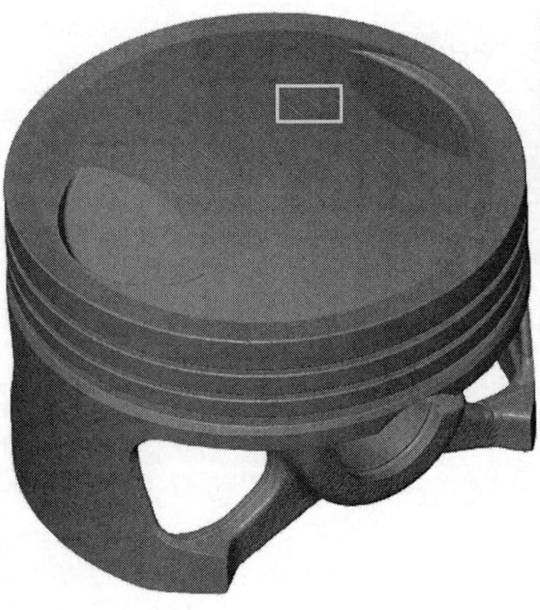

Dual-grid generation: 7 sec.
Marching cubes: 29 sec.
(including file IO)
Mesh triangles: 24 millions
STL size = 6 × octree size
by using
12 cores: Xeon E5-1650 3.6 GHz

10

Result of Isosurface Extraction : Aluminum Step Cylinder

- Reasonable time for mesh generation, but accuracy is not high because if CT artifacts.

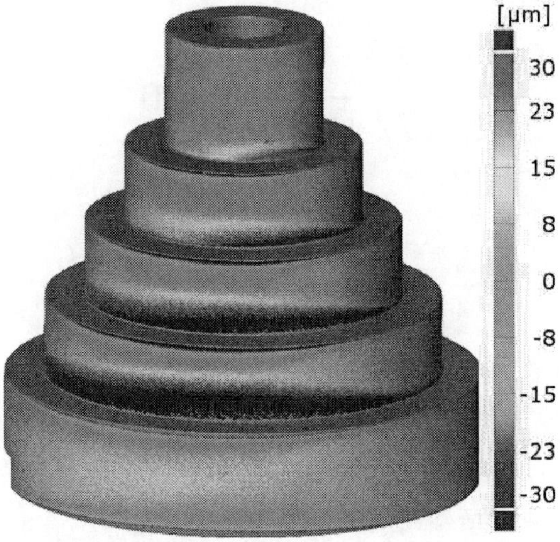

Colored by deviation from CMM cylinders

Dual-grid generation: 32 sec.
Marching cubes: 55 sec.
(including file IO)
Mesh triangles: 60 millions

by using
12 cores: Xeon E5-1650 3.6 GHz

11

Contents

1. Reconstruction of octree CT volume
2. Isosurface extraction on octree CT volume
3. Gradient-maxima surface extraction

12

Gradient-maxima Surface for Improving Measurement Accuracy

- The surface point is defined as a local maximum of the gradient-norm $\|\nabla f\|$ along the gradient-direction

$$g \equiv (\nabla f\,)^t\, \nabla\|\nabla f\| = \frac{(\nabla f\,)^t(\nabla^2 f)(\nabla f)}{\|\nabla f\|^2} = 0$$

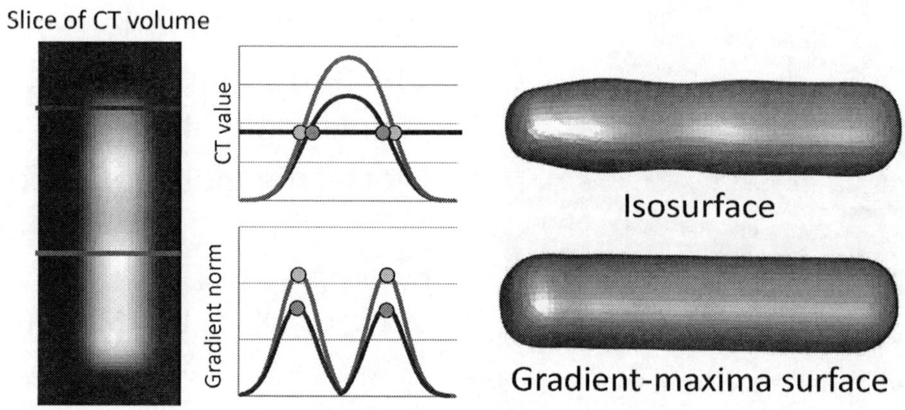

13

Estimation of Gradient ∇f and Hessian $\nabla^2 f$ on Octree CT Volume

- Integration of derivatives of projection data
  - Analytical derivatives of FDK-formula
  - Grid-free estimation on CT volume

$$\frac{\partial^{\alpha+\beta+\gamma}}{\partial x^\alpha \partial y^\beta \partial z^\gamma} f \cong \frac{1}{2}\int_0^{2\pi} P^{-(\alpha+\beta+\gamma+2)}\,(\cos\theta)^\alpha(-\sin\theta)^\beta \frac{\partial^{\alpha+\beta+\gamma}}{\partial u^{\alpha+\beta}\partial v^\gamma} S\, d\theta$$

[(x, y, z): coord. of CT volume, (u, v): coord. of projection iamge、P: magunification]

$$\nabla f = \begin{pmatrix} f_x \\ f_y \\ f_z \end{pmatrix} = \frac{1}{2}\int_0^{2\pi} \alpha^3 \begin{pmatrix} \cos\theta\, S_u \\ -\sin\theta\, S_u \\ S_v \end{pmatrix} d\theta$$

$$\nabla^2 f = \begin{pmatrix} f_{xx} & f_{xy} & f_{xz} \\ & f_{yy} & f_{yz} \\ & & f_{zz} \end{pmatrix} = \frac{1}{2}\int_0^{2\pi} \alpha^4 \begin{pmatrix} \cos^2\theta S_{uu} & -\sin\theta\cos\theta\, S_{uv} & \cos\theta\, S_{uv} \\ & \sin^2\theta S_{uu} & -\sin\theta\, S_{uv} \\ & & S_{vv} \end{pmatrix} d\theta$$

14

Extraction of Gradient-maxima Surface

1. Label inside / outside by thresholding CT values
   - The initial surface is the isosurface

2. Resolve inconsistency to $g = 0$ by iterative local re-labeling
   - Move the surface $g = 0$ along dual-grid edges

15

Result of Gradient-maxima Surface Extraction: Aluminum Step Cylinder

- The accuracy was improved, but it is noise sensitive because of high-order derivatives

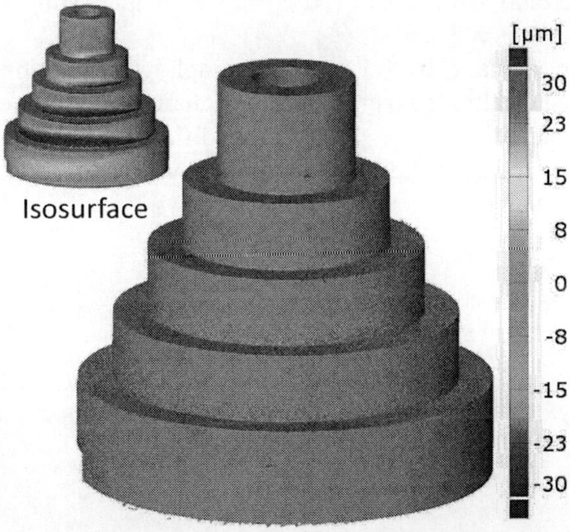

Derivative estimation: 2304 sec.
Dual-grid generation: 32 sec.
Marching cubes: 72 sec. (including file IO)
Mesh triangles: 67 millions

by using
1 × GeForce GTX 1080
12 cores: Xeon E5-1650 3.6 GHz

Colored by deviation from CMM cylinders

16

Optimization of soft X-ray tomography by compressed sensing

Juan WU [1], Mirna LEROTIC [2], Slava BEREJNOV [3], Darija SUSAC [3],
Juergen STUMPER [3], Adam HITCHCOCK [4]
[1] Guangzhou Haozhi Imaging Technology, Guangzhou, China
[2] 2nd Look Consulting, Hongkong, China
[3] Automotive Fuel Cell Cooperation (AFCC) Corporation, Burnaby, Canada
[4] McMaster University, Hamilton, Canada

Contact e-mail: 2006wuair@163.com

Abstract

Compressed sensing (CS) methods have relatively recently been applied to tomographic reconstruction algorithms, providing faithful 3D reconstructions with a small number of projection images to reduce artifacts arising from the missing wedge. In CS, signals are assumed to be sparse in some transform domain, then a linear matrix method is used for analysis.

CS has been applied for scanning transmission X-ray microscopy (STXM) tomography. STXM is used for the measurement of radiation sensitive materials such as biological samples and polymers due to the relatively low radiation damage. STXM tomography provides 3D chemical mapping based on the X-ray absorption properties of the sample. Reconstructions by weighted back-projection (WBP), the simultaneous iterative reconstruction technique (SIRT), and CS are compared. The comparisons show that CS can provide improved reconstruction quality relative to WBP and SIRT, with increasingly pronounced advantages as the angular sampling is reduced.

With CS, the reduction of radiation damage by STXM tomography to radiation sensitive materials have been demonstrated. The cathode layer in proton exchange membrane fuel cell (PEMFC) have been measured in 3D under different radiation condition. By limiting the number of tilt angles to 14, it is possible to measure the 3D structure of ionomer at room temperature while causing minimal radiation damage.

DIR 2019

McMaster University

OPTIMIZATION OF SOFT X-RAY TOMOGRAPHY BY COMPRESSED SENSING

Juan Wu

HAO ZHI
Guangzhou Haozhi Imaging Technology Co., LTD.

Adam Hitchcock (McMaster University)
Mirna Lerotic (2nd Look Consulting)
David Shapiro (Advanced Light Source, Berkeley Lab)
Viatcheslav Berejnov, Darija Susac, Juergen Stumper (AFCC)

CONTENT

Technique:

- STXM: Scanning Transmission X-ray Microscopy
  - STXM tomography
- Soft X-ray Ptychography
  - Ptychography tomography

Reconstruction:

- Compressed sensing method

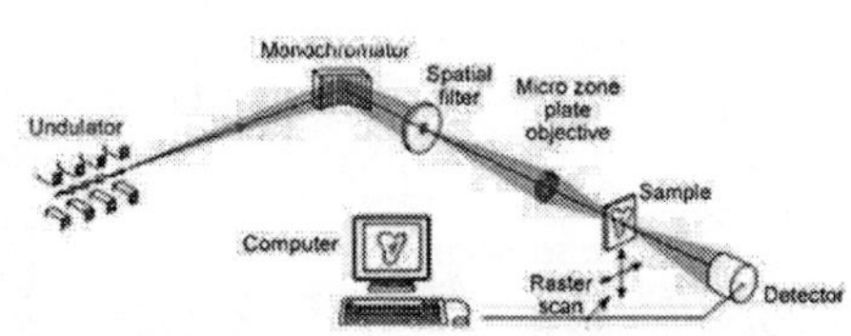

MOTIVATION

PEMFC Proton exchange membrane fuel cells

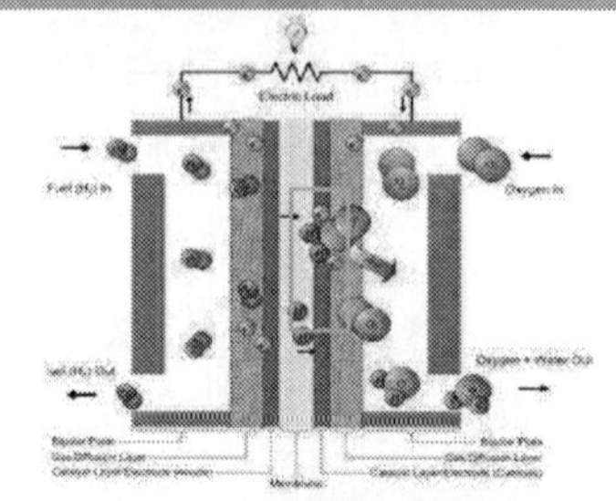

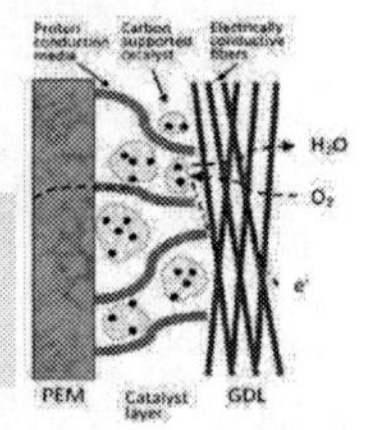

Pt particles
Carbon support particles
Ionomer (perfluorosulfonic acid-PFSA)

https://www.slideshare.net/ReddyPrasad4/comparison-of-fuel-cell-based-hybrid-vehicles-with-other-alternative-systems
*R. Devanathan, **Energy Environ. Sci.**, 1 (2008) 101.*
*S. Litster and G. McLean, **J. Power Sources**, 130 (2004) 61*

SCANNING TRANSMISSION X-RAY MICROSCOPY

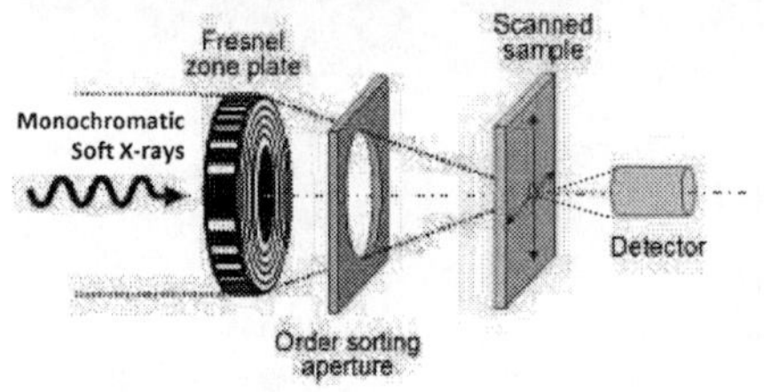

- Imaging: 30-50 nm resolution
- NEXAFS: 0.05-0.1 eV resolution
- Sample thickness: 150 nm - 1 μm
- Applications: bacteria, biological materials, radiation-sensitive materials, et al.

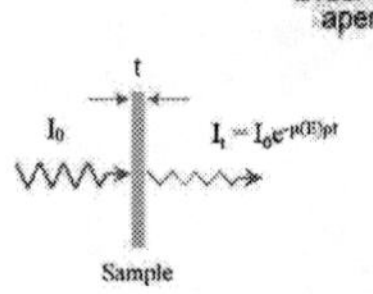

$OD = -\ln(I_t/I_0) = \mu(E).\rho.t$

μ - mass absorption coefficient (cm^2/g)
ρ -density (g/cm^3)
t - sample thickness (cm)

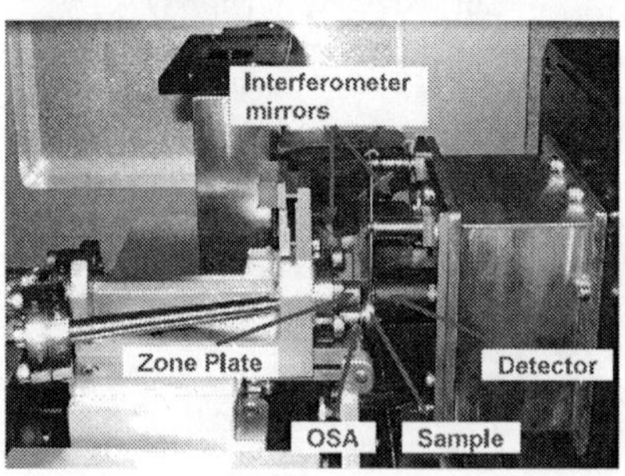

http://exshare.lightsource.ca/sm/Pages/Endstations.aspx

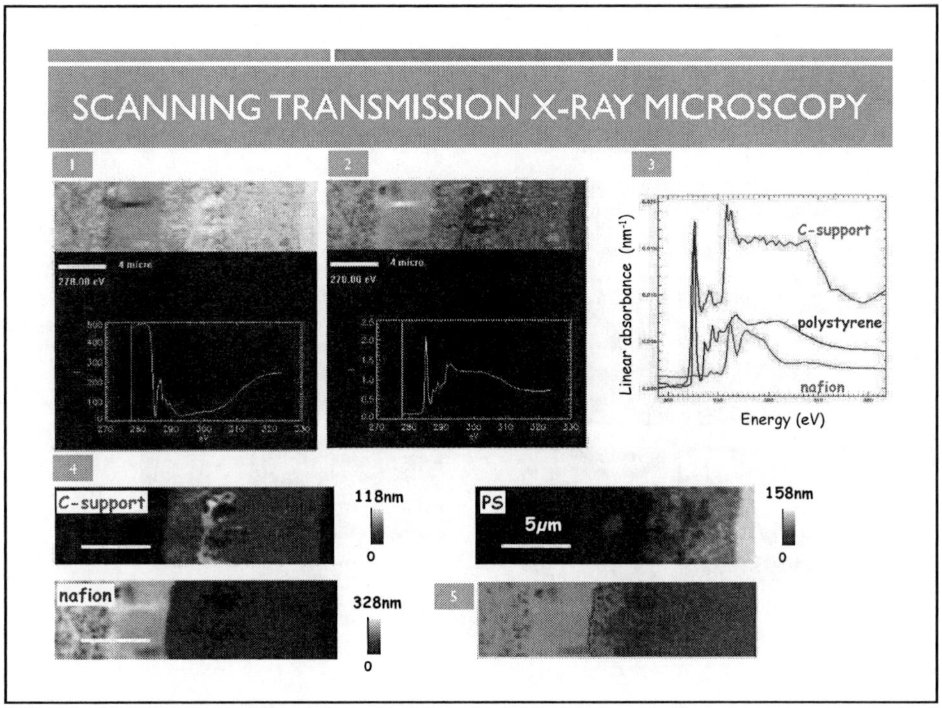
SCANNING TRANSMISSION X-RAY MICROSCOPY
1
2
3
4 micro
4 micro
Linear absorbance (nm-1)
C-support
polystyrene
nafion
Energy (eV)
4
C-support
118nm
0
PS
5μm
158nm
0
nafion
328nm
0
5

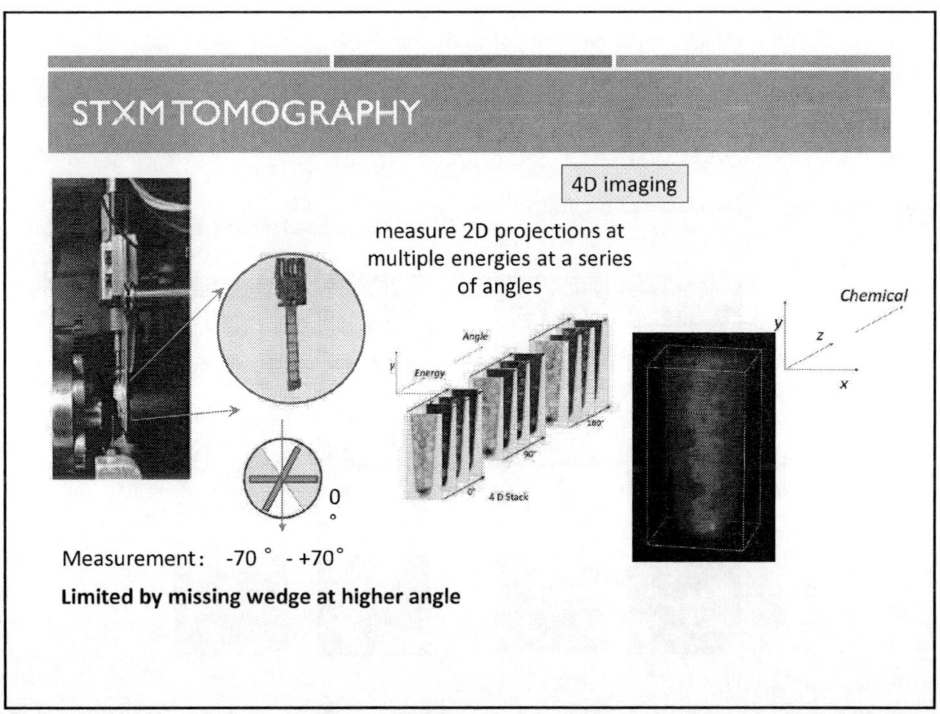
STXM TOMOGRAPHY
4D imaging
measure 2D projections at multiple energies at a series of angles
Chemical
y
z
x
Angle
Energy
0°
90°
180°
4 D Stack
0°
Measurement: -70° - +70°
Limited by missing wedge at higher angle

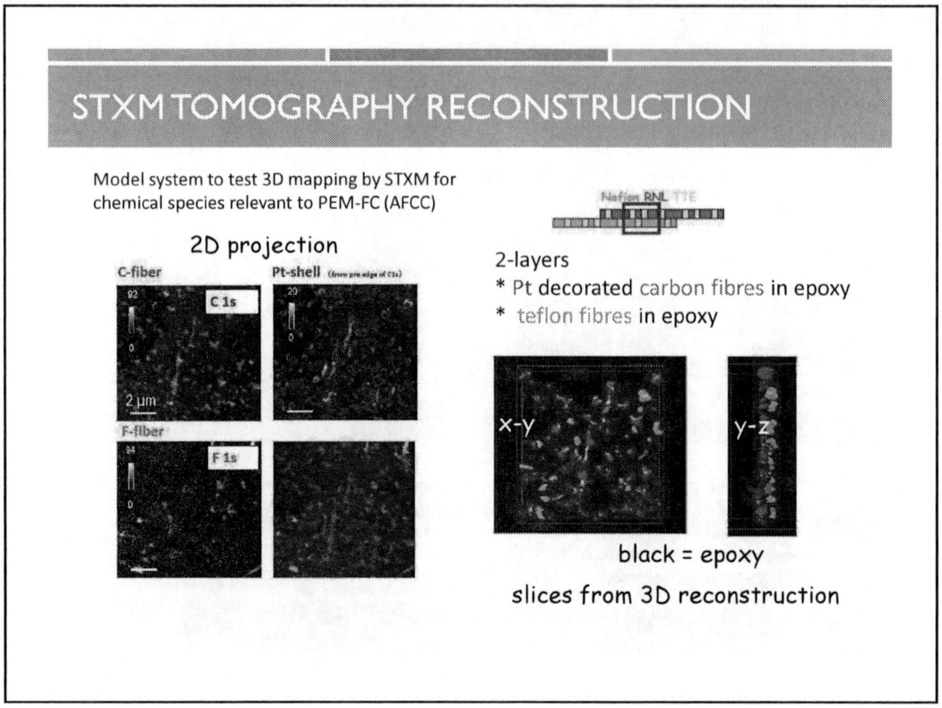
STXM TOMOGRAPHY RECONSTRUCTION
Model system to test 3D mapping by STXM for chemical species relevant to PEM-FC (AFCC)
2D projection
C-fiber
Pt-shell
C 1s
2 µm
F-fiber
F 1s
2-layers
* Pt decorated carbon fibres in epoxy
* teflon fibres in epoxy
x-y
y-z
black = epoxy
slices from 3D reconstruction

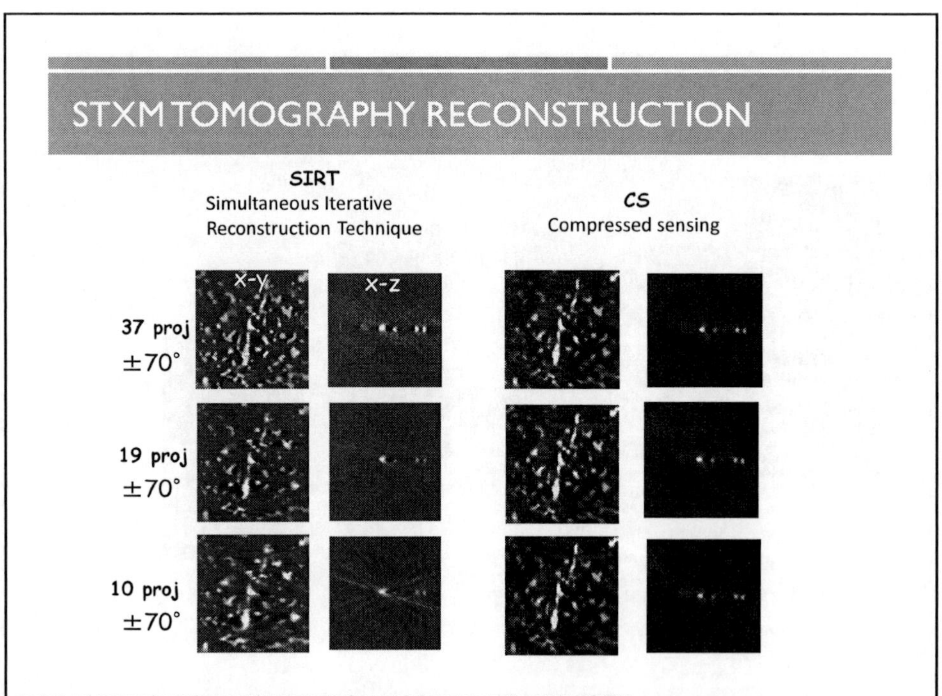
STXM TOMOGRAPHY RECONSTRUCTION
SIRT
Simultaneous Iterative Reconstruction Technique
CS
Compressed sensing
x-y
x-z
37 proj ±70°
19 proj ±70°
10 proj ±70°
Lerotic, M. et al, Compressed sensing for tomographic reconstruction of sparse projection x-ray spectromicroscopy data, XRM2016

STXM TOMOGRAPHY RECONSTRUCTION

CS reconstruction results: sparse-views

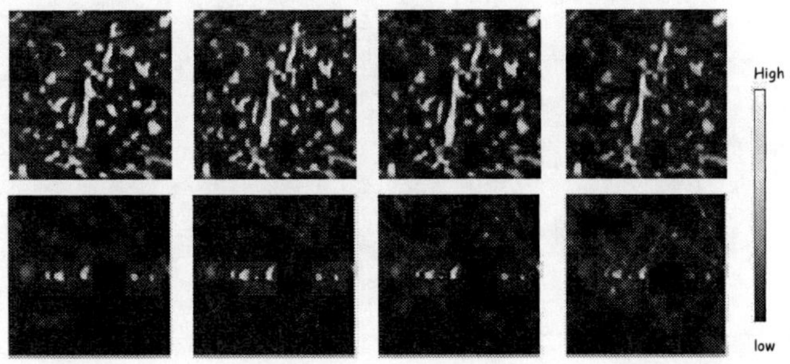

10 proj | 8 proj | 6 proj | 4 proj

Tilted range ±56°

STXM TOMOGRAPHY RECONSTRUCTION

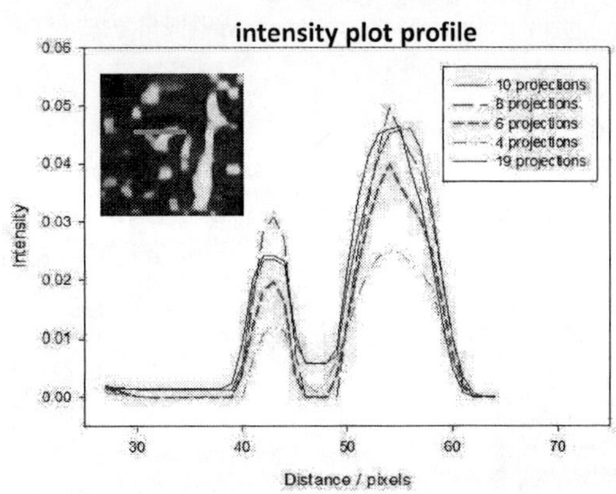

Radiation:
6*4 ms = 24 ms
< 30 ms !!!

EVALUATING RADIATION DAMAGE IN 3D

Fewer angles 45 → 14
Lower flux 8 → 2 MHz
Defocus 30 → 50 nm (step size)

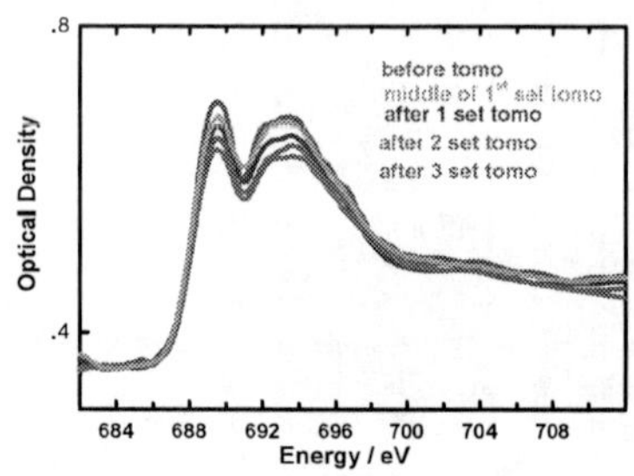

Ionomer chemical maps 2D

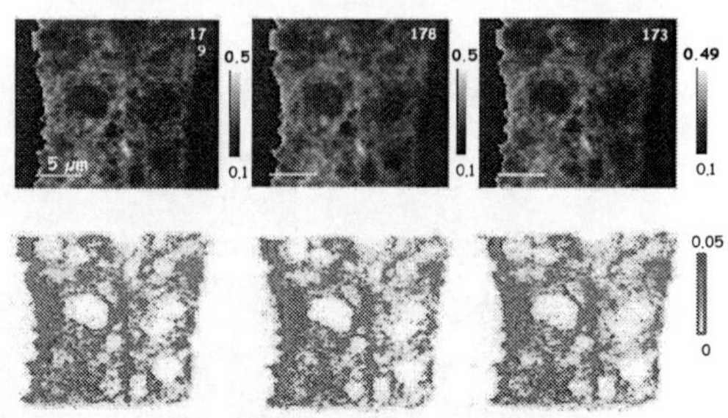

Ionomer chemical maps 3D @ 0°

Radiation dose	**3 MGy**	**7 MGy**	**11 MGy**
Ionomer volume fraction	**31%**	**29%**	**28%**

<<130~400 MGy (before)

SOFT X-RAY PTYCHOGRAPHY

Ptychography – coherent diffraction imaging method
Analysis of diffraction patterns greatly improves spatial resolution, typically to 1/10th of the spot size

- Add Area Detector to STXM
- Record Diffraction Pattern
- Overlap sampled spots
- SHARP Reconstruction algorithm

3 nm - record soft X-ray spatial resolution !

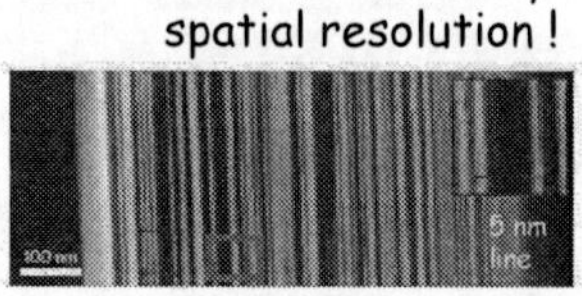

Shapiro et al. Nature Photonics 8 (2014) 765

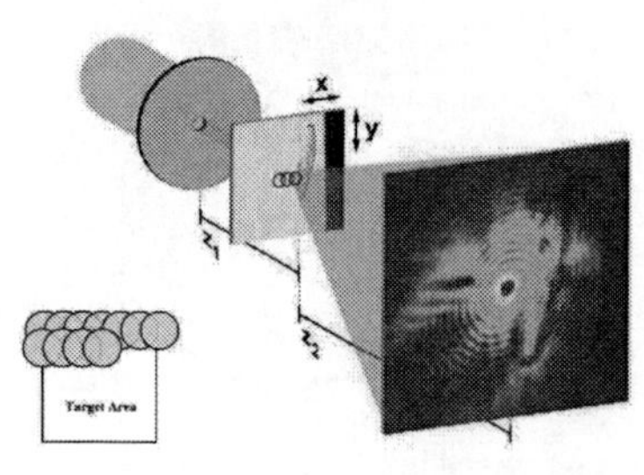

J.M. Rodenburg et al. Phys. Rev. Lett. 98 (2007) 034801.

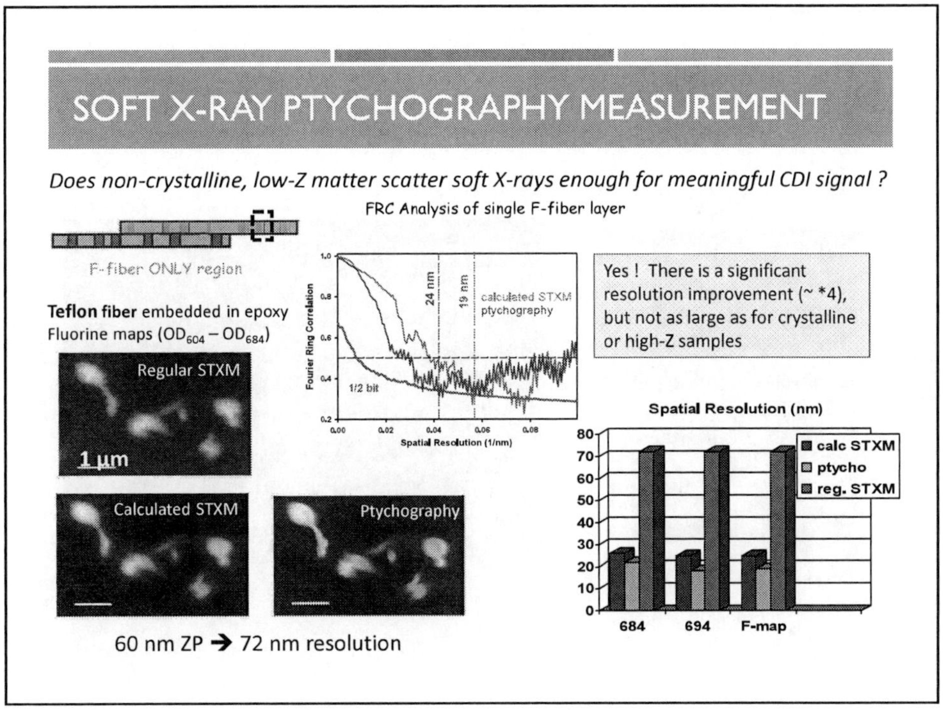
SOFT X-RAY PTYCHOGRAPHY MEASUREMENT
Does non-crystalline, low-Z matter scatter soft X-rays enough for meaningful CDI signal ?
FRC Analysis of single F-fiber layer
F-fiber ONLY region
Teflon fiber embedded in epoxy
Fluorine maps (OD604 – OD684)
Regular STXM
1 µm
Calculated STXM
Ptychography
60 nm ZP → 72 nm resolution
24 nm
19 nm
calculated STXM
ptychography
1/2 bit
Fourier Ring Correlation
Spatial Resolution (1/nm)
Yes ! There is a significant resolution improvement (~ *4), but not as large as for crystalline or high-Z samples
Spatial Resolution (nm)
calc STXM
ptycho
reg. STXM
684
694
F-map

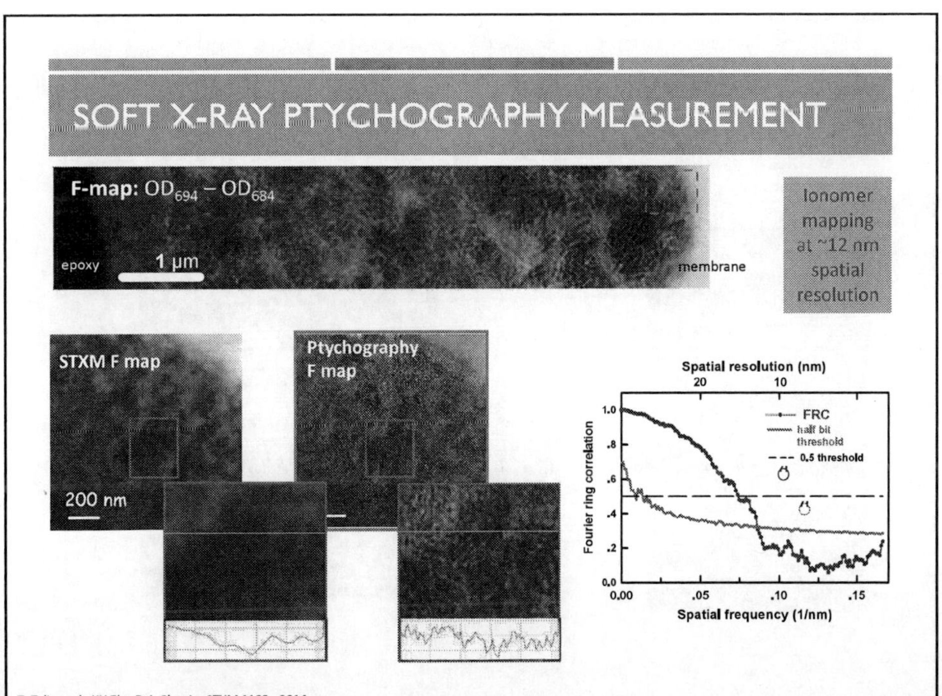
SOFT X-RAY PTYCHOGRAPHY MEASUREMENT
F-map: OD694 – OD684
epoxy
1 µm
membrane
Ionomer mapping at ~12 nm spatial resolution
STXM F map
200 nm
Ptychography F map
Spatial resolution (nm)
FRC
half bit threshold
0,5 threshold
Fourier ring correlation
Spatial frequency (1/nm)
T. Tyliszczak, XH Zhu,D.A. Shapiro STXM 1102, 2014

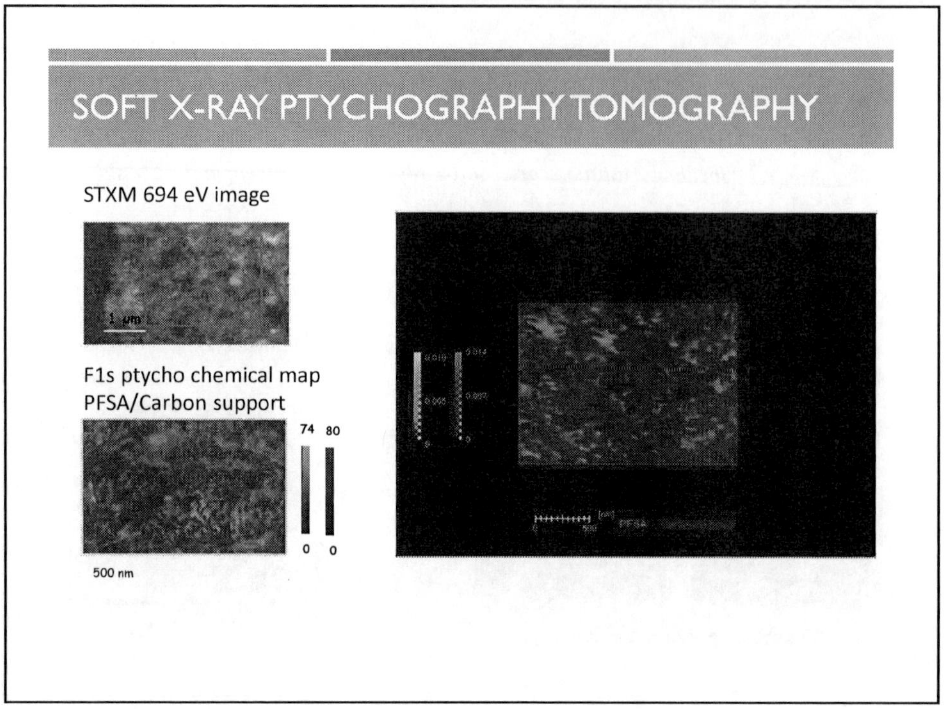
SOFT X-RAY PTYCHOGRAPHY TOMOGRAPHY
STXM 694 eV image
1 μm
F1s ptycho chemical map
PFSA/Carbon support
74
80
0
0
500 nm

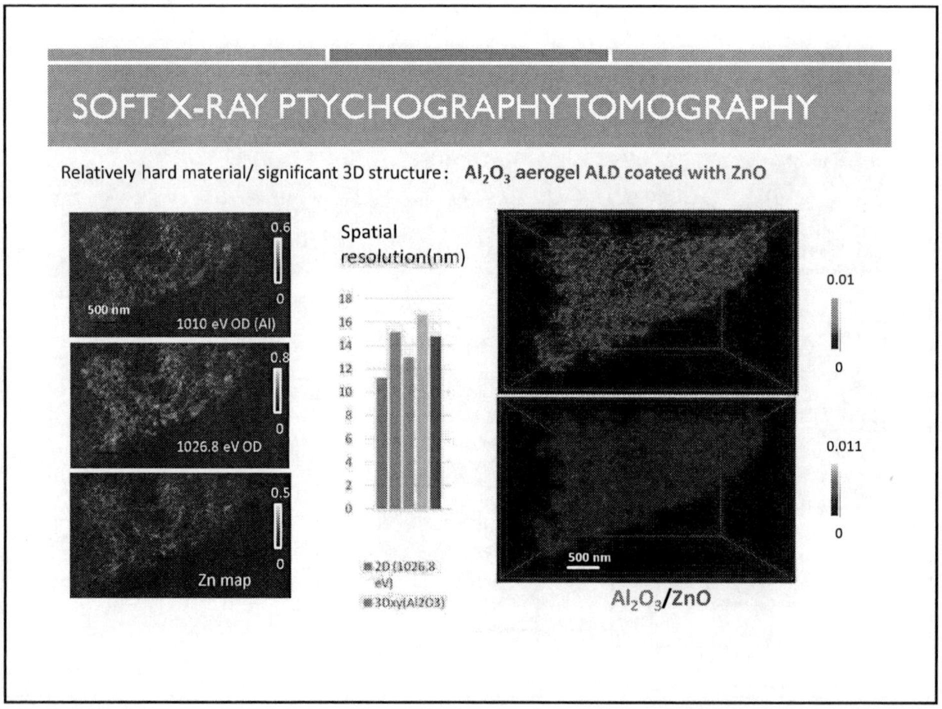
SOFT X-RAY PTYCHOGRAPHY TOMOGRAPHY
Relatively hard material/ significant 3D structure: Al_2O_3 aerogel ALD coated with ZnO
0.6
0
500 nm
1010 eV OD (Al)
0.8
0
1026.8 eV OD
0.5
0
Zn map
Spatial resolution(nm)
18
16
14
12
10
8
6
4
2
0
2D (1026.8 eV)
3Dxy(Al2O3)
0.01
0
0.011
0
500 nm
Al_2O_3/ZnO

SOFT X-RAY PTYCHOGRAPHY TOMOGRAPHY

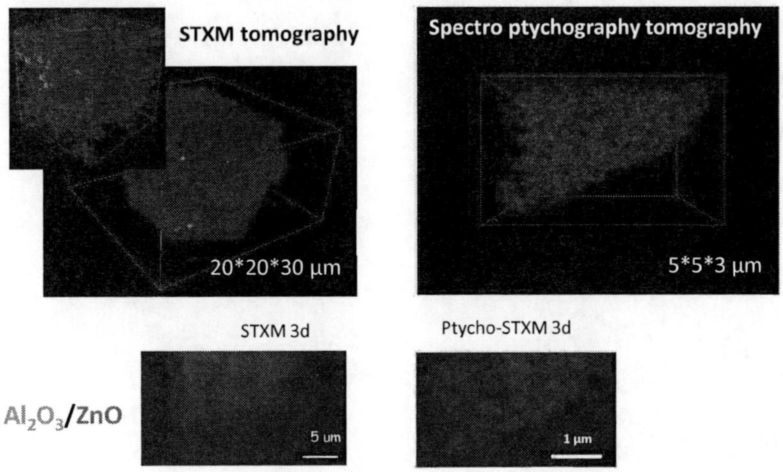

Zn 2p & Al 1s spectro- tomography of FIB sample

Sample courtesy: J.Lee et al (LLNL)

OVERVIEW

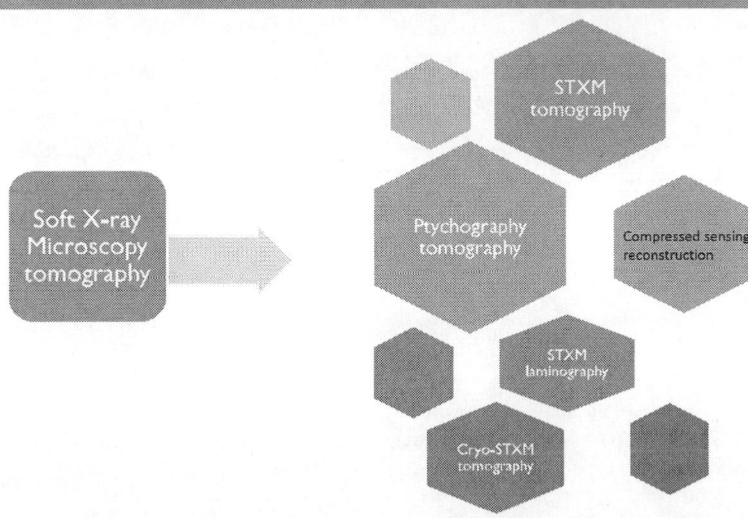

ACKNOWLEDGEMENTS

Adam Hitchcock (Mcmaster University)
Marcia Reid (McMaster University)
Paul Midgley, Rowan Leary (Cambridge)
Tolek Tyliszczak, David Kilcoyne (ALS)
Jian Wang (CLS)
Martin Obst (Beyreuth University)

Canadian Light Source / Centre canadien de rayonnement synchrotron

Funding

ALS ADVANCED LIGHT SOURCE

CaRPE-FC Catalysis Research for Polymer Electrolyte Fuel Cells

NSERC CRSNG

HAO ZHI

McMaster University

AUTOMOTIVE FUEL CELL COOPERATION

Canada Foundation for Innovation / Fondation canadienne pour l'innovation

Thank you!

Questions ?

Three-dimensional in-situ growth surveillance of bulky SiC crystals

Michael SALAMON [1], Matthias ARZIG [2], Peter J. WELLMANN [2], Norman UHLMANN [1]
[1] Fraunhofer Development Center for X-ray Technology, Fürth, Germany
[2] Crystal Growth Lab, University Erlangen-Nuremberg, Erlangen-Nueremberg, Germany
Contact e-mail: michael.salamon@iis.fraunhofer.de

Abstract. The integrated design of modern Computed Tomography systems, optimized to achieve the required environmental conditions for a scan, such as stable Temperature or low vibration has helped to widespread the CT method along non-destructive testing and coordinate measurement applications. For in-situ analysis of processes this integrated design is often limiting the application to processes with relatively low peripheral requirements and small footprint. In general, the in-situ process is adapted to the CT using an experimental environment. In cases a productive environment is required the CT has to adapt to the process and its environment. The complex structure of a physical vapor transport (PVT) reactor represents such a case. Inside an evacuated crucible heated inductively to approx. 2400 ° C, the deposition of silicon and carbon atoms takes place on a seed crystal. Over a period of about 100 to 200 hours, crystal of about 20 - 30 mm in height and currently 150 mm in diameter are grown. The growth kinetics depends strongly on the temperature field and further processes characteristics. The geometry of the interior of the crucible and the growth interface of the crystal are relevant parameters that can be observed during the process by means of computed tomography. The publication presents the integration of a CT system into a PVT crystal growing process with a crystal diameter of 75 mm and the challenges related to the absorption characteristics of the bulky crystal and the quality losses resulting from the applied region of interest scanning.

Introduction

The recent results of process visualization of physical vapor transport (PVT) crystal growth of 75 mm SiC crystal boules using computed tomography (CT) provide beneficial insights into the process [1]. Analysis of features like geometry of the boule, position of the evolving facets and transformation of the SiC source powder from a particulate package to porous needle type structure is allowing to further improve the simulation used for parameterization of the growth process [2]. Thus a main aspect of our research work is to improve the quality of the CT results conquering the harsh environmental conditions of the monitored process [3]. The aspects addressed in the following section concerns the influence of the furnace structure superposing the scanned region of interest and the possibilities to compensate the effects.

1. Material and Methods

In this section the principle setup introduced by Wellmann et al. [1] is further discussed in terms of absorption of the object and of the peripheral superposing structures of the furnace.

1.1 Scan object and periphery

The object to be scanned is a SiC boule with a diameter of 75 mm inside a graphite crucible with a diameter of approx. 130 mm. The surrounding periphery extends the all-over size of the scanned setup to approx. 400 mm. The scan is to be performed during the PVT growth process of the crystal with a growth rate of approx. 100 µm/h. The growth process requires a temperature of approx. 2300 °C inside the crucible induced by an inductive furnace based on a water cooled induction heating coil surrounding the growth chamber.

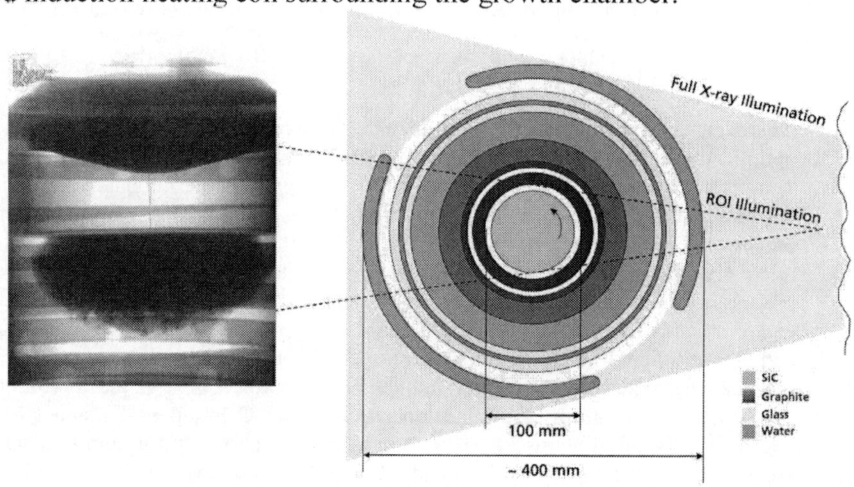

Fig. 1. Resulting Radiograph of the furnace with installed crucible (left) and the schematic setup of layers to be penetrated by the X-ray beam (right). The dotted fan line corresponds to the illuminated area realized within this setup. The yellow marked area indicates a fraction of the fan needed to illuminate the whole furnace, resulting in a relatively long Source to Detector Distance.

1.2 The scan approach: Region of Interest (ROI)

The CT on the large furnace (see cross section in Fig. 1) requires two major adaptions: on the one hand, a large detector field of view (FOV) is needed and on the other hand, a large footprint of the CT system is required in order to illuminate and project the object with a minimum magnification onto the limited FOV. A medical CT system could be suitable for this task if it was possible to switch the imaging orientation from vertical to horizontal slicing. Since the interesting region is located in the centre of the furnace and not in the peripheral layers it is obvious to focus the imaging just on the inner part. Using an ROI approach as presented by Van Gompel et al. in [3] allows to use a smaller detector FOV and a much smaller source to detector distance than in standard CT preserving the footprint of the furnace. Furthermore, the tangential X-ray beam paths along the cylindrical outer layers would require an higher intensity X-ray source, which is not intended in order to keep the requirements on the radiation shielding low. The ROI is focused on the relevant inner part of the setup marked by the dotted triangle in Fig. 1. In a conventional ROI approach, the rotation of the entire setup would harm the image quality due to the interchanging peripheral structures in each projection. The structural information of the periphery would lead to a slight degradation of the Signal to Noise Ratio (SNR) affecting the image quality. In our setup this effect is minimized by rotating only the relevant inner part while the periphery is remaining in position. This has the benefit that the periphery of the furnace may be considered as a static pre-filter adding a static not interchanging structural information into the reconstructed volume. Ideally, the whole superposing structures may be homogenized during gain and offset calibration, except for the spectral energy shifts behind different penetration lengths. The reconstruction of the ROI scan is performed using a cosine padding algorithm suppressing all peripheral structures acquired in the truncated ROI by a cosine function to an area outside the ROI. Fig. 2 shows the different scans derived from a phantom designed corresponding to the furnace structure and material types. The ROI scan in addition to the padding based reconstruction delivers the closest results compared to the full object acquisition.

Fig. 2. The result of the full furnace scan (left), the standard ROI scan with conventional reconstruction (centre) and the result of the padding based reconstruction approach (right).

1.3 The gain-offset correction

The purpose of the gain-offset correction is to homogenize the pixel behaviour of the detector and to compensate for intensity gradients resulting from the X-ray cone beam. Standardly, a two point correction is performed acquiring an offset image without X-ray signal and a gain image with X-ray signal but without any object in the beam path to compensate just for the detector and X-ray beam inhomogeneity. Multigain calibrations are performed at different X-ray intensity levels in order to improve the detectors homogeneity further. For the simple case of a linear calibration each pixel $S_{x,y}(g)$ is calibrated by the following formula:

$$S_{x,y}(g) = \left(\frac{(m_{gain}-m_{offset})}{S_{x,y}(gain)-S_{x,y}(offset)}\right) * (S_{x,y}(image) - S_{x,y}(offset)) + m_{offset} \quad \textbf{eq. 1}$$

In conventional CT applications, the gain-offset correction is done on a daily basis to ensure a sufficient image quality at each time. In the crystal growth application the acquisition of new correction images is limited to the offset image because the peripheral furnace structure is permanently installed into the X-ray beam and cannot be detached to acquire the air image. Bad correction images may diminish the image quality, the superposition of even small static inhomogeneity can be observed by ring-type concentric artefacts in the XY plane of the reconstructed volume. Thus, it is preferred to update the images regularly considering the inclusion of the furnace structure into the correction signal. Fig. 3. shows the corresponding images for offset, gain and the gain+periphery.

Fig. 3. Different type of calibration images. Offset correction image (left), gain correction image acquired before installation of the furnace (4 years ago. centre), gain+periphery correction image acquired just before crucible installation for the growth process (right).

The gain images without peripheral structures have been acquired before installation of the furnace. Their age of several years implicates to use more actual gain calibration images accepting the additional furnace structures. This gain+periphery image contains the peripheral structures but can be acquired before every single growth process while the crucible is not installed yet.

1.4 The Electromagnetic Compatibility

Another aspect for image quality is related to the electromagnetic compatibility (EMC) of the X-ray components to the inductive furnace. The inductive furnace emits an electromagnetic field with a radio frequency of 5-6 kHz at a power between 8-10 kW. After installation of the X-ray components a major influence on the detector system appeared. After optimization of the detector housing some interference signals remained in the projection images. The interference occurs at the readout electronics and affects the image line by line, leading to a fluctuating intensity between each line. Due to the regular behaviour of the interference, a convolutional approach has been chosen, eliminating the intensity fluctuation. In Fig. 4 the EMC affected raw data, the convolved result and the unaffected data imaged without heating is shown. The filter kernel is adjusted correspondingly to the intensity of the artefacts, reducing them successfully.

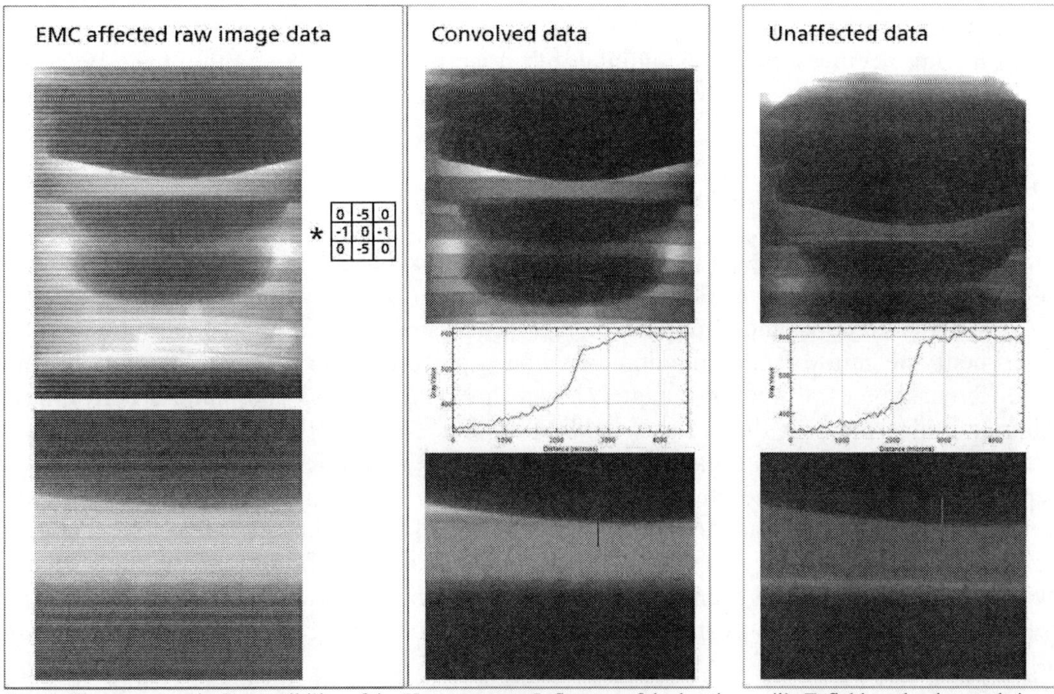

Fig. 4. Electromagnetic Compatibility of the X-ray system. Influence of the heating-coil's E-field on the detected signal (left). Result of the convolution approach applied to the affected raw data (centre). Reference image acquired after growth process, without RF influence (right).

2. Experiments

This section focuses on the experiments performed to evaluate the different offset-gain calibration methods and their influence on the aimed Computed Tomography.

2.1 Basic settings of the Computed Tomography

The scan parameters are adjusted corresponding to the maximum available ratings of the X-ray tube as shown in table 1 and to a reasonably short scan duration minimizing the growth related blurring during the scan. A maximum growth of ¼ voxel is aimed to not be exceeded during one CT scan. The static pre-filtration of 4 mm Al represents mainly the shielding applied to the X-ray system in order to reduce the electromagnetic and thermal interferences of the furnace.

Table 1. Scan parameter of the in-situ Computed Tomography

Parameter	
Energy	125 kV
Static pre-filtration	4 mm Al
Current	350 μA
Integration time	4000 ms
Number of angles	200 / 400
Pixel size	127 μm
Pixel matrix	1488 x 1850
Binning in reconstruction	2
Magnification	~2
Scan time	~ 13 / 27 min
Scan procedure	Stepwise

The complex peripheral structure to be penetrated require a quite long exposure time of 4 s for a single projection limiting the number of projections achievable within the typical scan time of 20 minutes. Based on that condition a binning factor of 2 is conducted enhancing the image quality.

2.2 Data acquisition, processing and feature extraction

The acquisition of the CT data was based on a stepwise rotation of the inner crucible inside the furnace. By means of Fraunhofer Volex 6 the single raw projection images were acquired without any correction and supplied for post-processing steps and reconstruction. In this experiment, the gain-offset calibrations have been performed on the same dataset but using the different type of correction images shown in Fig. 3. On the one hand, a conventional calibration incorporating the old gain calibration image and on the other hand a calibration incorporating the up-to-date gain+periphery image were applied. Further post-processing steps like geometrical calibration and reduction of electromagnetic field interferences are performed on the data in advance to the Feldkamp [4] type reconstruction. The analytical comparison of the results is done using representative features in both reconstructed volumes. By use of the Fraunhofer VolumePlayerPlus it is possible to synchronize both volumes and to extract the same regions by navigating a line- or rectangular-tracker in just one of the two volumes.

3. Results

In this section the applied correction methods and the CT scan approach are evaluated regarding the resulting image quality after calibration. The different calibration approaches show a strong influence on the image quality. The heating coil of the furnace is most representative due to its strong superposition and local intensity distribution. In the 2D radioscopy shown in Fig. 5 the effects are well visible. In the standard calibration approach (Fig. 5, left) the detector is corrected to a good homogeneity but the coil structures superpose the low absorption regions intensively. The displayed grey value profile shows a typical signal drop related to the coil structures of approx. 40 % (light green arrows). In the gain+periphery result (right) the superposition of low absorption regions is reduced, the grey value drops are strongly suppressed resulting in a better homogeneity of signal in this local region (light green arrow). On the other hand, another effect appears diminishing the image quality in the regions of the measured crystal: the mostly homogenous crystal structure is superposed by coil structures which are contained in the calibration image (dark red arrows in the right image). This effect indicates that behind the bulky crystal the contrast resulting from the coil superposition is completely lost and thus the structural information in the gain-periphery image is superposing the region instead.

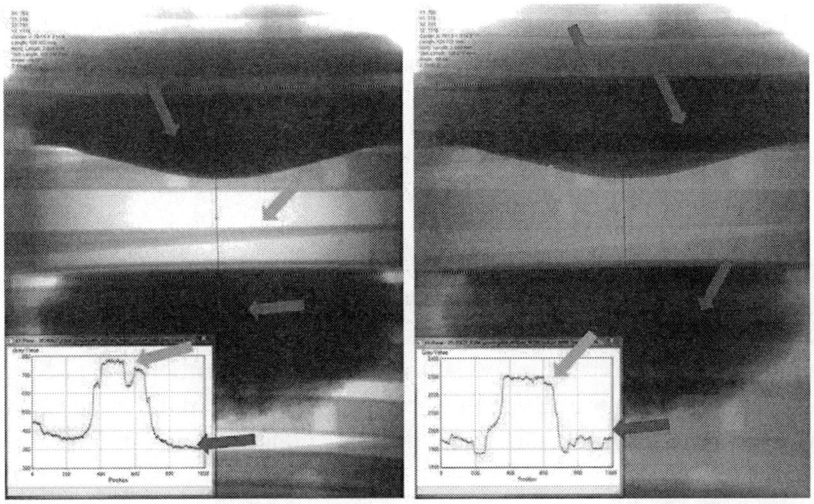

Fig. 5. 2D radiograph calibrated with the standard gain image acquired prior to the installation of the furnace (left), and calibrated with the gain+periphery image calibrated after installation, but with detached crucible (right)

In the reconstructed CT volume the same effect can be observed. In Fig. 6 the CT volumes are compared in the YZ plane, corresponding to the plane shown in the radioscopy. Whereas the low absorption region is reconstructed with a minimum inhomogeneity, the bulky crystal is superposed by fan shaped structures in the gain+periphery result (right). The standard calibration shows the opposite behaviour, the bulky crystal is displayed homogeneous while the low absorption area contains the fan shaped structures (left).

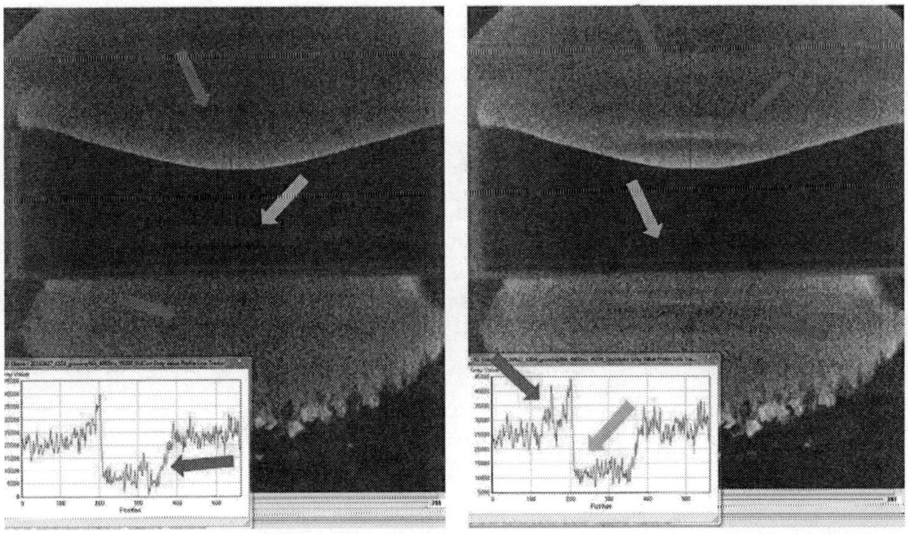

Fig. 6. Resulting effect in the reconstructed volumes. Standard calibration approach (left), gain+periphery approach (right).

4. Discussion and Outlook

The adaption of a Computed Tomography system to the crystal growth process is well feasible with today's available CT techniques and allows important insights into the growth process of SiC crystals using the PVT process. The potentials for improvement of the quality by optimized scan and reconstruction approaches are high, enabling a wider use of the X-ray technique. The results of the investigation on the gain-offset calibration show that the penetration capability of the applied X-ray energy is far too low to penetrate the bulky crystal and thus the contrast features expected to arise from the superposition of the heating coil are not visible. Ideally, if these superposing structures are missing in this region due to a total X-ray shadow produced by the bulky crystal the structures contained in the gain+periphery calibration image would also be cancelled due to the term (2) in equation (1).

$$(S_{x,y}(image) - S_{x,y}(offset)) = 0 \qquad \text{eq. 2}$$

In reality this is not the case due to the Gaussian X-ray intensity distribution and especially due to the amount of scattered radiation superposing the absorption signal $S_{x,y}(image)$. In a regular application with a most homogeneous gain image without superposing structures with high gradients and contrasts this effect is nearly unnoticed, in our case it is limiting the gain+periphery calibration totally. The standard gain calibration with old gain calibration images delivers better results especially for the interesting crystal regions as shown by the comparison in Fig. 6.

The future work will focus on the better compensation of the occurring artefacts using the gain+periphery calibration approach. The superposition of the gain image structures might be a possible indicator for scattered radiation intensity and helpful to even compensate for it in the reconstructed volume and so enhance the image quality.

Acknowledgments

This work has been funded by the German Science Foundation (DFG) under the contract number UH246/4 and WE2107/12.

References

[1] P.J. Wellmann, G. Neubauer, L. Fahlbusch, M. Salamon, N. Uhlmann, Cryst. Res. Technol. 50, No. 1, 2–9 (2015) / DOI 10.1002/crat.201400216

[2] P. J. Wellmann, Semicond. Sci. Technol., in press https://doi.org/10.1088/1361-6641/aad831 (2018)

[3] G.V. Gompel, G. Tisson, D. V. Dyck, J. Sijbers "A new algorithm for 2D region of interest tomography", Proc. SPIE 5370, Medical Imaging 2004: Image Processing, (12 May 2004); https://doi.org/10.1117/12.536685

[4] I. A. Feldkamp, L. C. Davis, J. W. Kress, Practical cone-beam algorithm; Journal of the Optical Society of America A: Optics and Image Science, and Vision 1 (1984) 612-619

XXL-CT capabilities for the inspection of modern Electric Vehicles

Michael SALAMON [1], Nils REIMS [1], Michael BÖHNEL [1], Katrin ZERBE [1], Michael SCHMITT [1], Norman UHLMANN [1], Randolf HANKE [1]
[1] Fraunhofer Development Center for X-ray Technology, Fürth, Germany
Contact e-mail: Michael.Salamon@iis.fraunhofer.de

Abstract. The automotive industry currently undergoes a significant transformation process from the conventional combustion technology to other drive technologies i.e. electric drives. The market of electric vehicles (EVs) has emerged thanks to the availability and performance boost of the Lithium Ion battery technology. The new technology enables on the one hand new car concepts, on the other demands a completely new power train technology. With rising production capacity of those new components and systems, the availability of characterization and inspection methods will become an important technology driver. One already intensively applied method is the Micro-Computed Tomography for the investigation of the microstructure and assembly in single battery cells. In combination with high energy X-ray sources, even full battery modules can be analysed with standard Computed Tomography methods. The XXL-CT at Fraunhofer EZRT made the next major step forward in allowing to inspect fully assembled vehicles three-dimensional for the first time. One unique benefit of this method is a non-invasive analysis of potentially damaged battery geometries even after a crash test, where a destructive analysis is of severe risk due to the high impact sensitivity of the Li-Ion batteries.

In this contribution, the XXL-CT method for scanning of EV's will be demonstrated by real scan results of a state of the art EV. Two scan geometries are evaluated, i.e. horizontal and vertical vehicle orientation. The focus lies on the massive battery package and its influence on the overall image quality.. Furthermore, other applications and the development status of the XXL-CT at Fraunhofer EZRT will be demonstrated and discussed.

Introduction

Since 2013 the Fraunhofer EZRT (Fürth, Germany) provides a world unique scale of Computed Tomography (CT) for novel, scientifically and commercially yet unexplored applications [1,2]. One of these major application fields is the digitalization of cars for purposes of deformation analysis after crash tests [3] or the analysis of prototype cars during different development phases i.e. endurance tests. The unimpaired non-destructive insight arise potential time and cost savings for the data gathering processes in research and development of modern car manufacturers motivating the Fraunhofer research. With shorter product cycles and growing variety of models, the process of crash simulation validation can benefit strongly from this technology in the near future. Also for the new trends in automotive industry like electrical vehicles (EV's) the XXL-CT enables uniquely new capabilities, like the analysis of battery modules and power train components even after real crash tests. With traditional tear down methods an injury of the Lithium-Ion batteries may lead to a burn down of the whole car limiting the application of those methods. The research of the Fraunhofer EZRT is dedicated to several topics in the domain of car CT scanning, starting with data acquisition over processing and reconstruction [4] up to the data conversion for certain user groups [5] and image evaluation [6]. The data acquisition techniques developed for car scanning are focussed in this publication.

1. Material and Methods

This section describes the arrangements and scanning methods established for car scanning at the XXL-CT system already described in [7]. These are the street orientation region of interest (SOROI) and the rocket orientation (VERTICAGE) arrangements. Both procedures have their advantages in different application fields and for the different car types.

1.1 Scan object

The object to be scanned is a state of the art electric vehicle (EV) allowing the proof of methodology of XXL-CT for new trends in automotive engineering. Contrary to the combustion technology, where high X-ray absorption occurs in local areas e.g. in gear transmissions and crankshafts, the absorbing regions are spread along large areas in EV's leading to higher influences on the image quality.
The scanned car is a Tesla Model S P85 (2014 version) in which the battery module is located flat under the passenger area. It consists of over 7000 Lithium Ion battery cells of type 18650. The single cells are arranged in 16 modules (bricks) aligned in two parallel columns alongside the car chassis. With a cell diameter of 16 mm and a matrix of approx. 14 x 32 cells in each brick a total area of approx. 1.7 x 1 m^2 and a height of 65 mm forms. Even though the average absorption coefficient of the battery module may be approximated to the absorption coefficient of Aluminum, its large dimension is limiting the penetration capability. A demand for an optimized scan orientation in CT is obvious.

1.2 Street Orientation Region of Interest scanning (SOROI)

The street orientation scan is the standard approach applied at the XXL-CT since beginning of operation [3,7,8]. For the street orientation scan the car is parked on two steel girders extending the size of the turntable to the needed footprint of the car. The steel structure allows adjusting the scan region by positioning the car decentered on the turntable. The overhanging section of the car shown in Fig. 1 is limiting the angular scan range to 216° in order to avoid collisions with the detector system. Due to the 4 m width of the detector, only a partial volume of the car can be reconstructed at once. For a full reconstruction of the car a second scan has to be performed with a rearranged car position. Beside the unfavorable aspect ratio resulting from the car geometry and the Region of Interest type of CT data the main advantage of the SOROI scan approach is the low preparation effort not even requiring to empty the fluids.

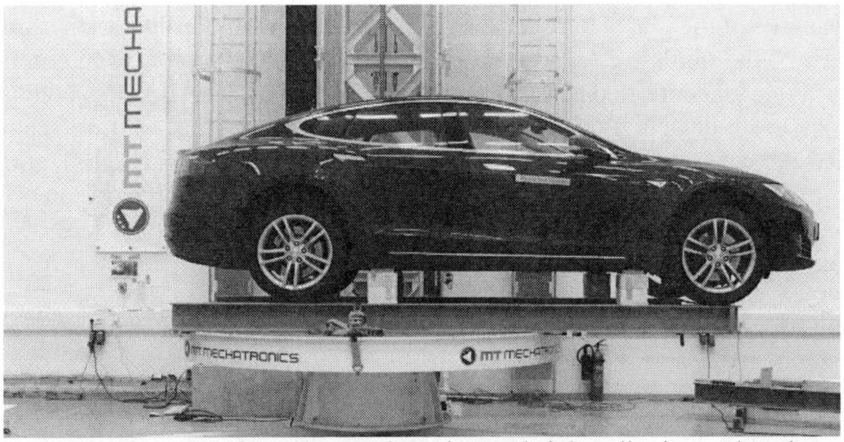

Fig. 1. Car positioning for street orientation scanning using steel girders allowing to place the car decentred onto the turntable.

1.3 Rocket Orientation full vehicle scanning (VERTICAGE)

The recently developed rocket orientation scan emulates a gantry type scanning known from the medical CT application where the patient is scanned bedded on a horizontal table. This allows minimizing the penetration paths inside the scanned object. The advantage is an enhanced image quality especially for all longitudinal structures like battery modules or long straight metal sheets. The realized approach requires the installation of the car in a structure called VERTICAGE specially developed for the purpose of tilting and scanning cars. The design of the rigid steel framework is optimized for low X-ray absorption in order to minimize artifacts from superposition of the frame structure. The size and load capacity is orientated to today's car market allowing to scan from low weight cars to heavy pickups with up to 3.200 kg weight. The fixation is done using load resistance assemblies at each wheel and at the undercarriage of the vehicle. For the manipulation and positioning of the VERTICAGE, the internal 10 t crane is used. The tilting procedure is performed using the fixation points at one side of the VERTICAGE and sliding the other end on a permanently installed bearing pedestal as shown in Fig. 2 (left).

Fig. 2. Car preparation for VERTICAGE scanning. The car is fixed to the platform of the verticage steel frame structure. Using a bearing pedestal, it can be tilted around the rotating point to the vertical orientation by the crane.

The disadvantage of the VERTICAGE scan approach is the high preparation effort requiring to empty few of the fluids e.g. window washer and optionally motor oil in combustion cars in order to avoid uncontrolled leaking in vertical orientation. Also, the conventional car battery needs to be detached to prevent the car from powering up during scanning. After having tilted the car into the required scan orientation, an additional stabilization period is required to prevent the car from sinking during the scan. The total preparation efforts for the VERTICAGE scanning approach are approx. 24 hrs.

2. Experiment

2.1 Scan preparation

Due to the limited field of view of SOROI scans and time available for the investigation, the measurements were focussed on the rear end of the Tesla S containing the most absorbing and therefore most relevant parts, the battery module and the electric drive. As mentioned before all, we emptied unsealed vessels to a minimum before tilting the car to the vertical orientation. We monitored the sinking of the car by means of a digital gauge positioned on a car frame.

2.2 Acquisition type and scan parameterisation

All measurements were performed with the same hardware arrangement of X-ray source, detector and manipulation system. The detailed scan settings are summarized in the following Table 1.

Table 1. Parameters of the CT-scans for SOROI and VERTICAGE measurement setups.

Parameter	**SOROI scan**	**VERTICAGE scan**
X-ray energy working point	7.9 MeV	7.9 MeV
Dose rate at 1 m distance	22 Gy/min	22 Gy/min
Integration time per line	24 ms	24 ms
Scanned height	1.44 m	2.812 m
Number of lines	2400	4687
Number of images	2500	1500
Resulting image acquisition time	**40 h**	**46.87 h**
Resulting volume	Ø 3.2 x 1.5 m^3	Ø 2.2 x 2.8 m^3

The chosen scan parameters are corresponding to a comparable total scan time (40 hours and 46.87 hours, respectively) and a similar effective volume size. The smaller envelope of the VERTICAGE results in a scan volume diameter of approx. 2.2 m and allowed using a lower number of projections, while the scan height have nearly been doubled to achieve a similar volume size as in the SOROI scan. Fig. 3 shows the different 2D projection images of the two scans. Both scans were acquired with the same vertical sampling of 0.6 mm resulting in an anisotropic voxel size of 0.4 x 0.4 x 0.6 mm^3.

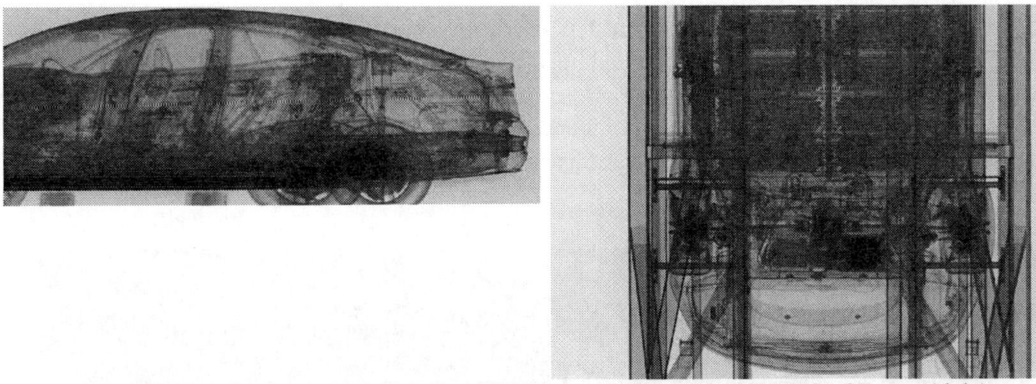

Fig. 3. 2D projections acquired during SOROI scan (left) and the VERTICAGE scan (right).

2.3 Image processing and data handling

The derived data undergoes several calibration and adjustment steps before it is reconstructed using Fraunhofer Filtered Back Projection reconstruction library. The raw dataset of the SOROI scan has a size of 111 GB and the one of the VERTICAGE scan has a size of 130 GB. The achievable volume size is even larger, thus the comparison has been performed on an coarse sampled voxel grid of 1 x 1 x 1 mm^3, resulting in approx. 30 GB of volume size for each reconstruction.

3. Results

3.1 SOROI scan results

An XY cross-section of the passenger cabin acquired using the SOROI approach is shown in Fig. 4. Due to the Region of Interest type of measurement, only the centre and back part of the car can be reconstructed accurately. The front part section in the left third of the image is just partially acquired during the scan, leading to inconsistent reconstruction data. As mentioned before, a full scan of the car requires the front and back part of the car to be scanned individually. Nevertheless, the reconstruction in this car region allows to precisely visualize the different metal sheets of the car body, different electrical and plastic assemblies and even the outer shape of the seat defined by the leather cover.

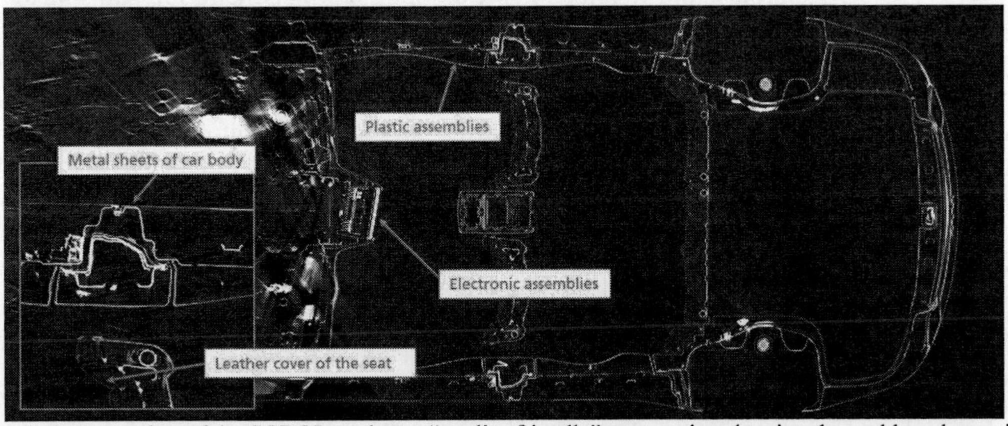

Fig. 4. Cross-section of the SOROI result at a "quality friendly" scan region showing the usable volume size and details detectable with high quality (sub-image left).

A closer look at the regions with highest penetration lengths shows the limitations of the SOROI approach especially for EV's. The cross-sections displayed in Fig. 5 (1, 2) are showing the expected typical artefacts deriving from too long penetration depths in the battery region and even in the massive electric drive and power supply section. The shape of the structural elements can be hardly reconstructed due to the limited information from all angular views.

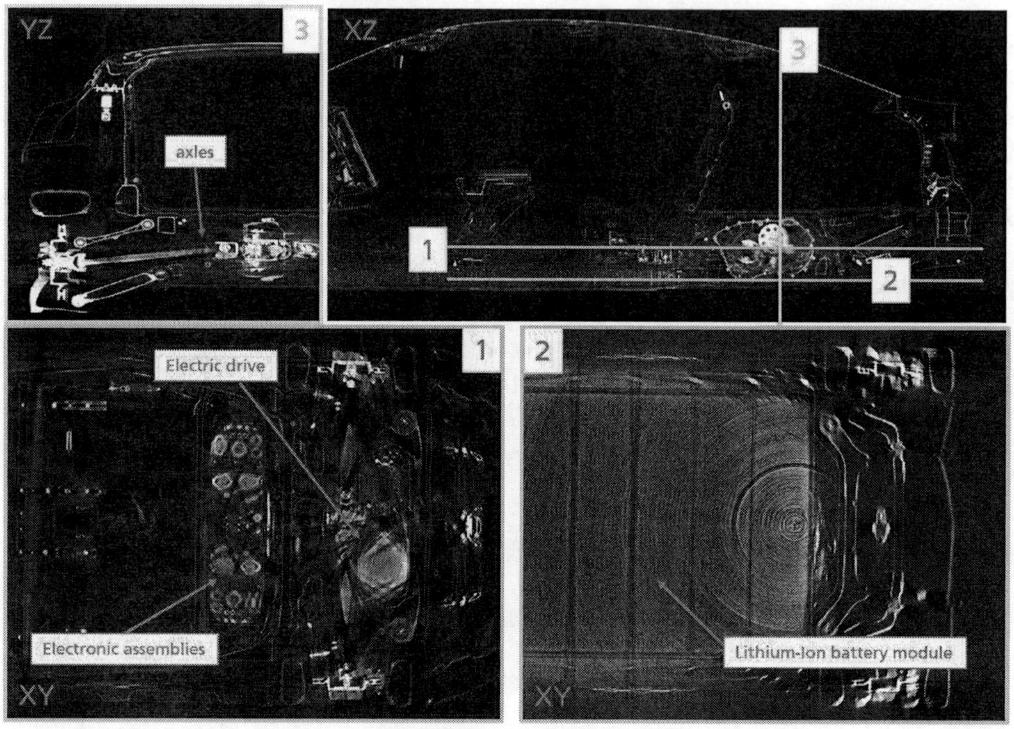

Fig. 5. Different cross-sections of less "quality friendly" regions showing the limitations of the SOROI approach for CT on EV's.

3.2 VERTICAGE scan results

The VERTICAGE approach allows reducing the amount of long penetration paths inside the scanned car. As shown in Fig. 6, the detection capability increase significantly, even in regions of the battery module and in the power supply and drive section. The high quality of the results allow for an analysis of the inner assembly of the battery module down to the single cells location in the brick. Dislocations due to crash or endurance tests during the simulation of lifetime are well detectable. The high achievable quality in the XZ planes allow also virtually disassembling subcomponents of the power train and electrical wiring.
Nevertheless, also the VERTICAGE approach has its limitations, which are visible at the axles shown in the XY cross-section in Fig. 6. The longitudinal structure and the in beam orientation of the dense material reduce the visibility drastically and leading to a worse signal than in the SOROI scan results. The necessity of an optimum car positioning before scanning is obviously visible by that result.

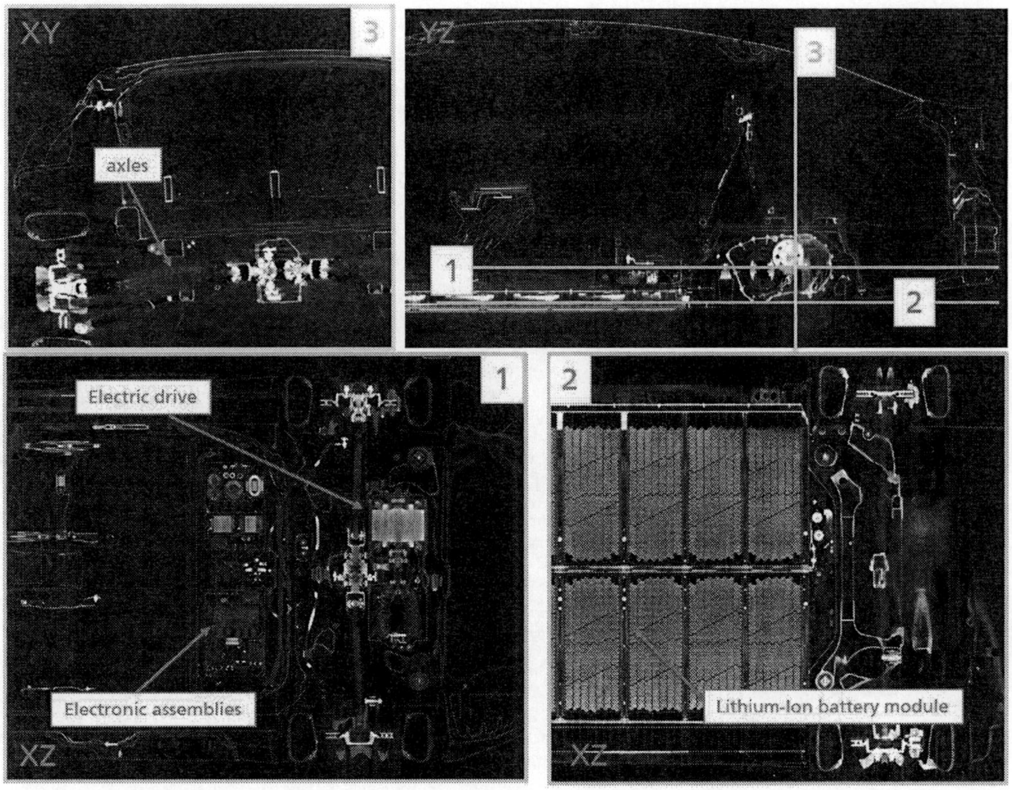

Fig. 6. Comparable cross-sections discussed in the SOROI results acquired using the VERTICAGE approach with far better image quality.

4. Discussion and Conclusion

The comparison of the SOROI and the VERTICAGE scan setups shows the advantage of the vertical approach especially for the regions with an unfavorable aspect ratio like the battery module. The highest demands on quality justify the large effort in car preparation, handling and scanning itself. However, the SOROI approach has still its advantages: besides the simple handling, shorter scan times at larger volume sizes can be achieved as long as too long penetration paths are avoided. For local area scans like the upper passenger cabin section and roof the SOROI approach is a reasonable choice. Further comparisons and analyses are aimed to be performed for combustion vehicles. A combination of VERTICAGE and SOROI scanning could help to further enhance the detectability, but the already high time effort for a single scan will not justify a second scan at the moment. Possibly, with new X-ray components and reconstruction methods, scan times can be shortened so that even two scans can be performed economically.
The VERTICAGE setup has proven to be highly beneficial for the investigation of cars motivating the future development at the Fraunhofer EZRT of a rotating Gantry system. The simplified handling and at the same time highest grade of image quality will open new markets and application fields for the revolutionary digitalization technology.

Acknowledgements

Co-financed by the free state of Bavaria

References

[1] N. Reims, A. Schulp, M. Böhnel, P. Larson, "An XXL-CT-scan of an XXL Tyrannosaurus rex skull", 19th World Conference on Non-Destructive Testing 2016

[2] Hassler U., Boehnel M., Errmann G., Osman A. (FHG), High energy CT scanning applied to large aerospace samples, Proceedings of the 3rd International EASN Association Workshop on Aerostructures, EASN, December 2013, http://www.easn.net

[3] G. A. Ciliberti, P. Janello, P. Jahnke, L. Keuthage, "Potentials of Full-Vehicle CT Scans within the Automotive Industry",Proceedings of the 19th World Conference on Non-Destructive Testing 2016

[4] N. Reims, T. Schoen, M. Boehnel, F. Sukowski, M. Firsching, „Strategies for Efficient Scanning and Reconstruction Methods on Very Large Objects with High Energy X-ray Computed Tomography" Proceedings, Conference on Industrial Computed Tomography (ICT) 2014.

[5] T. Fuchs, F. Sukowski, C. Schorr, T. Schön, S. Schröpfer, U. Haßler, T. Hofmann, N. Reims, M. Firsching, K. Tigkos, V. Heusinger, S. Moser, R. Langkemper, S. Nau, "High-Energy 3-D X-ray Tomography for Container Inspection", 8th Future Security 2013

[6] U. Hassler, N. Reims, M. Böhnel, T. Grulich, M. Rehak, M.Oettl, „Segmentation Approach for 'Full-Car'-CT Applications ", 3rd International Conference on Tomography of Materials and Structures (ICTMS) 2017

[7] M. Salamon, M. Boehnel, N. Reims, V. Voland, M. Schmitt, S. Kasperl, N. Uhlmann, R. Hanke, "XXL-CT, unique CT capabilities for industrial applications", 5th Conference on Industrial Computed Tomography, 2014

[8] Projektübersicht BMW Wendler Restauration, http://www.deutsches-museum.de/forschung/forschungsbereiche/sammlungen/restaurierungsforsch/bmw-328-wendler/

Application of a compact X-ray source with small focal spot using a 950 keV linear accelerator

Norihito MATSUNAGA [1], Tomoya SATO [1], Atsushi YAMADA [1], Masayuki ZAIKE [1], Toshinori TSURUYA [1], Tadashi HATANO [2], Masashi YAMAMOTO [2], Kazuaki SUZUKI [1]

[1] Nikon Corporation, Yokohama, Japan

[2] Accuthera Inc., Kawasaki, Japan

Contact e-mail: Norihito.Matsunaga@nikon.com

Abstract. A compact X-ray source with small focal spot using a 950 keV linear accelerator (compact 950 keV LINAC) has been developed. 9.3 GHz (X-band) radio frequency is used for accelerating electrons in order to make the size of X-ray source smaller compared with lower radio frequency such as 3.0 GHz (S-band). Size and weight of the developed X-ray source are W 600 mm × D 800 mm × H 500 mm and 230 kg, respectively.

A focal spot size of conventional X-ray source using LINAC has been about 1- 3 mm (FWHM). By activating a focusing mechanism for electron beam, small focal spot (0.24 mm in FWHM of Line Spread Function) is successfully achieved at the electron beam energy of 950 keV. Clearer transmission images using a rectangular groove sample are obtained with small spot setup compared to conventional one.

In addition, CT images of a turbine blade and an engine block are obtained using compact 950 keV LINAC with small spot setup. From this result, we can understand that our developed CT system using compact 950 keV LINAC is effective for industrial non-destructive inspection.

Therefore, we are convinced that this compact 950 keV LINAC is widely useful for industrial non-destructive inspection of large structures because of compactness, high transmission ability and spatial resolution.

Introduction

High energy X-ray CT system with X-ray source around maximum energy 1 - 10 MeV has been used for non-destructive inspection for large structures made of iron, aluminium or copper [1, 2]. In such systems, LINAC method using the acceleration tube is often used as the high energy X-ray source, because of high energy electron beam of about 1 - 10 MeV can be obtained stably without the risk of discharge inside the X-ray source. For example, it is used for non-destructive inspection of automobile parts and shape evaluation of combustion chambers of rocket engines [1, 2]. In such a situation, it is certain that high energy X-ray CT system with high spatial resolution is required in the near future. However, it was difficult to achieve a small X-ray spot or a small X-ray beam diameter less than 1 - 3 mm in a conventional X-ray source using LINAC method.

Recently, we have developed compact 950 keV LINAC with small X-ray focal spot by installing a focusing mechanism for electron beam, and experimental transmission images have been obtained for several industrial samples [3]. This time, by newly obtaining CT images for several industrial samples, we demonstrated that high transmission ability and

spatial resolution of our CT system using compact 950 keV LINAC with small focal spot of FWHM 0.24 mm. Therefore, we are convinced that this compact 950 keV LINAC with small focal spot is widely useful for industrial non-destructive inspection of large structures because of compactness, high transmission ability and spatial resolution.
This paper is structured as follows. In Sec 1, we explain the achieved specification of compact 950 keV LINAC and CT system. In Sec 2, we predict spatial resolution of our CT system, and In Sec 3, we demonstrate high spatial resolution by using a rectangular groove sample. In Sec 4, we show CT images of a turbine blade and an engine block, and explain about high spatial resolution and transmission ability. Finally, we summarize our result in Sec 5.

1. Specification of Compact 950 keV LINAC and CT System

Figure 1 (a) shows a photograph of the developed X-ray source. This X-ray source has two main features. First, by installing a focusing mechanism in X-ray source, this X-ray source is designed to have a smaller focal spot than a conventional one, in order to enable higher resolution imaging than a conventional LINAC. If the spot size is defined as FWHM of Line Spread Function of the X-ray focal spot distribution, this X-ray source can be narrowed to 0.24 mm compared to the switchable conventional size 2.06 mm. Target electron currents of small spot and conventional spot are about 40 mA and 130 mA at a pulse peak, respectively. Second, this source is designed to be compact because it is expected to be used in various ways. Size and weight are W 600 mm × D 800 mm × H 500 mm and 230 kg, respectively. The radio frequency used in the acceleration tube is 9.3 GHz (X-band) to make the system compact.

(a)

(b)

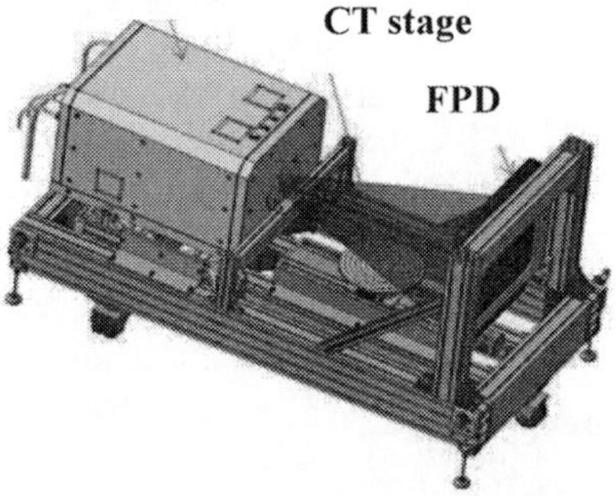

Fig.1 (a) Photograph of compact 950 keV LINAC. (b) CAD figure of CT system.

Figure 1 (b) shows a CAD figure of our CT system using compact 950 keV LINAC. This system has high precision shift and rotating stage for CT imaging, and high performance FPD (PerkinElmer: XRD1621CN3EHS. pixel pitch and total imaging area are 200 μm and 409.6 mm × 409.6 mm, respectively) is used for high energy imaging. X-ray spot and scintillator plane distance FFD is designed to be about 1030 mm to enable use for high resolution imaging of large structure.
Specification is summarized in Table 1. Note that small spot and conventional spot setup are swichable by activating a focusing mechanism of X-ray source. Since target electron current depends on X-ray spot size, this means that brightness of obtained X-ray image can be changed by changing the X-ray spot size.

Table 1. Specification of compact 950 keV LINAC and CT system

Item	Specification
Radio Frequency	9.3 GHz (X-band)
Maximum X-ray Energy	950 keV
X-ray Focal Spot Size (FWHM of Line Spread Function)	Small : 0.24 mm Conventional : 2.06 mm ※ Switchable
Target electron current	Small : 40 mA Conventional : 130 mA ※ Pulse peak
Size and weight of Compact 950 keV LINAC	W 600 mm × D 800 mm × H 500 mm 230 kg
FFD of CT system	1030 mm
FPD pixel pitch FPD imaging area	200 μm 409.6 mm × 409.6 mm

2. Prediction of Spatial Resolution

In the evaluation of the spatial resolution of CT system, it is important to understand the focal spot and the detector blur.

Figure 2 shows a schematic figure of the measurement configuration of the focal spot distribution and the detector blur distribution. In order to obtain two Line Spread Functions, we adopted the edge method using a tungsten cylinder (Diameter 35 mm, height 50 mm) for simplicity of alignment. In the case of a focal spot measurement, the projection magnification M of a tungsten cylinder is set at 6.64. On the other hand, projection magnification M of the detector blur case is set at 1.1 using CT stage.

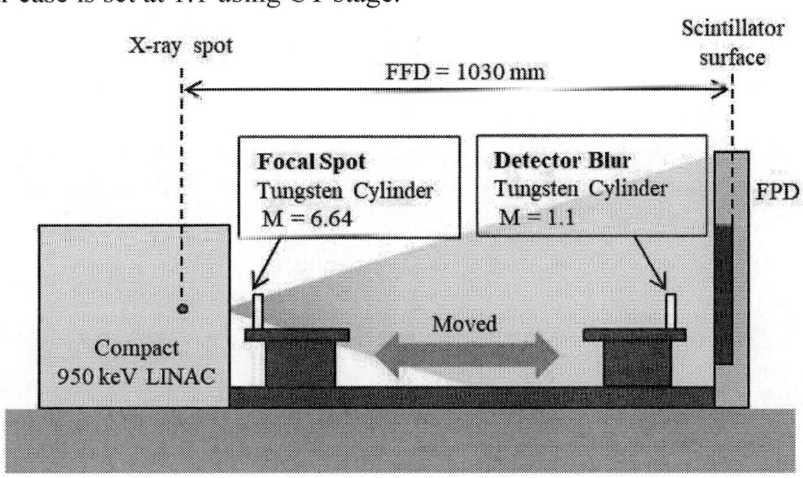

Fig. 2. Schematic figure of the measurement configuration of the focal spot and the detector blur, where M is the projection magnification of a Tungsten cylinder.

Figure 3 (a) shows measurement results of focal spot distributions, where the blue line corresponds to a focal spot distribution with conventional setup, and the red line corresponds to a focal spot distribution with small spot setup using a focusing mechanism. FWHM values of Line Spread Function are about 2.06 mm and 0.24 mm, respectively. The spatial resolution of X-ray imaging system depends on X-ray spot size [4], therefore, this result suggests that the spatial resolution of X-ray image can be improved about 8.6 times by activating a focusing mechanism of compact 950 keV LINAC.

On the other hand, Figure 3 (b) shows measurement results of detector blur distributions, where the blue line corresponds to a detector blur distribution with conventional setup, and the red line corresponds to a detector blur distribution with small spot setup. Estimated FWHM values are same about 0.5 mm.

(a) (b)

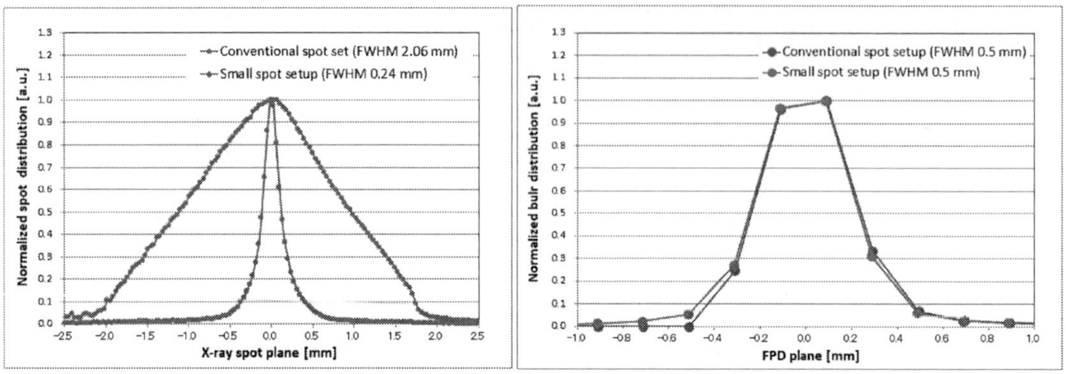

Fig. 3 (a) X-ray focal spot distributions. (b) Detector blur distributions.

By using results of Fig 3 (a) and (b), we can estimate the Geometric Unsharpness which is one of the indices of spatial resolution. The Geometric Unsharpness Ug is defined by

$$U_g = \frac{1}{M}\sqrt{((M-1)\Delta s)^2 + (\Delta x)^2}.$$

Where Δs and Δx are correspond to an X-ray focal spot size and a detector blur size, respectively. M is the projection magnification of a sample. Figure 4 shows estimation results of Ug, where the blue line corresponds to an Ug distribution with conventional setup, and the red line corresponds to an Ug distribution with small spot setup. From the result of the red line, we can understand that the Geometric Unspharpness Ug of small spot setup is between 0.2 mm (X-ray spot side) and 0.5 mm (FPD side), and spatial resolution of small spot setup is significantly improved compared to conventional one, where FOD is the distance between X-ray spot and sample position.

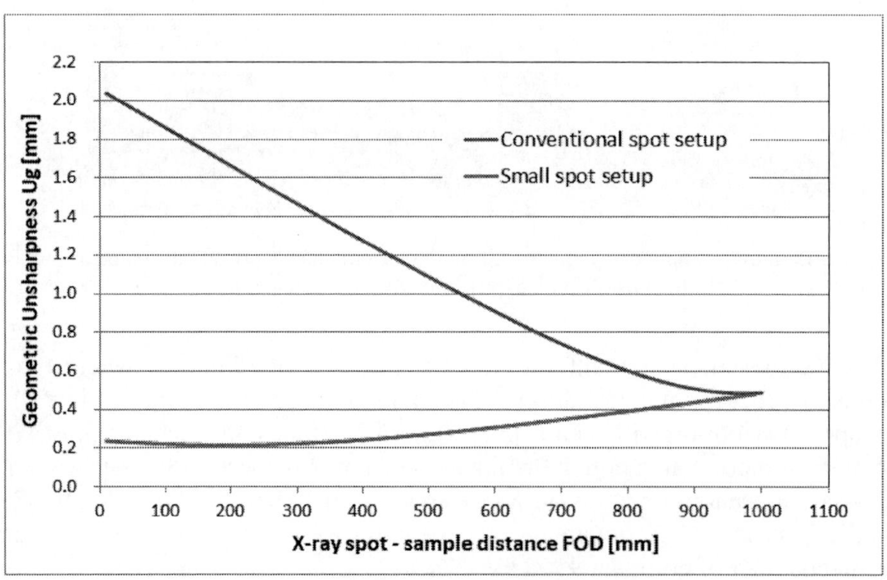

Fig.4. Estimation result of the Geometric Unsharpness Ug.

3. Demonstration of High Spatial Resolution using A Rectangular Groove

The sample for evaluation of spatial resolution is observed using compact 950 keV LINAC. Figure 5 shows images of a rectangular groove, where (a) a schematic figure, (b) a transmission image with conventional setup (M=6.72, FOD=153 mm), (c) a transmission image with small spot setup (M =6.72, FOD=153 mm), (d) horizontal plot of (b) and (c), and (e) is a magnified partial plot of (d) (see dashed green circle in (d)). By comparing the blue line and the red line in (d) and (e), we can say that spatial resolution with small spot setup is much improved compared to conventional one. In particular, size of two groove structures of left side in Figure 5 (a) is width 0.1 mm and width 0.2 mm, respectively. From this result, we can understand that groove structure of width 0.2 mm can be resolved with sufficient contrast (visibility is 12 %) by activating a focusing mechanism.

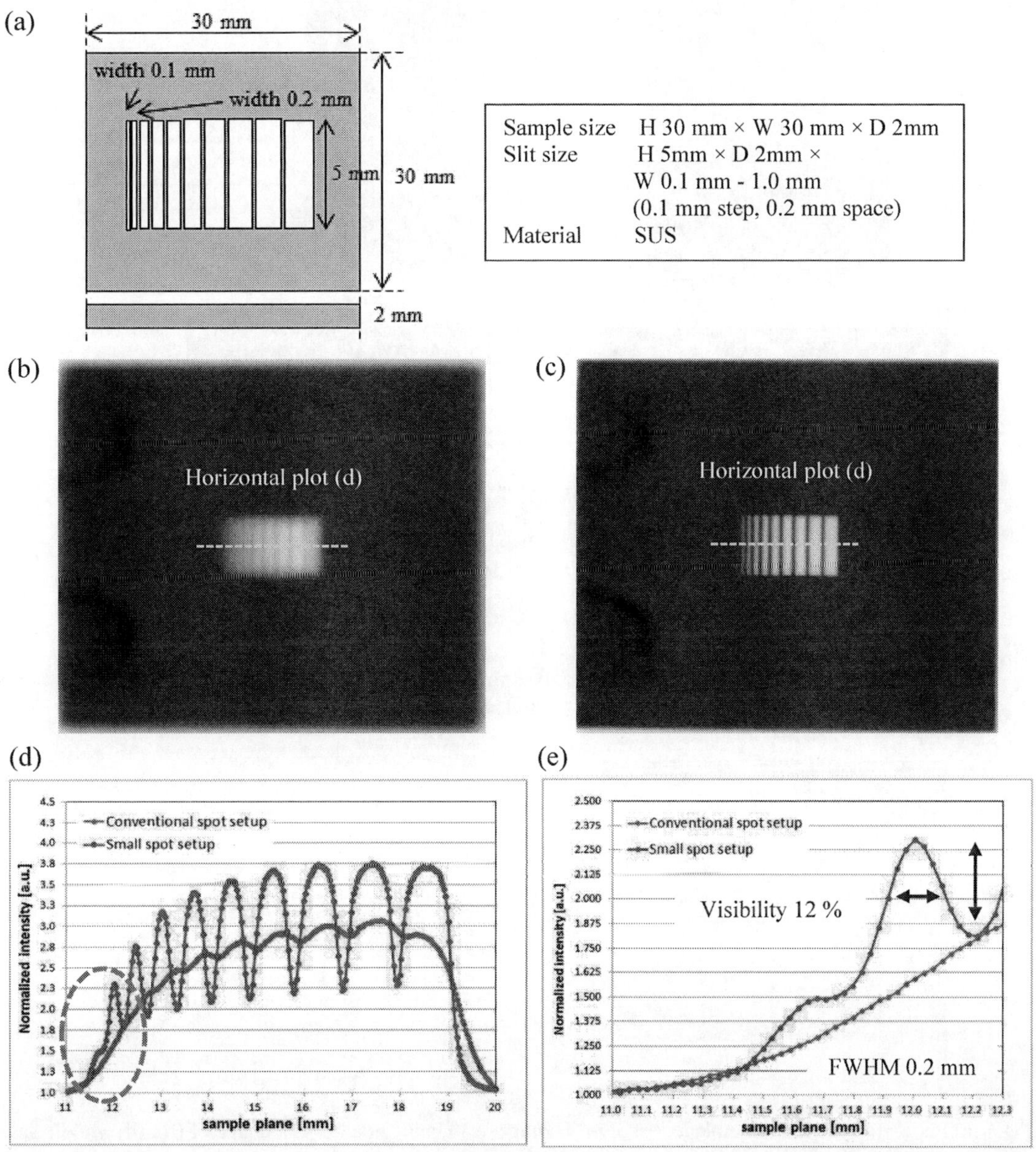

Fig. 5 (a) Schematic figure of a rectangular groove sample. (b) Transmission image with conventional setup. (c) Transmission image with small spot setup. (d) Horizontal plot of (b) and (c). (e) Magnified partial plot of (d).

4. CT Measurement for Industrial Samples

Several industrial samples are measured using compact 950 keV LINAC. Figure 6 shows CT images of a turbine blade, where (a) a photograph, (b) a 3D-CT image with small spot setup (M=3.08, FOD=334 mm, voxel size of X,Y and Z are same 0.065 mm), (c) a upper side 2D-CT image of (b), (d) a middle side 2D-CT image of (b),(e) a magnified partial 2D-CT image of (c), and (f) is a horizontal plot of (e). From the result of (c), (e) and (f), we can understand that hole structures of about FWHM 0.4 mm can be sufficiently resolved by activating a focusing mechanism. This spatial resolution is consistent with the evaluation result of Geometric Unsharpness Ug in Fig 4.

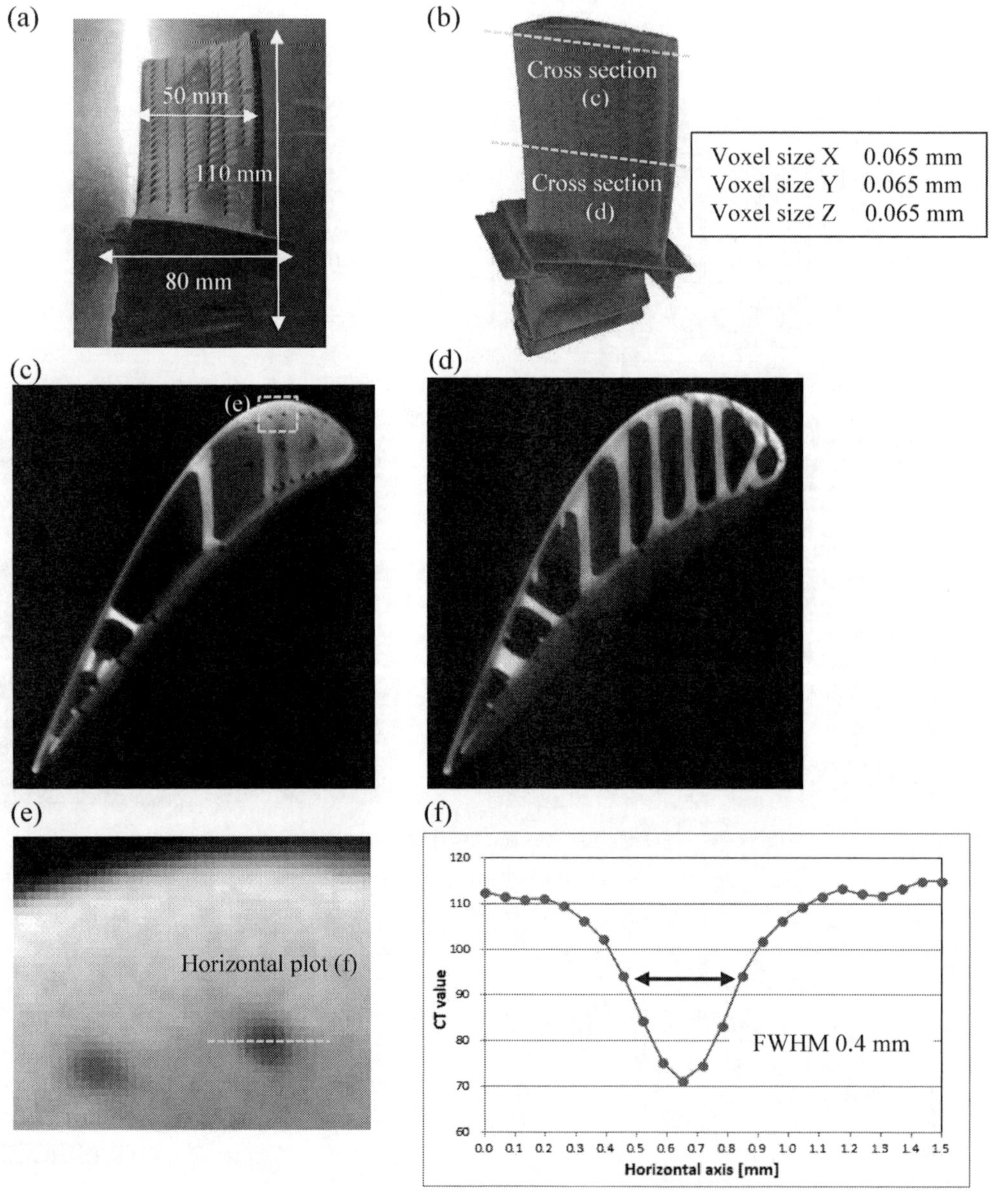

Fig.6 (a) Photograph of a turbine blade. (b) 3D-CT image. (c) Upper side 2D-CT image of (b). (d) Middle side 2D-CT image of (b). (e) Magnified 2D-CT image of (c). (f) Horizontal plot of (e).

Figure 7 shows CT images of an engine block, where (a) a photograph, (b) a 3D-CT image with small spot setup (M=1.31, FOD=786 mm, voxel size of X,Y and Z are same 0.153 mm), (c) a 2D-CT image of sliced cylinder part in (b), (d) a 2D-CT image of sliced thick part in (b),(e) a magnified partial 2D-CT image of (c), and (f) is a magnified partial 2D-CT image of (d). From results of (e) and (f), we can understand that internal structures of an engine block can be clearly confirmed. Especially, sizes of internal blowholes in (f) are roughly estimated about FWHM 2 mm - 3 mm, therefore this result suggests that our CT system using compact 950 keV LINAC is useful for industrial non-destructive inspection of large structures because of high transmission ability.

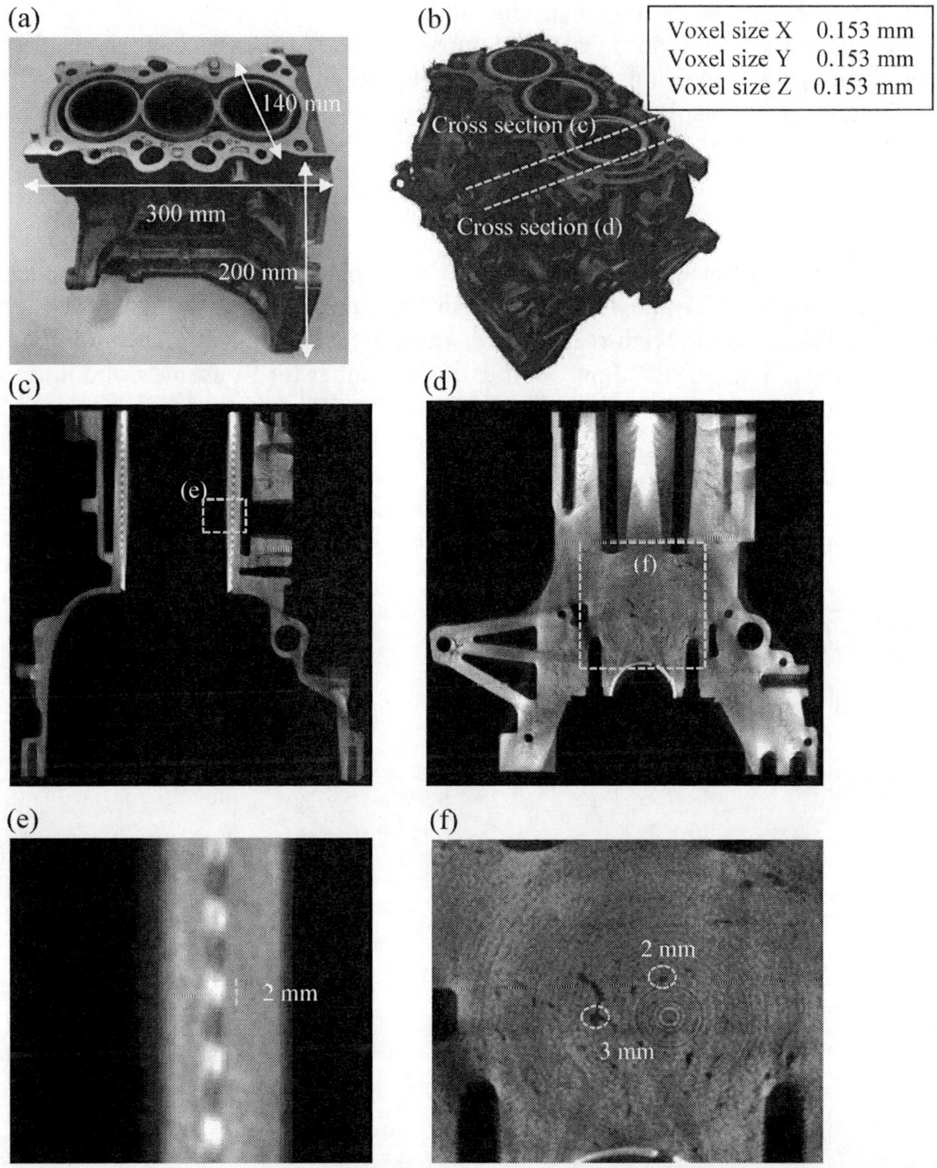

Fig.7 (a) Photograph of an engine block. (b) 3D-CT image. (c) 2D-CT image of sliced cylinder part in (b). (d) 2D-CT image of sliced thick part in (b). (e) Magnified partial 2D-CT image of (c). (f) Magnified partial 2D-CT image of (d).

5. Summary

A compact 950 keV LINAC with small focal spot of FWHM 0.2 mm and CT system had been developed. By obtaining the X-ray spot and the detector blur size, we estimated the Geometric Unsharpness Ug and demonstrated high spatial resolution with small spot setup using a rectangular groove sample. In addition, CT images of a turbine blade and an engine block were measured using compact 950 keV LINAC with small spot setup. In the case of a turbine blade, it is demonstrated that spatial resolution is consistent with estimated Ug in Sec 2, and in the case of an engine block, we obtained the suggestion that our CT system using compact 950 keV LINAC is effective for industrial non-destructive inspection of large structures because of high transmission ability.
Therefore, we are convinced that this compact 950 keV LINAC is widely useful for industrial non-destructive inspection of large structures because of compactness, high transmission ability and spatial resolution.

References

[1] M.Salamon, et al., High energy X-ray imaging for application in aerospace industry, 4th International Symposium on NDT in Aerospace 2012 -Tu.4.A.2.

[2] H.Takagi, et al., High-resolution X-ray CT Image Analysis Technology for the Deformation of the Main Combustion Chamber for Reusable Sounding Rocket Engines, Journal of JSNDI Vol.65, No.6, pp.277-286 (2016).

[3] N.Matsunaga, et al., Development of 950 kV X-ray source with small focal spot using a linear accelerator, 9th Conference on Industrial Computed Tomography (iCT2019).

[4] S.Carmignato, W.Dewulf and R.Leach Editors, Industrial X-ray Computed Tomography, 1st edition, Springer International Publishing AG, 2018.

Implementation of a beam hardening correction method for mono material parts using a linearization technique

Andreas Michael MÜLLER [1], Lorenz BUTZHAMMER [1], Tino HAUSOTTE [1]
[1] Institute of Manufacturing Metrology (FMT), Friedrich-Alexander-Universität Erlangen-Nürnberg (FAU), Erlangen, Germany
Contact e-mail: andreas.mueller@fmt.fau.de

Abstract. Common X-ray sources for industrial computed tomography (CT) exhibit a polychromatic spectrum. While the absorption of radiation by the measurement object is dependent on the X-ray energy, the commonly used reconstruction algorithms assume an absorption coefficient under monochromatic conditions. In the context of industrial CT, the material of a measurement object is oftentimes known, which makes the application of a linearizing beam hardening correction method possible for parts made out of a single material. In this work, we present an efficient setup and data processing pipeline to determine and manage experimental X-ray absorption models for mono material parts. This pipeline estimates the correlation between attenuation and penetration length using a cone shaped specimen and subsequently corrects the grey values of the projections of a measurement object of the same material as the specimen. In contrast to commonly used step wedges, the special specimen makes it possible to sample the absorption curve with a large number of grid points, while using a standard measurement operation mode for the CT system. The correction method makes it possible to measure the tooth flanks of a steel gear wheel, which could otherwise not be measured due to beam hardening artefacts.

Introduction – Beam Hardening and Correction Concepts

The physical interaction of electromagnetic radiation with matter causes the radiation to be attenuated, which is described by the Lambert-Beer law in Eq. (1).

$$I = \int_0^{E_{max}} I_0(E) \cdot e^{-\int_0^l \mu(E,\rho,Z,x)dx} dE \tag{1}$$

The intensity I_0 of the photons is attenuated along their path through all the volume elements of a penetrated object with thickness l dependent on the attenuation coefficient $\mu(E, \rho, Z, x)$, which is dependent on the energy of radiation E as well as on the density ρ and the atomic number Z of the volume elements. X-ray computed tomography (CT) in general assumes a straight-line propagation and an energy-independent absorption coefficient when applying standard reconstruction techniques like the commonly used FDK algorithm [1] for cone-beam CT. The generation of monochromatic X-rays with sufficiently large intensity to be used in industrial computed tomography is currently not possible (except for rare and expensive synchrotron accelerators). Consequently, X-ray tubes are almost exclusively used to generate the radiation required for the imaging by accelerating electrons onto a target. This process generates a wide distribution of polychromatic photons, which are then used in the CT imaging process. The energy dependent attenuation of X-rays during penetration of matter causes an increase of the average photon energy, which is called "beam hardening". X-rays from an identical source passing the same location within a measurement object from different directions consequently show different energy distributions, which leads to different intensity deviations from an ideal monochromatic absorption curve for the same object location. This causes the occurrence of beam hardening artefacts, which can prevent the correct determination of the surface of the measurement object by examination of the volume data [2].

One commonly used method to suppress the effects of beam hardening is the pre-filtration of all generated X-rays. This causes a smaller energy distribution of used photons during the measurement process at the cost of lower X-ray intensities and thus a worse signal-to-noise ratio of the measurements. In the last decades, a lot of different correction methods have been proposed and implemented by the scientific community, X-ray device manufacturers and software companies. Exemplary classifications can be found in [3–5]. Those methods range from dual-energy imaging [6], over volume data post-processing and projection data adjustments and can vary greatly regarding the available and used information of the measurement setup and can oftentimes be limited to certain use cases [2]. An overview over methods to compensate metal artefacts in general can be found in [7, 8].

This contribution has implemented a version of the commonly known linearization technique, which operates on the assumption that for monochromatic radiation, a linear relation exists in the sense of Eq. (2) between X-ray penetration lengths and observed attenuated intensities, if the measurement object exhibits a homogeneous composition. The concept of the linearization describes the approach to transform a measured polychromatic projection value from its nonlinear absorption curve onto the corresponding mono-energetic projection value representing the same penetration length [2, 9].

$$-\ln\left(\frac{I}{I_0}\right) = \mu x \qquad (2)$$

The focus was put on presenting a robust, fast and easily adjustable method to repeatedly perform a pre-defined inspection task of steel gear wheels with known material composition. It will be shown, that the measurement task can be fulfilled by the implementation of the beam hardening correction (BHC) routine into the measurement chain and thus expand the fields of applications of the used CT system to steel parts with minimum additional effort.

1. Measurement Task and Used Measurement Setup

The measurement task was defined as the dimensional characterization of a wire-eroded, straight-toothed steel gear wheel (tooth tip diameter 19 mm, thickness 8 mm, material 16MnCr5) using the industrial cone beam CT Zeiss Metrotom 1500. Goal of the examination was the evaluation of the gear tooth flanks. Because of that, the gear wheel was manually positioned within the measurement volume such that the rotation axis of the gear wheel matched the rotation axis of the rotatory stage. This orientation was chosen to ensure the same measurement uncertainty characteristics for each tooth flank. Nonetheless, this resulted in larger average X-ray penetration lengths compared to an upright orientation of the gear wheel and this also meant, that there were no projections showing the contour of the tooth system. Consequently, the criterion for success was defined as being able to evaluate the geometric properties of all tooth flanks. The measurement settings were chosen as follows: tube voltage 225 kV, tube current 200 μA, nominal X-ray spot size 45 μm, detector gain 16x, detector integration time 1000 ms, geometric magnification approx. 19.6, voxel size approx. 10 µm, number of projections 2000. A projection consists of 2038 x 2046 pixels with a pitch of 200 µm and value range of 16 bit. All subsequently mentioned data processing steps were performed using the corrected projections, which also included a flat field correction (Zeiss Metrotom OS). No additional pre-filtration (besides a filter of 0.6 mm aluminium built into the X-ray tube) was used. The built-in ring artefact reduction method was disabled in order to simplify the following data processing pipeline. The tomographic reconstruction was done using the filtered (Shepp-Logan filter kernel [10]) back projection algorithm provided by Siemens CERA XPlorer 3.0.3 [11]. It was not possible to determine the surface of the measurement object in the areas between the gear teeth from the reconstructed volume using VGStudio Max 3.2.5 (VGS), despite experimenting with numerous different algorithm parameters[1]. We also unsuccessfully tried filtering the volume data before the surface determination with Gaussian filters of various cubic kernel sizes using VGS, as well as filtering each projection before reconstruction with a Gaussian filter (sigma = 5 pixel) using MATLAB. Fig. 1 shows a volume data slice of the reconstructed volume. It is clearly observable, that the areas between neighbouring teeth exhibit incorrect grey values, which is caused by beam hardening as well as X-ray scatter artefacts. The measurement task was set to quantitatively evaluate the geometry of the gear teeth of many gears of the same kind in order to determine impacts caused by wear and tear. Because of that, the subsequently described data processing pipeline was set up in order to correct the beam hardening artefacts, which prevented a robust and stable surface determination of these measurements.

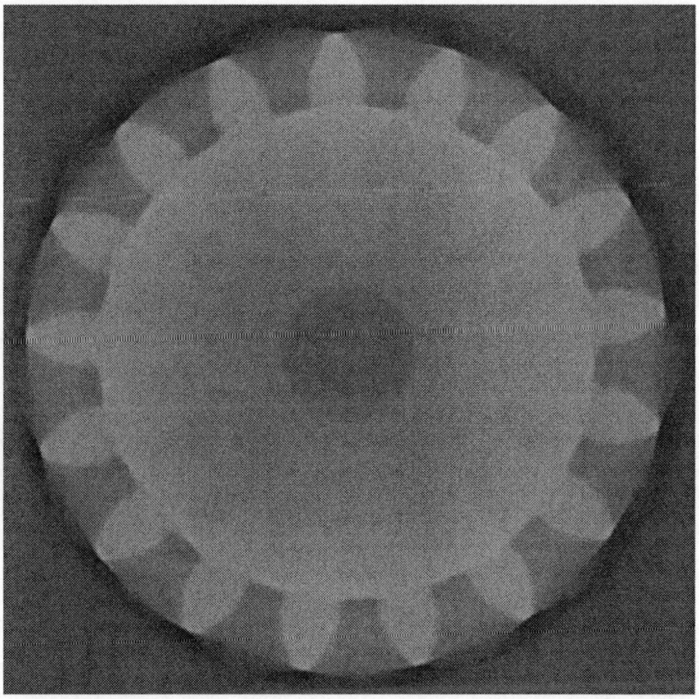

Fig. 1: Visualization of a single volume data slice (z = 630 voxel) of the gear measurement. The beam hardening and X-ray scatter artefacts visible between the gear teeth made it impossible to determine the object surface. The colour maps contain the 16 bit value range resulting after reconstruction.

[1] The presented results in this article are not to be understood as an evaluation of the software performance. Therefore, no detailed descriptions regarding the exact settings applied for the surface determination are given.

2. Data Processing Pipeline for Beam Hardening Correction

In the following chapter, the complete data processing pipeline is described. The main focus was put on implementing a robust, fast and flexible beam hardening correction method for a given CT system and work pieces of known composition. The complete approach is characterized by the following steps:

- I. Design and manufacturing of a simple, cone-shaped specimen with homogeneous absorption properties, preferably by using the same material as the measurement object of interest (gear wheel).
- II. Performing a complete CT scan using the same CT settings as for the gear wheel measurement.
- III. Determining the position and orientation of the measured cone from the determined surface using geometry element regression analysis in VGS.
- IV. Extracting the absorption curve by determination of the corresponding cone penetration lengths for each detector pixel in each captured projection.
- V. Correction of the projection grey values of the gear wheel measurement using the absorption curve and the associated correction values.
- VI. Reconstruction of the corrected projection stack and evaluation of the result.

2.1. Cone shaped Specimen

Usually, a single projection (radiography) of a calibrated normal is used in order to determine the correlation between X-ray penetration length and radiation attenuation [2]. In general, this approach can have several downsides, including:

- Usage of only discrete X-ray penetration lengths,
- Disregarding of the cone beam geometry, which can lead to slightly incorrect X-ray penetration lengths,
- Limited statistical robustness of the measured correlation due to the evaluation of only a small number of observations,
- Necessity of a separate CT system operation program in order to record single projections with adequate image corrections (e.g. flat field correction). These corrections are usually proprietary, meaning corrections equal to those performed within the measurement software cannot be performed.

In order to remedy these possible downsides, a cone-shaped specimen (Fig. 2) was manufactured, which can directly be mounted onto the perforated plate of the rotatory stage. Consequently, the rotation axis of the specimen is then aligned parallel with the rotation axis of CT rotatory stage.

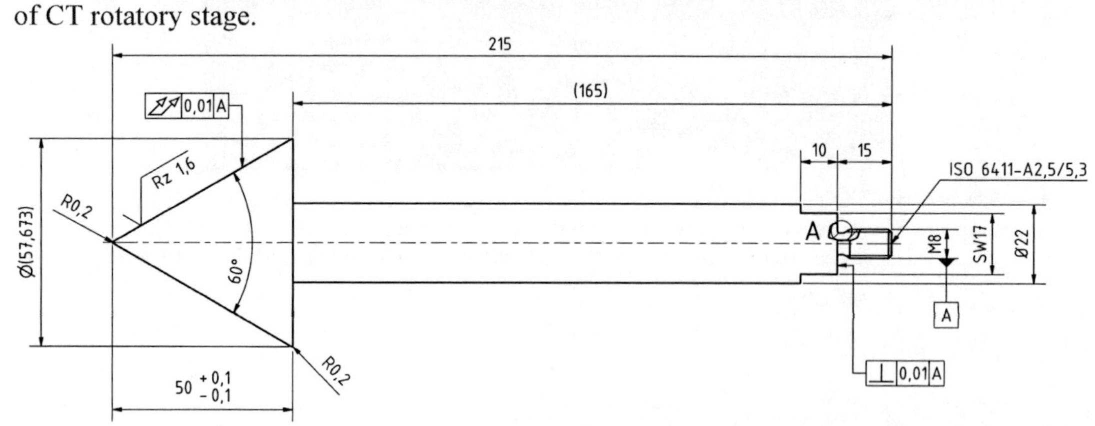

Fig. 2: Technical drawing of the used cone-shaped specimen, which was then mounted in a standing position onto the rotatory stage of the CT, resulting in a parallel alignment of both rotatory axes.

The rotation symmetry of the specimen makes it possible to correctly determine the cone shell surface using locally adaptive CT surface determination routines after the reconstruction, because no object edges are hidden in any projections. This condition is also not affected by the material of the specimen. A correct reconstruction of the cone shell surface would still be possible in case of nearly complete radiation attenuation at the detector. The penetration lengths are not limited to discrete values by the specimen (between the cone tip radius and the maximum cone diameter). This specimen can then be positioned and measured using the same CT settings as for the gear measurement. The evaluation of a large number of projections can then result in a statistically very robust determination of the absorption curve (see subsequent chapter). Additionally, turned parts can be manufactured very accurately and cost efficiently, which makes it feasible to produce additional parts with different material compositions if needed.

For the implementation of the linearization technique it is crucial to use materials with matching absorption properties for both the measurement object and the specimen. In this contribution, we chose the material for the specimen as X5CrNi18-10, because of its availability and its corrosion properties. Despite the different chemical compositions of both alloys (Table 1), the resulting attenuation coefficients are almost identical (Fig. 3).

Table 1: Chemical composition in mass percent of the alloys of the gear wheel (16MnCr5 [12]) and the specimen (X5CrNi18-10 [13]). The remaining part of the alloys consist of iron (Fe).

mass percent p^M	C	Si	Mn	S	Cr	Ni	ρ_{Alloy} in g/cm³
16MnCr5	0.16	0.2	1.1	0.035	0.9	-	7.81
X5CrNi18-10	0.07	1.0	2.0	0.03	18.0	9.5	7.90

The underlying calculations shown in Eq. (3) [14] determine the absorption coefficient $\mu(E)$ for a certain energy E by adding up the weighted mass attenuation coefficients $\mu(E)\,/\,\rho$ (from the database of the national Institute of Standards and Technology NIST [15]) for each of the n elements of the alloy chemical composition in mass percent p^M (Table 1). Linear interpolation was used between the discrete sampling points of the NIST database.

$$\mu(E) = \rho_{alloy} \cdot \sum_{i=1}^{n} \left(\frac{\mu(E)}{\rho}\right)_i \cdot p_i^M \tag{3}$$

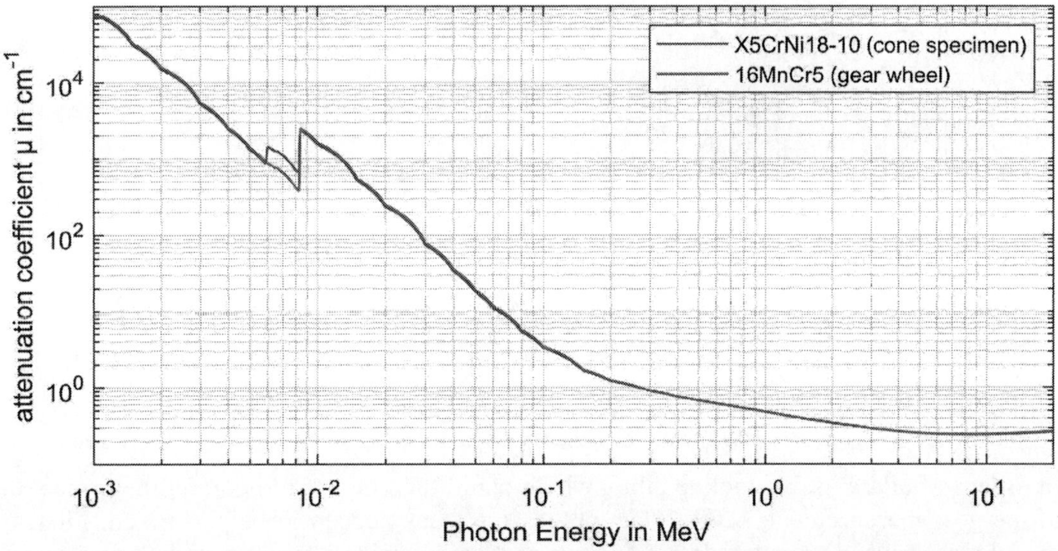

Fig. 3: Calculated absorption coefficients of the gear (16MnCr5) and the specimen (X5CrNi18-10).

2.2. Extraction of the Absorption Curve and Generation of a Correction Lookup-Table

To obtain the reference data, the object was mounted onto the rotatory stage of the CT and a full scan was performed. The settings regarding the X-ray tube and the detector were chosen to match those of the gear measurement. After the tomographic reconstruction of the measurement consisting of 1500 projections, a geometric regression analysis with a cone model ("cone fit") was performed onto the segmented volume data. Both steps (surface determination and cone fit) were carried out using VGS. Now, a full 3-D model of the CT scan can be set up in an arbitrary coordinate system (we used the Siemens CERA 3.0.3 world coordinate system (WCS), as described in [11]). The cone model can then be placed into the scene by using the geometric parameters of the regression analysis (Fig. 4).

The generation of the 3-D scene makes it possible to apply a ray tracing routine in order to determine the cone penetration lengths associated with each detector pixel. For that, the connection vectors for each pixel to the X-ray source were tested against the analytical cone model, determined by the VGS fit, using the intersection test described in [16]. The usage of the analytical description of the cone was used mainly for performance reasons, as this approach was significantly less computationally demanding than using the triangulated model resulting from the surface determination. This can be justified, because the turned part was manufactured very accurately. The correlation of the value of each pixel and the calculated penetration lengths then formed the desired relationship between intensity attenuation and penetration length. This approach was then repeated for each of the 1500 projections, thus laying the foundation for a statistically very robust absorption model. At this point, the absorption model consisted of more than 10^9 grid points.

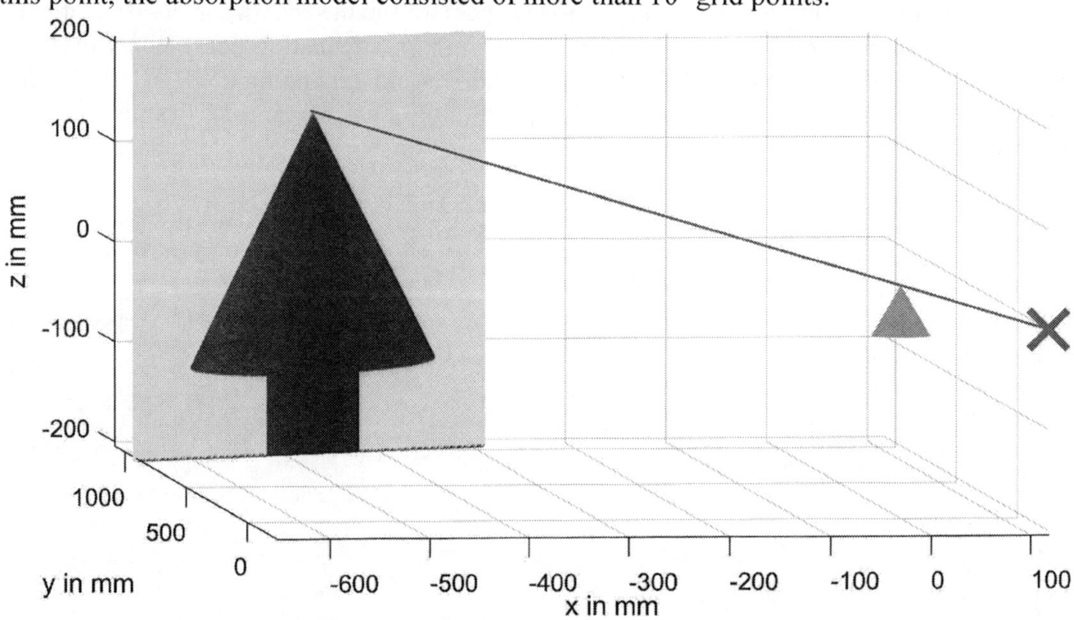

Fig. 4: Geometric setup for the intersection setup with X-ray source (red cross), the measured cone geometry (green cone) and the corresponding projection (RGB colour map for better visualization). The red line connects source with the cone tip up to the detector, the figure is true to scale.

In order to build a usable lookup table, which could then be used to correct the projections of the gear measurement, several data processing steps needed to be performed. First, the model was reduced to data points, which were located inside a target cone half angle of 1°. This excludes the faulty influence of the cone curvature onto the calculated grid points, which increases rapidly with decreasing intersection angle between the test ray and the cone shell surface. Secondly, all model grid points, which exhibited a cone penetration length of smaller than 250 µm were excluded from the model. This was done because of the limited detector pitch and because the model showed degenerative behaviour for penetration lengths below that point. These two measures reduced the number of model points to $3.1 \cdot 10^7$. Lastly, the model points needed to be binned in order to acquire unique x/y grid points to increase the subsequent regression analysis. The absorption values are represented by discrete values in the 16 bit value range, but for each of those values there possibly existed more than one penetration length. Thus, a unique penetration length for each discrete absorption value was calculated by the mean value of those distances within less than three scaled median absolute deviations from the median of those distances [17].

Following the data binning, the logarithms $-\log I/I_0$ of the grey values were taken ($I_0 = 2^{16} - 1$) and the absorption model was finalized by piecewise defined polynomial regression analyses. This was mainly done because of the requirement that the model must include the data point (0/0), which represents the maximum intensity on the logarithmic scale at zero X-ray penetration length, as shown in Fig. 5. Because of that condition on the solution of polynomial regression analysis, the solution of a constrained linear least-squares problem in the form of Eq. (4) was necessary [18].

$$\min_{x} \frac{1}{2} \|C \cdot x - d\|_2^2 \; such \; that \; A_{eq} \cdot x = b_{eq} \tag{4}$$

Here, C represents the Vandermonde matrix, up the desired degree n of the polynomial regression model in Eq. (5). The variables x and d represent the value pairs of the binned absorption curve. The linear equality constraint of (0/0) is realized by the definition of A_{eq} and b_{eq}, as shown exemplarily in Eq. (5). Here, the superscript index 1 stands for model part 1, which requires the equality constraint (0/0).

$$\begin{gathered} C(x_1, x_2, \dots, x_n) = \begin{pmatrix} x_1^{n-1} & \cdots & x_1^1 & x_1^0 \\ x_2^{n-1} & \cdots & x_2^1 & x_2^0 \\ \vdots & \ddots & \vdots & \vdots \\ x_n^{n-1} & \cdots & x_n^1 & x_n^0 \end{pmatrix} \\ A_{eq}^1 = (0^{n-1} \quad \cdots \quad 0^1 \quad 0^0), \, b_{eq}^1 = 0 \end{gathered} \tag{5}$$

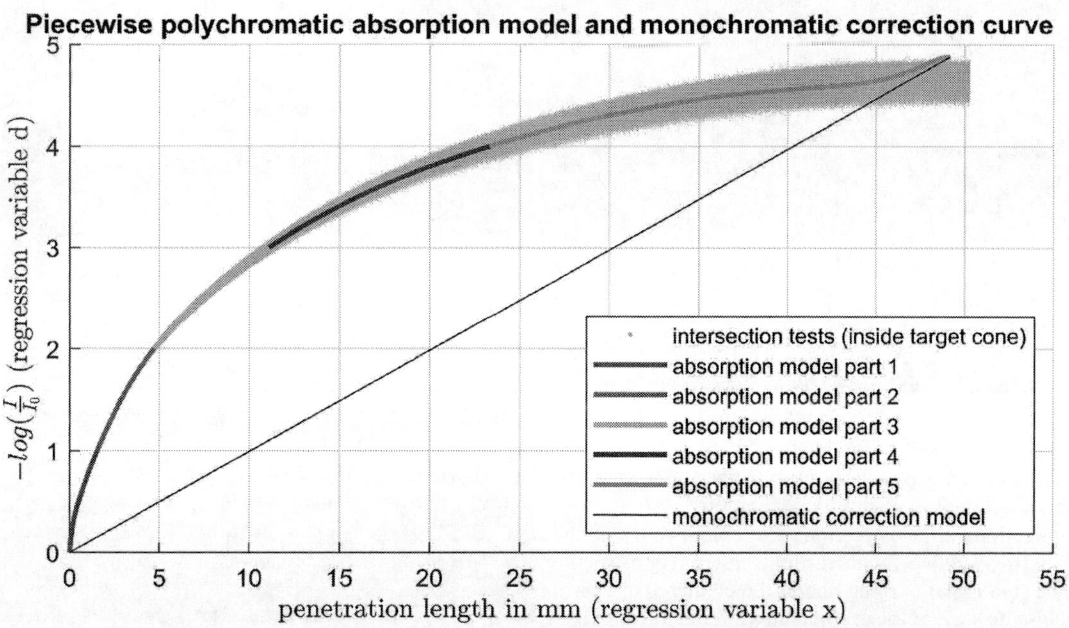

Fig. 5: Absorption model represented by piecewise defined polynomial functions with C^0 continuity based on the binned intersection test data satisfying the target cone half angle condition of 1°and a minimum penetration length of 250 µm. Material X5CrNi18-10; X-ray tube settings 225 kV, 200 µA.

For numerical robustness, the regression was performed piecewise, with subsequent polynomials satisfying C^0 continuity. Thus, the constraint $A_{eq} \cdot x = b_{eq}$ for the following curve satisfies the last point of the previous curve. Fig. 5 shows the final regression results. Finally, all required features to create a correction lookup table were prepared. The correction for a projection value is dependent on the definition of a monochromatic correction curve (Fig. 5). That means, that the corrected value of a pixel value is determined by the value of the monochromatic correction model for the same penetration length, which is defined by the absorption model. After the correction of all projections, the tomographic reconstruction was performed again, using Siemens CERA XPlorer 3.0.3 and the same settings as for the previous reconstruction.

All data processing steps but the correction of the projection itself, can be pre-computed, which reduces the actual required data processing steps for the correction of any measurement to the highly performant application of lookup table operations. The operations required for the application of a 16 bit lookup table are basically only limited to the I/O speed of the used data connection.

3. Results

The correction of the grey values resulted in a visible reduction of beam hardening artefacts (Fig. 6, left), in comparison to the uncorrected data (Fig. 1). Nonetheless, the image exhibits a high noise level, which prevented a successful surface determination. Because of that, the projections were subject to an additional Gaussian filtering (MATLAB, sigma = 5 pixel) after the application of the BHC. This operation resulted in an increased contrast in the volume data between the measurement object and the background (Fig. 6, right).

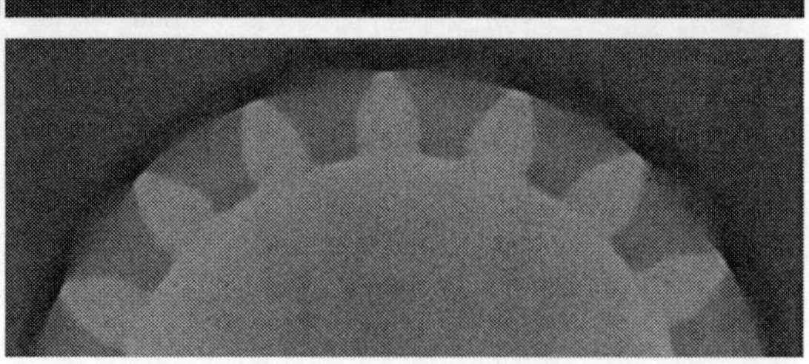

Fig. 6: Visualization of a volume slice (z = 630 voxel) of the corrected measurement utilizing the proposed method (left). Correction result (right, same slice) with an additional gauss filter (sigma = 5 pixel) of the projections after correction. The colour maps contain the 16 bit value range resulting after reconstruction. The small image (right) shows the unaltered top part of the same slice of the original measurement from Fig. 1.

The surface determination of the reconstructed volume of the post-filtrated projections (Fig. 6, right) lead to satisfying results when creating a nominal-actual comparison against the CAD model of the gear wheel (Fig. 7). The absolute deviations in the middle part of the tooth flanks do not exceed 15 µm. Note that due to the lack of a reference measurement, it is impossible to distinguish between geometric work piece deviations and measurement errors. Similar results, which means slightly noisier but also more accurate with respect to the CAD model, could be achieved by filtering of the volume data (according to Fig. 6, left; filtering result not shown) using a Gaussian filter of size 15 in VGS, followed by the surface determination and nominal-actual comparison (Fig. 8).

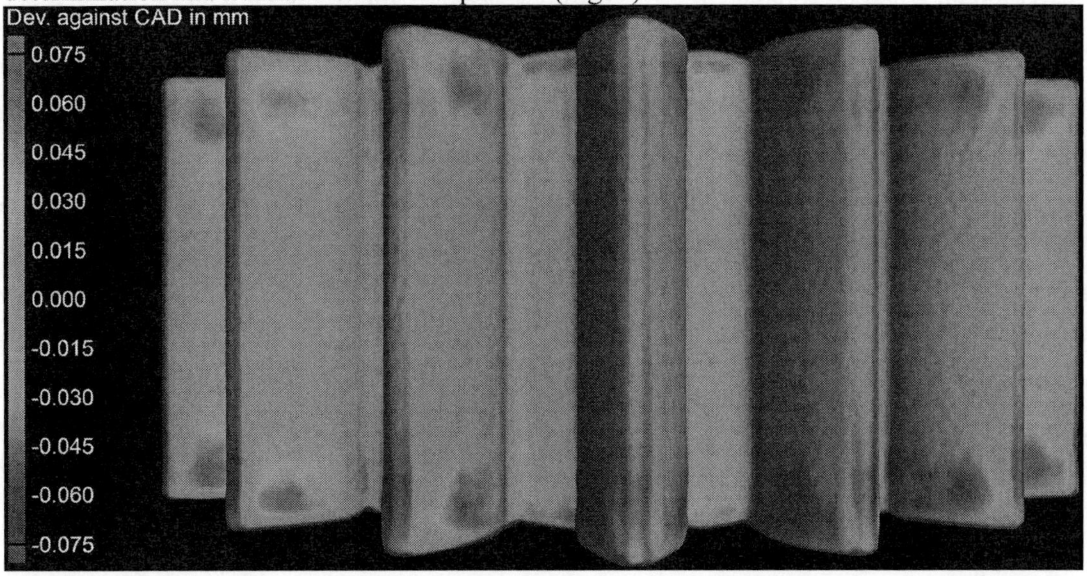

Fig. 7: Nominal-actual comparison of the reconstructed volume using the post-filtered projections (Fig. 6, right) against the nominal CAD model.

Fig. 8: Nominal-actual comparison of the reconstructed volume against the nominal CAD model using the BHC corrected projections (Fig. 6, left), followed by a volume data Gaussian filter (size = 15).

4. Discussion

We could show, that it was possible to significantly improve the CT volume data quality by estimating the faulty influence of X-ray beam hardening onto the measurement result. After the correction, surface determination was possible and a nominal-actual comparison with the CAD model could be calculated. This was not possible without the correction method. The investigation showed, that the problem of beam hardening is closely related to a low signal-to-noise ratio (SNR) of the reconstructed volume data, which can also impair a successful surface determination. In this contribution, we did not take into account the effect of X-ray scattering, which has a comparatively high influence for scenarios with low SNR. We are aware of the fact, that our calculated absorption model (Fig. 5) is also corrupted by X-ray scatter effects, which are inherently different for the gear wheel measurement. Nonetheless, this effect on the volume data could successfully be reduced by the introduction of an additional Gaussian filter onto the corrected projections (Fig. 7) and the volume data (Fig. 8), which lead to satisfying results.

5. Summary and outlook

The implementation of the beam hardening correction method discussed in this paper focused on a robust and easy to implement procedure, which can be utilized for any industrial CT. The determination of the absorption curve for a specific material was carried out exploiting a standard CT measurement of a cone-shaped specimen, which does not require any additional operating states of the CT system. From that point on, the data processing pipeline is straightforward and easy to implement for the specific use case. The quality of the results can further be increased by Gaussian filtering of the corrected projections or the reconstructed volume.

Future examinations will focus on examining the portability of the procedure on other measurement tasks, as well as on other materials, such as aluminium. Additionally, the validity of the introduced filtering operations regarding the measurement uncertainty must be proven, in order to be used by default within the data processing pipeline.

Acknowledgments

The authors would like to thank the German Research Foundation (DFG) for supporting the research project "FOR 2271 process-oriented tolerance management based on virtual computer-aided engineering tools" under grant number HA 5915/9-1 and for the financial support of the acquisition of the CT system "Zeiss Metrotom 1500" through Grant No. 324672600. The authors thank the Institute of Manufacturing Technology (LFT, FAU Erlangen-Nürnberg) for providing the steel gear wheel used for demonstration purposes in this article.

References

[1] L. A. Feldkamp, L. C. Davis, and J. W. Kress, "Practical cone-beam algorithm," *Journal of the Optical Society of America A: Optics and Image Science, and Vision*, vol. 1, no. 6, pp. 612–619, 1984.

[2] S. Kasperl, "Qualitätsverbesserungen durch referenzfreie Artefaktreduzierung und Oberflächennormierung in der industriellen 3D-Computertomographie," Dissertation, Friedrich-Alexander-Universität Erlangen-Nürnberg (FAU), 2005.

[3] F. E. Boas and D. Fleischmann, "CT artifacts: causes and reduction techniques," *Imaging in Medicine*, vol. 4, no. 2, pp. 229–240, 2012.
[4] M. Kachelrieß, "Reduktion von Metallartefakten in der Röntgen-Computer-Tomographie," Dissertation, Friedrich-Alexander-Universität Erlangen-Nürnberg (FAU), Erlangen, 1998.
[5] A. C. Kak and M. Slaney, *Principles of computerized tomographic imaging*. New York: IEEE Press, 1988.
[6] R. E. Alvarez and A. Macovski, "Energy-selective reconstructions in X-ray computerised tomography," *Phys. Med. Biol.*, vol. 21, no. 5, pp. 733–744, 1976.
[7] L. Gjesteby *et al.*, "Metal Artifact Reduction in CT: Where Are We After Four Decades?," *IEEE Access*, vol. 4, pp. 5826–5849, 2016.
[8] M. Katsura, J. Sato, M. Akahane, A. Kunimatsu, and O. Abe, "Current and Novel Techniques for Metal Artifact Reduction at CT: Practical Guide for Radiologists," (eng), *Radiographics : a review publication of the Radiological Society of North America, Inc*, vol. 38, no. 2, pp. 450–461, 2018.
[9] G. T. Herman, "Correction for beam hardening in computed tomography," *Phys. Med. Biol.*, vol. 24, no. 1, pp. 81–106, 1979.
[10] L. A. Shepp and B. F. Logan, "The Fourier reconstruction of a head section," *IEEE Trans. Nucl. Sci.*, vol. 21, no. 3, pp. 21–43, 1974.
[11] Siemens Healthcare GmbH, *CERA: User Manual (Part 1 – CERA API)*. Version 3.0, 2015.
[12] STAUBER GmbH Metalltechnologie, *1.7131 - Werkstoff 1.7131 Datenblatt.* [Online] Available: https://www.stauberstahl.com/werkstoffe/17131-werkstoff-datenblatt/. Accessed on: May 31 2019.
[13] STAUBER GmbH Metalltechnologie, *1.4301 - Werkstoff 1.4301 Datenblatt.* [Online] Available: https://www.stauberstahl.com/werkstoffe/14301-werkstoff-datenblatt/. Accessed on: May 31 2019.
[14] C. Niggemann, *Wissensbasiertes System zur Verbesserung dimensioneller Messungen mittels Röntgen-Computertomografie,* 1st ed. Aachen: Apprimus-Verl., 2012.
[15] NIST - International Institute of Standards and Technology - U.S. Department of Commerce, *X-Ray Mass Attenuation Coefficients.* [Online] Available: https://physics.nist.gov/PhysRefData/XrayMassCoef/tab3.html. Accessed on: May 31 2019.
[16] D. Eberly, *Geometric Tools: Intersection of a Line and a Cone.* [Online] Available: https://www.geometrictools.com/Documentation/IntersectionLineCone.pdf. Accessed on: May 31 2019.
[17] The MathWorks, Inc., *rmoutliers: Detect and remove outliers in data.* [Online] Available: https://www.mathworks.com/help/matlab/ref/rmoutliers.html. Accessed on: May 31 2019.
[18] The MathWorks, Inc., *lsqlin: Solve constrained linear least-squares problems.* [Online] Available: https://www.mathworks.com/help/optim/ug/lsqlin.html. Accessed on: May 31 2019.

Optimization of Image Quality and Throughput at CT Scans of Large Parts

Oliver BRUNKE [1], Eberhard NEUSER [1], Alexander SUPPES [1]
[1] GE Sensing & Inspection Technologies GmbH, Wunstorf, Germany

Contact e-mail: alexander.suppes@bhge.com

Abstract

For production process control of complex and expensive automotive, aerospace or 3D printed components, Computed Tomography (CT) becomes the technology of choice for many inspection and metrology tasks. A major challenge is the increased cycle time requirement with high inspection depth.

In our paper we will present improved methods to deal with imaging artefacts to bring CT to the production floor, including advanced scatter correction technique, next generation high dynamic X-ray detectors and x-offset CT.

Scattering of X-rays is main factor for such artefacts in CT. While state of the art scatter reduction just simulates scatter, the new method is really measuring the scatter portion of that specific sample and minimizes it for every individual voxel. We will show comparison results between up to 100 times slower fan beam CT as well as conventional, scatter radiation impacted cone beam CT and advanced scatter reduced cone beam CT. The new method will not only substitute fan beam minifocus CT in many inspection and metrology applications: as case studies prove, 300 kV microCT scans can now be applied for CT tasks which normally require investment in more expensive 450 kV high energy CT.

A key factor for excellent CT results is detector performance. We will show the next industrial 16” X-ray detector generation with up to 10x improved efficiency and sensitivity compared to state of the art 200 µm pixel DXR detectors. This allows 2-3x cycle time increase without an image quality impact, allowing fully automated high throughput CT inspection on the production floor.

A further challenge for CT manufactures is the increased demand to scan bigger and complex shaped parts which require alternative scanning and reconstruction methods beyond standard circular scanning. We will demonstrate how alternative rotation axis offset based scanning methods can improve scanning volume without significant cycle time impact.

BAKER HUGHES
a GE company

Optimization of Throughput and Image Quality at CT Scans of Large Parts

Eberhard Neuser, Oliver Brunke, Alexander Suppes,
GE Sensing & Inspection Technologies GmbH, Wunstorf, Germany

International Symposium on
Digital Industrial Radiology and Computed Tomography
Fürth, Germany, July 3rd 2019

Content

1. Motivation
2. Improved X-Ray Source Technology
3. Improved X-Ray Detector Technology
4. Improved CT Scanning Procedure
5. Conclusion

Motivation

How to optimize or improve Throughput & Image Quality in CT ?

Tube Technology	→	High|Flux target
Detector Technology	→	Dynamic 41|100&200
CT scanning procedure	→	Offset|CT

High|flux target X-ray source technology
How does an X-ray source work?

An beam of electrons is accelerated across a vacuum gap into a tungsten target

- <1% of incident energy converted to useful x-rays
- >99% of energy converted to heat

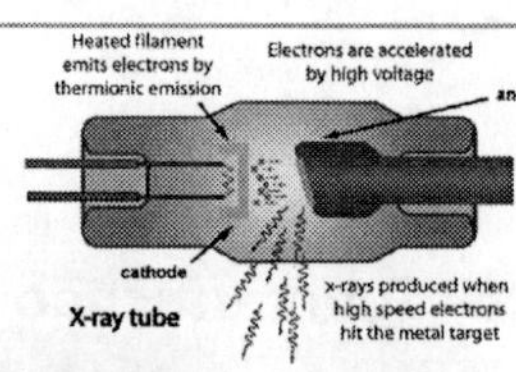

http://www.arpansa.gov.au/radiationprotection/basics/xrays.cfm

Rocket Science

X-ray Tube Science!

Fission Reactor Core
1-2 W/mm²

Rocket Nozzle Interior
10 W/mm²

Plasma Torch
20 W/mm²
S. Borkar, Intel 1999

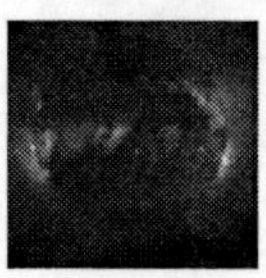

Sun's Surface
60 W/mm²

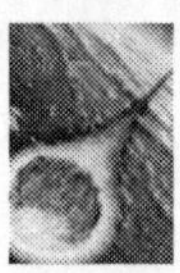

Meteor Entry
100-500 W/mm²

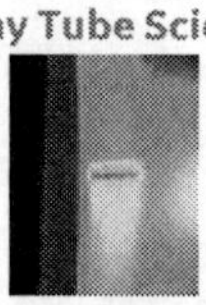

X-Ray Tube
Focal Spot ~1000 W/mm²

X-ray source development is a <u>*thermal management*</u> challenge

- *... beyond rocket science*
- **Better heat dissipation = faster scan times**

High|flux target X-ray source technology
Example Open Source

- GE patented advanced X-ray target **based on material design**, not on a mechanical rotating anode concept
- Higher flux density due to optimized thermo-physical performance by **combining tungsten and diamond** in the target design
- Designed for lab and production process control **microCT** applications
- By **doubling the X-ray inspection power from 50 to 100W without enlarging the focal spot size**, the high-flux|target provides better CT quality with less image noise, enabling users to get more accurate results or to scan twice as much

High|flux target X-ray source technology
Higher productivity & performance

conventional target - 7W power

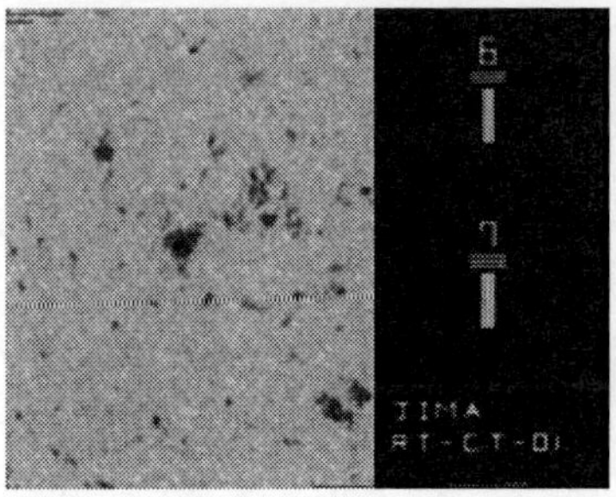

SNR: low 11
Resolution: high 6 µm

conventional target - 14W power

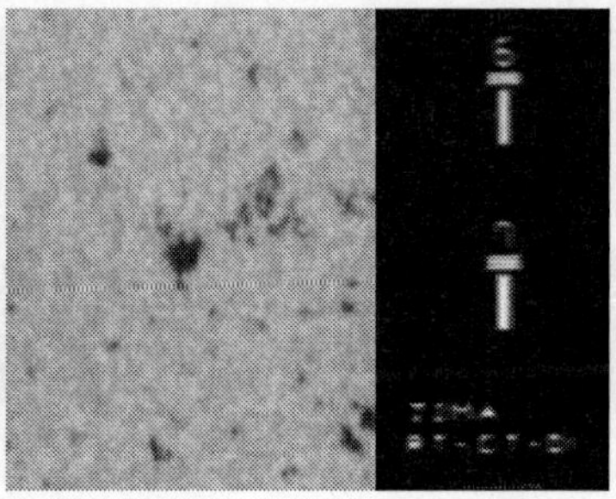

SNR: high 23
Resolution: low 14 µm

high-flux|target – 14 W power

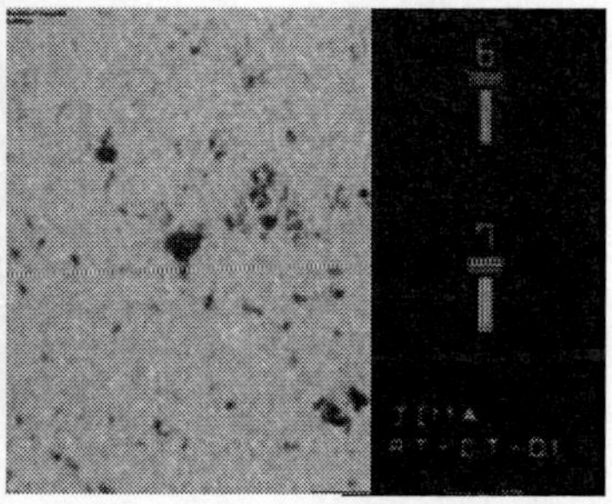

SNR: **high** 23
Resolution: **high** 6 µm

Improved Detector Technology
dynamic 41|100 & 200

Key features & benefits

GE proprietary 16" X-ray detector family with 100µm & 200 µm pixel size (16/4 Mpixel)

Next generation electronics, photodiode and scintillator design for improved efficiency and sensitivity

optimized for long-term reliability at industrial use (high kV, mA @ high duty cycle)

Improved Detector Technology
CT Case Study: 1993 Video Camera

Test Object:
- Hi8 video camera
- Chip: 1/3" CCD sensor
- Resolution: 470 000 pixel
- Data storage type: Magnet tape

CT Scan:
- Scanner: phoenix vtomex m 300 with scatter|correct technology
- detector: dynamic 41|100
- X-ray parameters U = 270kV, I = 400 µA
- 2x 4000 projections

CT dataset:
- voxelsize: 35/70µm
- Dataset size: ~2500*3500*7000 (dynamic41|100)
- Reconstruction time: ~35 min

Improved Detector Technology
CT Case Study: 1993 Video Camera

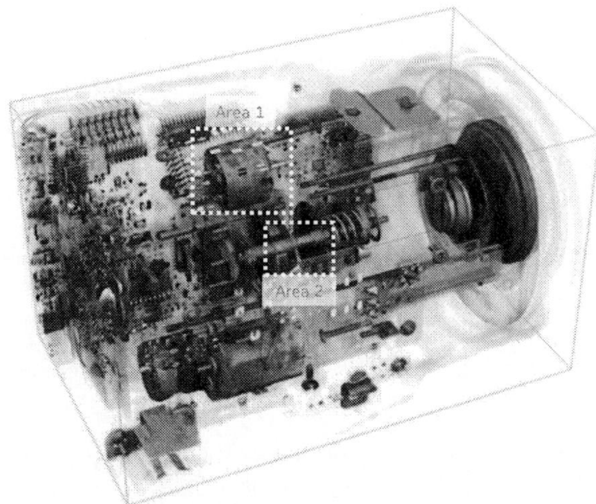

Area 1: Zoom Motor
„Resolution test“

Area 2: Driveshaft Optical Zoom
„SNR test“

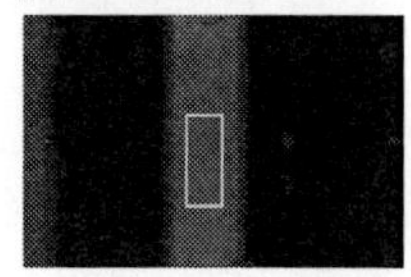

Improved Detector Technology
CT Case Study: dynamic41|100 vs DXR250

Area 1: Zoom Motor „Resolution test“

DXR250
60 min

baseline

Dynamic41|100 2x2 binning
30 min

2x faster @ same resolution

Dynamic41|100
60 min

Same speed as baseline @ 2x higher resolution

Improved Detector Technology
CT Case Study: dynamic41 vs DXR250

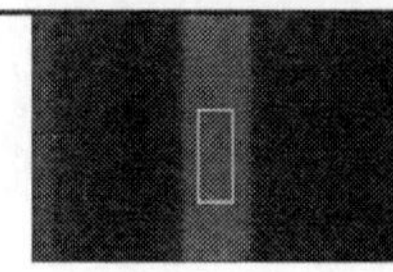

Area 2: Driveshaft Optical Zoom: „SNR test"

Detector	dynamic 41|100		dynamic 41|200	DXR250
Scan Time	60 min	30 min	60 min	60 min
Voxelsize	70 µm (2x2 binning)	70 µm (2x2 binning)	70 µm (native)	70 µm (native)
SNR	161	136	206	154

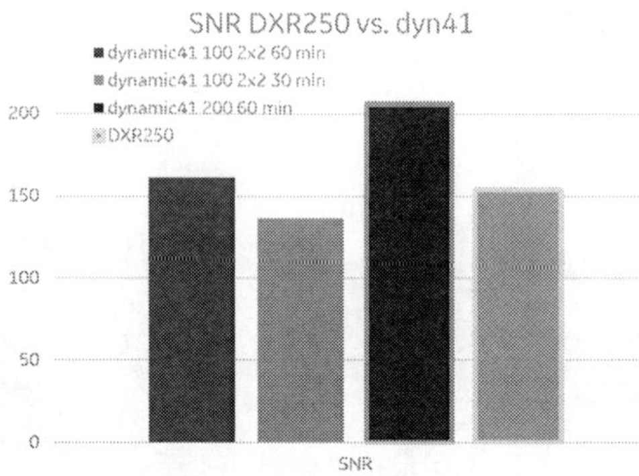

Conclusion: dynamic41 allows ~1.2-1.7x **better image quality** (SNR) at same speed or 1.5-3x **higher speed** at same image quality (SNR)

BAKER HUGHES a GE company

Improved CT scan procedure
Offset|CT

Offset|CT

Scanning of larger parts

Avoiding larger detector or

Virtual detector enlargement

→ same spatial resolution

→ 2x faster compared to virtual detector concept

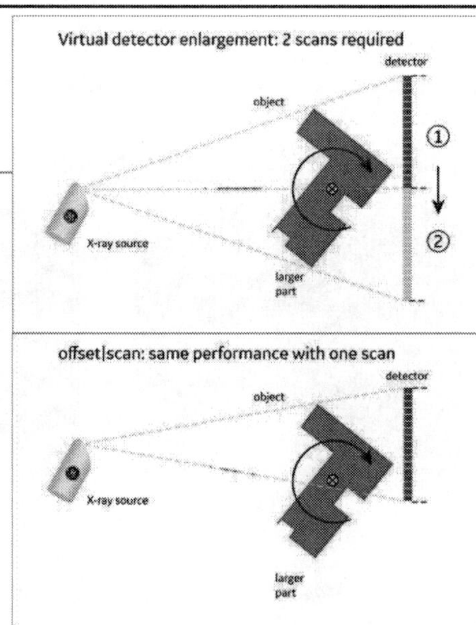

BAKER HUGHES a GE company

Improved CT scan procedure Offset|CT – 3D results

Automotive Aluminum part

Length 400mm

Scanned @ 260kV

3000 projections

Voxelsize 112µm

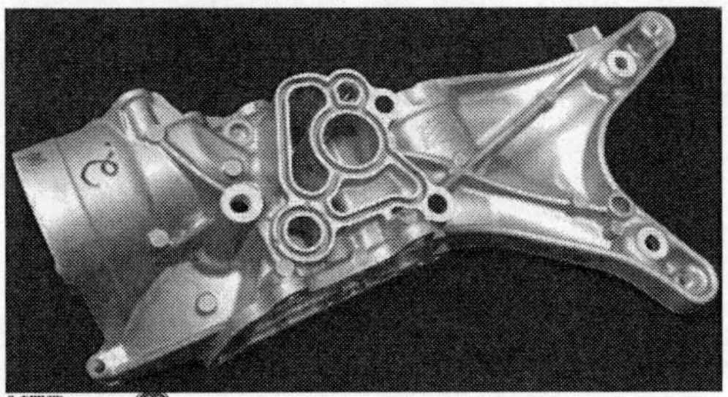

Set of Projections – 360° movie

Improved CT scan procedure | Offset|CT – 3D results

Automotive Aluminum part

Length 400mm

Scanned @ 260kV

3000 projections

Voxelsize 112µm

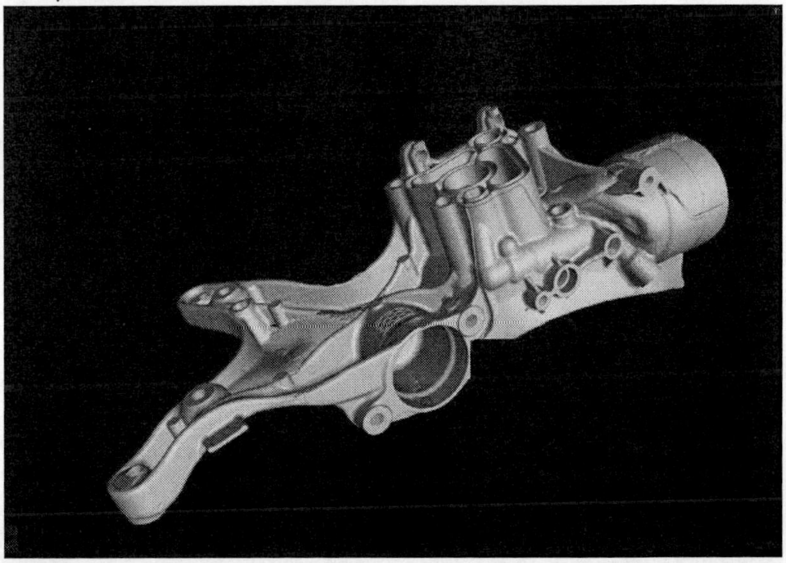

Improved CT scan procedure
Offset|CT – 3D results

Cross section

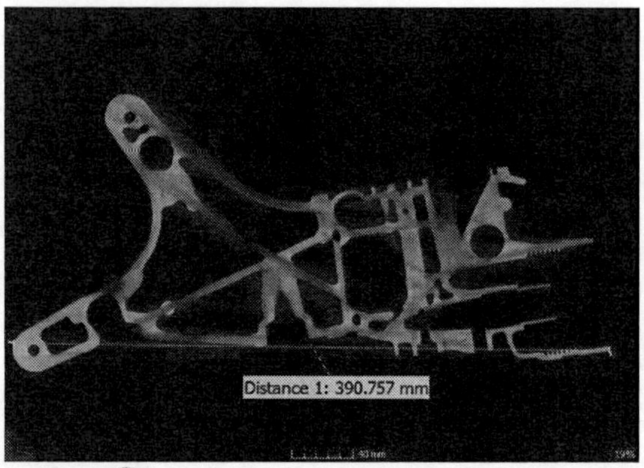

Resolution detail

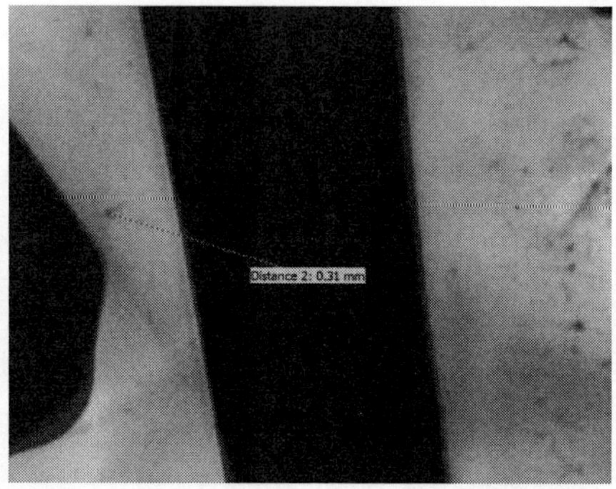

Conclusion

Optimization of Throughput and Image quality in CT scanning have been shown by implementation of improved

- **X-Ray source technology**
- **Detector technology**
- **CT scan procedure**

Towards a full physics radiography model

Andreas SCHUMM [1], Anthony TOURON [2], Romain JONCHIÈRE [3], Caroline VIENNE [2]
[1] EDF, Moret sur Loing, France
[2] CEA, Saclay, France
[3] EDF, Saclay, France

Contact e-mail: andreas.schumm@edf.fr

Abstract

The current state of the art in radiographic modelling, as implemented in CIVA and MODERATO, models the main photon-matter interactions in an energy range from roughly 50keV to about 10MeV, including Rayleigh scattering, Compton scattering and pair production. However, at very low and very high energies, other effects such as fluorescence and Bremsstrahlung grow more important and add additional contributions to the final image in terms of scatter, in particular within the detector, but also within the part to be inspected. These additional interactions require the electrons being produced by photoelectric absorption, Compton scattering and pair production to be handled, and the electron cascade with its additional interactions to be treated accordingly. Electrons being the main contributors to the latent image in argentic film radiography, this "true physics approach" has the potential to significantly improve radiographic modelling. However, as the electron mean free path length is extremely small compared to photons, this process is in general considered prohibitively expensive.

In an effort to enhance the accuracy of modelling results, CEA and EDF engaged a 3 year project to identify essential photon/electron interactions, and to implement them in a computationally manageable way. As the electron cascade implies a severe penalty on the entire modelling chain, a parallel effort was made to search for efficient variance reduction schemes. In this presentation, we present the current state of affairs, and give an outlook on what might be available in terms of truly physical modelling in an upcoming version of CIVA.

Towards a full physics radiography model

Andreas Schumm, Romain Jonchière (EDF R&D)
Anthony Touron, Caroline Vienne (CEA)

Towards a full physics radiography model

Outline

1. A look at the current state of the art in radiographic modelling: The photon shower
2. The electron cascade in the inspected object and the detector
3. Dealing with the performance penalty

Andreas Schumm, Romain Jonchière (EDF R&D)
Anthony Touron, Caroline Vienne (CEA)

A particle transport problem

The photon shower

a) Direct radiation
b) Photoelectric absorption
c) Rayleigh scattering
d) Compton scattering
e) [pair production]

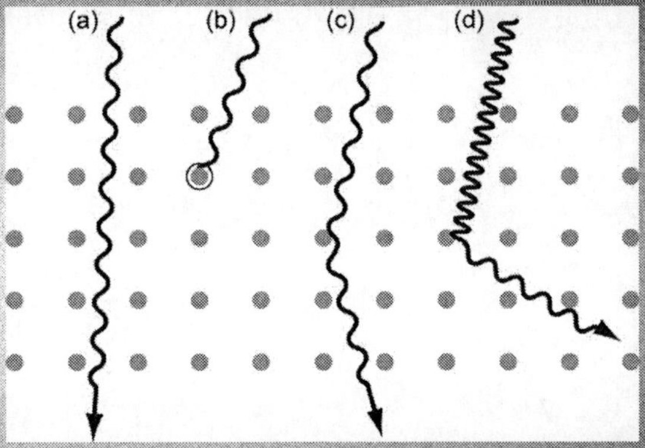

A particle transport problem

The photon shower including secondary radiation

a) Direct radiation
b) Photoelectric absorption + e +γ
c) Rayleigh scattering
d) Compton scattering + e
e) [pair production] e- + e+ $\rightarrow \gamma + \gamma$
f) (b,d,e): Bremsstrahlung e $\rightarrow \gamma$

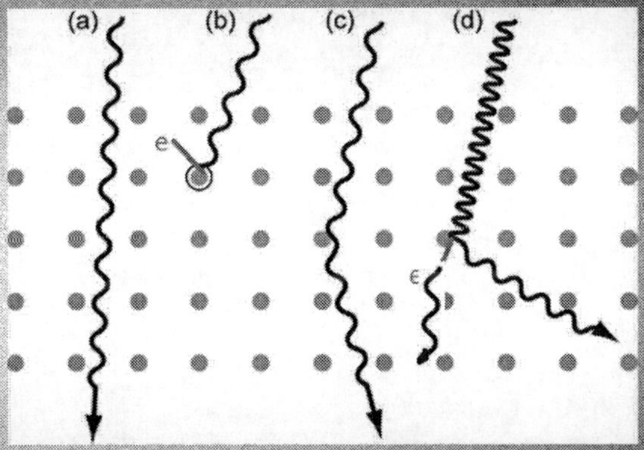

Ionisation

Photoelectric Absorption	Compton Scattering	Pair production
$E_e=E_\gamma-E_{bind}$	$E_e=E_\gamma-E_\gamma'$	$E_e \approx E_\gamma-2*0.511MeV$
Steel: (see table below) $\Rightarrow E_e \approx E_\gamma$!	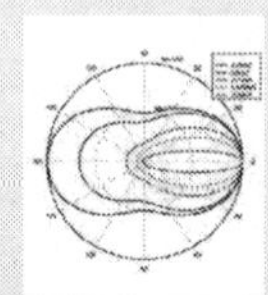 $E_e \approx [0..E_\gamma]$	Significant only >1MeV

Steel:

K	L1	L2	L3
7keV	0.8keV	0.7keV	0.7keV

Fluorescence after Photoelectric absorption

- Ejection of an electron with $E=E_{inc}-E_{bind}$
- Emission of a fluorescence photon with $E=E_k-E_{Lx}$

Fe:

K	L1	L2	L3
7keV	0.8keV	0.7keV	0.7keV

Pb:

K	L1	L2	L3
88keV	16keV	15keV	13keV

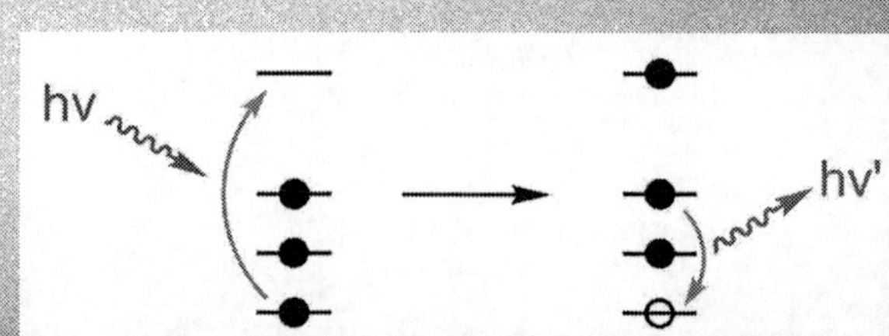

$\Rightarrow$Significant only for materials with high Z (as used in filters)

Secondary radiation – when does it matter?

In the inspected part

1. At moderate to high energy (Bremsstrahlung)
2. If the detector does not contain a filter (fluorescence)

In the detector

1. Always

Study the contribution of fluorescence and Bremsstrahlung by comparison with reference codes (Penelope and MCNP)

Secondary radiation – when does it matter?

Photon spectra

Cobalt source
60mm Fe block
200u lead filter/screen

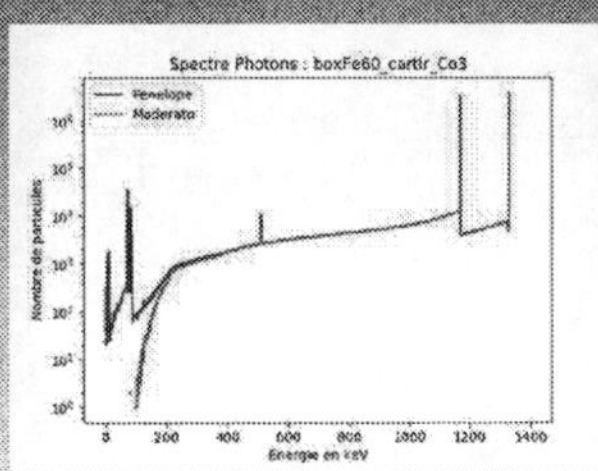

Selenium source
20mm Fe block
100u lead filter/screen

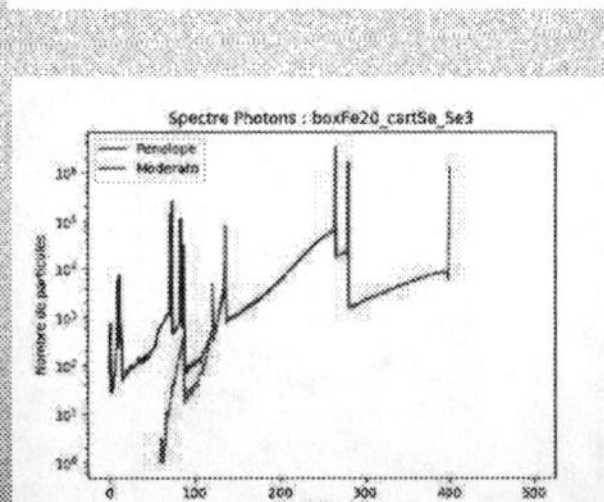

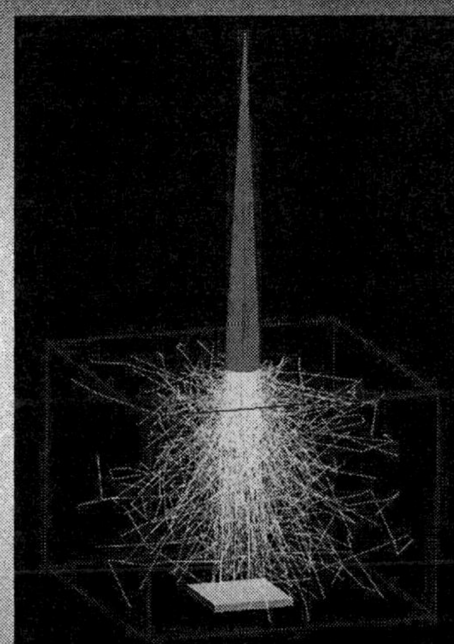

Secondary radiation – when does it matter?

Photon spectra

Iridium source
40mm Fe block
Without screen/filter

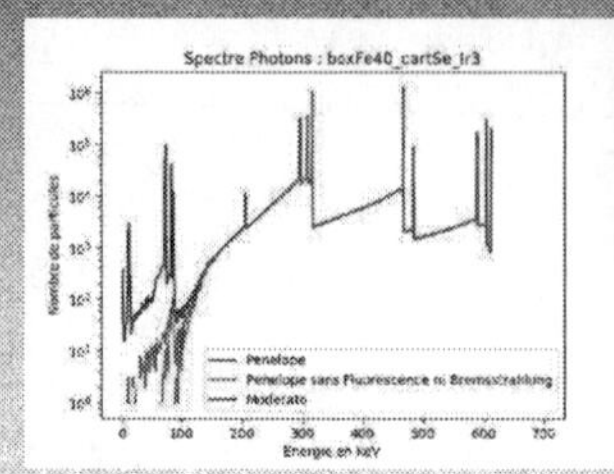

With lead screen/filter

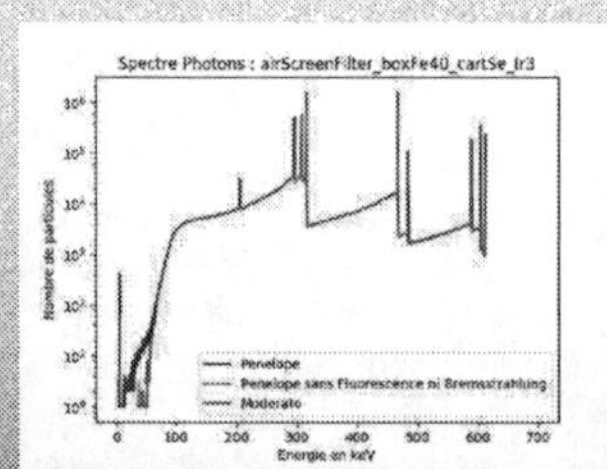

Fluorescence created within the Fe part is filtered

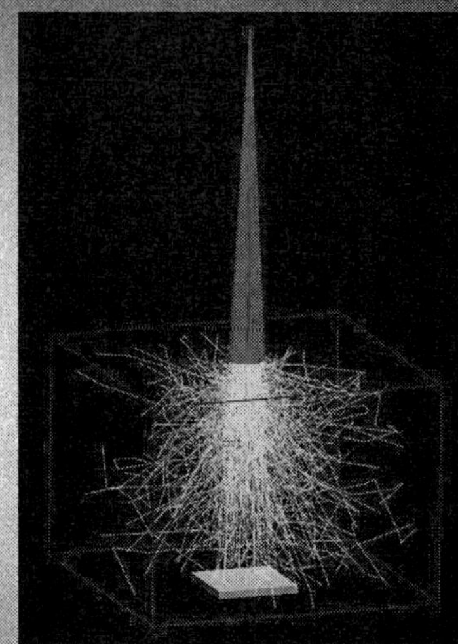

Fluorescence within lead screen

Photon paths for „pencil beam“
100keV monochromatic

5mm Fe

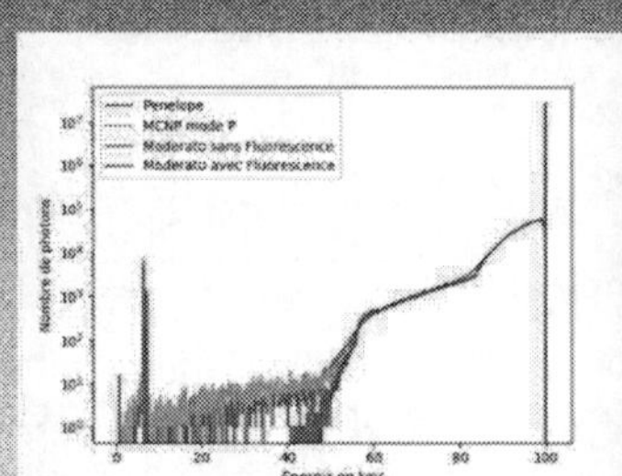

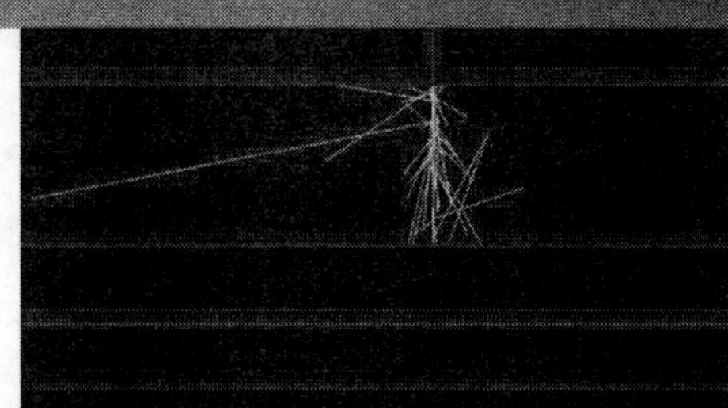

5mm Pb

Bremsstrahlung?

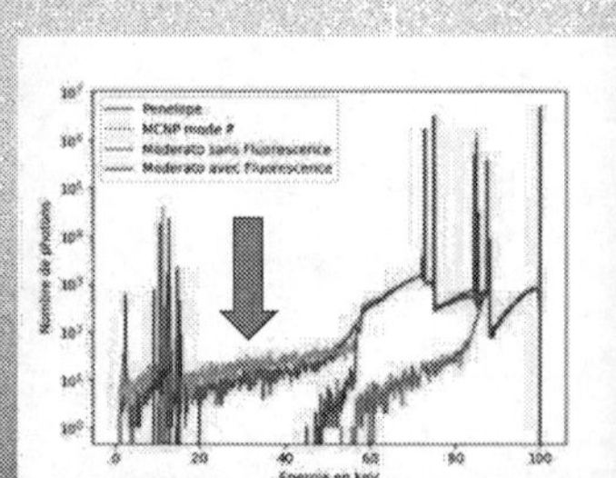

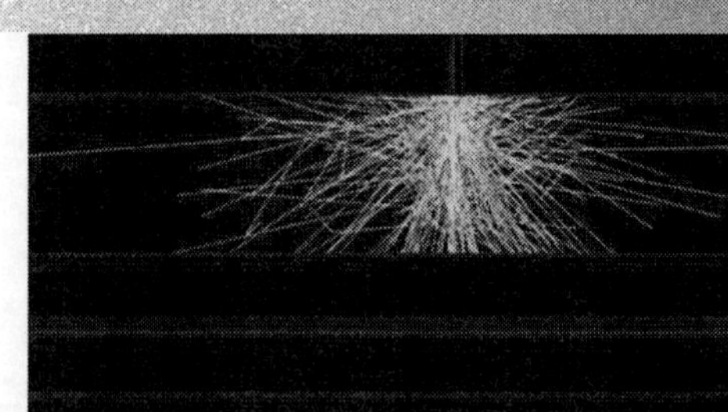

Fluorescence implementation

1. Determine the electron shell (K, L_1, L_2, L_3) where the photoelectric absorption takes place
2. Simulate atomic relaxation which produces fluorescence rays

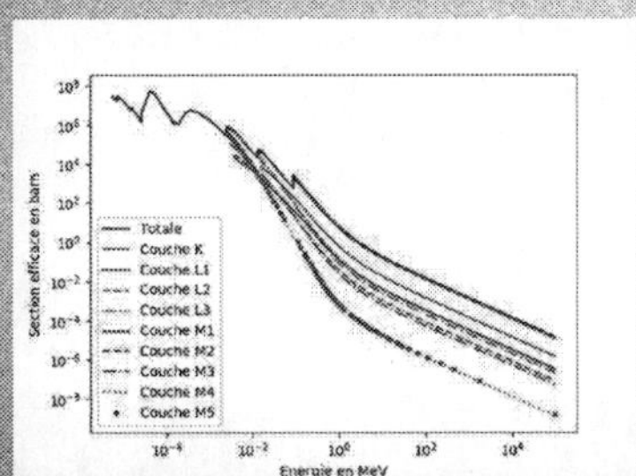

IAEA database EADL97

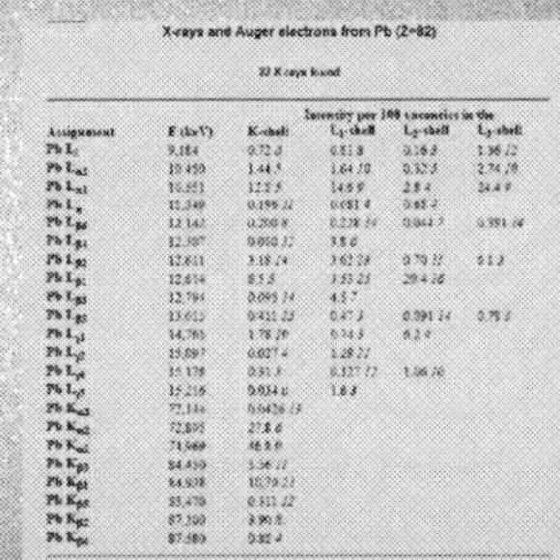

NuclearData database

Fluorescence implementation

Speed penalty:

Without fluorescence	Fluorescence in filter	Fluorescence in filter and part
1	+2%	+9%

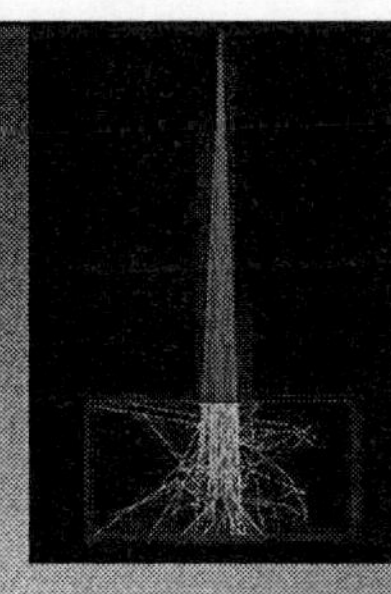

Error w/r Penelope:
- No ionisation by electrons (only photons)
- No Bremsstrahlung

$K_{\beta 2}$	$K_{\beta 1} + K_{\beta 3}$	$K_{\alpha 1}$	$K_{\alpha 2}$	L_β	L_α
15 %	-1 %	6,9 %	6,2 %	21 %	30 %

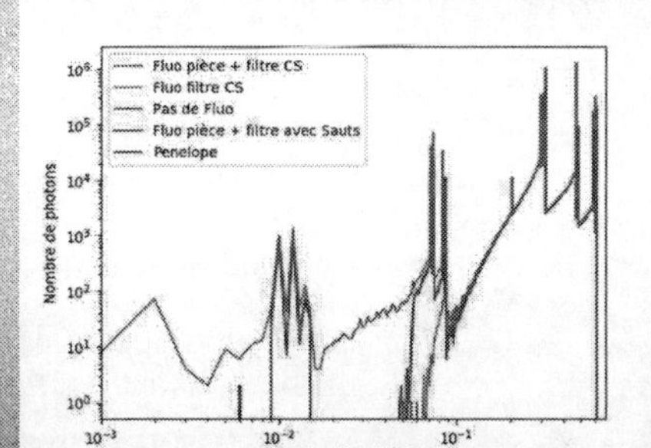

Bremsstrahlung

Electrons passing close to atomic cores are slowed down due to the electric charge of the core.
A fraction of this energy loss takes place in the form of photon emission. The effect is stronger for high Z materials.

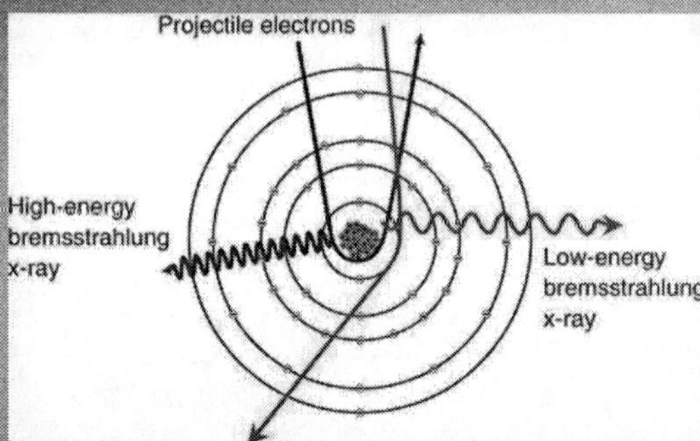

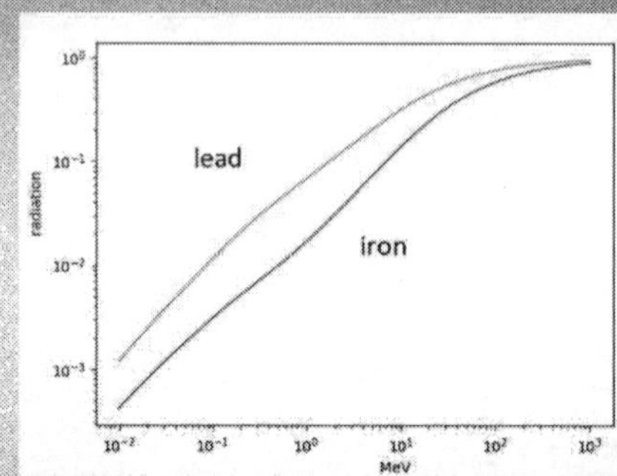

Fraction of electron energy converted to gamma radiation (NIST)

Bremsstrahlung

A rigorous implementation:

1. Would model the electron cascade (with a very small free path length)
2. Determine the photon emission due to kinetic energy loss at each scattering event

Prohibitive

A simplified implementation

1. Does not follow the electron cascade
2. Supposes immediate photon emission upon electron emission
3. Photon emission in the direction of the electron (known from PE/Compton/PP event)

→ Electrons can not reach the detector

Bremsstrahlung implementation [1]

1. Determine $E_{electron}$ and direction from scattering event
2. Determine electron stopping power from X-Ray cross section data to calculate number of photons emitted
3. Determine energy (from cross section data)

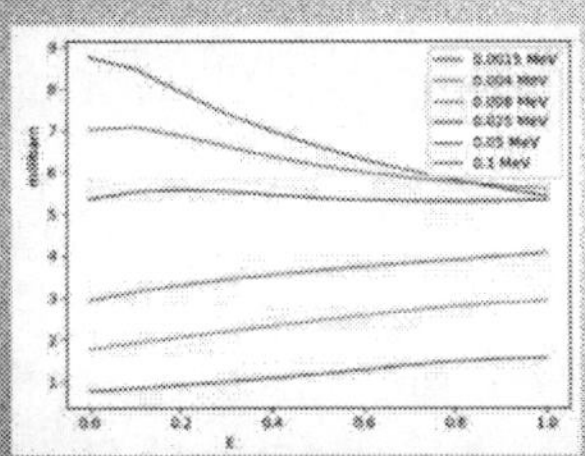

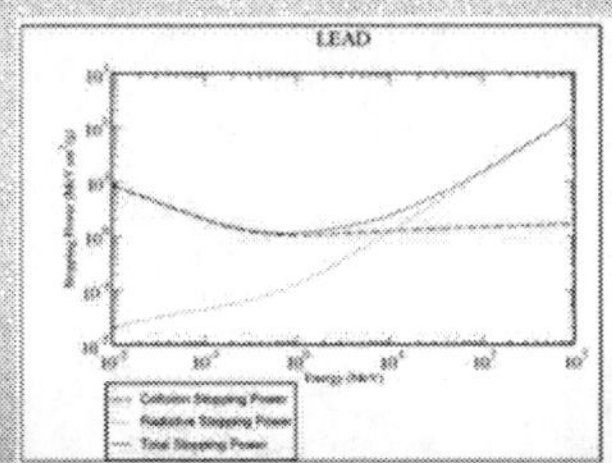

[1] Kaltiaisenaho, T. (2016). Implementing a photon physics model in Serpent 2
[2] Atomic Data and Nuclear Data Tables (Seltzer & Berger, 1986)

Bremsstrahlung

Evaluation (6MeV monochromatic source

- Results close to MCNP mode P (without electron transport)
- Slight over-estimation compared to Penelope

Potential improvements:
- Model photon emission direction
- Positrons due to pair production are handled like electrons – apply correction

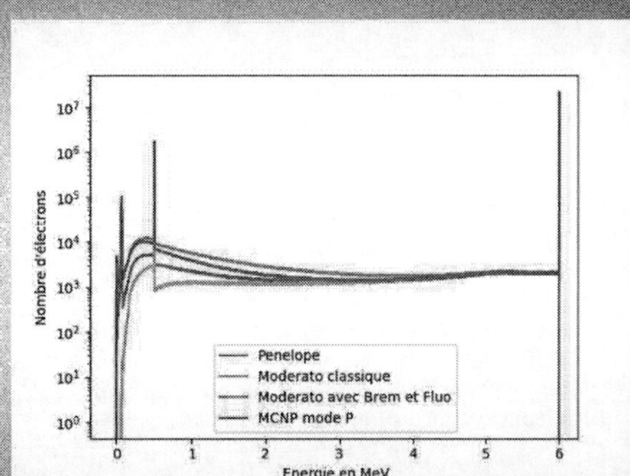

Dealing with the performance penalty

Characterising performance (FOM)

1. CPU time for an identical number of photons is not a good criterion: With variance reduction schemes, more is achieved with less
2. Gains depend on the average number of free path lengths (≈ build-up)
3. A convergence indicator based on the average error is

Dealing with the performance penalty

Successfully Implemented variance reduction schemes

1. Implicit capture (within the filter)
2. Particle projection
3. Forced interaction (within the cartridge)

Evaluated and discarded techniques:

- Russian roulette
- Adjoint methods
- Systematic splitting (combined with russian roulette)

Dealing with the performance penalty

Implicit Capture

A particle entering a highly absorbing material has a low probability of surviving.

Idea: reduce the probability of photoelectric absorption to explore the geometry further

Method: Replace absorption events by scattering events, and use a corrective weight

Idea: Separation of absorption and scattering cross section

$$\sigma_{total} = \sigma_A + \sigma_S$$

- Monte-Carlo sampling only on the scattering cross section
- Correction of weight $wcorr = \sigma S / \sigma total$

Dealing with the performance penalty

Split and forced interaction in intensifying screens

w

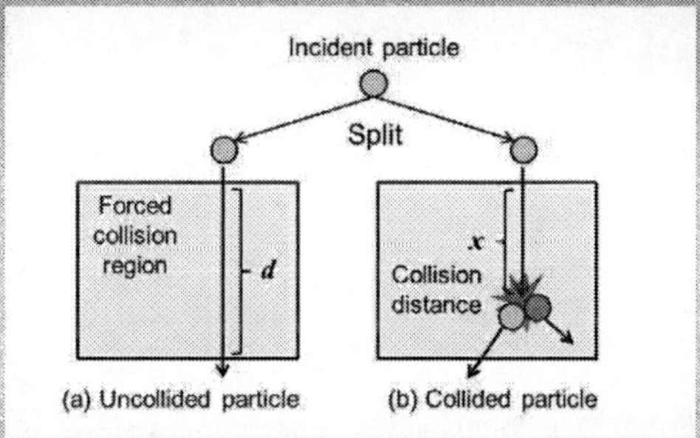

1-w

$w = \exp(-d.\ u)$ $x = -(1/u)\log(1 - \xi(1 - \exp(-d.\ u)))$

Dealing with the performance penalty

Projection

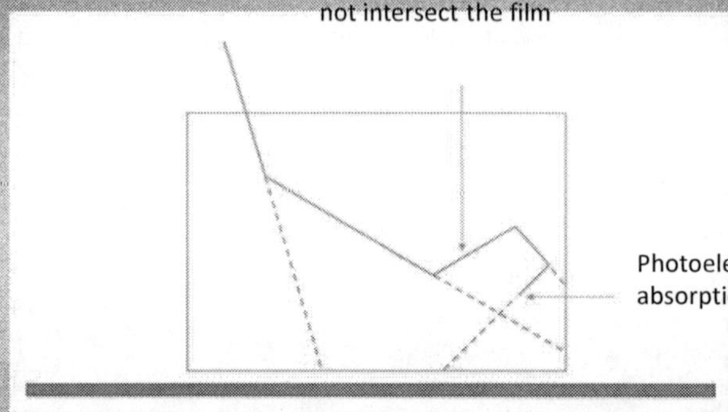

Idea: The ray-tracing algorithm always determines all intersections. If the detector is intersected, split.

Dealing with the performance penalty

Evaluation of speed-up due to variance reduction

Step wedge 5-10mm, 200keV tube

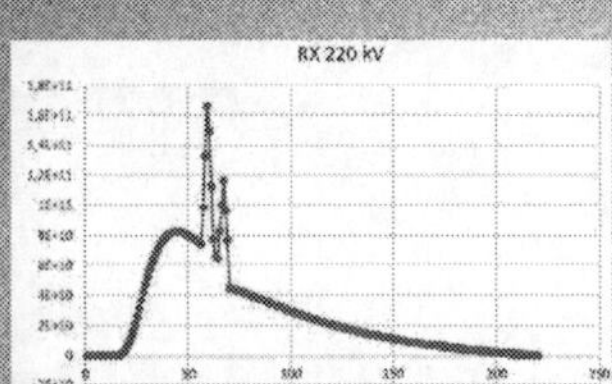

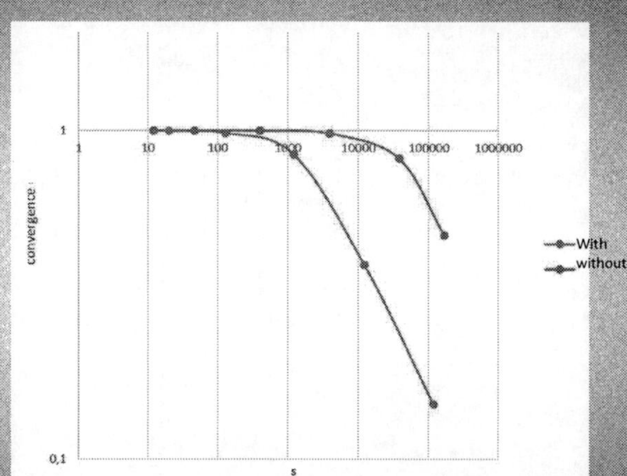

Conclusions

- Secondary radiation may make a difference
- Fluorescence in the detector can be modelled with negligible performance penalty
- Bremsstrahlung within the component can be modelled without an explicit electron cascade

Perspectives

- Full cartridge model with secondary radiation
- Evaluation on real world examples

Perspectives

- Decrease of exposure time as a function of screen thickness
- Increase of intrinsic unsharpness due to fluorescence in lead screens
- Contribution of Bremsstrahlung to Build-Up

Thank you

Determination of Metrological Structural Resolution using an Aperiodic Spatial Frequency Standard

Florian WOHLGEMUTH [1], Matthias FLEẞNER [2], Tino HAUSOTTE [1]
[1] Institute of Manufacturing Metrology, Friedrich-Alexander-University Erlangen-Nürnberg, Erlangen, Germany
[2] Volume Graphics GmbH, Heidelberg, Germany
Contact e-mail: Florian.Wohlgemuth@fmt.fau.de

Abstract. The metrological structural resolution is both an important indicator of a measurement system's metrological scope (also in comparison to other measurement systems) as well as an indicator for low pass filtering that typically reduces probing errors. For computed tomography systems, it is defined in the VDI/VDE guideline 2630 Part 1.3 as 'the size of the smallest structure that can still be measured dimensionally'. The determination procedure suggested in this guideline includes the measurement of multiple calibrated microspheres which has proven expensive and time-consuming in practice. Therefore, alternative approaches to measuring the metrological structural resolution are currently being proposed and researched. Ideally, these are applicable to a wide range of measurement systems.

In prior works of the Institute of Manufacturing Metrology, an aperiodic spatial frequency standard was presented. Qualitatively, it could be shown that with this standard, the expected effects can be detected. This work presents and discusses the quantitative determination of the metrological structural resolution using this standard. The determination is based on the calculation of a spatial frequency dependent amplitude transfer function and a subsequent fit with a Gaussian function. In this work, the effect of different cutoff frequencies for the fit is studied and influence factors on the MSR values are demonstrated.

1. Introduction

The metrological structural resolution for computed tomography (CT) systems is defined in the VDI/VDE guideline 2630 Part 1.3 as 'the size of the smallest structure that can still be measured dimensionally' [1]. This abstract definition is accompanied by the suggestion to measure multiple calibrated microspheres until the measurement deviation exceeds a threshold provided by the manufacturer. The diameter of the smallest microsphere that is still measureable within the provided threshold is the structural resolution of the CT system. This determination procedure has proven unattractive in practice due to the required measurement time as well as the cost of multiple calibrated microspheres. As the metrological structural resolution can both serve as an important indicator of a measurement system's metrological scope as well as indicate low pass filtering (which typically reduces probing errors in acceptance testing [2]), an alternative approach to the determination of the metrological structural resolution is of high interest. It is advantageous if this alternative approach is also applicable to a wide range of measurement systems, as not only the capabilities of different

CT systems can be compared, but also a wide range of different sensors can be compared regarding their metrological properties.

There are projection- and volume-based resolution characterisation methods like the modulation transfer function [3, 4, 5]. From the point of view of dimensional metrology, those characterisation methods are unsatisfying because they cannot easily be transferred to all dimensional measuring systems (e.g. as other systems do not provide comparable projection or volume data). But even for CT systems, these characterisation methods cannot cover all influences relevant for dimensional metrology. The result of a dimensional measurement using a CT system is influenced by all measurement chain steps (projection acquisition, reconstruction, surface determination and subsequent dimensional evaluation (e.g. fitting of a standard geometry)). A metrologically sound resolution quantification needs to compass all measurement chain steps before the dimensional evaluation to be sensitive to all potential influence factors on the dimensional measurement result.

In the field of CT, different determination methods have been proposed for the structural resolution (e.g. [6, 7]). In prior work at the Institute of Manufacturing Metrology, an aperiodic spatial frequency standard (ASFS) was presented. Qualitatively, it could be shown that with this standard, the expected effects can be detected [8, 9]. In this contribution, we want to discuss a quantitative determination method in detail.

2. Determination Approach

In system theory, a system's transfer function $f(\vec{x})$ describes its properties by connecting system input $i(\vec{x})$ with the system output $o(\vec{x})$:

$$o(\vec{x}) = \int f(\vec{x} - \vec{y})\, i(\vec{y}) d\vec{y} \tag{1}$$

This convolution can be rewritten as a multiplication in Fourier space, yielding a simple determination formula for the system's transfer function:

$$\tilde{f}(\vec{k}) = \frac{\tilde{o}(\vec{k})}{\tilde{\imath}(\vec{k})} \tag{2}$$

The basic idea behind the metrological structural resolution determination using the aperiodic spatial frequency standard is to characterise a measurement system by its transfer function. The aperiodic spatial frequency standard (see Fig. 1) is a rectangular cuboid of dimension 4 mm x 4 mm x 26 mm. It is made of titanium and manufactured using laser ablation with one of the longer side faces featuring aperiodic structures. The structures are constant along the shorter 4 mm direction of the side face and vary without a reproducible pattern along the longer 26 mm side. The total length of the aperiodic structure is 25 mm, leaving 0.5 mm of even surface on each end of the standard that are useful e.g. for tactile profilometer measurements for initial positioning of the stylus. For the determination of the metrological structural resolution, the CT system's transfer function is determined for profile lines extracted along the standard (compare section 3), yielding a 'one dimensional metrological structural resolution'.

The aperiodic spatial frequency standard is designed to have amplitudes over a broad frequency range in order to effectively sample a good range of the transfer function. Typical structure sizes are in the order of 100 µm which is also the order of structural resolutions that can be determined using this standard. The maximum inclination is 45° to allow for a wide range of optical measurements. The design spectrum of the aperiodic spatial frequency standard (based on a profile line taken from the CAD) is shown in Fig. 2.

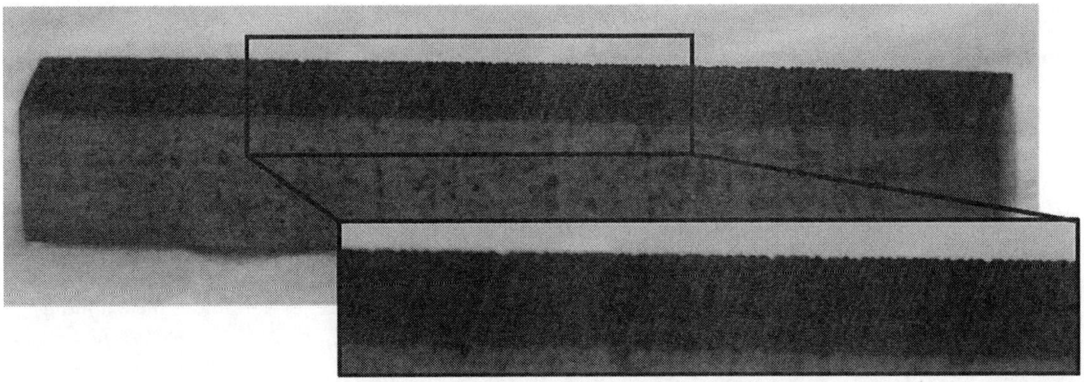

Fig. 1. Photograph of the aperiodic spatial frequency standard.

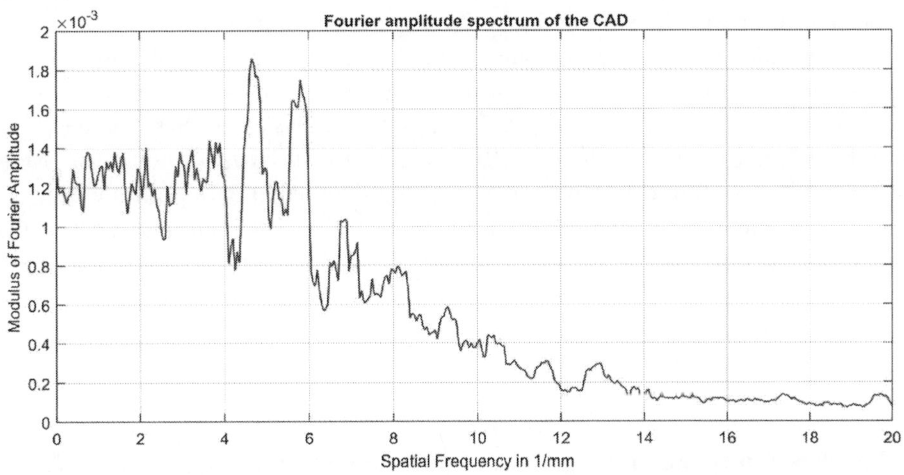

Fig. 2. Aperiodic spatial frequency standard CAD spectrum.

3. Methodology

For the determination procedure using the aperiodic spatial frequency standard, the same profile is measured both with a reference measurement system of nominally significantly higher resolution as well as with the measurement system to be characterised. For the case of CT, a tactile profilometer can serve as reference measurement system. Alternatively, if simulations using a CT digital twin are used, the simulation mesh can serve as reference data set. The main idea is to perform a Fourier transform of reference and measurement profile and then determine the transfer function by analysing the ratio of the measurement and the reference profile amplitudes. In the following, the single data processing steps are described in detail. Fig. 3 shows a conceptual summary.

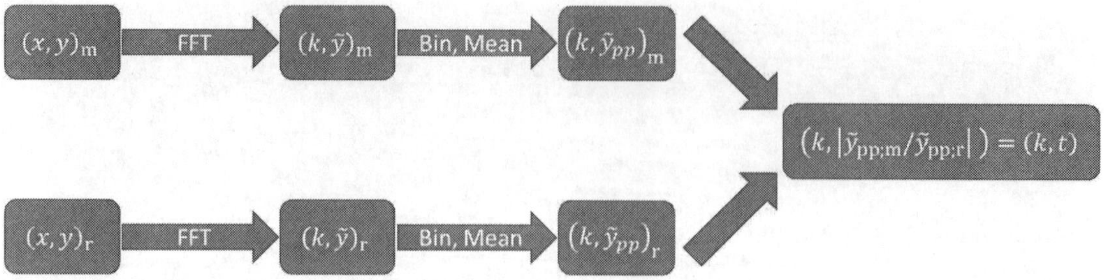

Fig. 3. Determination of the transfer function (k,t) (k is the spatial frequency variable, t the amplitude transfer at that frequency) from the (aligned) measured profile values $(x,y)_m$ and a reference profile $(x,y)_r$.

3.1 *Alignment*

The measurement data both for reference measurement system and measurement system is a 2D data set of a profile. Due to the working principle of some measurement systems, a reasonable alignment of the profile direction and the coordinate system the profile was measured in is not always guaranteed. To solve this, the measurement data (x,y) is fitted with a robust (bisquare weighted [10]) linear fit $y = p_1 \cdot x + p_2$ in MathWorks Matlab 2019a. The constant offset is then subtracted from each data point ($y \mapsto y - p_2$) and subsequently, the data set is rotated with an angle of $-\arctan(p_1)$ to ensure that the x coordinate represents the coordinate along the profile and y the profile height at that x value. To ensure comparability, this alignment procedure is performed for data sets from any measurement system.

3.2 *Fourier Transform*

The measurement and reference measurement data after alignment result in a profile of the measurement $\{(x_{1;\text{measured}}, y_{1;\text{measured}}), \dots, (x_{n;\text{measured}}, y_{n;\text{measured}})\}$ as well as a reference profile $\{(x_{1;\text{reference}}, y_{1;\text{reference}}), \dots, (x_{n;\text{reference}}, y_{n;\text{reference}})\}$. In this study, a constant profile length of 25.2 mm is used for the CT measurement data. In general, a comparable profile length for different measurement systems should be chosen to guarantee comparability of the profiles and spectra obtained. The y-values of both profiles are then transformed to the spatial frequency domain. The result are two complex valued amplitude vectors $\{\tilde{y}_{1;\text{measured}}, \dots, \tilde{y}_{n;\text{measured}}\}$ and $\{\tilde{y}_{1;\text{reference}}, \dots, \tilde{y}_{n;\text{reference}}\}$ as well as two corresponding spatial frequency vectors $\{k_{1;\text{measured}}, \dots, k_{n;\text{measured}}\}$ and $\{k_{1;\text{reference}}, \dots, k_{n;\text{reference}}\}$. In the following, only the complex amplitude (and not the phase) of the FFT is of interest. Thus, the spectral data contains redundant information and it suffices to consider the positive spatial frequency part with indices $\{1, \dots, m\}$.

3.3 *Binning and Filtering*

Due to unavoidable differences in the sampling positions $\{x_{1;\text{measured}}, \dots, x_{n;\text{measured}}\}$ and $\{x_{1;\text{reference}}, \dots, x_{n;\text{reference}}\}$ (at least due to limited numerical precision, but also due to different sampling of different measurement devices), the measurement and reference spatial frequency vectors $\{k_{1;\text{measured}}, \dots, k_{m;\text{measured}}\}$ and $\{k_{1;\text{reference}}, \dots, k_{m;\text{reference}}\}$ are not identical. To obtain a transfer function from the amplitude spectra $\{|\tilde{y}_{1;\text{measured}}|, \dots, |\tilde{y}_{m;\text{measured}}|\}$ and $\{|\tilde{y}_{1;\text{reference}}|, \dots, |\tilde{y}_{m;\text{reference}}|\}$, their ratio has to be calculated at the same spatial frequencies (compare equation (2) on page 2). To obtain their values at identical spatial frequencies, the amplitude spectra are rebinned with a fixed spatial frequency bin width Δ_{bin}. In this study, 0.05 mm^{-1} was used as bin width. The binning process yields two binned amplitude spectra $\{|\tilde{y}_{1;\text{measured};\text{b}}|, \ldots, |\tilde{y}_{h;\text{measured};\text{b}}|\}$ and $\{|\tilde{y}_{1;\text{reference};\text{b}}|, \ldots, |\tilde{y}_{h;\text{reference};\text{b}}|\}$ with one corresponding frequency vector $\{0, \ldots, (h-1) \cdot \Delta_{\text{bin}}\}$. In principle, a transfer function can already be calculated from this data. In practice, there is a lot of noise on these binned amplitude spectra. To suppress that, a moving mean filter is used to smoothen the spectra. For this study, a filter width of 9 bins was used. After the application of this filter, the result of this data processing are two binned and filtered, post-processed amplitude spectra - $\{|\tilde{y}_{1;\text{m};\text{pp}}|, \ldots, |\tilde{y}_{h;\text{m};\text{pp}}|\}$ for the measurement and $\{|\tilde{y}_{1;\text{r};\text{pp}}|, \ldots, |\tilde{y}_{h;\text{r};\text{pp}}|\}$ for the reference.

3.4 Transfer Function and Fit

The transfer function at the spatial frequencies $\{0, \ldots, (h-1) \cdot \Delta_{\text{bin}}\}$ is calculated as $\{|\tilde{y}_{1;\text{m};\text{pp}}|/|\tilde{y}_{1;\text{r};\text{pp}}|, \ldots, |\tilde{y}_{h;\text{m};\text{pp}}|/|\tilde{y}_{h;\text{r};\text{pp}}|\}$. This transfer function now serves to characterise the system response for the spatial frequency range. Empirically, a Gaussian characteristic of this transfer function could be observed for the CT systems at the Institute of Manufacturing Metrology. Thus, the transfer function is fitted with a Gaussian function of the form:

$$f(k) = \exp\left(-\ln(2) \cdot \frac{k^2}{k_c^2}\right) \quad (3)$$

The transfer function has a limit of 0 for $k \to \infty$, signifying that arbitrarily small structures cannot be resolved. Additionally, the transfer function is set to be 1 (signifying an amplitude transfer of 100 %) for $k = 0$ by definition.

To fit the data points at all spatial frequencies $\{0, \ldots, (h-1) \cdot \Delta_{\text{bin}}\}$ is usually not constructive as the nominal spectrum's information content (i.e., its amplitudes) drop drastically at a certain frequency (compare Fig. 2), leading to high noise levels due to the division by a small reference measurement amplitude. Therefore, it is useful to set a cutoff frequency for the fit. We discuss the influence of this cutoff frequency and its choice in Section 4.4. Our usual choice for the cutoff frequency in this study is 13 mm^{-1}.

k_c is the spatial frequency for which the transmission characteristic of the system drops to 50 %. The metrological structural resolution l_{MSR} is calculated as the corresponding length:

$$l_{\text{MSR}} = \frac{1}{k_c} \quad (4)$$

The choice of the 50 % transmission criterion is based on the analogy to other filters used in dimensional metrology [11] and agrees with current standardisation in other dimensional metrology fields [12]. Future research might lead to the conclusion that p% transmission are a more suitable measure to represent a dimensional metrological system's capabilities. In this case, the new value for the structural resolution would be given by:

$$l_{\text{MSR}} \cdot \sqrt{\frac{\ln\left(\frac{p}{100}\right)}{\ln\left(\frac{1}{2}\right)}} \quad (5)$$

4. Results

4.1 Setup

Reference measurements were carried out using both a tactile profilometer (Taylor Hobson (Leicester, England) Form TalySurf Serie 2 PGI, 2 µm diamond tip stylus) and an optical focus variation system (Alicona (Raaba/Graz, Austria) Infinite Focus G4 using a 50 x objective). The profilometer measurement was repeated four times, the Alicona measured a surface of 26.45 mm x 0.24 mm. Due to the working principle of the Alicona, inclinations can only be measured up to a certain acceptance angle. This causes gaps in the measured surface that have no height information. As 332 parallel profiles can be extracted from the surface data, the profile with the fewest gap points was chosen for further use as reference profile. For this profile, the ratio of gaps is 1.7 %. The missing data points from this profile were interpolated linearly for the Fourier transform.
The titanium ASFS (compare Section 2 and Fig. 1) was measured using a Carl Zeiss Industrielle Messtechnik GmbH (Oberkochen, Germany) Metrotom 1500 225kV (2048 x 2048 pixel detector with 200 µm pixel size). Measurements were performed using the parameters given in Table 1. Measurement M1 is a measurement with a small nominal spot size to minimise spot blurring. M2 represents a typical choice for a good measurement while measurement M3 is a measurement with high unsharpness and a shorter measurement time due to a large focal spot (high tube power).

Table 1. Measurement parameters

Measurement Number	**M1**	**M2**	**M3**
	small spot	normal spot	large spot
Voltage/kV	200	200	200
Current/µA	91	182	546
Filter	0.5 mm Cu	0.5 mm Cu	0.5 mm Cu
Magnification	6.54	6.54	6.54
Voxel Size/µm	30.56	30.57	30.56
Nominal Spot Size/µm	18	36	109
Number of Projections	1600	1600	1600
Binning	1 x 1	1 x 1	1 x 1
Integration Time/ms	2000	2000	667
Detector Gain	16	16	16
Image Averaging	none	none	none

The obtained volume data from the CT measurements was furtheron processed using VGSTUDIO MAX 3.2.5. The surface was determined using the local adaptive surface determination with a search radius of 4 voxels from the iso50 surface (iso50 value determined automatically by VGSTUDIO MAX). It was then registered onto the CAD using a Best Fit registration and a measurement template consisting of 11 parallel line fits along the standard (compare Fig. 4) was applied. These lines are spread with a spacing of 0.3 mm and an edge distance of 0.5 mm. To ensure that no points on the surface are excluded from the fit, a search distance of 0.2 mm and a maximum gradient of 90° were used as fitting parameters. The fit points of those lines were exported as csv-files and processed in MATLAB as described in section 3 to obtain values for the metrological structural resolution. The intent behind using 11 line fits was to detect noise. For the purpose of this paper, we evaluated only one line per measurement and we always evaluated the same line on the standard for all measurements.

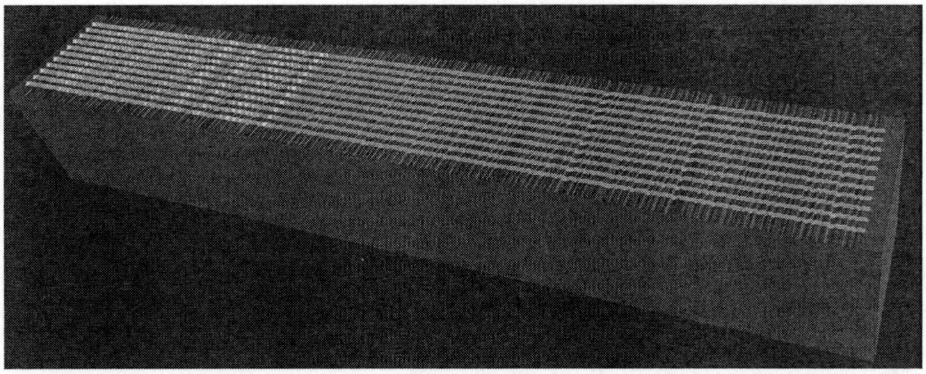

Fig. 4. Measurement template defined on the CAD file in VGSTUDIO MAX showing the 11 parallel lines for the profile extraction from the standard.

4.2 *Influence of the Focal Spot Size*

Using the same profilometer measurement as reference, the metrological structural resolution (MSR) for measurement M1 is 117 µm, the MSR value for M2 is 151 µm and the MSR for measurement M3 is 209 µm. The transmission spectrum and the Gaussian fit for measurement M2 are shown in Fig. 5. The MSR determination can thus clearly register the loss of information due to the increased focal spot smearing.

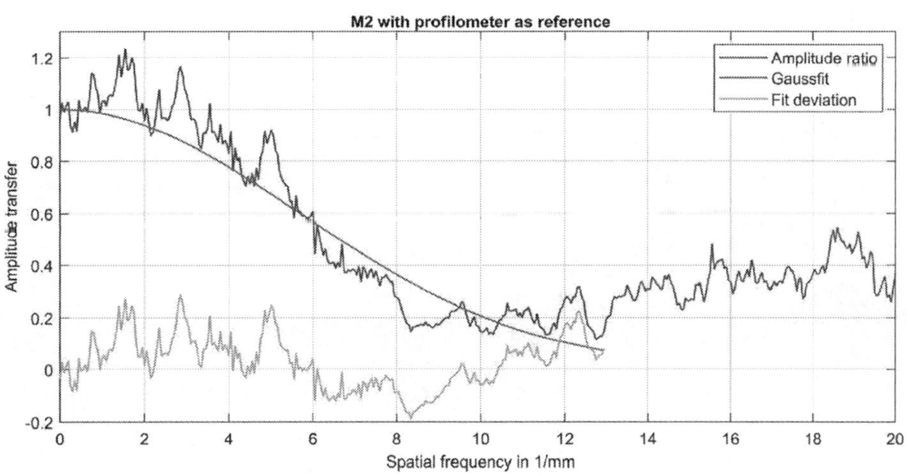

Fig. 5. Transmission spectrum and Gaussian fit with fit deviation for measurement M1.

4.3 *Choice of Reference Measurements*

The four profilometer measurements show a small variation concerning their amplitude spectra – especially when compared to the difference between a profilometer measurement spectrum and a focus variation measurement spectrum (compare Fig. 6 and Fig. 7). The differences are likely to be caused by the different transmission spectra that characterise the profilometer and the focus variation, respectively. Due to the lack of a higher resolution reference measurement, no spectrum can be regarded as inherently 'better' or 'more true' than the other.

The MSR values determined using all five reference spectra (four profilometer spectra, one focus variation spectrum) are in the range of 113 μm to 118 μm, 149 μm to 153 μm and 207 μm to 215 μm for measurements M1, M2 and M3, respectively. The span of those values (5 μm, 4 μm, 8 μm) can be seen as a minimum value for the reference measurement's contribution to the uncertainty of the MSR determination. We however believe that there are other influence factors with significantly higher contributions to the uncertainty of the determined MSR value and that thus both profilometer and focus variation measurements can be seen as valid reference measurements.

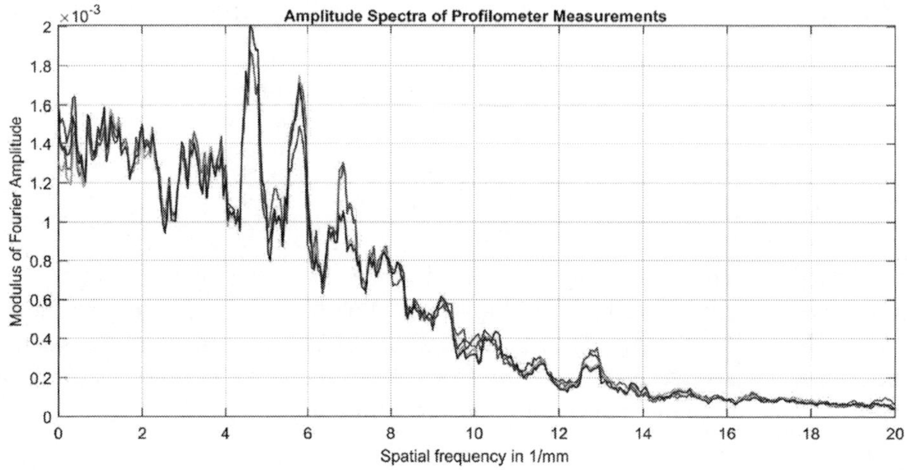

Fig. 6. Fourier amplitude spectra of the four different profilometer reference measurements.

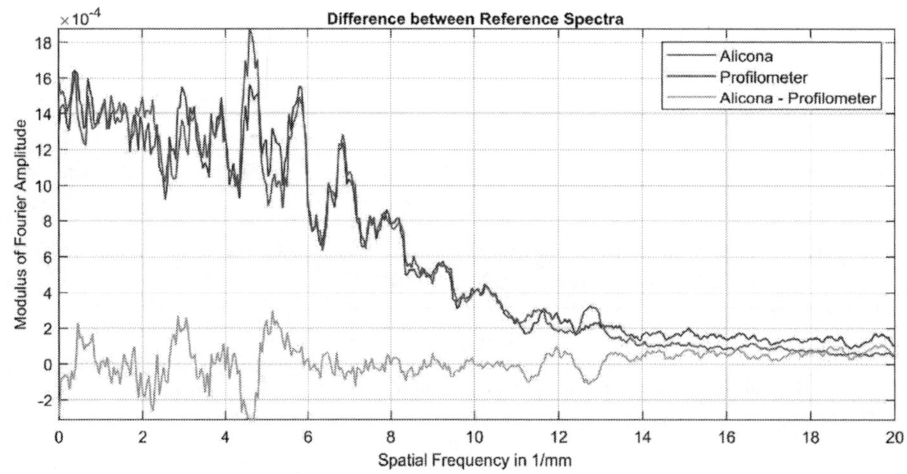

Fig. 7. Fourier amplitude spectra of a profilometer reference measurement and the focus variation reference measurement along with the amplitude difference.

4.4 *Influence of the Fit Cutoff Frequency*

As already mentioned in Section 3.4, the Gaussian fit uses only spectrum amplitudes up to a certain maximum spatial frequency to minimise the impact of high frequency noise. The choice of input data for the fit also influences the fit result and thus the determined MSR value. Therefore, the choice of the fit cutoff frequency (13 mm^{-1} in this work) is important. We chose this value due to the nominal frequency spectrum (Fig. 2) and the reference measurement spectra (Fig. 6 and Fig. 7). All these spectra have a last amplitude peak shortly before 13 mm^{-1} and drop to lower amplitudes at higher frequencies. As we think that the division by small reference amplitudes is the main reason for non-negligible high frequency 'noise' amplitudes distorting the measurement, basing the choice on this criterion seems reasonable and should guarantee comparability. We believe that the fit cutoff frequency needs to be a design parameter set for each specific determination standard.

We further believe that the cutoff frequency value should be significantly larger than the critical frequency k_c (compare equation (3) and (4)) as otherwise, the MSR value is determined from an extrapolation. A cutoff frequency of 13 mm^{-1} corresponds to 77 µm which is thus the minimal MSR that could be determined.

Fig. 8 shows the dependence of the determined MSR value on the cutoff frequency used. In particular, there are pronounced differences for frequencies above ~15 mm^{-1}. In this spatial frequency region, the Alicona reference spectrum has significantly higher reference amplitudes (compare Fig. 7). Thus, the transfer spectrum values are significantly lower, not shifting the Gaussian fit function to higher spatial frequencies (and thus, lower MSR values). This again indicates that our idea of small reference amplitudes being problematic should hold. Between 8 mm^{-1} and 15 mm^{-1}, the MSR values determined with both reference measurements agree to 2 µm.

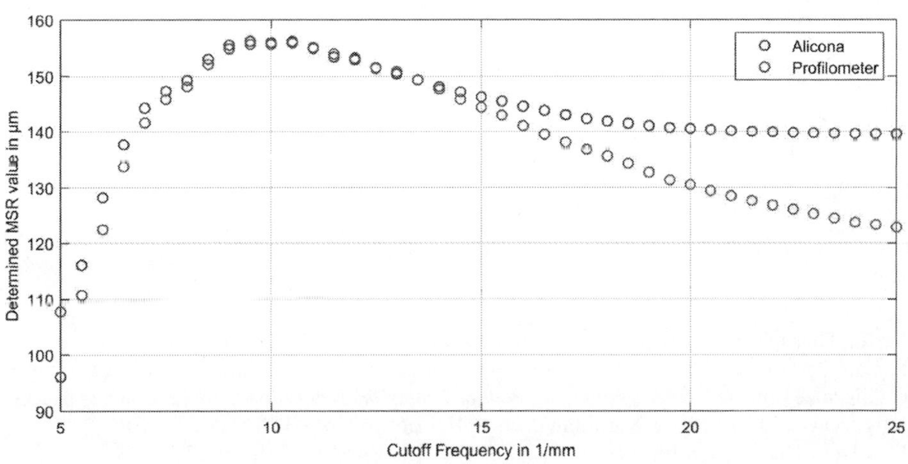

Fig. 8. Determined MSR value for measurement M1 and both reference measurements depending on the cutoff frequency used.

4.5 *Influence of Meshing*

For many metrology applications, the determined surface is converted to a mesh (e.g. in STL format) prior to subsequent processing. It is thus of interest to inspect the influence of this meshing on the structural resolution. For this purpose, the measurement volume of M1 was imported into VGSTUDIO MAX and subsequent to surface determination, the surface was converted to three different meshes using the presets 'Ray-based fast', 'Ray-based normal' and 'Ray-based precise'. Then, the MSR was determined for the volume surface as well as for the mesh surfaces. The results are summarised in Table 2. As reference, the first profilometer measurement was used.

For measurement M1, the preset 'Ray-based fast' results in a similar loss of metrological resolution as a spot of double size (compare Section 4.2). The 'Ray-based normal' and 'Ray-based precise' presets result in small differences.

Table 2. MSR values for different meshing

Measurement on:	**Volume surface**	**Mesh with Preset Ray-based fast**	**Mesh with Preset Ray-based normal**	**Mesh with Preset Ray-based precise**
M1:MSR value in µm	117	152	119	120

5. Conclusion

The quantitative determination of metrological structural resolution values using the aperiodic spatial frequency standard (ASFS) was described in detail. The determination comprehends the calculation of a Fourier spectrum, the determination of a transfer spectrum and a Gaussian fit. It could be shown that the determination method using this standard is able to detect changes in resolution due to spot size changes and is further able to detect loss of resolution due to coarse STL-meshing. The choice of the right reference measurement and a suitable cutoff frequency for the Gaussian fit was discussed.

Acknowledgements

The Zeiss Metrotom 1500 CT system was subsidised by the German Research Foundation DFG under grant 324672600. The authors thank Philipp Kaller (Institute of Manufacturing Metrology) for the tactile profilometer measurements and Andrej Torba (Institute of Manufacturing Metrology) for the focus variation measurements.

References

[1] Verein Deutscher Ingenieure e.V., "VDI/VDE 2630 Part 1.3," Düsseldorf, 2015.

[2] S. Carmignato and E. Savio, "Metrological performance verification of coordinate measuring systems with optical distance sensors," *International Journal of Precision Technology*, vol. 2, no. 2/3, p. 153, 2011.

[3] DIN Deutsches Institut für Normung e. V., "DIN EN 16016-3," 2012.

[4] DIN Deutsches Institut für Normung e. V., "DIN EN ISO 15708-3," 2018.

[5] ASTM International, "ASTM E1695-95 (2013), Standard Test Method for Measurement of Computed Tomography (CT) System Performance," 2013.

[6] F. A. Arenhart, V. C. Nardelli, and G. D. Donatelli, "Comparison of surface-based and image-based quality metrics for the analysis of dimensional computed tomography data," *Case Studies in Nondestructive Testing and Evaluation*, vol. 6, pp. 111–121, nov 2016.

[7] J. Illemann, M. Bartscher, O. Jusko, F. Härtig, U. Neuschaefer-Rube, and K. Wendt, "Procedure and reference standard to determine the structural resolution in coordinate metrology," *Measurement Science and Technology*, vol. 25, no. 6, p. 064015, Apr. 2014.

[8] M. Fleßner, M. Blauhöfer, E. Helmecke, A. Staude, and T. Hausotte, "CT measurements of microparts: Numerical uncertainty determination and structural resolution," in *SENSOR 2015, Nürnberg, 19.-21.05.2015*, 2015.

[9] M. Fleßner, N. Vujaklija, E. Helmecke, and T. Hausotte, "Determination of metrological structural resolution of a CT system using the frequency response on surface structures," in *MacroScale 2014 - Recent developments in traceable dimensional measurements, Wien, 28.-30.10.2014*, 2014.

[10] *MathWorks MATLAB Documentation: Least-Squares Fitting*, https://uk.mathworks.com/help/curvefit/least-squares-fitting.html (22. 05.2019).

[11] M. Krystek, "ISO-Filter für die Fertigungsmesstechnik," *tm - Technisches Messen*, vol. 76, no. 3, pp. 133–159, jan 2009.

[12] DIN Deutsches Institut für Normung e. V., "DIN EN ISO 25178-600 (Entwurf)," 2017.

X-ray back scatter techniques for additive manufacturing

Uwe ZSCHERPEL [1], Maryam NAZARZADEHMOAFI [1], Mingchuan YUAN [1], Anja WASKE [1]
[1] BAM Berlin, Berlin, Germany
Contact e-mail: uwez@bam.de

Abstract

X-ray back scatter imaging is rarely applied compared to classical X-ray projection imaging. 20 years ago the company Philips developed "COMSCAN", a first application case in the aircraft industry, which allowed even a depth resolution using back scatter imaging. The company AS&E in Boston offers back scatter imaging solutions for the security market. Their principle is to scan the object with a highly collimated X-ray needle beam from one side only and detect the backscattered radiation by a large area detector side by side with the collimation wheel. A new prototype is investigated at BAM for application and optimization in non-destructive testing. As modern industrial application field in-situ testing in additive manufacturing is targeted. The accessibility of the printed part during the production process is very limited. This prevent the application of a two sided X-ray inspection or Computed Tomography, were an rotation of the object is required to acquire projections from 360 degrees. An important advantage for the X-ray back scatter technique are also the materials used in additive manufacturing (polymers, ceramics, light metals like Aluminum or Titanium). These materials with lower density and lower Z values give better scatter signals than metals with higher densities and Z values. The back scatter intensity decreases with increasing density and Z value of the material. But the requirements on spatial resolution and contrast sensitivity are more stringent for non-destructive testing of additive manufactured parts compared to the security area. In NDT sizes of indications smaller than 1 mm have to be detected clearly. The investigation of these limits on a state-of-the-art prototype for X-ray back scattering using rotating collimated X-ray needle beams is a part of the BAM project "ProMoAM". The contribution shows first results of the optimization for NDT and the achieved application limits for several example cases.

Sicherheit in Technik und Chemie

BAM

Bundesanstalt für Materialforschung und -prüfung

X-RAY BACK SCATTER TECHNIQUES FOR ADDITIVE MANUFACTURING

U. Zscherpel, M. Nazarzadehmoafi, M. Yuan, A. Waske

Federal Institute for Materials Research and Testing (BAM), Berlin
Division 8.3 „Radiological Methods"

DIR2019, Fuerth, 03.07.2019

CONTENTS

BAM

Introduction

X-ray back scatter techniques
- Starting point: Applications for security
- Various detection principles

Prototype „ModBx" of AS&E, Boston
- Construction and conditions for work
- Example images and optimization of image quality

Applications for additive manufacturing
- First results
- Additional test objects
- Integration into AM equipment?

Applications for Security

BAM

Aims of imaging by X-ray back scattering:

- Mobile application with **one-sided** measurement access only
- Detection of weapons, explosive materials and contraband in baggage and freight containers
- Identification of dangerous liquids in complex environment

X-ray back scatter techniques: various principles

BAM

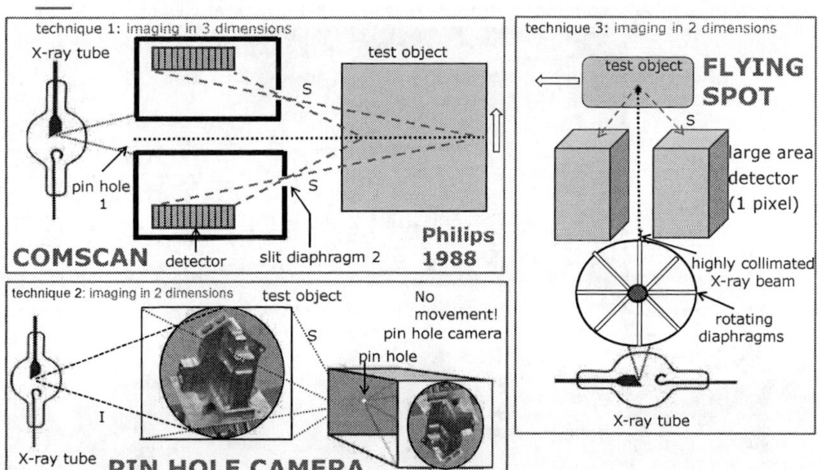

Example image by single slit diaphragm

BAM

back scatter image

Experimental parameters:
- 420 kV, 10mA, 3 min
- Diaphragm slit opening: 0.9 mm
- Source object distance: 61cm
- Object diaphragm distance: 47 cm
- Diaphragm detector distance: 22 cm

- Damages in the plate can be identified
- Intruded water between surface and core can be visualized
- Structural integrity can be displayed, **but only limited field of view**

AXISS: mobil X-ray back scatter technique by American Science & Engineering

AS&E

X-ray source: 140 kV, rotating anode with 600 W continuous power

Mobile system with 2D scanning: rotating collimator & horizontal sweep

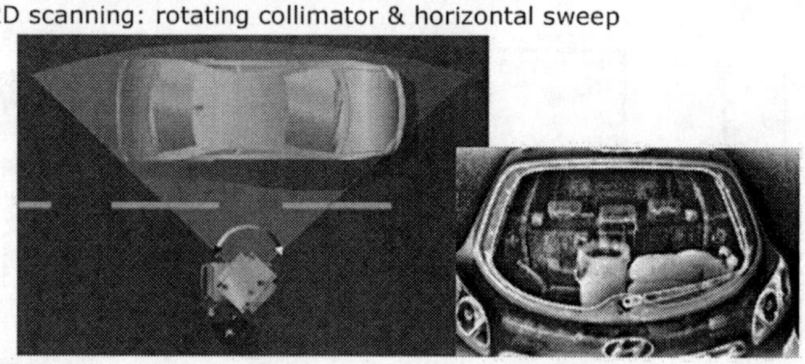

Field of view: vertical 42°, source adjustable by +/- 22° top/bottom:

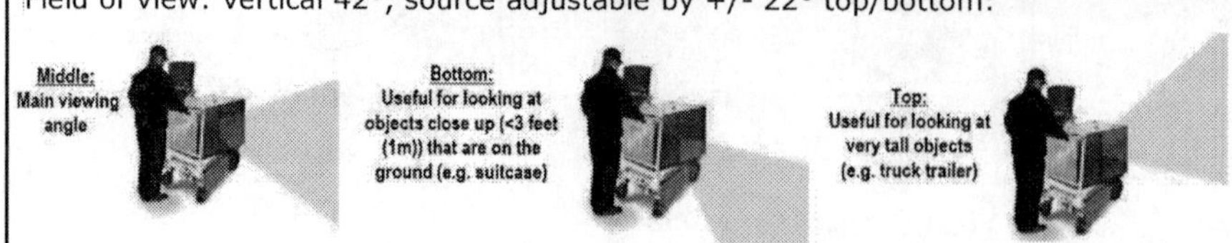

Back scatter prototype "ModBx" by AS&E in BAM 8.3

BAM

0.1 mSv/h in beam in 1m distance

Front view on collimator wheel:

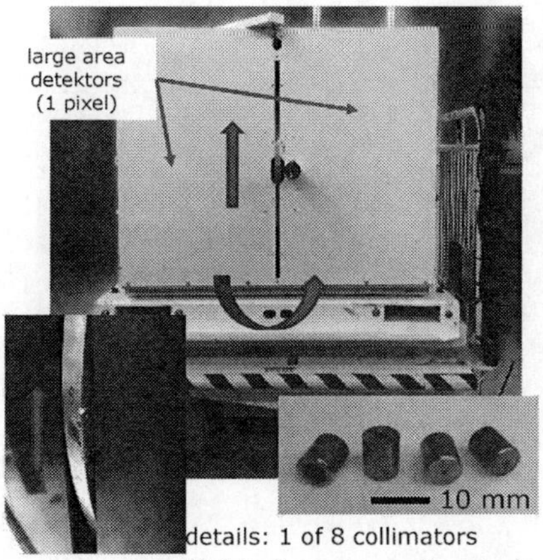

details: 1 of 8 collimators
(0.1 to 2 mm inner diameter)

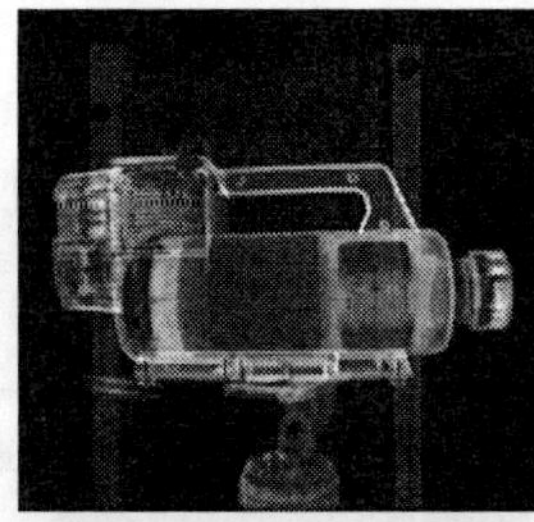

40 cm
distance

200 kV flash tube
(plastic case)

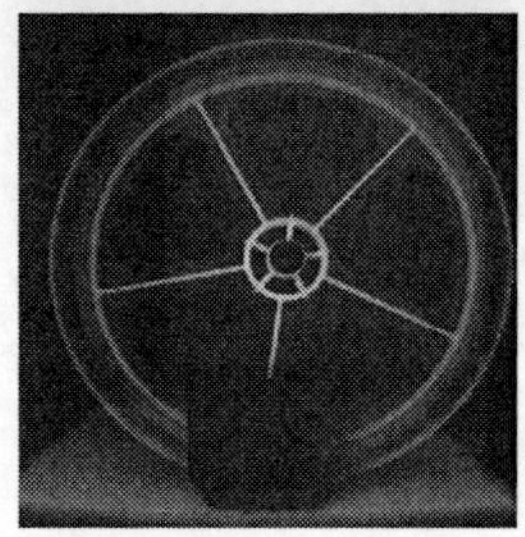

40 cm
distance

40cm
Plastic wheel

Example images: flying spot back scattering, back scatter pin hole camera & penetration image with flash tube

BAM

Back scattering with flying spot collimator: 0.5 mm needle beam diameter

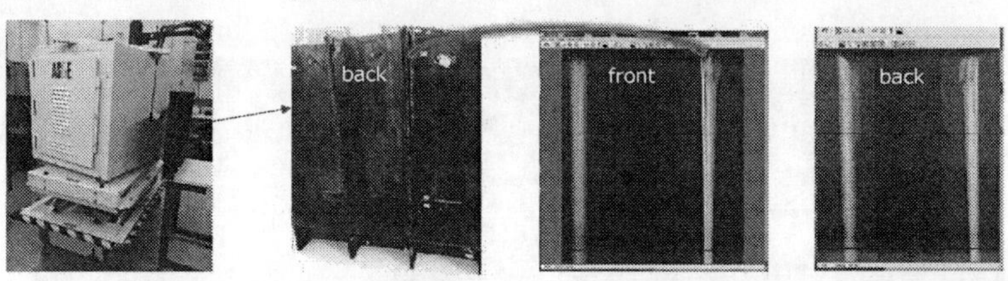

Back scatter camera with single slit: 0.5 mm

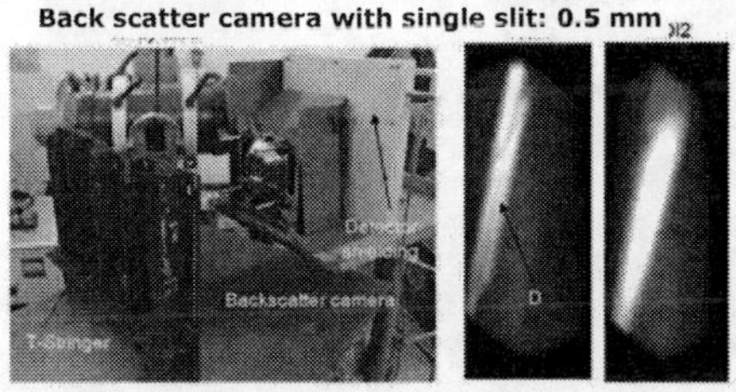

Penetration with flash tube: pixel 0.15 mm

Optimization of back scatter prototype "ModBx": stack of Aluminum plates with hole IQIs

BAM

side view

7.7 mm

IQIs stacked on 8 plates

front view
76.8 × 100 × 1.8 mm^3

IQI hole Ø: 0.15 - 2 mm

back view

Optimization: higher X-ray energy → higher scatter signal

BAM

pixel size: 0.3 mm
collimator: 0.5 mm
D = 130 mm

70 kV	100 kV	140 kV

Better image quality with higher X-ray energy

70 kV	100 kV	140 kV
SNR= 6	SNR= 15	SNR= 35
GV= 282	GV= 2062	GV= 6902
SR_b = 0.4 mm	SR_b = 0.4 mm	SR_b = 0.4 mm
CNR=3.5	CNR=11	CNR=18

Basic spatial resolution SR_b: Duplex wire IQI according to ISO 19232-5

BAM

X-ray voltage 140 kV, 0.5 mm Collimators
D = 130 mm

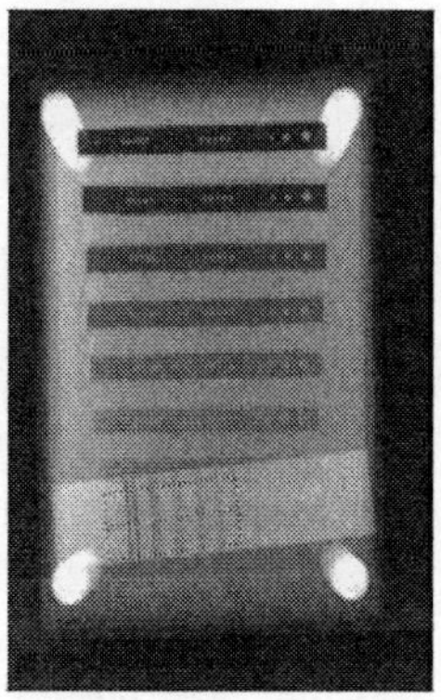

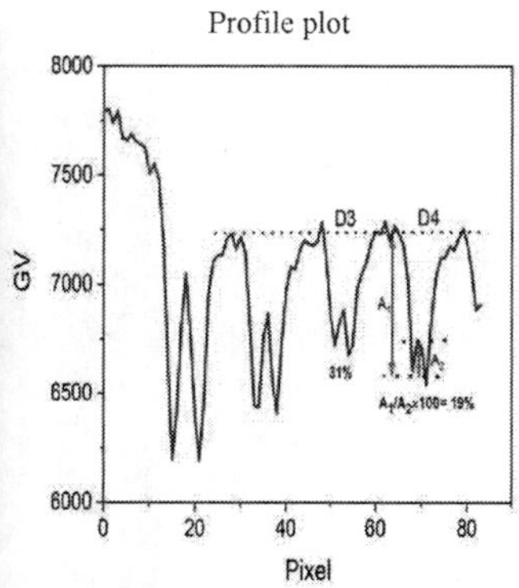

Basic spatial resolution
ISO 19232-5
D4
↓
SR_b = 0.4 mm
at
0.3 mm pixel size

0.1 mm collimators: SR_b better, but signal too low!
0.25 mm collimators: built, testing in progress...
Wolframet drilled, not eroded!

Optimization: SNR & spatial resolution with Duplex IQI

BAM

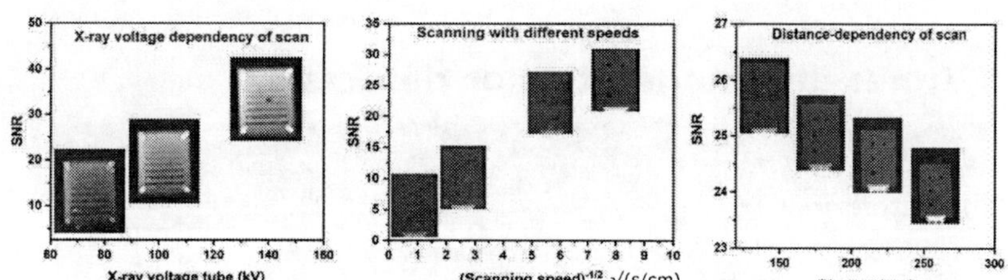

large pixels

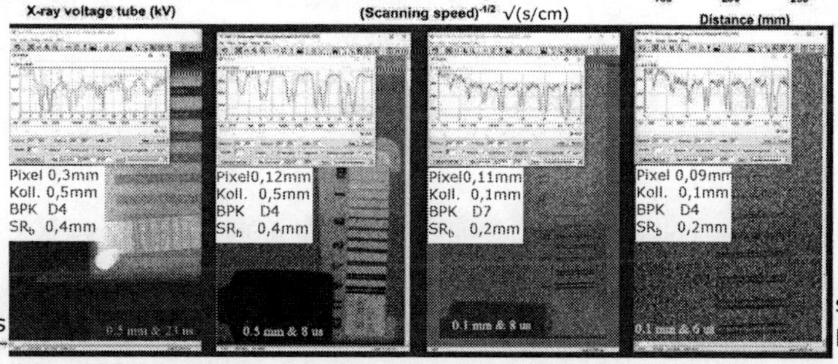

small pixels

Enhancement of basic spatial resolution by reduced collimator holes and higher sampling frequency but reduction of SNR

BAM

140 kV, 4.2 mA, 600 W

Collimator hole: 0.5 mm	Collimator hole: 0.1 mm

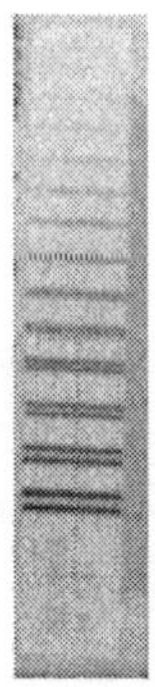

ISO 19232-5

D4

↓

SR_b = 0.4 mm

- Pixel size = 0.12 mm

ISO 19232-5

D7

↓

SR_b = 0.2 mm

- Pixel size = 0.09 mm

➡ Mechanical tolerances limit SR_b to 0.2 mm!

Accentance levels in the aircraft industry for parts manufactered by AM

BAM

- Inner inhomogenities or defects -

Imperfection or defect	Acceptance criteria for **internal** characteristics for metallic 3D printed parts		
	Level A	B	C
Cracks	Not permitted		
Gas cavities	No line porosity is accepted		
	< 200 µm		**< 500 µm**
Solid inclusions	< 200 µm		< 500 µm
Lack of fusion	Not permitted		

[3] Airbus Spec: AIPS 01-04-020 „Additive Manufacturing –Powder Bed Fusion

See talk by Testia: http://www.spacetechexpo.eu/site/presentations/ non-destructive-testing-on-metal-am-parts-and-structures.pdf

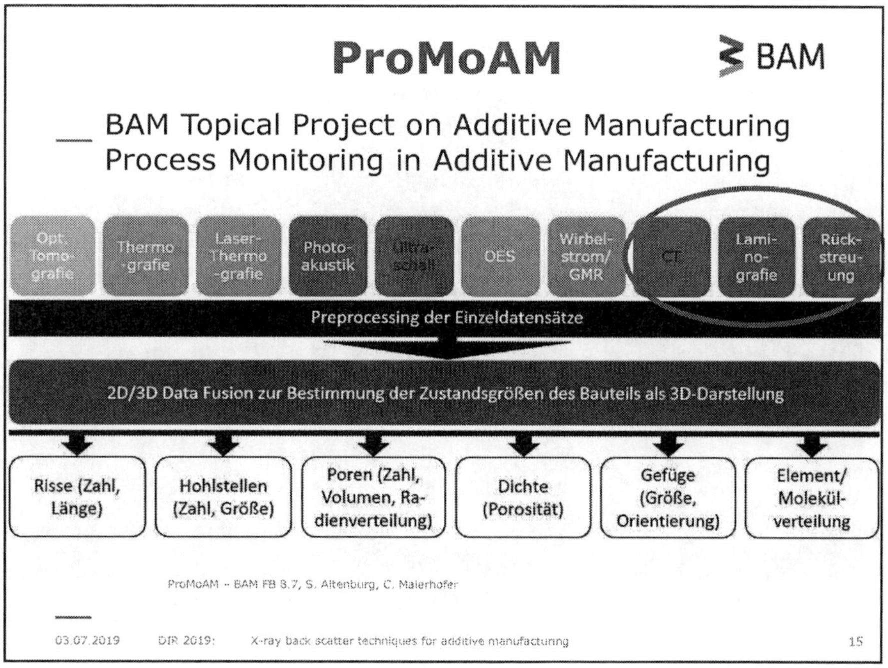
ProMoAM
BAM
BAM Topical Project on Additive Manufacturing
Process Monitoring in Additive Manufacturing
Opt. Tomo-grafie
Thermo-grafie
Laser-Thermo-grafie
Photo-akustik
Ultra-schall
OES
Wirbel-strom/ GMR
CT
Lami-no-grafie
Rück-streu-ung
Preprocessing der Einzeldatensätze
2D/3D Data Fusion zur Bestimmung der Zustandsgrößen des Bauteils als 3D-Darstellung
Risse (Zahl, Länge)
Hohlstellen (Zahl, Größe)
Poren (Zahl, Volumen, Ra-dienverteilung)
Dichte (Porosität)
Gefüge (Größe, Orientierung)
Element/ Molekül-verteilung
ProMoAM – BAM FB 8.7, S. Altenburg, C. Maierhofer
03.07.2019 DIR 2019: X-ray back scatter techniques for additive manufacturing 15

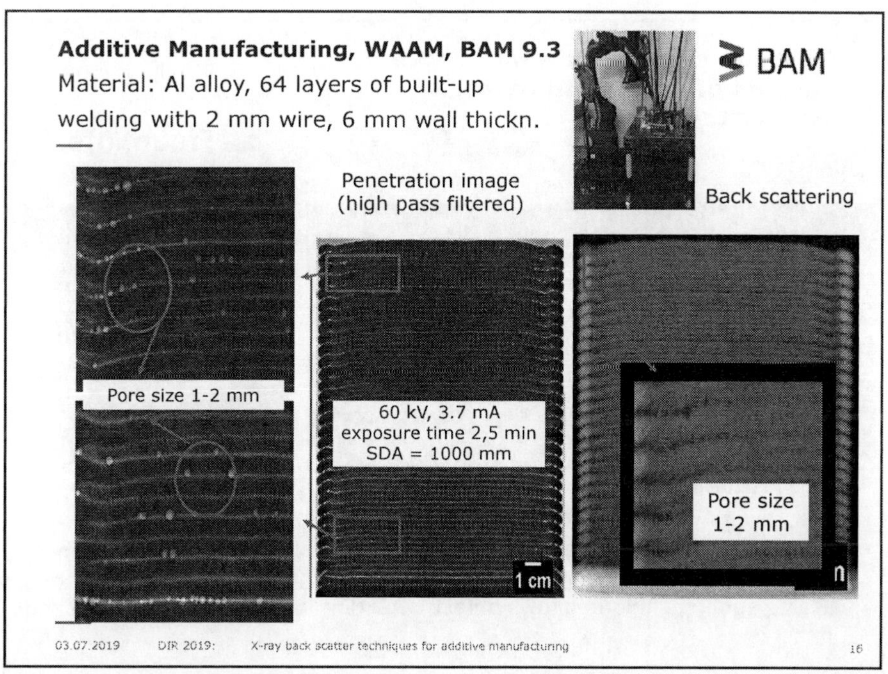
Additive Manufacturing, WAAM, BAM 9.3
Material: Al alloy, 64 layers of built-up welding with 2 mm wire, 6 mm wall thickn.
BAM
Penetration image (high pass filtered)
Back scattering
Pore size 1-2 mm
60 kV, 3.7 mA
exposure time 2,5 min
SDA = 1000 mm
1 cm
Pore size 1-2 mm
03.07.2019 DIR 2019: X-ray back scatter techniques for additive manufacturing 16

Test samples of Fraunhofer IWS

BAM

Fraunhofer

AGENT3D
ADDITIV GENERATIVE FERTIGUNG

Contact at IWS: Samira Gruber

Starting point STL-File:

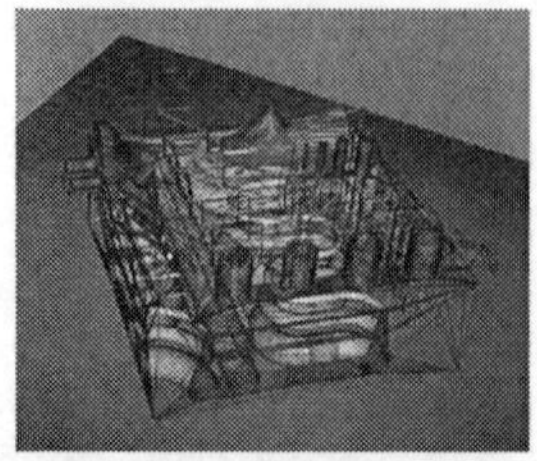

length: 40 mm
width: 40 mm
hight: 25 mm

SLM 3D prints from powder of different materials:

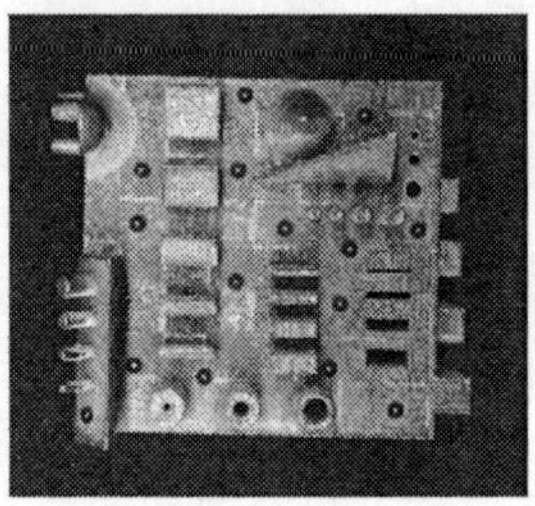

AlSi10Mg

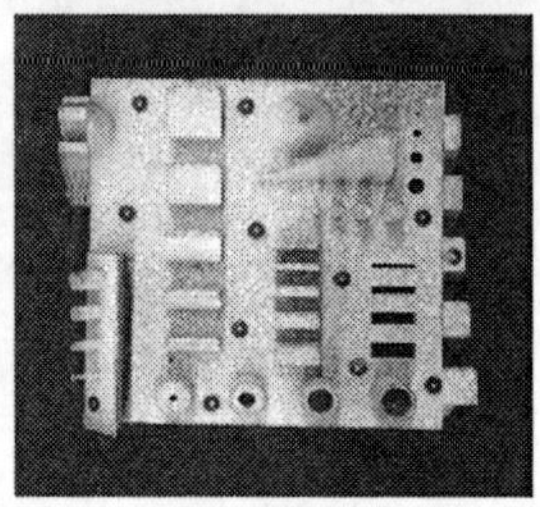

Ti-6Al-4V

INC718

Test samples of Fraunhofer IWS Results of inspection by back scatter technique

BAM

AGENT3D
ADDITIV GENERATIVE FERTIGUNG

Fraunhofer

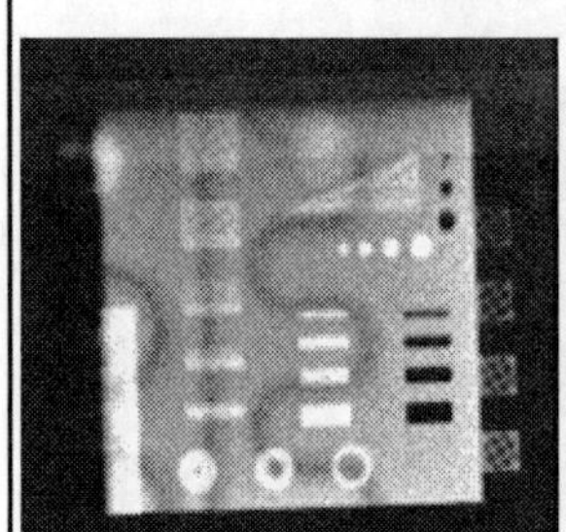

AlSi10Mg

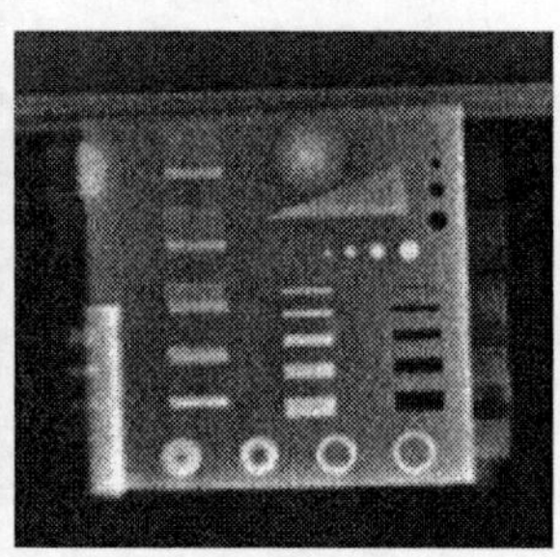

Ti-6Al-4V

INC718

- Best results with light metals (Compton scattering highest)
- Detection inner structures depending on material (attenuation by steel requires higher X-ray energy)
- "Flying spot" technique allows defect detection in real time above 0.2 mm

Conclusions

BAM

- Image quality in back scattering:
  - long scanning time (down to 1 min/cm)
  - small Object Collimator Distance
  - high X-ray voltage (max. 140 kV)
  - low material attenuation,
    light materials are the best scatterer
- Basic spatial resolution limited by mechanical tolerances of collimator wheel, if collimator holes too small, only signal loss. Optimum for ModBX: 0.25 mm
- Integration into AM set-up? Place in front of open printer!
- Further investigations within "ProMoAM" of BAM

AS&E prototype

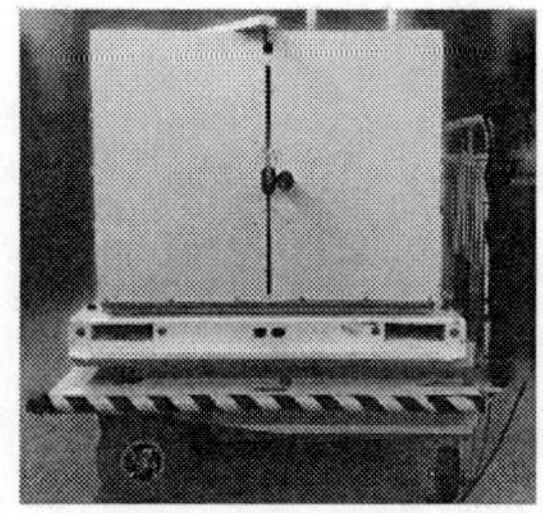

OUTLOOK: further miniaturization 2016: test of mobile X-ray back scatter technique Limitation up to 70 kV (10W), today 120 kV reached!

BAM

AS&E Mini-Z

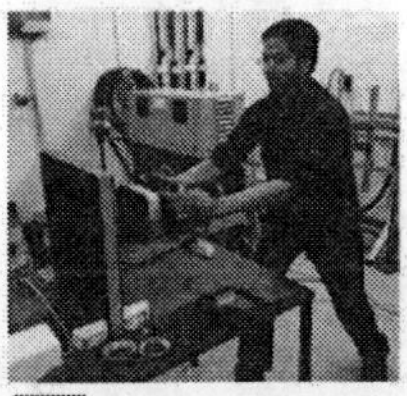

Tests at BAM

LuFo IPrO 2016 Project with Airbus

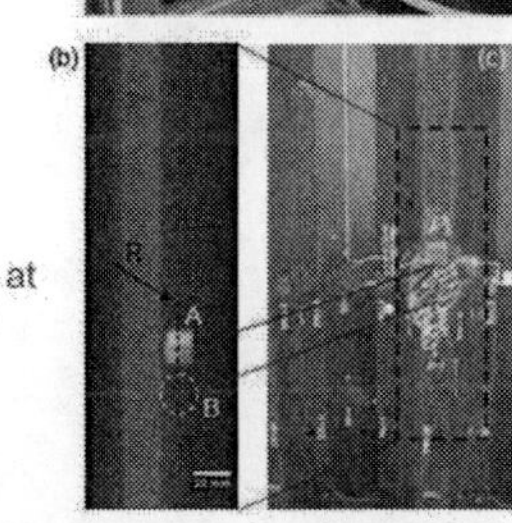

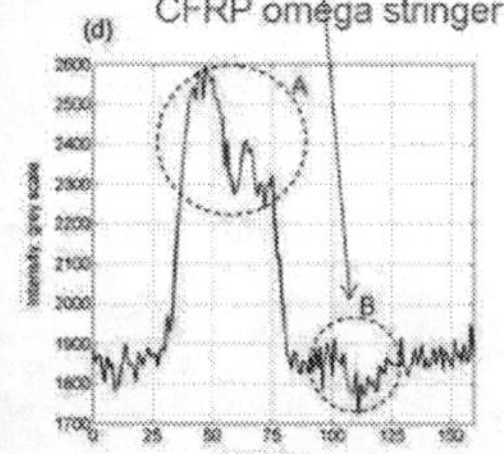

BAM

The END

BAM

Next steps:

- Test of ModBx prototype for in-situ measurements
  - External detector panels
  - Optimization of Collimator size (Contrast – Resolution)
- Investigation of identical AM test objects from Al, Ti and Inconel (Fraunhofer IKS Dresden)
- Ex-situ measurements of ProMoAM test objects
- Development of scenarios for combination of ModBx prototype direct in the AM printers of BAM 9.3 / 9.4

externe
Detector panels

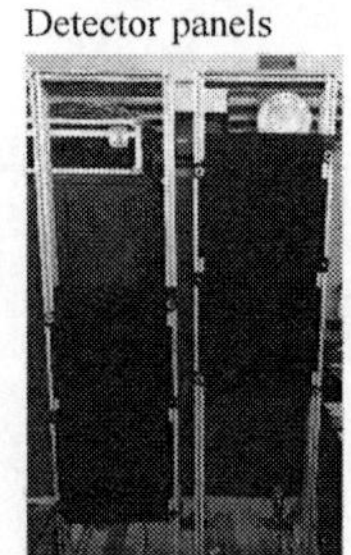

X-ray back scatter techniques

BAM

- Nomovement of object necessary
- One-sided access to object sufficient
- Suitable for typical AM materials (plastics, ceramics, metalls)
- In-situ measurements possible!

AS&E prototype

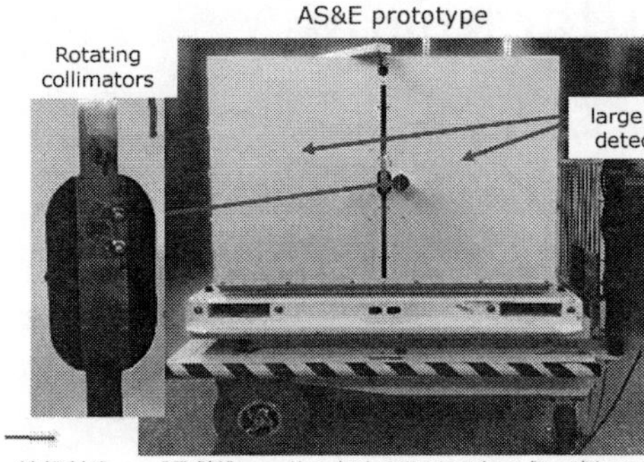

Basic spatial resolution (0.5mm coll.)

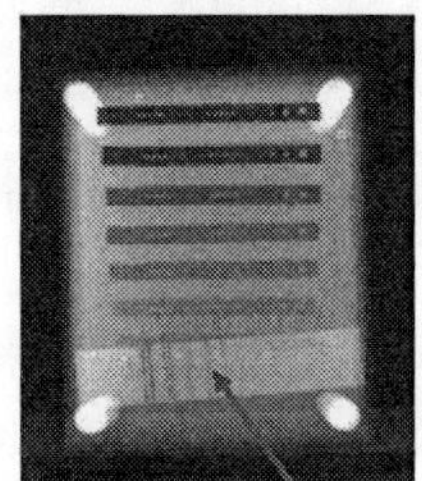

Ø = 400 µm

Aluminum plate

size: 60 × 100 × 3 mm³
hole Ø = 8, 6, 5, 4, 2, 1 mm

BAM

scans with different scanning speeds (pixel size 0.29 mm):

Speed: fast	Speed: 1/4×	Speed: 1/32×	Speed: 1/64×
SNR= 2	SNR= 6	SNR= 18	SNR= 26
GV= 3376	GV= 3607	GV= 3631	GV= 3651
CNR=0.8	CNR=1.8	CNR=5.5	CNR=7.6

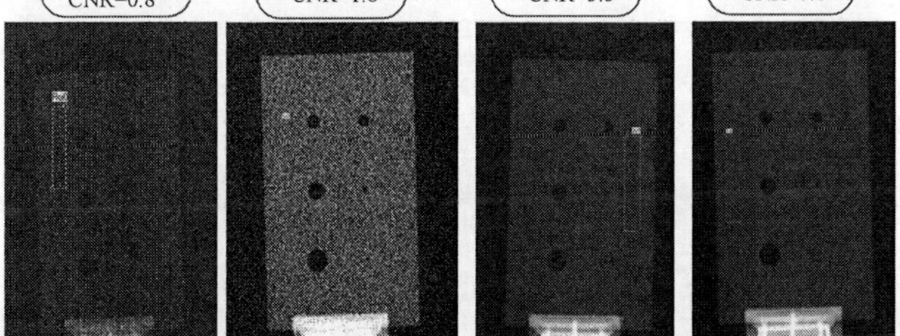

Image quality increases with scanning time

X-ray and Computed Tomography as a tool for quality assurance and process optimization in the field of additive manufacturing

Philip SPERLING[1], Gabriele MAEURER[1], Nils ACHILLES[1]
[1] Yxlon International, Hamburg, Germany

Contact e-mail: Gabriele.Maeurer@hbg.yxlon.com

Abstract

Computed Tomography (CT) is becoming an increasingly important technique for quality control of additive manufactured (3D printed) parts. Complex AM parts with a variety of inner structures, require additional, more precise testing methods to detect smallest defects in µm scales and to ensure the reproducibility of the part. This paper will demonstrate how CT contributes to the quality control of additive manufactured parts and how it improves the entire processing chain.

Additive manufacturing (AM) is one of the most revolutionary processes in recent years, making a dramatic impact on the industrial market. Also known as 3D printing, AM is a manufacturing technique that builds objects layer by layer. The materials range from polymers and metals to composites and multi-materials. This fast-evolving technology is increasingly adopted by many industries and allows completely new product designs and functionalities. However, the more complex the design, the more challenging the quality control will become.

Many additive manufacturers and designers have relied on industrial computed tomography (CT) as non-destructive testing (NDT) method for quality assurance throughout the R&D and production processes. The 3D volumetric density map that is produced from a high number of 2D X-ray projections, allows insight into the type and location of defects in an AM part and thus assesses the quality of the part. In addition, R&D engineers also use CT as tool for the optimization of the production process, i.e. to ensure the stability of the process parameters and the consistency of the part quality. CT provides valuable information about the design, the raw materials, critical defects and the verification of geometric parameters.

In this paper several application examples from different industries will be discussed to show challenges, problems and typical defects throughout the AM production. It will demonstrate how industrial CT can contribute to the improvement of the AM processes with regard to analysis and inspection for quality assurance, metrology and process optimization.

The Future of Manufacturing: Industry 4.0

Additive Manufacturing (AM) is one of the key technologies with the potential of revolutionizing the current manufacturing processes as it allows individualization and customization of products. This will bring a shift from mass production to single-item production, and from centralized production to local on-the-spot production. AM allows new designs and additional functionality of products which are currently not possible with existing constructions and manufacturing techniques.

1. Additive Manufacturing – A Growing Industry

Starting out mainly as a prototyping technology, additive manufacturing has gained tremendous momentum for use in industrial applications during the last 5 years. Conventional manufacturing techniques can produce a great variety of shapes and designs, however, designers are bound to predetermined patterns by production constraints, moulding forms and/or material behaviour. Additive Manufacturing can overcome this and take design and production to the next level.

One of the greatest benefits is of AM is the combination of different methods, materials, forms, shapes and properties resulting in parts with higher complexity and new product features. For example, shapes with a scooped out or hollow center can be produced as a single piece, without the need to weld or bond individual components together. Also, weight can be reduced by internal structural design changes with the advantage of being stronger. The hip cup as in Fig.1 is an example from the medical field and shows a novel design of an implant that is impossible to produce when using conventional precision casting. AM allows the exact fit of such cups in the size and shape to a person's individual hip and thus generates the most precise match prior to an operation. The rough surface with specific grid structures serve for better in-growth results and to avoid rejection reactions.

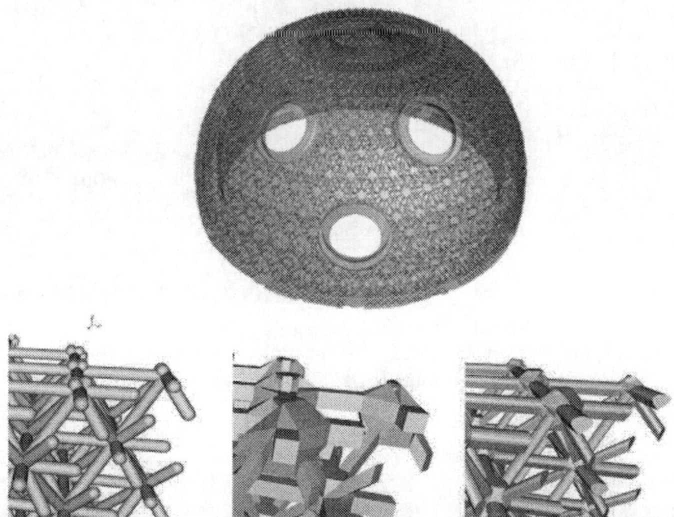

Fig.1 Additive manufacturing offers new levels of individualization and complexity. Here: hip cup implant (medicine) and possible surface grid structures, © AMPower GmbH & Co. KG

2. CT Technology as quality inspection tool in the AM Process Chain

While AM has opened up more freedom to product designers, the inspection methods have to be adapted to the new constructions as well. Little is yet understood what defects can occur in an AM manufactured parts and about the root cause of such effects. NDT techniques, and among them particularly CT, seems to be an efficient tool to uncover them, thus providing an indication on what went wrong in the manufacturing process. The easy insight into the inner structures of a part makes it a particularly useful method.

2.1 Powder Analysis

Like in any other production process the individual steps along the AM process include upstream, in-situ and downstream processes and have to be closely monitored. Various techniques are already applied for in-situ monitoring, mainly as integral or optional part of commercially available 3D printers. For upstream and downstream processes CT is a beneficial tool which provides valuable insights. Prior to the actual printing the analysis of the base AM materials (i.e. powder, granulate, filament) is essential as little variances in properties can have a huge effects or even ruin the printed results. CT, for example, provides information on metal powder, i.e. internal porosity (open or closed), particle morphology (volume, surface area, sphericity) and the presence of impurities.

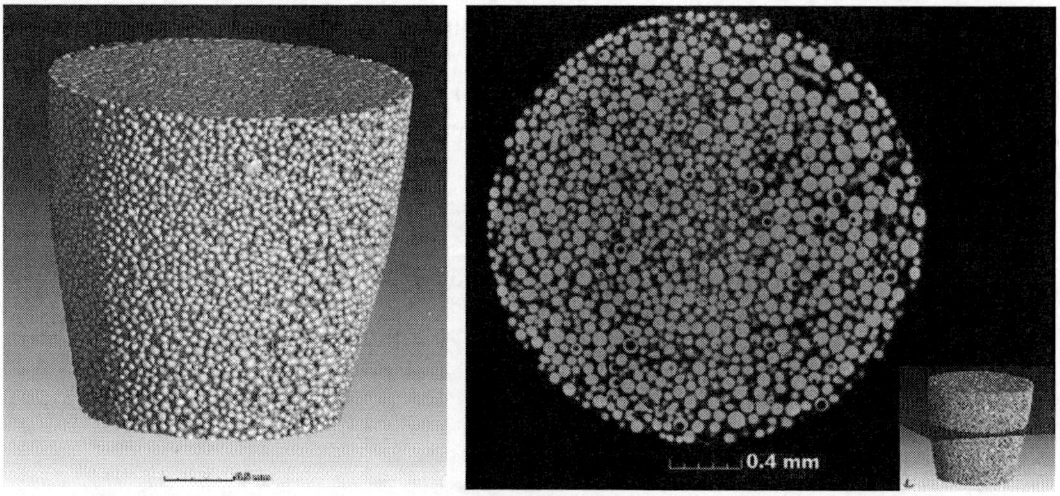

Fig.2. Powder analysis with CT for particle morphology, porosity and presence of impurities.© Stellenbosch University

2.2 CT as Quality Assurance Tool

More widely CT is used as part of the AM downstream processes where it helps to ensure the quality of printed parts. Typical defects include porosity, lack of fusion, balling, excessive surface roughness or micro-structural issues that can be easily detected with CT. Other challenges such as density variation, embedded features, organic part design, dissimilar metals or materials and surface finishes are also addressed with CT within a relatively short amount of time. Detecting errors inside an object without destroying it (i.e. in a metallographical analysis) is an essential benefit of this technique.

Fig.3 shows an AM demonstrator that was produced to uncover typical quality issues and this way allows conclusions on the printing process. In this part inside channels were printed and their shape was inspected with CT. In this example, cracks have formed between the channels due to residual stress during the 3D printing process. These cracks might lead to part failure when exposed to high pressure fluid or gas. Another defect is the residue powder in the inner channels of this part resulting in congestions (grey areas) and could be a reason for reduced flow capacity or complete blockage.

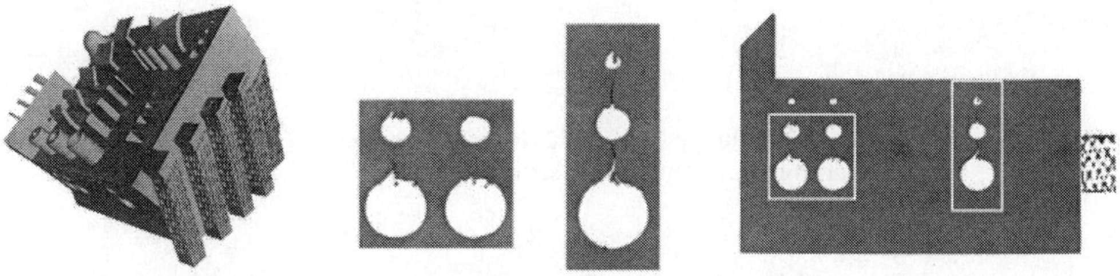

Fig.3 AM-demonstrator. CT detected cracks in the channels that formed due to residual stress during 3D printing. © Fraunhofer IWS / AGENT – 3D e.V.

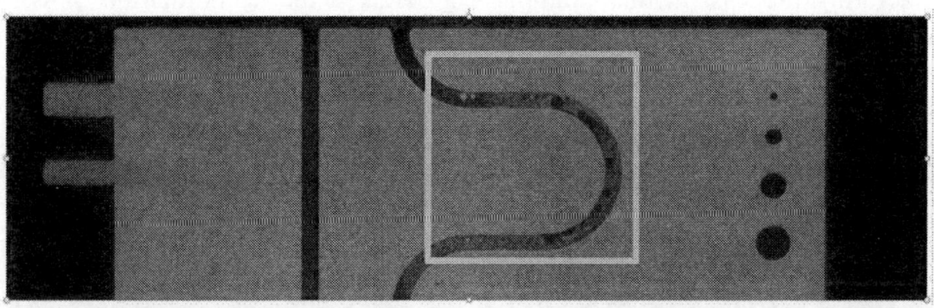

Fig. 4 AM-demonstrator. CT detected powder residue inside the channel that might prevent continuous liquid or gas flow. © Fraunhofer IWS / AGENT – 3D e.V.

2.3 CT as tool for measurements

Based on the CT reconstruction images it is possible to conduct precise measurements of internal structures which is an additional benefit for quality assurance in additive manufacturing. The helicopter bell crank, for example, was designed based on bionic principles with the goal to reduce the weight of the original part while increasing torsional and bending stiffness. X-ray and CT helped to investigate the coaxiality and symmetry of this part thus ensuring the reproducibility. CT was used to look for typical defects that can occur during the printing process. A comparison of a CAD model with CT data allows additional information, such as the location of possible deviations. Due to the new bionic design, the weight of the bell crank was reduced by 35% and answers the requirements of the aerospace industry for stronger, but lighter materials.

Fig.5. Bell Crank:
conventional design

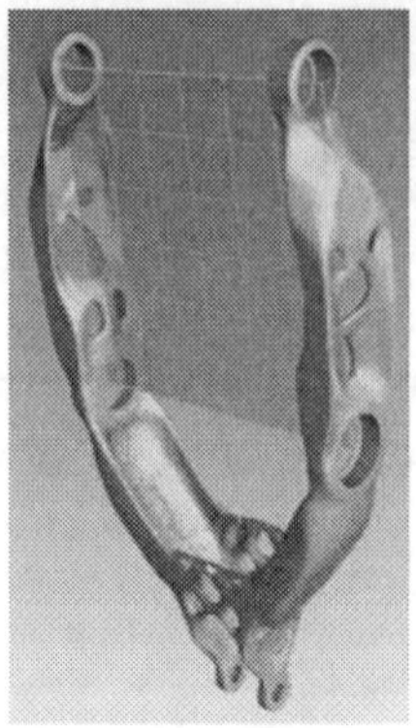

Fig. 6. Bell Crank:
bionic design

Fig.5. and **Fig.6.** Conventional and additively manufactured helicopter bell crank (optimized by bionic design) - 35% saving in weight, © Fraunhofer IAPT

2.4 CT and optimization of the AM process

Not all aspects of additive manufacturing are fully understood yet, therefore researchers are trying to identify critical factors in the printing process and their effects on the printed parts. Industrial CT helps to determine such relationship and allows conclusions on the manufacturing process. In this example CT detected the accumulation of porosities at a constant distance from the rim of the printed structures (around the channels and close to the part borders). Researchers are investigating these effects and trying to correlate them with the printing parameters. Limiting or avoiding such errors will optimize the printing process resulting in reduced risk, cost, and manufacturing time. The CT scan data help to predict specific effects based on their images and allow a fast and easy feedback to increase productivity.

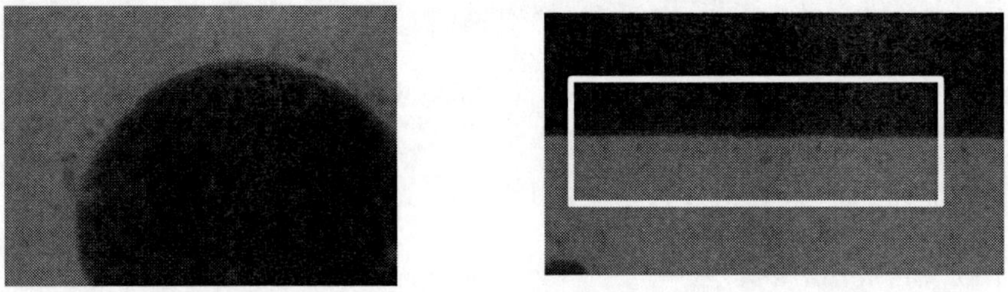

Fig. 7 AM-demonstrator. CT detected porosities at constant distances around channels and close to part borders. © Fraunhofer IWS / AGENT – 3D e.V.

3. Conclusion

Computed Tomography (CT) is a useful NDT method for the investigation of additive manufactured parts. Complex inner structures resulting from new design possibilities can be easily detected and measured with CT. Typical errors in the printing process include porosities, lack of fusion and balling effects. CT also allows conclusion on the manufacturing process by identifying critical factors in the printing process.

X-ray CT offers many benefits for quality control of 3D printed parts, among them routines of defect recognition, CAD (target-actual) comparisons and analyses of raw materials. It also provides feedback on possible erroneous printing parameters and contributes to faster development processes for new AM parts and their properties.

Preliminary Research on Laser Plasma Accelerator based Inverse Compton Scattering X-ray Imaging

Yongshun XIAO [1], Bo WANG [1], Yue MA [1], Zhiqiang CHEN [1]
[1] Department of Engineering Physics, Tsinghua University, Beijing, China
Contact e-mail: xiaoysh@mail.tsinghua.edu.cn

Abstract. Industrial CT system offers high precision non-destructive detection and capturing of the geometry of objects, especially for complex objects with hidden or difficult accessible surfaces. Modern micro focus X-ray CT system is used for high precision dimensional metrology. However, due to the energy and power limitations of the micro focus X-ray source, the micro-CT system currently performs high-precision metrology only for small objects of low-density materials. This paper will introduce the advantages of 3D CT imaging technology for accurate imaging complex structures inside objects, analyze the research and development trend of high-precision CT, and introduce the research progress of Tsinghua University in improving industrial CT resolution. Furthermore, preliminary research on laser plasma accelerator based inverse Compton scattering X-ray Imaging will be introduced. The novel micro CT system based on inverse Compton scattering radiation have the advantage of both high energy as well as small focal size. The design of micro-focus high-energy industrial CT based on new laser driven radiation source and its application prospects in high precision dimensional metrology of industrial components will be described.

Introduction

Because its strong ability to penetrate objects, X-ray has been widely used in medicine and industry for the detection of human bodies or objects, since its discovery in 1895. Later in 1967, Hounsfield designed and developed the first computed tomography (CT) machine, which laid a solid foundation for the extensive application of CT technology in non-destructive testing.

Micro CT, compared with traditional CT, can achieve a focal size of tens of microns, or even a few microns, which in turn has high accuracy with the resolution of micron level. At present, it is widely used in medicine, materials science, archaeology, agriculture and other fields[1]–[4], for its excellent qualities in quantitative detection. In industrial area, micro CT is currently possible to perform high-precision metrology of small objects of low-density materials[5].

However, due to the limitation of modern X-ray sources, it's hard to achieve small focus size and high radiation energy in one single source. Though the X-ray sources based on synchrotron radiation facilities can reach higher energy as well as small focus size, the number of those sources impede daily tests in industrial manufacture.

A novel kind of X-ray source based on inverse Compton effects has been under research in many facilities, which makes use of the inverse Compton effects between high brightness laser and high energy electrons.

The energy of generated photon can be expressed as

$$E_\gamma = \frac{E_L(1-\beta\cos\theta_1)}{1-\beta\cos\theta_2+\frac{E_L(1-\beta\cos(\theta_2-\theta_1))}{E_e}} \quad (1)$$

In which $\beta = \sqrt{1-\gamma^{-2}}$ and $\gamma = \frac{E_e}{0.511}$, and E_L is the energy of laser photon and $E_e$is the energy of electron, θ_1 is the angle between the direction of the laser photon and the electron and θ_2 is the angle between the direction of the generated photon and the electron. Thus, increasing the energy of laser and electron can get X-ray with higher energy. Many laboratories have achieved this kind of source and tried to use it in imaging.

H. Toyokawa presented in 2008, a Laser-Compton scattering(LCS) X-ray beam has been developed for industrial applications[6]. With the 800-MeV electron from storage ring, the LCS beam can be tuned up to 20MeV. And X-ray phottographs and Ct images of concrete specimen have been obtained. The focal size of the X-ray is 0.11 mm. To achieve such a high energy, this focal size is reasonable.

In 2012, K. Ta Phuoc et al, France, achieved an all-optical Compton gamma-ray source, which used same laser pulse for electron acceleration as well as inverse Compton by using so-called plasma mirror[7]. The source size was less than 3 um and the energy could reach several hundred keV. One radiograph of USB flash device was token and the internal structure could be clear seen.

In 2015, A. Döpp et al, showed one single-exposure X-ray imaging of a clock, by using the all-optical Compton source. Electron beams with 150 MeV peak energy were used to generate X-ray with a peak energy of 500 keV and about 50 percent energy spread at full width at half maximum (FWHM)[8]. Changhai Yu et al, in 2016, achieved tunable quasi-monochromatic MeV gamma-rays with high brightness.

1. Design of System

In Tsinghua University, we are researching on high-energy X-ray high-precision industrial CT. The inverse Compton source based on the Laser Plasma Accelerator can produce X-rays with energy up to 100keV and focus size of few um. And now the radiograph of line-pair card of 10lp/mm has been obtained and the line pairs can be clearly distinguished.

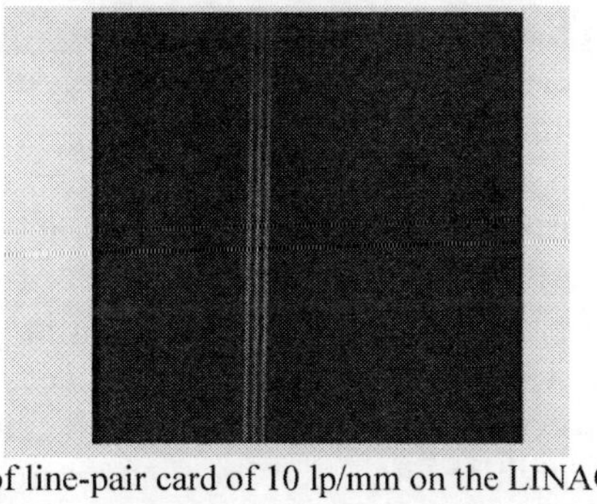

Fig. 2. Radiograph of line-pair card of 10 lp/mm on the LINAC Compton system

As the equivalent beam width of the CT system is expressed as:

$$BW = \frac{\sqrt{d^2+[a(M-1)]^2}}{M} \quad (2)$$

In which, d is the size of detector pixel, a is the focal size and M is magnification. M can be expressed as:

$$M = \frac{SD}{SO} = \frac{SO+OD}{SO} \quad (3)$$

SD is the distance between X-ray source and the detector, OD is the distance between the object and the detector, and SO is the distance between the source and object. Thus, equation can also be transformed as:

$$BW = \sqrt{(d\frac{SO}{SO+OD})^2 + (a\frac{OD}{SO+OD})^2} \quad (4)$$

Some simulations are done to show the relation of BW and the detector pixel size as well as the relation of BW and the magnification. The result is shown in Fig.1. In the simulation the size of X-ray focus is assumed as 2 um.

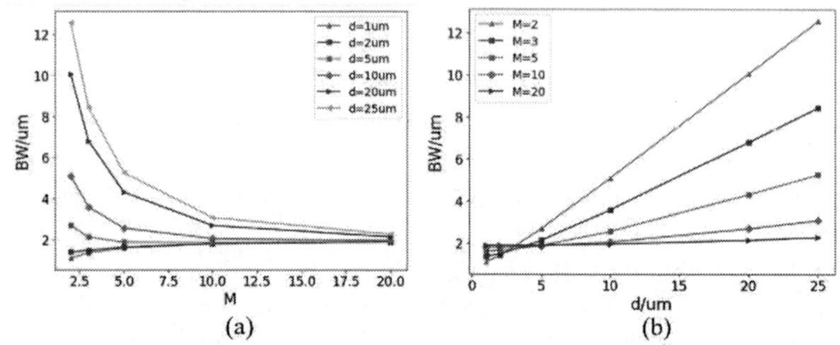

Fig. 2. The relation of BW and M as well as BW and d

As seen from Fig.2 when the magnification is fixed, the BW increases as the detector pixel size increases. And the BW is almost directly proportional to d, when d is much larger than a. When the pixel size is fixed, the change of magnification larger than 10 times doesn't have much impact on BW. Thus the traditional design of the micro CT system is to make the magnification as big as possible. But it is not feasible for inverse Compton source since this kind of source has small emission angle, which means object needs to be far away enough to be covered by beams and large magnification means lower dose in every detector pixel as well as large system design.

To sum up, when applying such a small focus X-ray source, the limitation of the resolution is the detector pixel size. And it is one of our goals to design and develop a suitable detector system to achieve high resolution

The length of LINAC decides the energy of electron can be accelerated. The all-optical Compton system uses laser plasma accelerator (LPC) instead of the LINAC. The LPC can accelerate the electron in a short length. Lawrence Berkeley National Laboratory has obtained 1GeV electron within 3.3centimeters[9], and it takes 64 meters for SLAC to achieve the same result. Thus If LPC is used for inverse Compton system, the whole system could be much more compact, and the blueprint for such a laboratory can be seen as Fig.3.

The inverse Compton source based on LPC is also called as all-optical Compton source, since the laser is used for acceleration as well as inverse Compton effect. The all-optical Compton source is compact, due to the small size of the LPC, thus it can be easily deployed in different situation with no need to depend on large accelerators, which can get higher X-ray energy than the LINAC ones in the same spaces. The energy of the X-ray can reach several MeV with the focal size of few micron, thus it could be used for non-destructive testing of metal workpiece with high-precision inner structure.

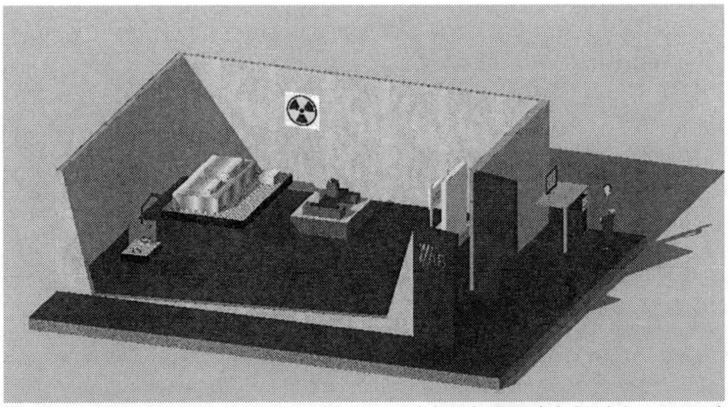

Fig. 3. Blueprint for high-energy X-ray high-precision industrial CT laboratory design

2. Reconstruction Algorithm

When applying the inverse Compton source or inverse Compton source in industrial CT system, the key point is to design a suitable reconstruction algorithm. There are two main problems. The first is when using such a small angle source to scan a big object, the scanning system needs to be carefully designed. The second one is the spectrum of the inverse Compton source varies with angle, as shown in Fig. 4. The energy of the photon in the central angle is around 0.8 MeV and the average energy decreases when the angle is away from the center.

To solve the first problem, a reconstruction algorithm based on re-binning has been attempted. The projection data was collected by using Translation-Rotation (TR) scanning method. TR method is a common method for scanning big object when Field-of-View (FOV) cannot cover the object. Then the projection data from different source was combined to be one single sinogram via re-binning method.

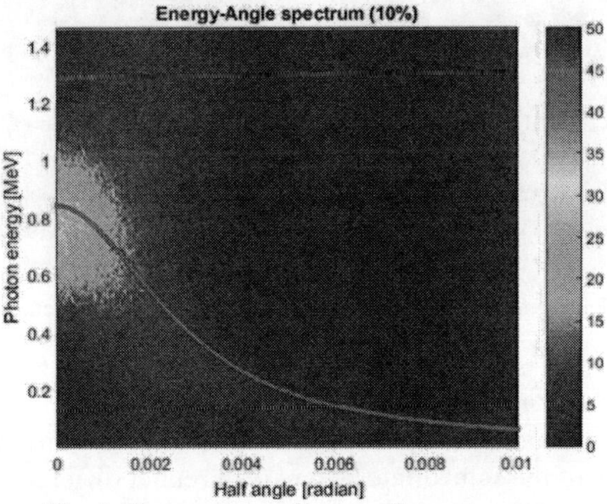

Fig. 4. The spectrum of inverse Compton source

Shown as Fig. 5, by using TR method several sinograms can be obtained when source in different positions. If those three sinograms are directly stitched together, the new sinogram cannot be used to reconstruct the image, since data in the new sinogram is not continuous. That's where the re-binning method works. The re-binning method change the projection data from different fan beams into parallel beam. And with the sinogram from parallel beam, the filtered backprojection (FBP) method can be used for reconstruction. The re-binned sinogram is shown as Fig. 6(a), and data in sinogram is continuous. Fig. 6(b) is the reconstructed image from Fig. 6(a) via FBP method and the detail of the 5 lp/mm line pairs can be clearly distinguished as Fig. 6(c).

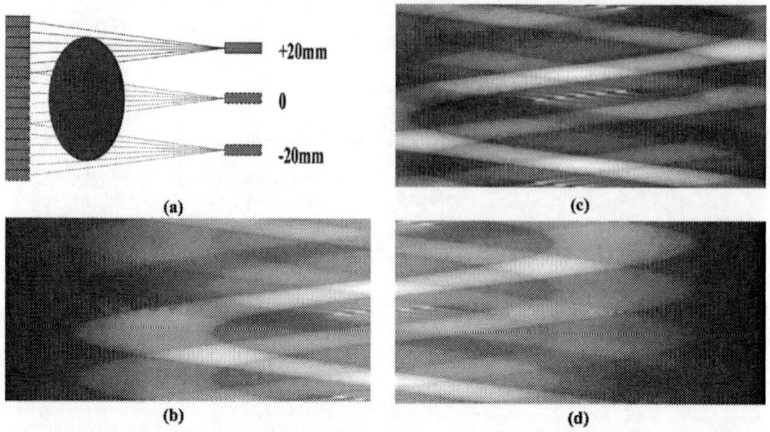

Fig. 5. The TR scanning system and projections from different source positions.

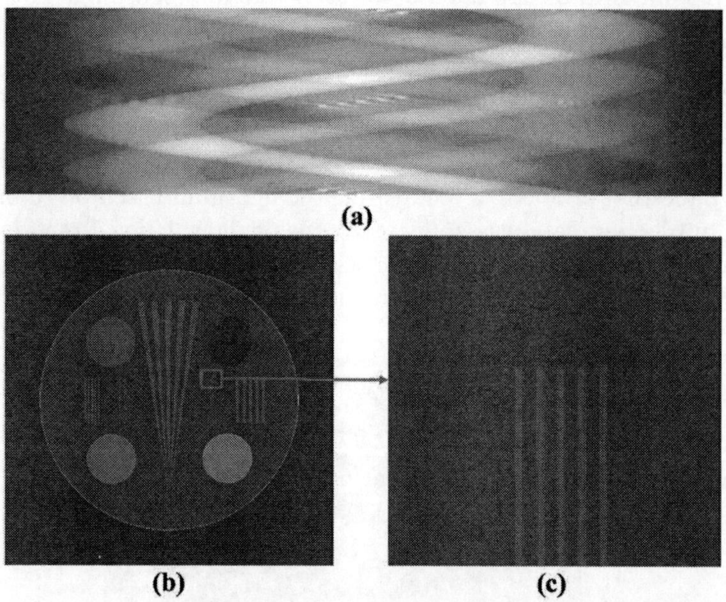

Fig. 6. The re-binned sinogram and reconstructed image as well as the detail of 5 lp/mm line pairs.

The result of Fig. 6 is from numerical experiment without considering the spectrum distribution of the source. But if the spectrum distribution of the source is taken into account, the direct reconstruction would introduce artifats into the reconstruction image because the attenuation coefficients of the same material are different at different energies. Thus, a novel method is needed to correct this kind of artifats.

Inspired by the idea of spectrum CT, the attenuation coefficient can be decomposed into two parts, when the energy is under 1 MeV.

$$\mu = \mu_{ph} + \mu_c = \rho_e(aZ^3 + b) \text{..............................} (5)$$

In the equation, μ is the attenuation coefficient obtained from the experiment and μ_{ph} is the contribution of photoelectric effect while μ_c is that of Compton scattering. And according to the empirical formulas, the equation can be transformed as the formula of ρ_e, the density of electron, and Z, the number of charges in nuclei, while a and b are constant only related to the energy of X-ray.

The a and b in above equation can be solved by scanning some known materials and be taken into account in different angles. Then a carefully designed iterative algorithm is applied to solve the ρ_e and Z, and the distributions of ρ_e and Z are obtained. At last, use the a and b in certain energy, such as 0.8 MeV, then the distribution of the attenuation coefficient of object is solved.

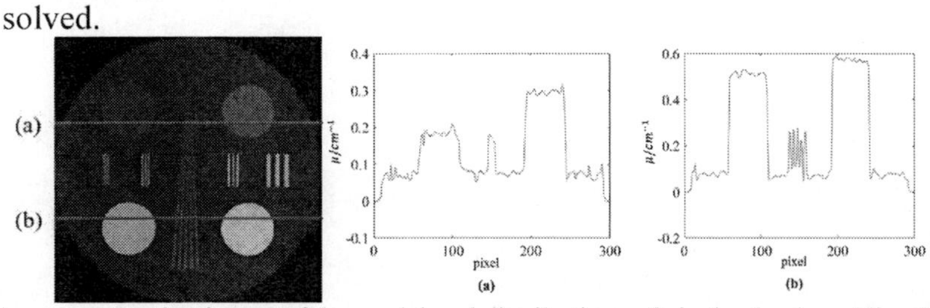

Fig. 7. The reconstruction image of the model and distributions of pixel value in red line (a) and (b).

Table 1. The attenuation coefficient comparison between reconstructed value and the ground truth

Material	μ/cm^{-1}	μ_{rec}/cm^{-1}
Fe	0.527	0.510
Al	0.185	0.190
Cu	0.592	0.582
Ti	0.298	0.289

As the comparisons of the attenuation coefficients of different materials shown in Table. 1. The attenuation coefficients of reconstructed image are close to the ground truth. And according to the distribution of attenuation coefficients in different red line, this kind of method can smooth the attenuation coefficients while recovering the value.

In summary, those two problems mentioned above when applying inverse Compton source in industrial CT have been solved by using re-binning method and artifacts recovery method. And if those two methods are combined, a novel reconstruction algorithm can speed up the progress of the application of inverse Compton source in industrial CT.

4. Conclusion

In this paper, we illustrate the micro CT for industrial usage. The novel micro CT system based on inverse Compton effect can have both high energy as well as small focal size. The inverse Compton system on LINAC has proved the feasibility of this kind of source when applying on imaging. A re-binning method and artifacts recovery method are proposed to solved two main problems when applying inverse Compton source in industrial CT. The future trend of micro industrial CT is smaller focal size and higher energy, and the all-optical Compton source is the choice.

References

[1] L. Schoeman, P. Williams, A. du Plessis, and M. Manley, "X-ray micro-computed tomography (μCT) for non-destructive characterisation of food microstructure," Trends in Food Science and Technology. 2016.

[2]J. Karch, B. Bartl, J. Dudak, J. Zemlicka, and F. Krejci, "Non-destructive imaging of fragments of historical beeswax seals using high-contrast X-ray micro-radiography and micro-tomography with large area photon-counting detector array," Micron, 2016.

[3]D. Panetta, "Advances in X-ray detectors for clinical and preclinical Computed Tomography," Nucl. Instruments Methods Phys. Res. Sect. A Accel. Spectrometers, Detect. Assoc. Equip., 2016.

[4]F. Peyrin, P. Dong, A. Pacureanu, and M. Langer, "Micro- and nano-CT for the study of bone ultrastructure," Curr. Osteoporos. Rep., 2014.

[5]J. Hiller, M. Maisl, and L. M. Reindl, "Physical characterization and performance evaluation of an x-ray micro-computed tomography system for dimensional metrology applications," Meas. Sci. Technol., 2012.

[6]H. Toyokawa, H. Kanada, T. Kaihori, M. Koike, and K. Yamada, "Application of high-energy photon CT system with laser-compton scattering to non-destructive test," in IEEE Transactions on Nuclear Science, 2008.
[7]K. Ta Phuoc et al., "All-optical Compton gamma-ray source," Nat. Photonics, 2012.
[8]A. Döpp et al., "An all-optical Compton source for single-exposure x-ray imaging," Plasma Phys. Control. Fusion, 2016.
[9]W. P. Leemans et al., "GeV electron beams from a centimetre-scale accelerator," Nat. Phys., 2006.

New MesoFocus NDT X-Ray tube concept

Adrian RIEDO [1]
[1] COMET AG, Flamatt, Switzerland

Contact e-mail: adrian.riedo@comet.ch

Abstract

A new tube concept with a variable, small focal spot is presented. The term MesoFocus means that the BF size ranges between the conventional MiniFocus and the open MicroFocus tubes.

The following features are achieved with a novel triode focusing concept:

- Adjustable focal spot size from 50 to 200mü (100 to 500mü) (ASTM E 1165-12)
- Constant tube power from 25% of the max. kV
- Homogeneous resolution in the entire radiation field, due to a 20° target angle

As functional models, unipolar tubes with a maximum voltage of 225kV were built. A resolution of <25µm (2D) and 30lp/mm (CT) could be demonstrated. Due to the steep target angle of 20°, the difference of the resolution in the entire radiation field is only one duplex wire. The wide voltage range with constant FS size and target power also enables the imaging of components made of lightweight materials.

The tests at different OEM's showed that such a MesoFocus tube concept is suitable for applications previously performed with open MicroFocus tubes. Thus, the advantages of a closed, robust metal ceramic X-Ray tube are combined with the high resolution of an open MicroFocus tube. Furthermore, that tube concept is scalable up to 600kV.

Due to the broad applicability of the MesoFocus tube concept, a wide range of applications (especially inline) can be addressed.

New MesoFocus NDT X-Ray tube concept

dir2019

Adrian Riedo, Senior Specialist

Topics

- Intended classification of the MesoFocus (MF) tube types
- Simulation based tube design
- Possible tube features and test results of the function models
- Application tests
- Summary and outlook

MesoFocus intended classification

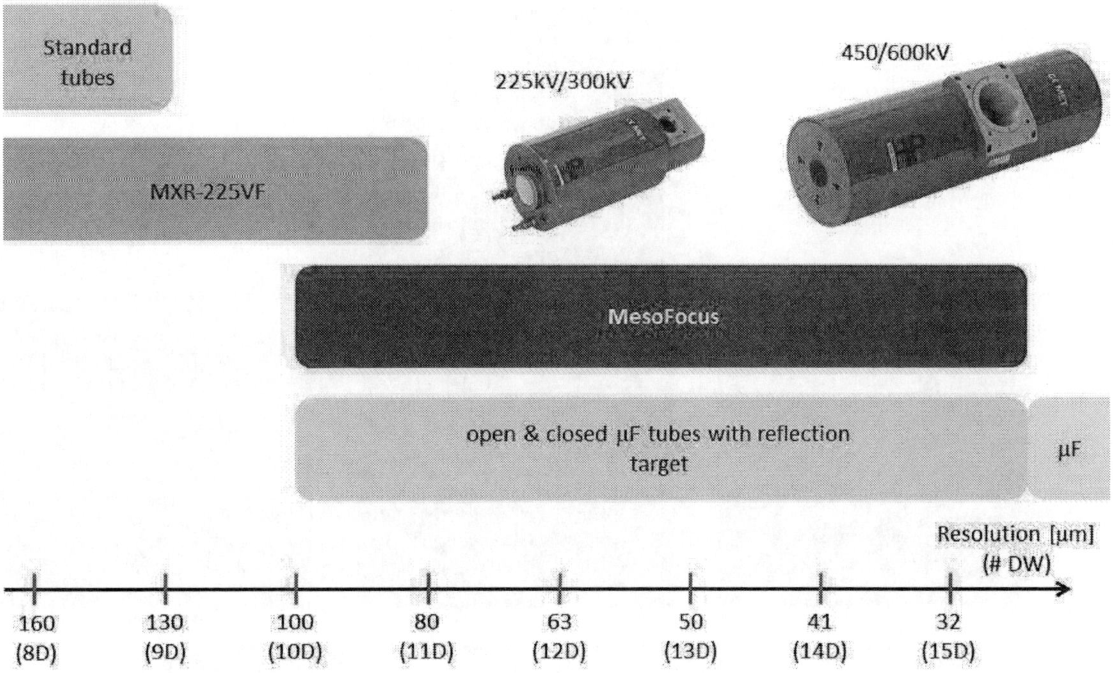

Simulation based tube design

FEA Study

Mechanical & electrical design:

- ❑ Heating current
- ❑ Temperature fields
- ❑ Thermal expansion
- ❑ Thermo-mechanical stress

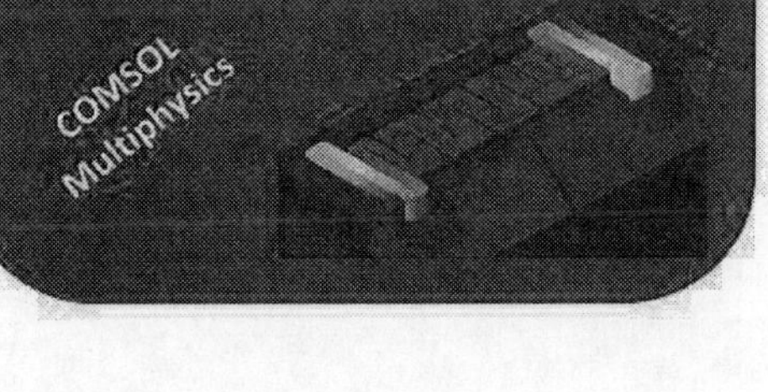

Electron trajectories:

- ❑ Current density at the emitting surface
- ❑ FS size as $f(U_{Grid})$
- ❑ Geometric tolerance analyzing

Optimization of the focusing geometry by FEA

FS size acc. ASTM E1165-12

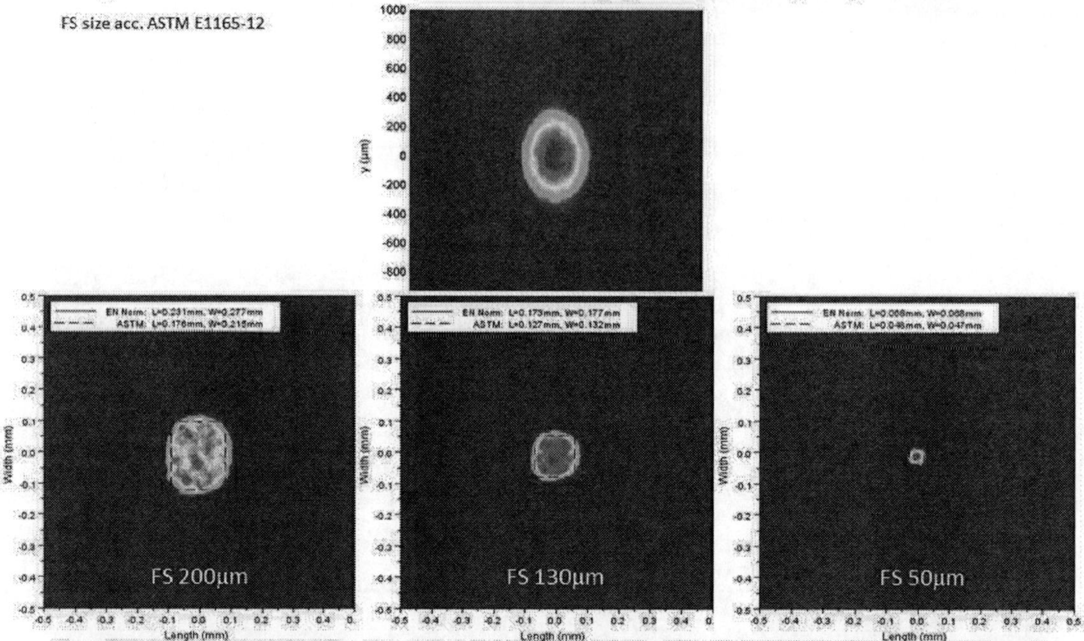

The result is a FS that can be varied from 50 to 200µm using the additional grid voltage.

Features of the MesoFocus (MF) tube concept

Small FS size:

- 50 to 200µm
- 100 to 500µm

possible FS ranges under discussion

High brilliance:

- 1W/µm FS size → 2W/µm achievable resolution

Wide kV range with full power:

- FS size and max. anode power are reached from 25% of max. kV.

Homogeneous resolution over the entire field of view:

- The target angle of >11° allows a homogenous resolution on the entire DDA surface.

Easy to use and robust:

- As reliable as any other standard X-Ray tube from COMET.

FS size & resolution

70kV, 0.71mA, FS-50, M=14

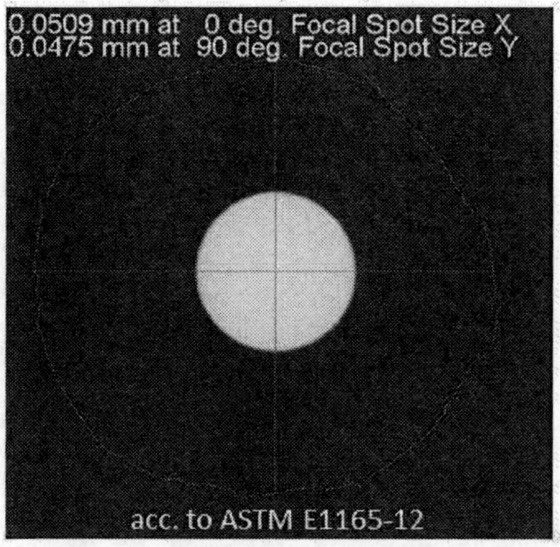

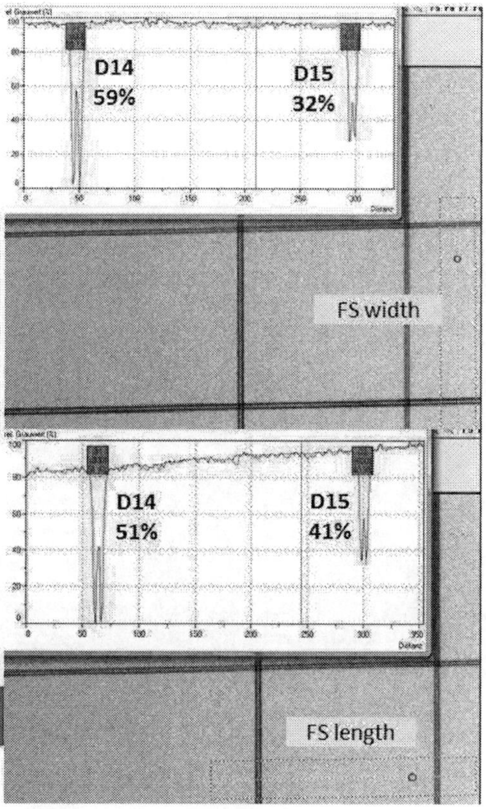

The clearly resolved D15 corresponds to an achievable resolution of < 30µm!

Resolution homogeneity

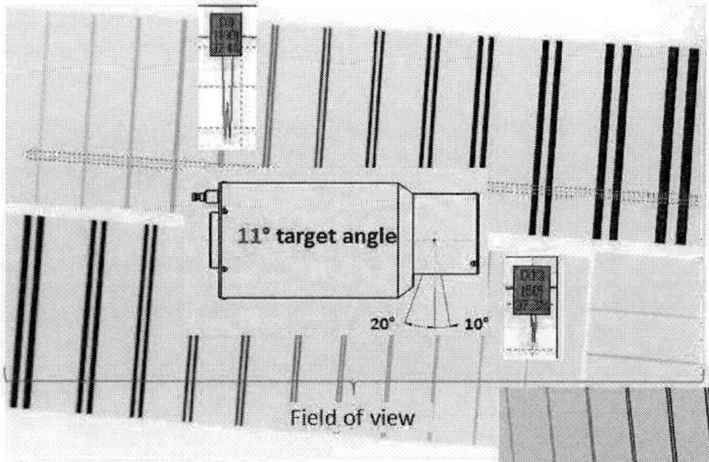

FS size: 100µm

Thanks to the target angle of e.g. 20° there is only one duplex wire of resolution difference at the entire DDA

Application tests at YXLON HH: Setup

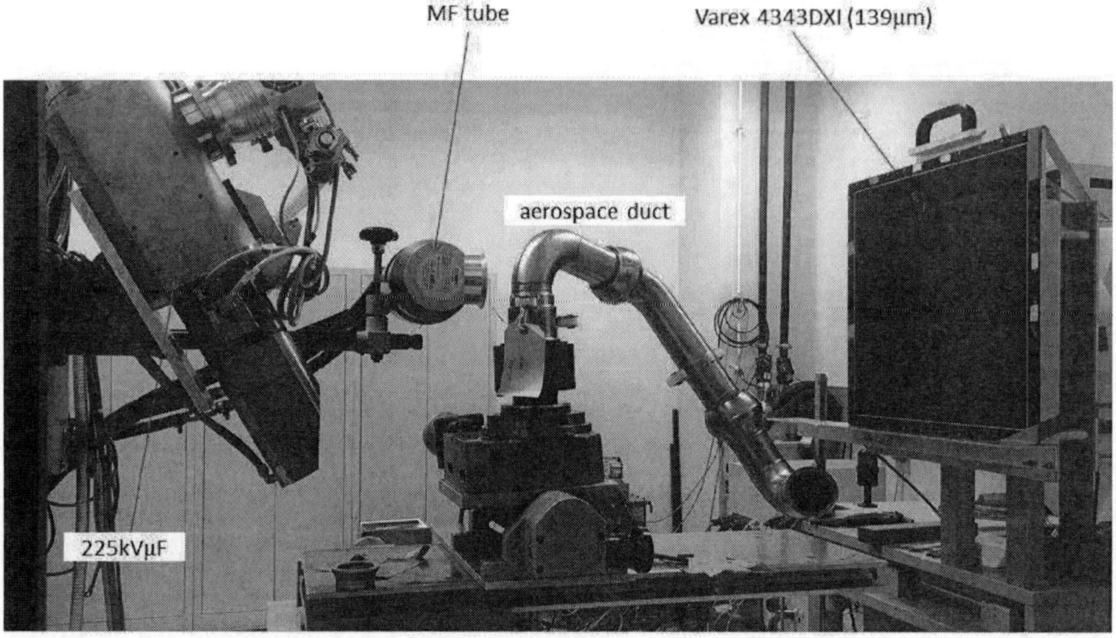

MF vs. µF: Inconel duct with flaw in weld

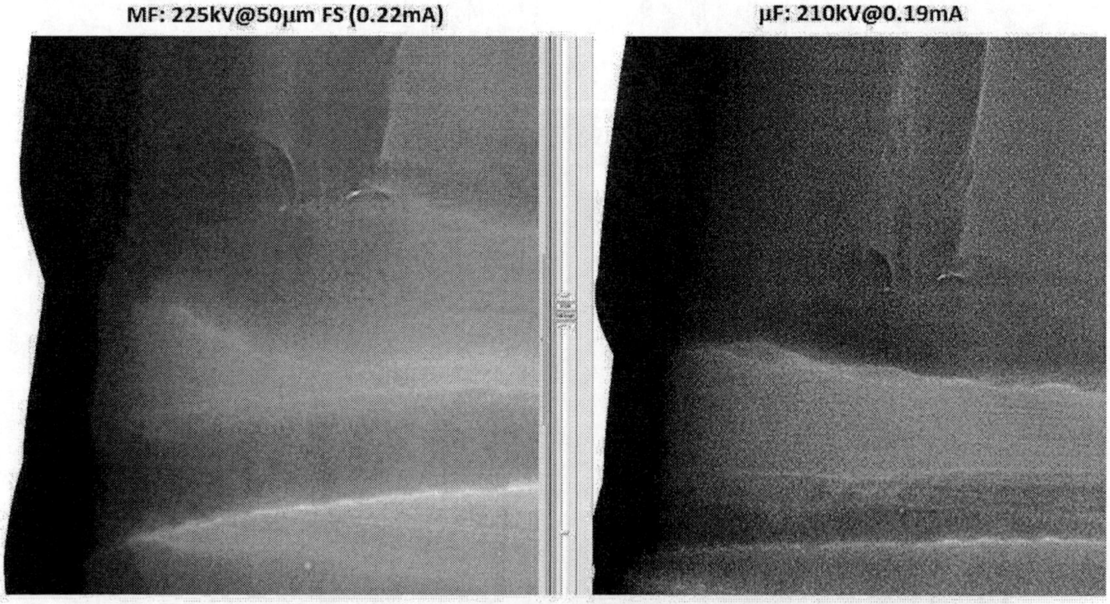

The MF tube provides the same detail recognition as the actual µF tube.

Turbine blade with 50 and 200μm FS

upper part: 225kV@50μm FS **lower part: 225kV@200μm FS**

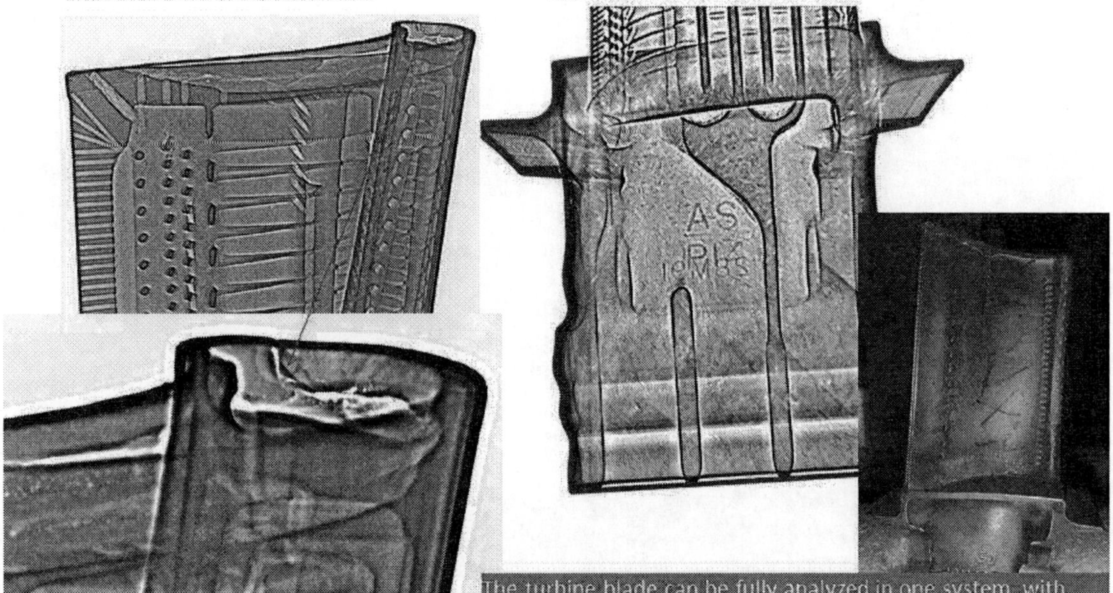

The turbine blade can be fully analyzed in one system, with two images. This saves costs, space and time.

3D printed pump rotor (SS)

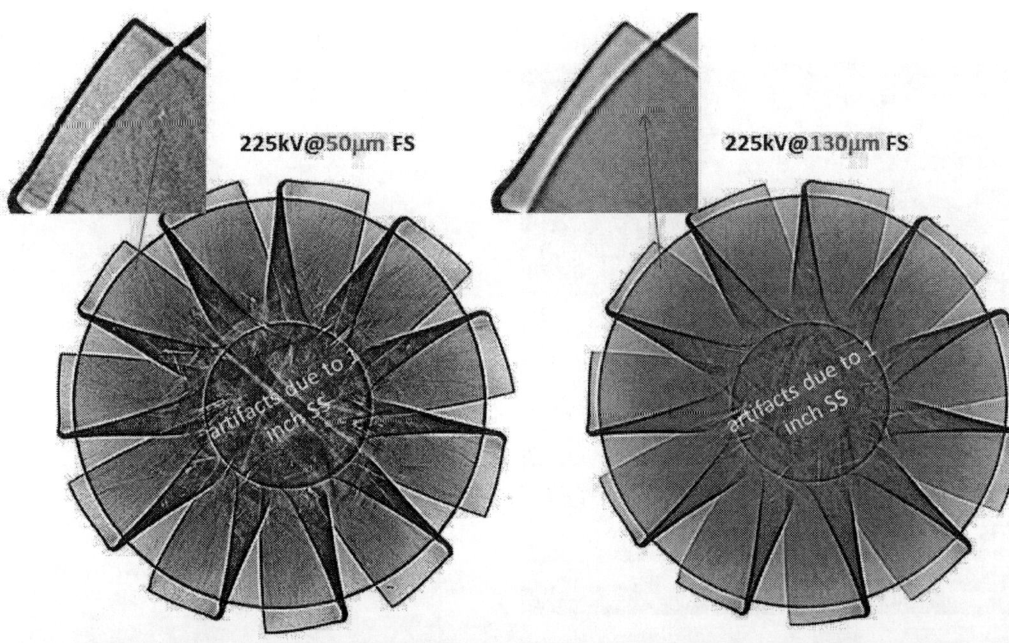

For most AM parts a resolution of 25μm is presumably necessary.

Lightweight materials: Carbon bicycle rim

60kV@50µm FS

Due to the full power at low energies, the MF tube is also suitable for light materials.

CT Scans

70kV, 0.71mA, FS-50, 2h, M=7, DDA: 150µm-Pitch

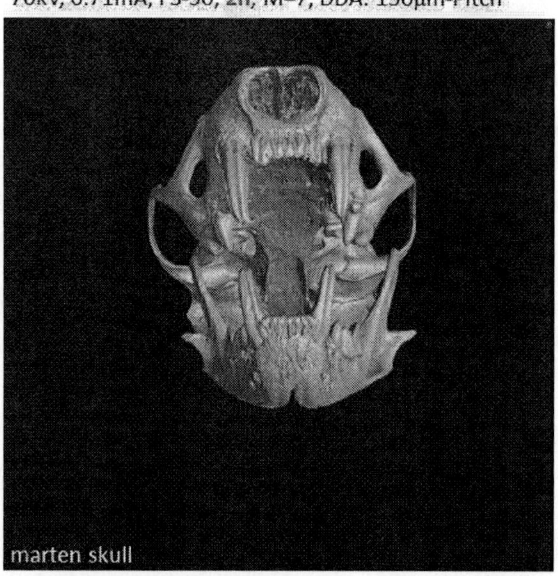

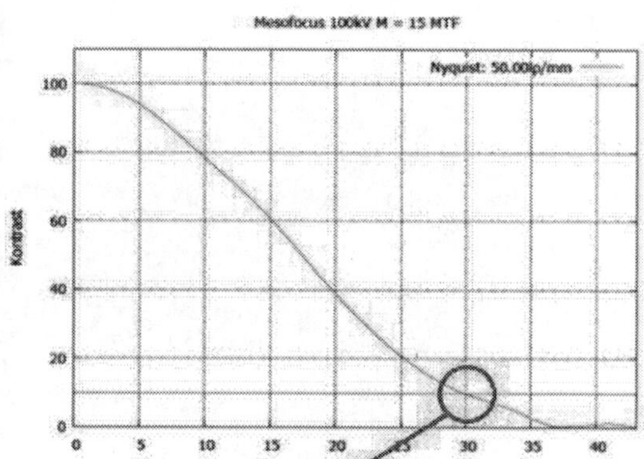

The high dose and BF stability of the MF tube allows it to be used for longer CT scans.

Summary and outlook

What's achieved with the function model tubes?

- ✓ The MesoFocus tube concept combines the reliability of a state of the art metal ceramic X-Ray tube with the high resolution performance of an open MicroFocus tube.
- ✓ A resolution of 25µm in 2D and 30lp/mm in CT was demonstrated.
- ✓ The FS size as well as the target power stays constant in the entire kV range.
- ✓ Due to the variable FS size a wide range of applications (incl. inline) can be addressed.
- ✓ That tube concept is scalable up to 600kV.

What's next?

- ☐ Complete the ongoing endurance tests.
- ☐ Define the range of the FS size with the highest market potential.
- ☐ Find the best compromise between target power and resolution homogeneity to determine the optimal target angle.
- ☐ Specify the voltage classes for a MF tube family.

X-Ray for money

Thank you for your attention

Evaluation of X-ray target materials to improve CT-based measurement of fiber orientations inside CF-SMC components

Jan TEUWSEN [1], Florian BITTNER [2], Jens Peter STEFFEN [3]
[1] Volkswagen AG, D-38436 Wolfsburg, Germany
[2] Fraunhofer Institute for Wood Research WKI, D-30453 Hannover, Germany
[3] X-RAY WorX GmbH, D-30827 Garbsen, Germany
Contact e-mail: jan.teuwsen@volkswagen.de

Abstract. Carbon fiber sheet molding compounds (CF-SMC) are increasingly used in automotive and aerospace industries. The accuracy of fiber orientation measurement depends on the quality of the computed tomography (CT) results. This is significantly influenced by signal and contrast obtained when imaging the low absorption materials. Using standard tungsten X-ray targets as available in most commercially available microfocus CT scanners appears to be a compromise in terms of contrast and flux.

The authors compare the X-ray spectra and imaging properties achieved using tungsten, copper, and silver X-ray targets. Projections and volume data sets obtained from scanning a CF-SMC sample component using different X-ray targets are compared and evaluated using common quantities like grey value distribution and signal-to-noise ratio. Finally, the different volumes are analyzed by using a common software module and fiber orientation tensors are compared.

1. Introduction

Over the last decades, compression molded discontinuous fiber composites (DFC) such as carbon fiber sheet molding compounds (CF-SMC) have been used extensively for interior and exterior, structural and non-structural composite applications in automotive and aerospace industry [1] [2]. High performance CF-SMCs, such as the materials used in this study (see section *2.1 Materials*), are characterized by a high delamination resistance, near quasi-isotropic in-plane stiffness, high out-of-plane strength and stiffness, and low notch sensitivity [3]. The epoxy-based CF-SMC HexMC has a very short curing time, leading to a 84 % shorter molding time and an overall process time reduction of 44 % for a large part (monocoque tub) compared to the same part produced in a resin transfer molding (RTM) process [4]. In comparison to the RTM process, the easy and time saving compression molding process leads to lower costs and also reduces the final amount of parts due to the chance of part integration (fasteners, inserts) [1]. Since CF-SMC is suitable for high volume production at low manufacturing costs and enables molding of complex three-dimensional (3D) geometries [1], lightweight components made of CF-SMC are used for many industry applications like window frames [2] [3], body panels, interior trims, seats, engine bay covers and braces [5], handles, air intakes, central tunnel claddings, inner monocoques, suspension control arms, and several other cockpit parts for example in the latest Lamborghini research and serial cars (Sesto Elemento, Veneno, Aventador, Huracán, and Urus) [3] [6] [7] [8].

Due to the part design and the manufacturing process a characteristic microstructure is induced in CF-SMC components. This process-induced microstructure is mainly characterized by a locally varying fiber orientation and fiber concentration determining the part's mechanical properties. The DFC materials used in this study consist of thermoset resins reinforced by transversely chopped carbon fiber tows ('strands', also called 'chips' or 'platelets' in literature) which are randomly distributed into a mat. The long carbon fibers are strongly aligned in longitudinal direction of the strands. Materials with those randomly oriented strands (ROS) are characterized by a high degree of heterogeneity (variability in intra- and inter-part structure on the meso- and macro-scale) and seek to reach quasi-isotropic mechanical properties [3] [9] [10]. Therefore, obtaining a realistic three-dimensional representation of the local fiber orientations in a CF-SMC part is key for a better understanding of the compression molding process, to be able to validate compression molding process simulations or to map the gained fiber orientation information into a structural mesh of an integrative (coupled) Finite Element Analysis (FEA) [11] [12]. In order to ensure a proper component design regarding mechanical requirements and for quality assurance of manufactured composite parts, the industry's endeavor is to determine the material's microstructure for large areas or ideally for an entire part [12].

In order to collect experimental 3D information about fiber orientation distributions of heterogeneous fiber reinforced materials (such as CF-SMC), X-ray computed tomography (CT) is widely used in industry due to the fact that it is easy to prepare samples and only requires a difference in the linear X-ray attenuation coefficients of the matrix and the reinforcement [13]. The inner structure/morphology of inhomogeneous materials like CF-SMCs can be investigated three-dimensionally by micro-CT (μ-CT), which is a high resolution X-ray computed tomography method, allowing an in-depth material characterization [13].

However, due to the lack of X-ray contrast between epoxy or vinyl ester resins and carbon fibers, μ-CT scans are so far not useable to scan larger CF-SMC components and still receive useful fiber orientation data. Obtaining CT data for an entire 3D part is always a trade-off between scan volume size, possible resolution (voxel size) and the required scanning time. Normally, attaining fiber orientations by CT scan data analysis of fiber reinforced polymer parts requires a finer scan resolution than the fiber's diameter, which by implication limits the scan volume size [14].

Useful μ-CT scans for CF-SMC parts need a sufficient resolution so that it is clearly distinguishable between strands and resin by grey value differences enabling a fiber orientation analysis with a common CT scan analysis software (e. g. VGStudio MAX, Volume Graphics GmbH, Heidelberg, Germany). The analysis algorithm within VGStudio MAX is intended to be used for the orientation analysis of discretely visible fibers [11] [15] [16] [17] [18]. Yet with correctly set parameters and with local relative density gradients between resin and fibers the image analysis principles are suited to be used for scans with meso-scale resolutions where no single fibers, only coarser structures like fiber strands and fiber bundles, respectively, are visible [12] [13]. Since microscopy shows that during compression molding the fibers within one strand flow and orient together, deforming yet remaining as an intact strand with locally highly aligned fibers, the density gradients of CT scanned ROS-based materials are sufficient for a determination of local average strand orientations even at a coarser scan resolution [11] [12] [13] [19]. In a CT scan of a ROS-based material the smallest density gradient is present in in-strand-direction (along the fibers), intermediate density differences are visible transverse to the strand orientation and the highest density gradient occurs normal to the strand direction [13].

Denos uses a CT scan with a meso-scale resolution of 53 μm (voxel edge length) to determine the heterogeneous internal microstructure of a 65 x 65 x 45 mm^3 strand-based long discontinuous fiber composite part [11] [20]. However, at that resolution and with the used CT scan settings it is not possible to distinguish between single carbon fibers (⌀ ≈ 7 μm) and the matrix or to discern strand boundaries (~ 100 μm thick) [11]. Although Denos' CT scan configuration is not able to represent distinct strand boundaries, it is still possible to receive a mean local fiber orientation [11].

Another common method to achieve bigger CT scan volumes at a reasonable resolution is to merge several scan volumes generating a digital twin of the scanned part and it's microstructure. Kravchenko merged 8 scans with a volume of 30 x 30 x 5 mm^3 each, using a scan resolution of 15 μm [13]. At that resolution the CT scan quality is high enough to discern between strands and suitable for reasonable fiber orientation analyses. Kravchenko uses an analysis mesh size of 0.7 x 0.7 x 0.1 mm^3, where each grid element contains about 13,000 voxels, which are used to determine a single orientation tensor from each of the measured orientation vectors by a grey scale analysis [13]. The finer analysis resolution in thickness direction better resolves the thin strands and enables gathering more detailed information about the local strand orientation changes.

The spectrum generated by the X-ray source significantly depends on the elemental composition of the used target material. Tungsten (W) is widely used as target material for microfocus X-ray sources. However, depending on the applied X-ray voltage and the absorption behavior of the sample material, alternative target materials might deliver beneficial spectrum characteristics that can improve CT measurement quality concerning the separation capability of fiber and matrix for fiber orientation analysis. A higher μ-CT scan quality, by means of a higher contrast between fiber and matrix, also enables to scan bigger composite part volumes, fulfilling industry demands, where the fiber orientation within a whole component is of interest.

To study the effect of copper (Cu) and silver (Ag) targets in comparison to a standard tungsten (W) target in μ-CT scans different carbon/epoxy composite samples are measured with different voxel resolutions. Furthermore, X-ray source resolution, grey value distribution as well as subjective image quality of the CT data are examined and comparative fiber orientation analyses are performed.

2. Experiments

2.1 Materials

The high performance CF-SMCs examined in this study are Hexcel's HexMC® and A. Schulman's Quantum AMC® 8593 HT. Both materials are designed for compression molding of complex shaped parts in a heated metal tool. HexMC® is a DFC which consists of unidirectional (UD) preimpregnated (prepreg) AS4/8552 carbon-epoxy tapes that are slit longitudinally and chopped transversely into strands and then randomly distributed into a mat [9] [10] [3]. Those ROS have nominal in-plane dimensions of 50 mm x 8 mm (2 in x 0.3 in) and a thickness of approximately 0.15 mm containing high strength carbon fibers impregnated by a fast curing Hexcel HexPly® M77 epoxy resin. The carbon fiber content is 62 % by weight, corresponding to 57 % fiber volume content and giving a material density of 1.55 g/cm^3.

A. Schulman's Quantum AMC® 8593 HT is a vinyl ester based SMC containing chopped 25 mm long PAN based 3 K carbon fiber tows [21]. The carbon fiber weight content is 50 %, giving a material density of 1.45 g/cm^3.

The samples analyzed in this study were cut out of plates (320 x 320 x 3.8 mm) (**Fig. 1** and **Fig. 2**) that were manufactured with a high mold coverage of 94 % (low- to no-flow conditions) in a 1000 ton (Dieffenbacher DCL 1000) compression molding machine at a temperature of 135 °C, a pressure of 110 bar, a closing speed of 16 mm/s, and a closing time of 480 s.

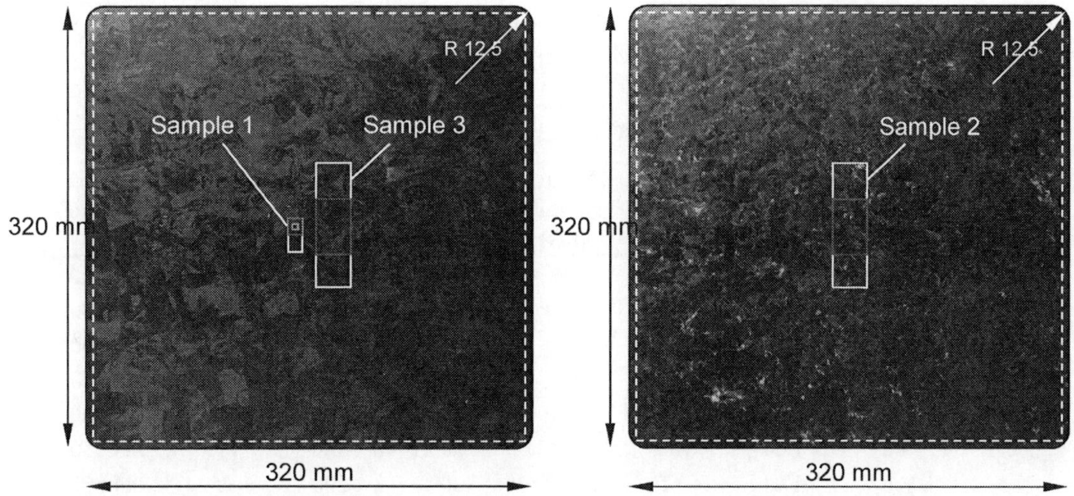

Fig. 1. Used SMC plates (left: HexMC; right: AMC), position of the 310 x 310 mm initial raw material charges (dotted white boxes) and location of cut out material samples (white boxes). Colored boxes denote CT analysis regions.

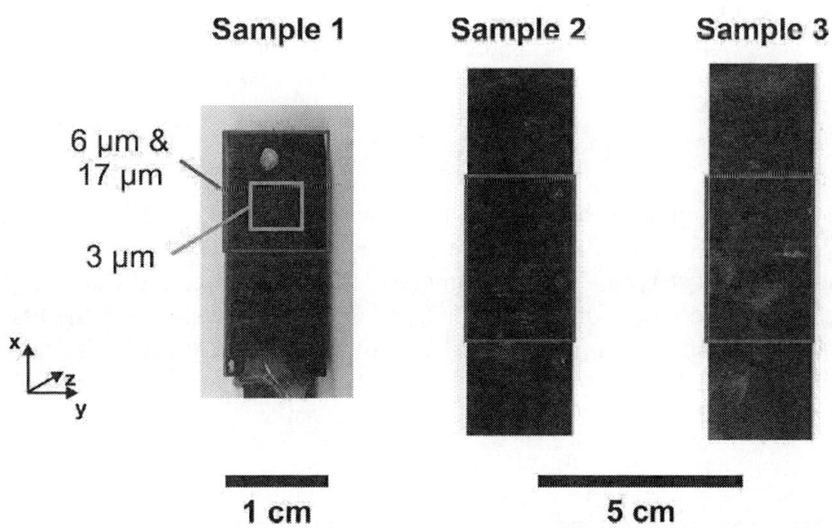

Fig. 2. Cut out material samples (colored boxes denote CT analysis regions with different scan resolutions)

2.2 Micro-CT Measurements

The CT measurements of three CF-SMC samples were performed on a CT-AlphaDuo device (Procon X-Ray GmbH, Sarstedt, Germany) operated by the Fraunhofer WKI, Hannover, Germany. The system is equipped with a 240 kV microfocus X-ray source XWT-240-TCHE Plus (X-RAY WorX GmbH, Garbsen, Germany) and a PaxScan® 2530DX detector (Varian Medical Systems, Inc., Salt Lake City, Utah, USA). While sample 1 was scanned with the three selected target materials and in different resolutions, for sample 2 and 3 only the target material was varied. The applied scan parameters are given in **Tab. 1**. To achieve comparable exposures of the X-ray images for the different targets the exposure time was adjusted. To maximize the scan resolution of the high aspect ratio samples scans at a voxel resolution of 6 and 17 µm were performed in helix mode. The scan of sample 1 at a resolution of 2.9 µm was performed in conventional axial mode. At the same resolution identical sample volumes were captured. For visualization and analysis of the volume data VGSTUDIO MAX 3.2 (Volume Graphics GmbH, Heidelberg, Germany) was used.

Tab. 1. Micro-CT measurement parameters

	Target material	Voxel resolution [µm]	X-ray voltage [kV]	X-ray current [µA]	FDD [mm]	Exposure time [ms]	Measuring time [min]	Approx. I_0 value [-]
Sample 1 (HexMC)	W	2.9	50	75	1000	8 x 3000	960	30,500
	W	6.3	65	140	600	8 x 300	135	35,000
	Cu	6.1	65	140	600	8 x 500	225	31,500
	Ag	6.1	65	140	600	8 x 300	135	36,700
	W	17.4	65	350	700	8 x 200	75	42,200
	Cu	17.4	65	350	700	8 x 300	115	33,250
	Ag	17.3	65	350	700	8 x 200	75	44,800
Sample 2 (AMC)	W	17.4	65	350	700	8 x 200	85	43,600
	Cu	17.4	65	350	700	8 x 400	170	44,000
	Ag	17.3	65	350	700	8 x 200	85	44,000
Sample 3 (HexMC)	W	17.4	65	350	700	8 x 200	90	42,800
	Cu	17.4	65	350	700	8 x 400	180	45,000
	Ag	17.4	65	350	700	8 x 200	90	43,600

2.3 JIMA Resolution Tests

To determine the effective image resolution of the selected setups, the JIMA resolution pattern RT RC-02B [22] was imaged at the same settings that were used for the particular CT scans. This includes target type, X-ray parameters, focus-to-detector distance (FDD), magnification, and exposure time for the single projection. The tested parameters and effective resolutions are listed in **Tab. 2**. The JIMA resolution was selected as the smallest pattern size showing at least 10 % variation of the estimated modulation transfer function (MTF) between the lines and spaces [23].

Tab. 2. JIMA resolution test parameters and determined resolution

Target material	X-ray voltage [kV]	X-ray current [µA]	FDD [mm]	Magni-fication	Exposure time [ms]	Effective resolution [µm]
W	65	140	600	80 x	8 x 300	7
Cu	65	140	600	80 x	8 x 500	7
Ag	65	140	600	80 x	8 x 300	7
W	65	350	700	95 x	8 x 200	15
Cu	65	350	700	95 x	8 x 300	15
Ag	65	350	700	95 x	8 x 200	15

2.4 X-Ray Spectra Simulations of Different Transmission Targets

The software *aRTist 2.10* published by Bundesanstalt für Materialforschung und -prüfung (BAM, Berlin) [24] was used to simulate the X-ray spectra of the three transmission targets applied in this study (**Fig. 3**). The simulation was performed for a voltage of 65 kV. At this energy the resulting tungsten spectrum does not exhibit the characteristic K-shell emission lines. The highest bremsstrahlung radiation in the considered energy range is delivered by the silver target which also gives the most significant contribution of the K-shell emission lines at 22 and 25 keV.

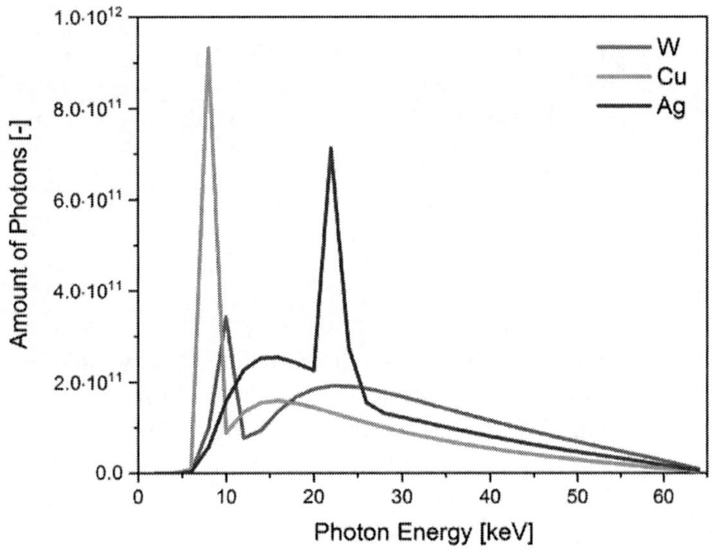

Fig. 3. Simulated spectra of tungsten target, copper target, and silver target

3. Results and Discussion

To investigate whether the different target materials have an effect on the ability to differentiate fiber and matrix, sample 1 was scanned with different voxel resolutions and fiber orientation analyses were applied to the same sample region. The fiber orientation tensors obtained from the scan with the W target at the highest resolution (2.9 µm) serve as reference for the other scan parameters. Samples 2 and 3 with slightly different compositions have larger sample dimensions than sample 1 and serve as demonstrators for application-orientated case studies.

In contrast to the standard W target a grey value analysis of the reconstructed volume data (**Tab. 3**) generally shows a shift of the grey values towards higher values by usage of Cu and Ag as target materials. Although the relative distance between the fiber and the matrix signals does not change significantly, the wider spreading of the grey value spectrum should principally improve the ability to separate fiber and matrix for fiber orientation measurements. For copper the higher grey values in exchange require higher measuring times due to the lower X-ray intensity.

JIMA resolution tests (cf. **Tab. 2**) have verified the resolution of the X-ray source for the different target materials and X-ray parameters. With 7 and 15 µm, respectively, it is comparable to the nominal voxel size of the CT scans. No difference in JIMA resolution is noticeable for the different target materials.

Tab. 3. Grey value distribution in dependence on target material

Sample	Target material	Voxel resolution [µm]	Avg. grey value matrix	Avg. grey value fiber	Relative difference grey value fiber/matrix [%]
Sample 1 (HexMC)	W	2.9	1127	1570	28.2
	W	6.3	1446	1737	16.8
	Cu	6.1	2489	2916	14.6
	Ag	6.1	2229	2644	15.7
	W	17.4	1308	1527	14.3
	Cu	17.4	1350	1520	11.2
	Ag	17.3	3366	3866	12.9
Sample 2 (AMC)	W	17.4	7160	8003	10.5
	Cu	17.4	8460	9564	11.5
	Ag	17.3	7790	8943	12.9
Sample 3 (HexMC)	W	17.4	2987	3376	11.5
	Cu	17.4	3600	4039	10.9
	Ag	17.4	3352	3845	12.8

The general effect of voxel resolution on detectable details is visible in **Fig. 4**, where sectional images at the same positions of sample 1 are compared for voxel sizes of 3, 6 and 17 µm. With a voxel resolution of 3 µm individual carbon fibers can be detected. The fiber orientation is clearly recognizable. This detail detectability at a voxel resolution of 3 µm gives reason to expect a plausible representation of the real fiber orientations of the sample, when a fiber orientation analysis is performed on these data.

When a voxel resolution of 6 µm is applied, individual carbon fibers are still visible, but the discrimination of fibers inside the strands deteriorates. For a resolution of 17 µm no single fibers are recognizable. The structure and orientation of the strands and their layered structure is still visible.

While the grey value gradients, which are detectable in a scan with a voxel size of 6 or 17 µm allow the execution of a fiber orientation analysis for this type of material, the loss of details unquestionably implicates an increased inaccuracy of the results. An evaluation of the accuracy of the obtained orientation vectors, that is required for the conduction of structural mechanics simulation, is out of the scope of this work. Therefore only a qualitative comparison of detail detectability and obtained fiber orientations in dependence on the target material is conducted.

Fig. 5 to **Fig. 8** compare sectional images of the different samples and voxel resolutions for the different target materials. Only slight differences in the image quality are observable. The scans of sample 1 acquired with the Ag target (**Fig. 5** and **Fig. 6**) appear to feature less noise than the scans with the W and Cu target. This is most notably in the areas occupied by matrix, which should be a homogeneous area without structures. While the noise level in all measurements of sample 2 and 3 (**Fig. 7** and **Fig. 8**) appears comparable, the edge contrast between strands and resin appears more distinct for the Cu and Ag target than for the W target. Both a lower noise level and a higher contrast between fiber and matrix would be beneficial for fiber orientation analyses.

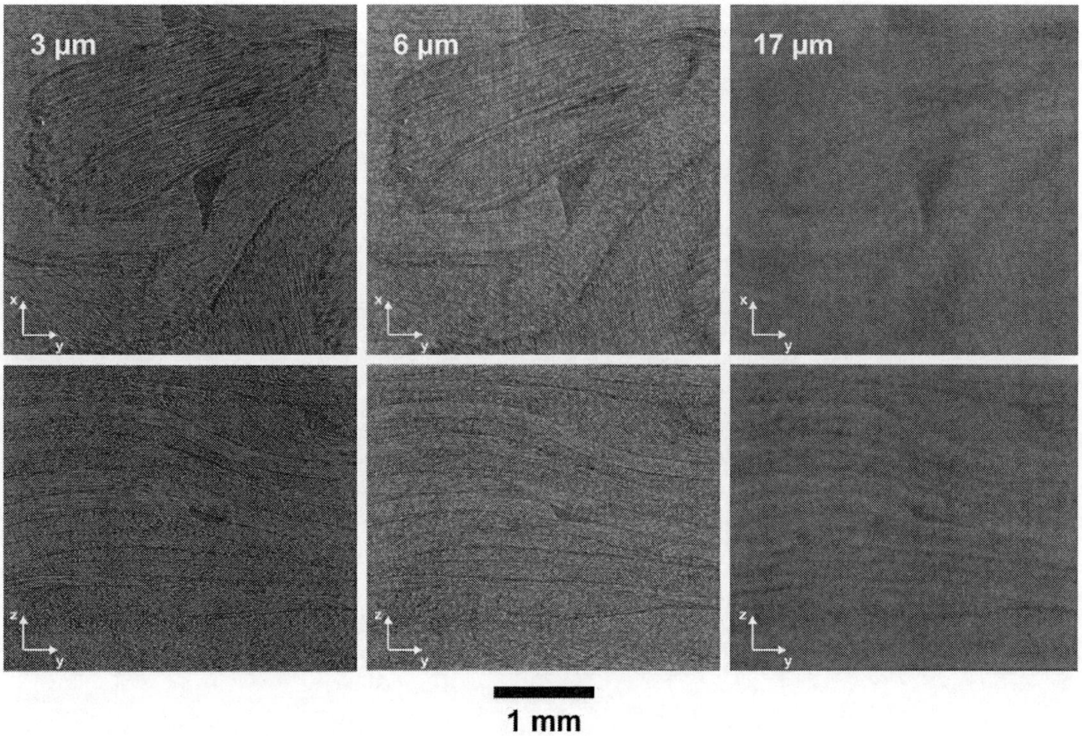

Fig. 4. Cross-sectional CT images of sample 1 at different voxel resolutions, W target; top: in-plane view, bottom: through-thickness view

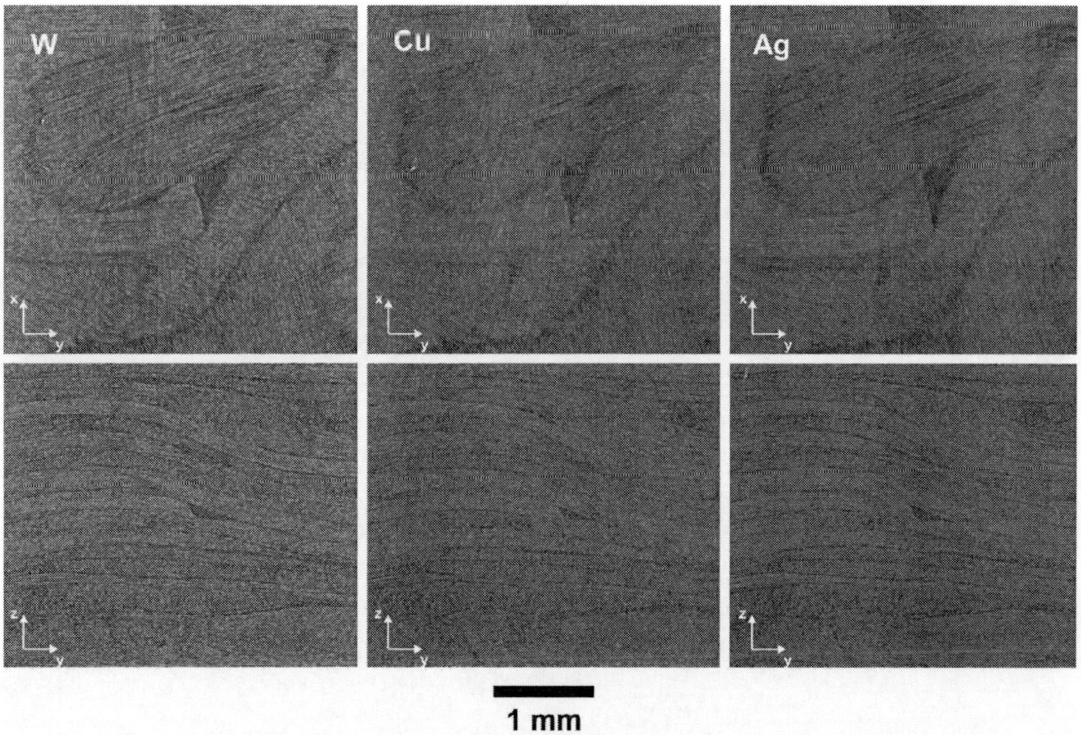

Fig. 5. Comparison of different target materials for sample 1 at a voxel resolution of approx. 6 µm; top: in-plane view, bottom: through-thickness view

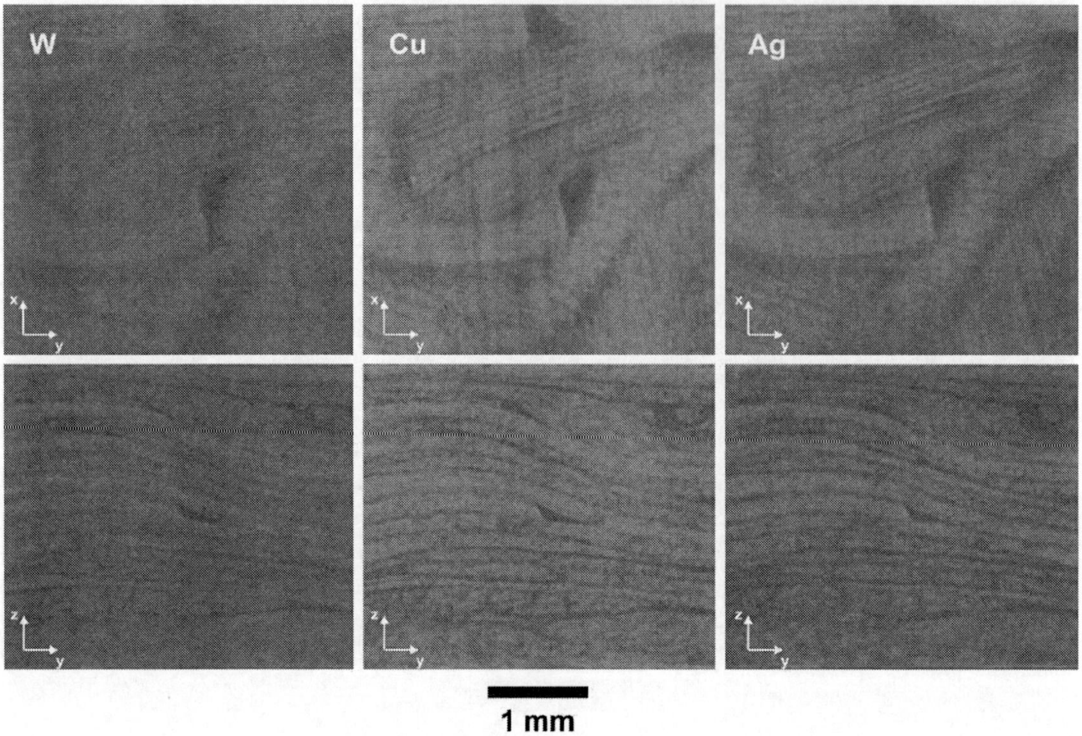

Fig. 6. Comparison of different target materials for sample 1 at a voxel resolution of approx. 17 μm; top: in-plane view, bottom: through-thickness view

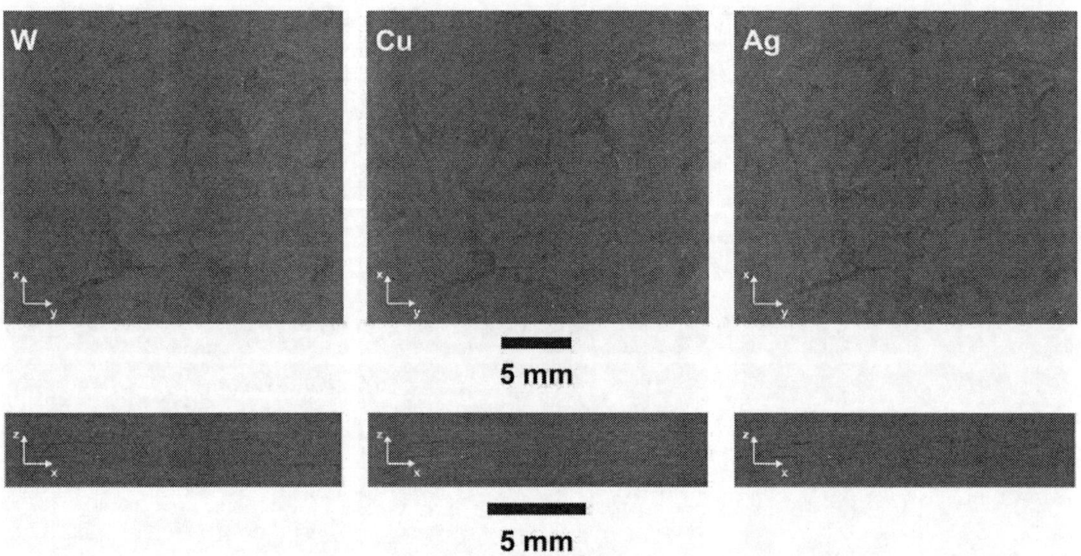

Fig. 7. Comparison of different target materials for sample 2 at a voxel resolution of approx. 17 μm; top: in-plane view, bottom: through-thickness view

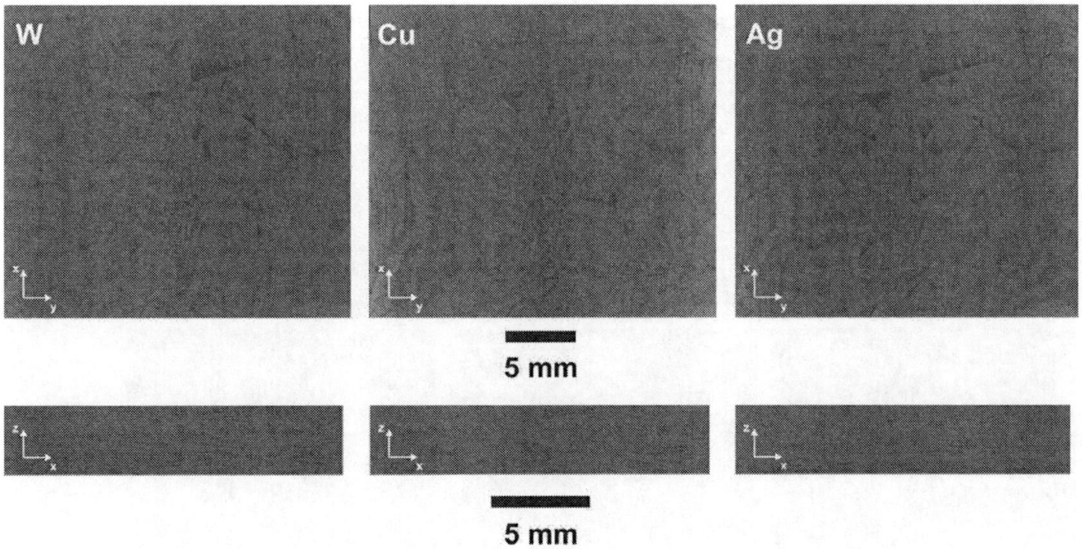

Fig. 8. Comparison of different target materials for sample 3 at a voxel resolution of approx. 17 µm; top: in-plane view, bottom: through-thickness view

Fig. 9 displays fiber orientation analyses of sample 1 based on the scans with the standard W target. While the color-coded illustrations of the analysis results yield a direct impression of fiber orientation, the mesh-based analysis with fiber orientation tensors per volume element produces results suitable for transfer into structural mechanics simulation software.

The color-coded illustration reveals a loss of precision that is caused by a lower voxel resolution. At the highest resolution of approximately 3 µm the course of the different fiber layers is reflected by the calculated fiber orientations. With decreasing voxel resolution the general course of fiber orientations is still described by the analysis, but a loss in detail can be observed. This is confirmed by the numerical results of fiber orientation summarized in **Tab. 4**. For the scans with the W target deviations of the fiber orientation tensors in comparison to the reference scan amount to max. 0.07. The scans of sample 1 with the Cu and the Ag target (**Fig. 10**) result in comparable deviations of the fiber orientation tensors. The relative distribution of fiber orientation components, with "xx" being the most pronounced direction, is determined correctly from all scans except the scan with the Cu target at a voxel resolution of 17 µm.

Fig. 11 and **Fig. 12** illustrate the fiber orientation analyses of samples 2 and 3. In comparison to sample 1 larger sample volumes have been scanned and the analysis regions are 6 times larger. Especially for non-homogeneous fiber-reinforced plastics, like CF-SMC, fiber orientation information over a representative sample volume is important to evaluate component quality and to have adequate input data for structural mechanics simulations. For this reason the determination of accurate fiber orientations from low-resolution CT scans is of high relevance.

Since no reference scans at high resolution exist for samples 2 and 3, only a qualitative evaluation of the results is possible. The depicted fiber orientations appear feasible for all target materials. **Tab. 5** and **Tab. 6** show minor differences of the calculated fiber orientation tensors, which are comparable in magnitude to the deviations observed for sample 1.

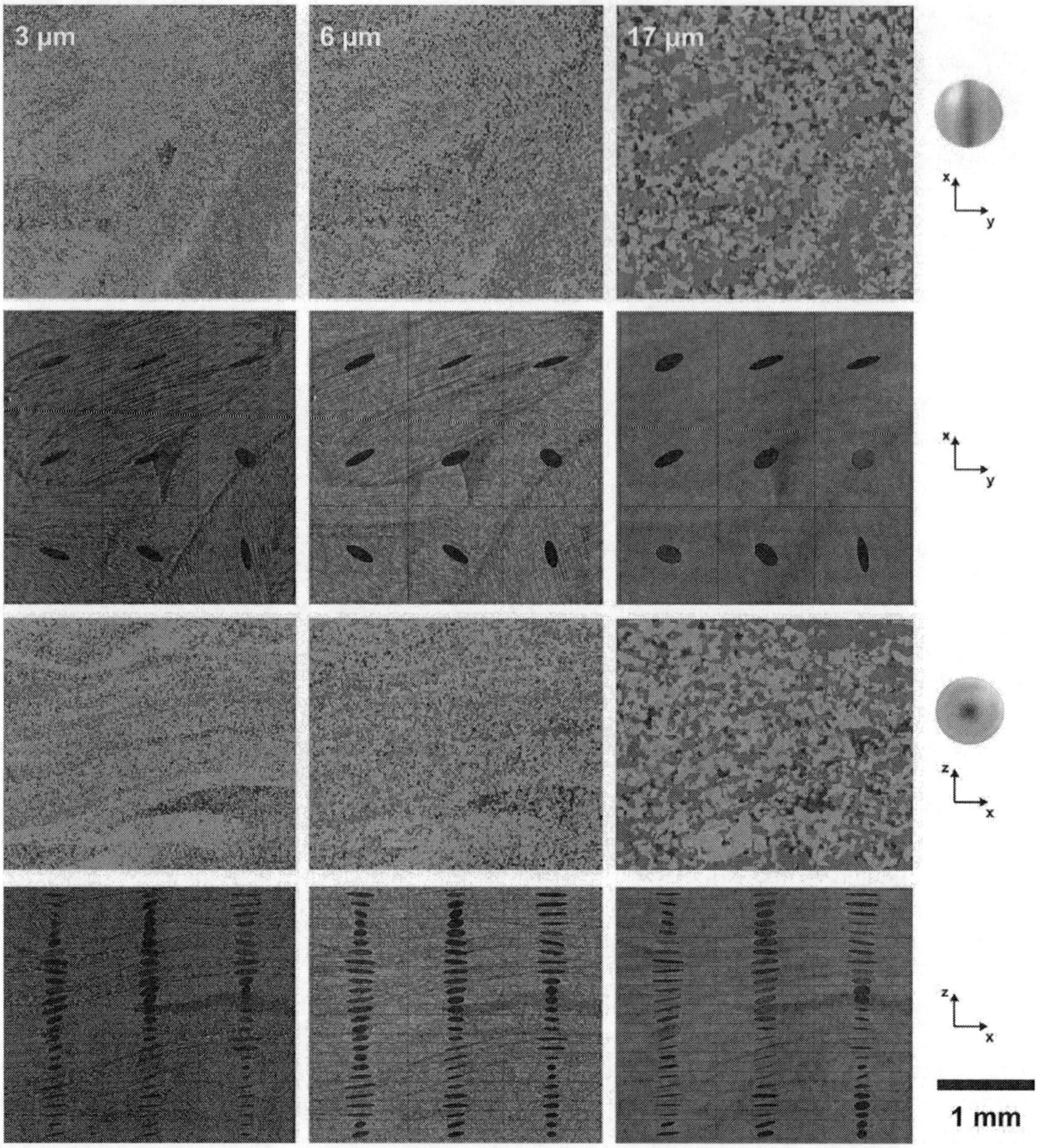

Fig. 9. Fiber orientation measurement of sample 1, W target. The color-coded images illustrate the local fiber orientations (compare globe colors for fiber orientation). The mesh-images represent the fiber orientation per unit cell. Each cell has dimensions of 1 x 1 x 0.1 mm³.

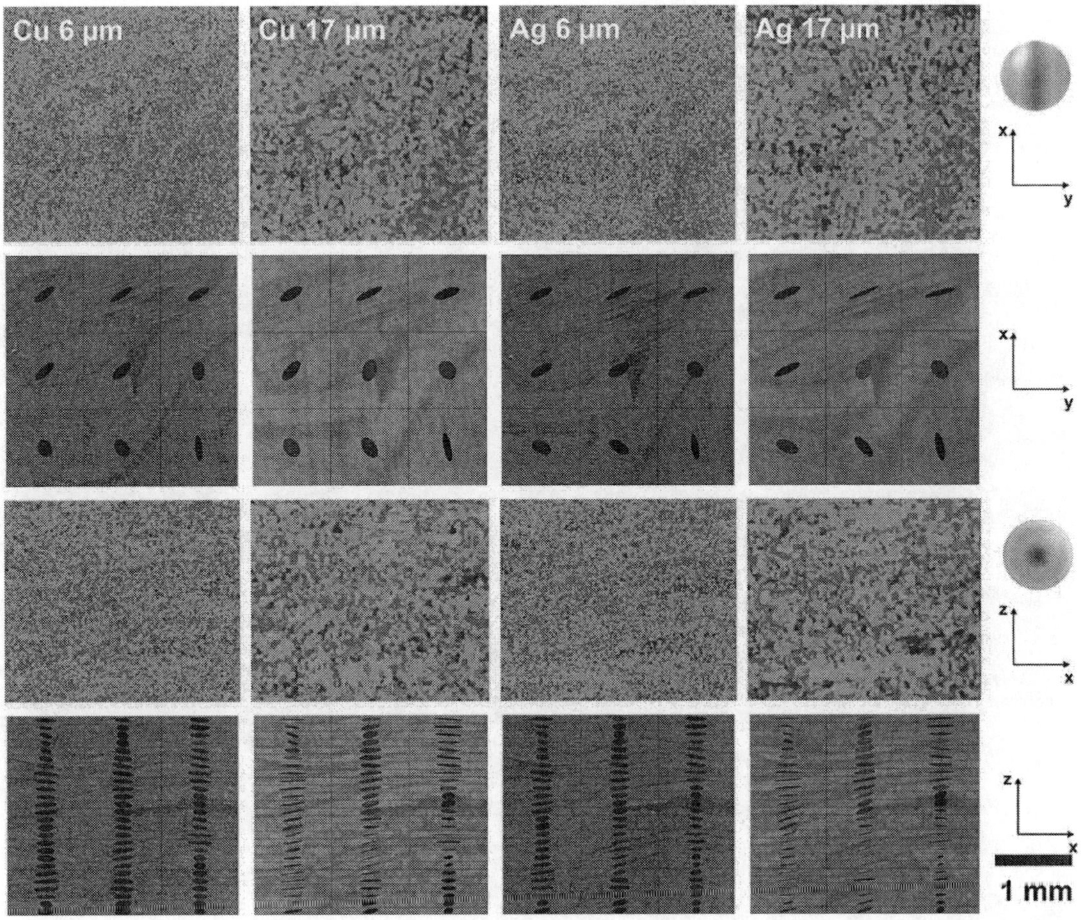

Fig. 10. Fiber orientation measurement of sample 1, Cu and Ag target. The color-coded images illustrate the local fiber orientations (compare globe colors for fiber orientation). The mesh-images represent the fiber orientation per unit cell. Each cell has dimensions of 1 x 1 x 0.1 mm³.

Tab. 4. Fiber orientation measurements of sample 1 (difference to 2.9 µm scan is given in brackets)

Target material	Voxel resolution [µm]	xx	yy	zz
W	2.9	0.36	0.54	0.10
W	6.3	0.39 (0.03)	0.52 (-0.02)	0.10 (0.00)
Cu	6.1	0.43 (0.07)	0.43 (-0.11)	0.14 (0.04)
Ag	6.1	0.39 (0.03)	0.49 (-0.05)	0.12 (0.02)
W	17.4	0.43 (0.07)	0.51 (-0.03)	0.06 (-0.04)
Cu	17.4	0.47 (0.11)	0.46 (-0.08)	0.06 (-0.04)
Ag	17.3	0.44 (0.08)	0.51 (-0.03)	0.05 (-0.05)

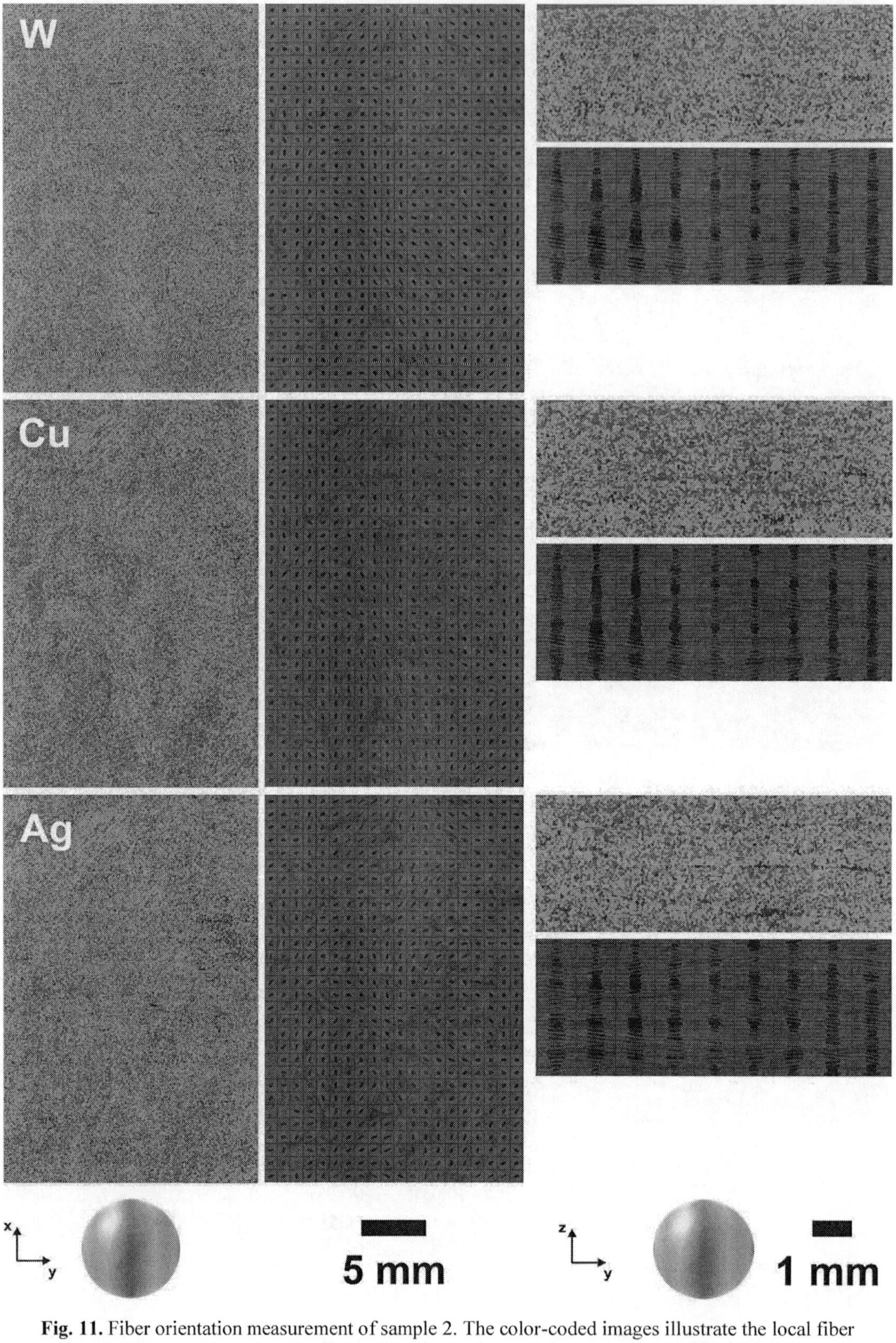

Fig. 11. Fiber orientation measurement of sample 2. The color-coded images illustrate the local fiber orientations (compare globe colors for fiber orientation). The mesh-images represent the fiber orientation per unit cell. Each cell has dimensions of 1 x 1 x 0.1 mm^3.

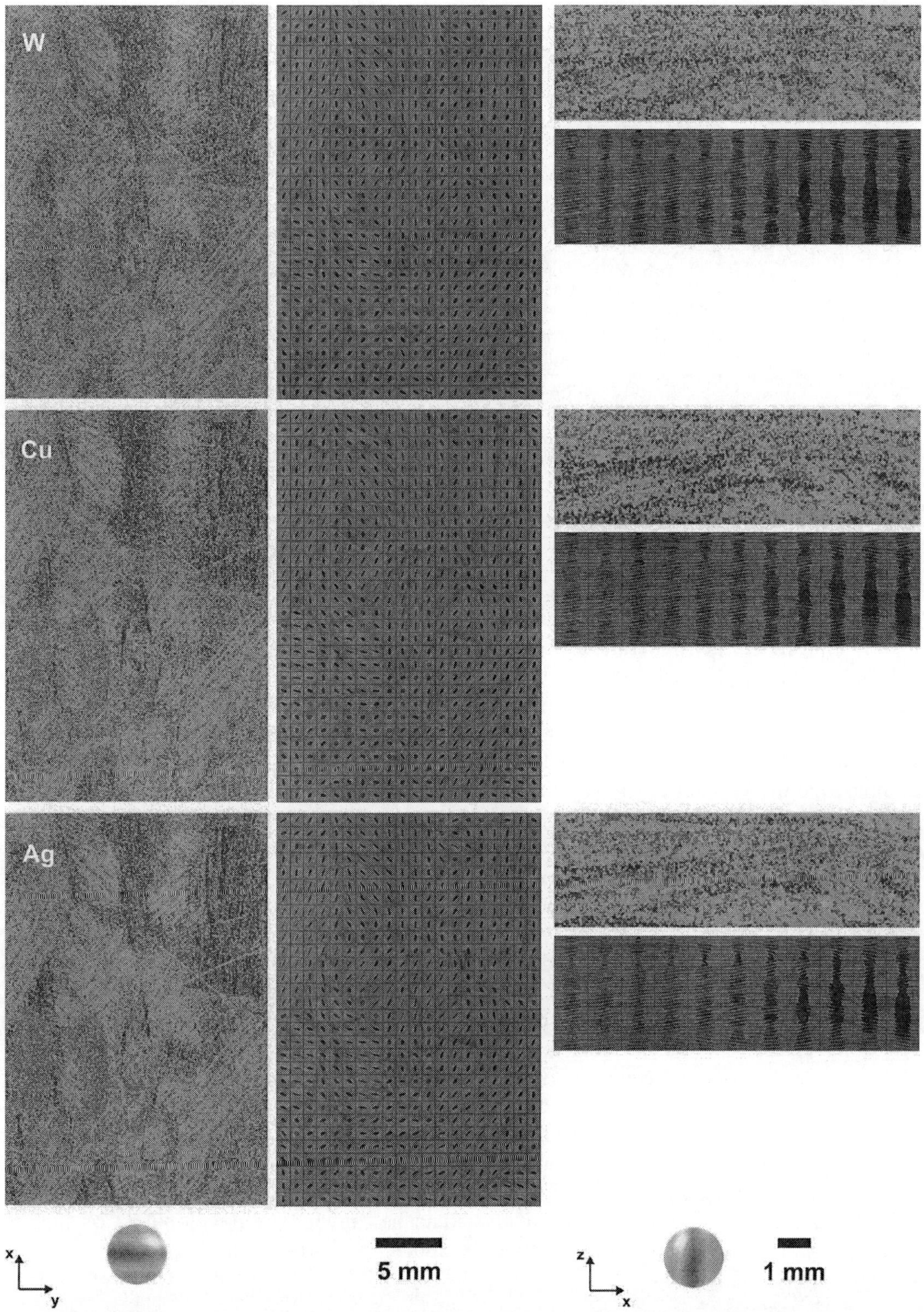

Fig. 12. Fiber orientation measurement of sample 3. The color-coded images illustrate the local fiber orientations (compare globe colors for fiber orientation). The mesh-images represent the fiber orientation per unit cell. Each cell has dimensions of 1 x 1 x 0.1 mm³.

Tab. 5. Fiber orientation measurements of sample 2

Target material	Voxel resolution [μm]	xx	yy	zz
W	17.4	0.53	0.43	0.04
Cu	17.4	0.51	0.40	0.09
Ag	17.3	0.45	0.48	0.07

Tab. 6. Fiber orientation measurements of sample 3

Target material	Voxel resolution [μm]	xx	yy	zz
W	17.4	0.54	0.40	0.06
Cu	17.4	0.53	0.41	0.06
Ag	17.4	0.49	0.47	0.04

4. Summary and Outlook

In this study three different X-ray target materials (W, Cu, and Ag) were compared regarding their performance in CT measurements of CF-SMC components.

All three target materials achieved equal resolutions using a JIMA test pattern. The CT measurements showed that Cu and Ag targets cause a shift of the grey values to slightly higher values in comparison to W target. Additional effects of the Cu and Ag target, which have to be approved, are enhanced edge contrast and, for the Ag target, improved noise levels. Because of a lower photon flux the Cu target requires about 1.5 times higher exposure times compared to the W and the Ag target to achieve comparable I_0 values.

In fiber orientation analysis no obvious advantages of the use of Cu and Ag targets could be detected. Generally the accuracy of fiber orientation analysis was observed to diminish with decreasing voxel resolution. Analyses based on scans with the Cu and Ag targets showed comparable differences to the reference high resolution scan as scans with the standard W target.

However, the general course of fiber orientations is described correctly also at a low voxel resolution of approximately 17 μm, which might be sufficient for some applications. At this voxel resolution samples with a diameter of approximately 20 mm and an even higher length have been scanned and fiber orientation analysis was performed.

Regarding the reference values for fiber orientation, a high resolution scan is assumed to deliver orientation values of high accuracy. However, no direct evidence for this assumption exists. To thoroughly evaluate the correctness of a fiber orientation analysis a reference sample of known fiber orientations would be required. Accordingly no final conclusion about the ability of the alternative target materials Cu and Ag to improve the μ-CT scan quality for fiber orientation analysis can be drawn. The development of suitable reference samples for SMC materials might be subject of further studies.

References

[1] Automobili Lamborghini S.p.A., "Technical Data Sheet Forged Composites.", 2017. Available: https://www.lamborghini.com/sites/it-en/files/DAM/lamborghini/forged/Forged%20presentation_EN.pdf. [Accessed April 23, 2019].

[2] The NORDAM Group, Inc., "Boeing 787 features composite window frames", *Reinforced Plastics,* vol. 51, no. 3, p. 4, 3 51 2007.

[3] M. Tuttle, T. Shifman and B. Boursier, "Simplifying Certification of Discontinuous Composite Material Forms for Primary Aircraft Structures", in *SAMPE*, Seattle, Washington, USA, 2010.

[4] L. De Oto, "Carbon Fiber Innovation for High Volumes - The Forged Composite", in *JEC Composites Show Paris 2011*, Paris, France, 2011.

[5] Hexcel Cooperation, "Hexcel Case Study: Audi R8 Carbon Fiber X-Brace", 2018. Available: https://www.hexcel.com/user_area/content_media/raw/HexcelCSAudiv7web(1).pdf. [Accessed April 25, 2019].

[6] B. Boursier, "New possibilities with HexMC, a high performance moulding compound", in *SAMPE Conference*, Paris, France, 2001.

[7] P. Feraboli, F. Gasco, B. Wade, S. Maier, R. Kwan, A. Masini, L. De Oto and M. Reggiani, "Lamborghini "Forged Composites®" Technology for the Suspension Arms of the Sesto Elemento", in *26th American Society for Composites (ASC) Annual Technical Conference*, Montreal, Quebec, Canada, 2011.

[8] Volkswagen AG - Group Research, Internal Documents, Wolfsburg, Germany, 2019.

[9] Hexcel Corporation, Stamford, Connecticut, USA, "HexMC User Guide", 2019. Available: https://www.hexcel.com/user_area/content_media/raw/HexMC_UserGuide.pdf. [Accessed April 23, 2019].

[10] Hexcel Corporation, Stamford, Connecticut, USA, "HexMC®-i Moulding Compound - Carbon Epoxy HexMC®-i / C / 2000 / M77." Data sheet, 2019. Available: https://www.hexcel.com/user_are a/content_media/raw/HexMCi_C_2000_M77_DataSheet.pdf. [Accessed April 23, 2019].

[11] B. R. Denos and B. R. Pipes, "Local Mean Fiber Orientation via Computer Assisted Tomography Analysis for Long Discontinuous Fiber Composites", in *Proceedings of the American Society for Composites: Thirty First Technical Conference*, Williamsburg, Virginia, USA, 2016.

[12] B. R. Denos, "Fiber Orientation Measurement in Platelet-Based Composites via Computed Tomography Analysis", vol. 10608588, Purdue University, Purdue: ProQuest Dissertations Publishing, 2017.

[13] S. G. Kravchenko, "Failure Analysis in Platelet Molded Composite Systems", West Lafayette, Indiana, USA: Purdue University, School of Aeronautics and Astronautics, 2017. PhD Thesis.

[14] S. C. Garcea, Y. Wang and P. J. Withers, "X-ray computed tomography of polymer composites", *Composites Science and Technology,* vol. 156, pp. 305-319, 2018.

[15] H. Shen, S. Nutt and D. Hull, "Direct observation and measurement of fiber architecture in short fiber-polymer composite foam through micro-CT imaging", *Composites Science and Technology,* vol. 64, no. 13, pp. 2113-2120, 2004.

[16] T. Riedel, "Evaluation of 3D fiber orientation analysis based on x-ray computed tomography data", in *Conference on Industrial Computed Tomography (ICT)*, Wels, Austria, 2012.

[17] S. Goris and T. A. Osswald, "Progress on the characterization of the process-induced fiber microstructure of long glass fiber-reinforced thermoplastics", in *Proc 16th SPE Automot Compos Conf Exhib (ACCE)*, Novi, Michigan, USA, 2016.

[18] D. Maier, T. Dierig, C. Reinhart and T. Günther, "Analysis of Woven Fabrics and Fiber Composite Material Aerospace Parts using Industrial CT Data", in *5th International Symposium on NDT in Aerospace*, Singapore, 2013.

[19] L. Brömstrup, "Fließpresssimulation kohlenstofffaserverstärkter Sheet Molding Compounds – Potentialanalyse der Einzelfasersimulation", Bachelor Thesis, Volkswagen AG, Wolfsburg, 2018.

[20] B. R. Denos, D. E. Sommer, A. J. Favaloro, B. R. Pipes and W. B. Avery, "Fiber orientation measurement from mesoscale CT scans of prepreg platelet molded composites", *Composites Part A: Applied Science and Manufacturing,* vol. 114, pp. 241-249, 2018.

[21] A. Schulman. Technical Data Sheet, "https://engineeredcomposites.aschulman.com/data-sheets", 2019. Available: https://cdn2.hubspot.net/hubfs/2015678/Engineered%20Composites%20/Documents/Product %20PDFs/Quantum/ASI-TDS-AMC%208593%20HT%20rev.pdf. [Accessed May 6, 2019].

[22] JIMA, "Japan Inspection Instruments Manufacturers' Association", Available: http://www.jima.jp/engli sh/assen-e.html. [Accessed May 2, 2019].

[23] N. Flay and R. Leach, "Application of the optical transfer function in X-ray computed tomography – a review", National Physical Laboratory (NPL), Report ENG 41, Teddington, Middlesex, United Kingdom, 2012.

[24] C. Bellon and G.-R. Jaenisch, "aRTist – Analytical RT Inspection Simulation Tool", in *DIR 2007 - International Symposium on Digital industrial Radiology and Computed Tomography*, Lyon, France, June 25-27, 2007.

At-line X-ray computed tomography of serial parts optimized by numerical simulations

Michael REITER [1], Christian GUSENBAUER [2], Reinhold HUEMER [2], Johann KASTNER [1]
[1] University of Applied Sciences Upper Austria, Wels, Austria
[2] Nemak Linz GmbH, Linz, Austria

Contact e-mail: michael.reiter@fh-wels.at

Abstract. Within this work, optimisation advances of an at-line X-ray computed tomography (XCT) inspection system at Nemak Linz are presented. The ZEISS VoluMax 1500 G2 is operated together with an ABB loading robot for automated inspection of serial parts as wells as prototypes of aluminium casted cylinder heads produced for the automotive industry. Typical scanning times for these high absorbing specimens are ranging from 40 to 400 seconds.

The defect detection software provided by the manufacturer is used to automatically identify and classify defects of serial parts and decide between "okay" and "not okay" parts. The ZEISS Automated Defect Detection (ZADD) software consists of anomaly detection, defect classification and a segmentation based on deep learning that requires an initial teaching phase.

The detection seems robust, but the artificial manufacturing of parts with critical defect types and sizes at specific locations and their experimental verification is challenging and very time consuming. This work presents an exemplary verification for one selected specimen and for one particular defect type with different sizes.

Besides real also simulated data with virtual defects are evaluated using the ZADD. Advantages of numerical simulation in combination with the presented defect detection are discussed in detail.

Introduction

In the automotive industry, the demand for 100 % part inspection by original equipment manufacturers, and the supply of parts with zero-defects, is increasing to minimize value loss to the industry. Nemak Linz – a supplier of casting parts – acquired an at-line X-ray computed tomography (XCT) inspection system from ZEISS in 2018 to fulfil these demands. At Nemak, XCT is used for two different scenarios. The first task targets for an automated serial inspection of cylinder heads with a part-to-part time of well below two minutes and the second task targets to robustly inspect prototypes in the development phase with scanning times of 7-8 minutes. Thus, the main differences of those two inspection tasks are throughput rates, achievable image quality, part handling, data evaluation and primary inspection goals. Table 1 shows a detailed comparison.

Table 1. Inspection tasks covered with the at-line inspection setup

Inspection task	Typical scanning times in [min]	Typical throughput per hour	Part handling	Data evaluation	Inspection goal
Serial inspection	1-2	30-60	Robot	Automatic	Defect detection and continuity of water jacket
Prototype inspection	7-8	7-8	Manual	Manual	Defect detection and dimensional metrology (check sand core shrinkage)

Although XCT data in the case of prototype inspection has to be inspected manually, the longer scanning times make this particular inspection task much more robust. Though, setting up a robust and reliable serial inspection based on limited data is difficult, especially if the ability to detect critical defects at critical positions has to be qualified. One way to achieve this is to produce physical parts with specific defect types and sizes at specific locations. This is in particular challenging as some defect scenarios only occur at specific casting situations or under critical casting parameters and may not be imitated artificially.
A more elegant way is to use a virtual engineering approach by using numerical simulations of the X-ray imaging process. This work shows the potential of numerical simulations to generate a virtual failure catalogue for a selected specimen, find the detection limits for different defect types, optimize XCT acquisitions and simplify the teaching process of the ZEISS Automated Defect Detection (ZADD) provided by the XCT manufacturer. Related work in the field of inline XCT inspection is presented in [1].

1. Methods

1.1 At-line X-ray computed tomography system

The scanning setup of the VoluMax 1500 at-line XCT is built around X-ray components that allow fast data acquisition. This is achieved by a Comet 225 kV dual-focus X-ray source with a maximum power of 1800 W and with a focal spot size of 0.4/1.0 mm combined with a Perkin Elmer XRD1620 flat panel detector with 33 ms minimal integration time, when operated in binning 2x2 mode. Continuous rotation of the rotary table ensures fast data acquisition. The X-ray source is equipped with a mechanical shutter for continuous failure-free operation and guaranteeing a long lifetime in 24/7 use.
The at-line system is supported by an automated part loading robot (see Fig. 1 right). Robot support is necessary to achieve constant and high throughput rates and to be able to handle several hundreds of pieces a day. Safety circuits of automation and XCT system are linked together, which means operation is only possible if both safety circuits are closed. The whole system is running within a robot cell with regulated access doors. The serial inspection process is shown in Fig. 2 in a simplified flowchart. Part handling and data acquisition steps are highlighted in blue. The data evaluation steps are highlighted in green and are described in detail in the following chapter. Note that, data acquisition and evaluation may run asynchronously.

Fig. 1. VoluMax 1500 with control panel in front (left) and ABB loading robot standing in the centre of the robot cell that is surrounded by a fence (right) at Nemak Linz

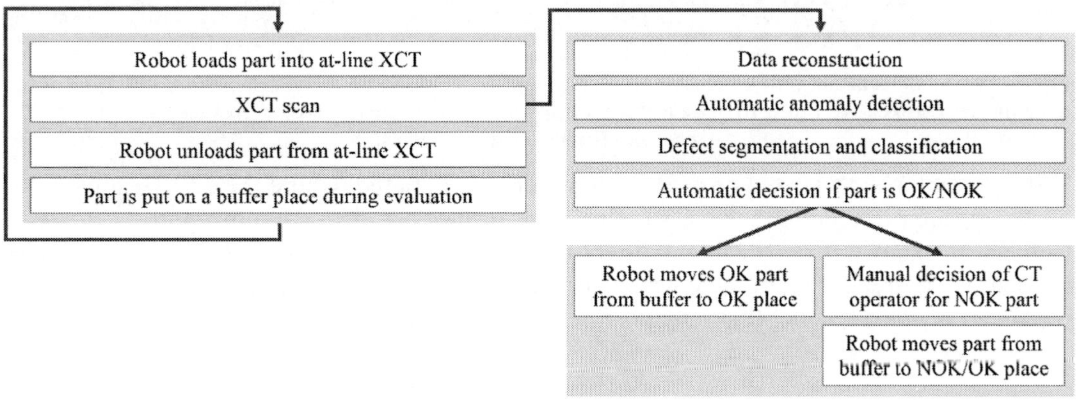

Fig. 2. Simplified workflow of the serial inspection process. Data acquisitions steps are highlighted in blue and data evaluation steps are highlighted in green.

1.2 Automated Defect Detection

The automated serial inspection is performed with the ZADD software that automatically segments and classifies defects. The configuration of this software for a specific evaluation task works in three levels.

In a first step, a so-called golden part is calculated based on XCT data of at least 10 "okay" parts that are within the specification. Therefore, the used scan parameters have to be constant and similar to the actual serial inspection parameters. This golden part serves as reference for all work pieces that are tested in the serial inspection process. Special local features (grey values, standard deviations etc.) of work piece and golden part are compared and analysed to detect deviations from nominal specimen properties.

In a second configuration step regions of interest have to be defined based on the golden part to be able to tolerate slight casting variations (e.g. position tolerances of sand cores) or varying casting characteristics (e.g. core prints, foundry marks), which avoids pseudo detections. If the comparison of local feature values to the golden part indicates local deviations above a certain threshold, defects are segmented to determine defect properties e.g. size.

Subsequently, the determined defect properties are classified and compared with specified limit values that may differ in different evaluation regions (e.g. water jacket).

Setting these local limits is the third configuration step of the inspection, which finally allows an automated decision between "okay" (OK) and "not okay" (NOK) part.

Typical macroscopic defects are shown in Figure 3 that can be found with the presented at-line inspection scenario.

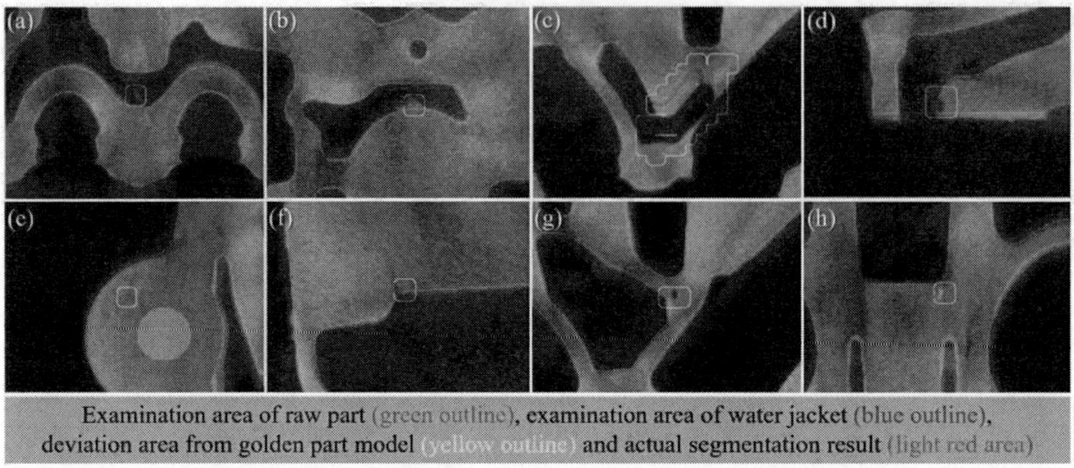

Fig. 3. Typical defects that can be detected with an at-line system used for serial inspection: (a) flash, (b) core failure, (c) broken sand core, (d) core inclusion, (e) porosity or shrinkage holes, (f) crack, (g) void or air bubble and (h) higher-dense inclusion

In the case of a NOK part, a post-decision process is initiated, where the XCT results of this NOK part are presented to the XCT operator in form of a ZEISS PiWeb-report.
Defects are shown from different directions and their position is visualised in a semi-transparent 3D rendering of the CAD. This visualisation enables the XCT operator to decide if the part is indeed NOK. If a defect is tolerable the XCT operator can declare a part as OK.

2. Experimental

2.1 Test specimen for serial inspection

The investigated cylinder head and cylinder heads in general are produced by the world leading dynamic casting process called Rotacast® process at Nemak. Rotacast® is a Nemak owned special permanent mould casting process for producing high quality parts. By a 180° rotation the mould is smoothly filled without turbulence creating an ideal temperature gradient that compensates the melt contraction and hence significantly reduces micro-porosity formation.
Systematic investigations within this work are done for a casted aluminium cylinder head with artificially prepared sand cores by sawing notches into the water jacket at selected positions. The notches result in wall-like aluminium structures in the casting that have a thickness of 1.2-1.35 mm and taper the water jacket. Sand core manipulations were made to achieve the following discontinuities in the water jacket:

- Position 1: partly closure of 75% at a single position
- Position 2: partly closure of 25% at a single position
- Position 3: partly closure of 25% at two positions in opposite direction
- Position 4: partly closure of 50% at two positions in opposite direction

Fig. 4 shows a 3D visualisation of the four water jackets and the corresponding manipulated sand cores used to cast the specimen.
The investigated cylinder head has a bounding box of roughly (200 mm x130 mm x400 mm) and is scanned by a stack of two cone-beam XCT scans with a voxel size of 313 μm. The X-

ray tube parameters for the serial inspection of this part are set to values slightly below the maximal voltage and current values to assure a safe 24/7 usage (reference to chapter 1.1) for the 1.0 mm big focal spot. Additionally, prefiltering of the spectrum with 1 mm of copper leads to a detectable dynamic range of intensities and avoids beam extinction. Since the ZADD is robust against artefacts caused by beam hardening and scattered radiation, no special emphasis is necessary to reduce those artefacts. The detector is operated in 2x2 binning mode with 33 ms integration time. Using the minimal integration time ensures minimal motion blur caused by the continuous rotation of the cylinder head during the acquisition.

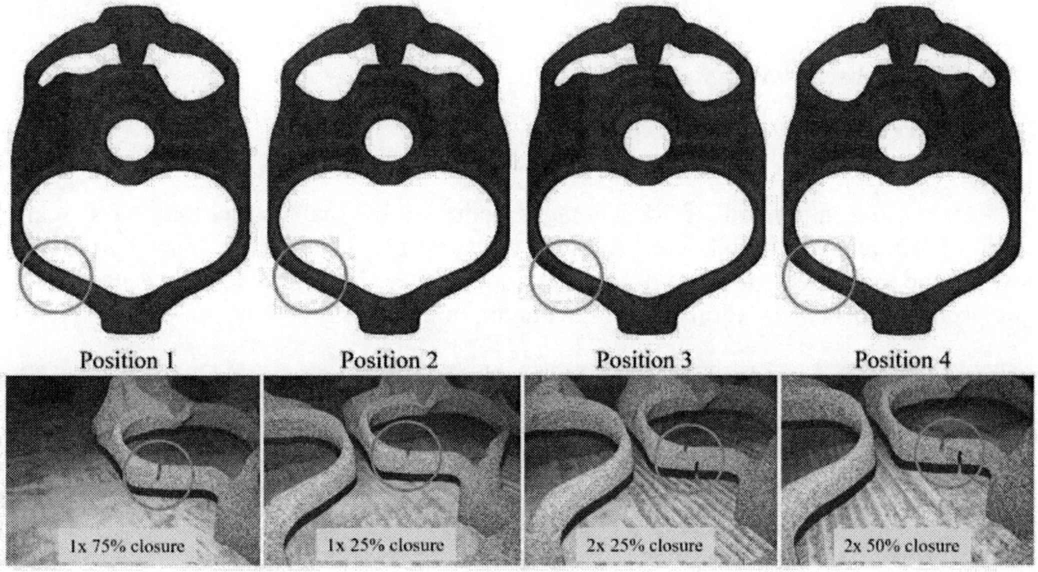

Fig. 4. The top image shows the corresponding CAD file from the water jacket and the images at the bottom different preparation scenarios of the water jacket sand core

2.2 Simulation study

Preliminary to the presented investigations, SimCT [2] has been extended to support the ZEISS VoluMax G2 XCT device. The tool models the complete acquisition chain from the generation of X-rays and the interactions of X-rays with matter to the detection of X-rays in a detector, while considering most relevant physical effects. Several experiments were made at Nemak Linz to model image noise, the energy dependent detector efficiency, scatter related effects as well as image blur by modulation transfer functions. This step is a prerequisite to simulate and create virtual datasets close to reality.

Within this work SimCT is used to investigate the number of acquired projection images and the effect on image quality. The final goal is to find a trade-off for the scanning time and the expected through put of the serial inspection. The simulation-based approach overcomes the issue of manufacturing parts with defined defect scenarios, which is very challenging.

The second investigation is a conceptual test, where SimCT is used to train and initialise the anomaly detection of the ZADD with a virtual golden part. For that purpose, 10 repeated numerical simulations are performed as realistic as possible of a defect-free and thus OK part. Not yet considered are uncertainties of the specimen placement by the robot and manufacturing tolerances [3]. Preliminary analyses have shown a high reproducibility of the specimen placement.

2.2.1 Virtual test specimen

A virtual replicate of the physically existing casting part described in section 2.1 is created with and without the discontinuities. The virtual replicate consists of a surface representation (triangulated meshes) and the aluminium alloy (mass weighted elements plus density), that is assumed to be homogeneously distributed within the closed surface.
Additionally, one virtual test specimen is created where the water jacked is tapered by walls with a thickness of only 0.62 mm. This thickness is represented by approximately two voxels in the acquired XCT data.

3. Results and discussion

3.1 Comparison of real and virtual XCT images

Fig. 5 shows a comparison of XCT images acquired by real and virtual XCT scans at Position 1 where the artificial defect (red arrow) tapers the water jacket. Visually the images show a good agreement in terms of artefact appearances, but simulated images are slightly more prone to noise at locations with high attenuation of X-rays.

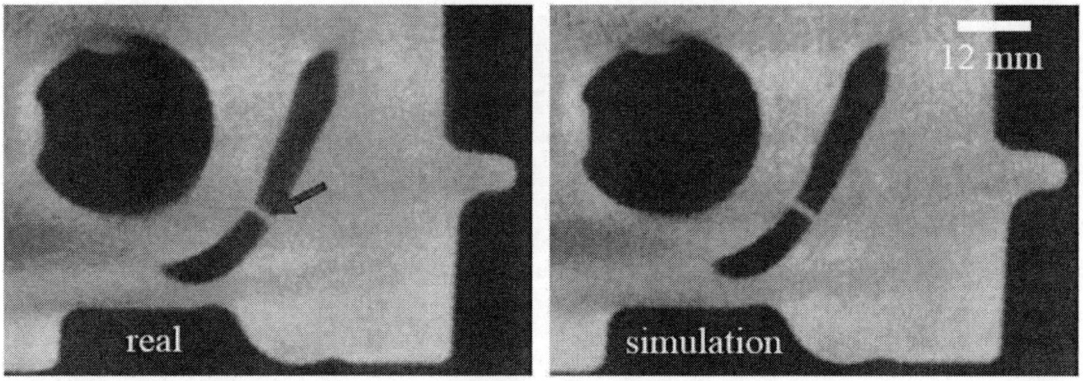

Fig. 5. Slice images of (left) real scan data with artificial defect marked red and (right) simulated replicate of the real scan.

3.2 Simulation-based investigation regarding number of projections and scanning time

Fig. 6-8 show results of a systematic simulation-based investigation of the needed number of projections, whereat the scan time increases proportionally with the number of projections.
Fig. 6 gives insight into the detectability of the artificial structure at Position 1 described in section 2.1. The contrast-to-noise ratio between air and aluminium flash, representing a discontinuity within the water jacket, increases with the number of projections. Nevertheless, 242 projections would be already enough to detect the flash within the water jacket. Less than 242 projections and even faster scans are not possible, since another limiting factor besides the minimal integration time is the maximal rotation speed of the rotary table. Results and trends are equal for even smaller structures in Fig.7, where the tapering wall is only two voxels in thickness.
With the current state of used source and detector models, simulated data is slightly more prone to noise, this means in real scans the detectability of the shown flashes would be slightly better.
On the contrary, Fig 8 shows a region of interest located close to the maximal radius of the permissible reconstruction volume. The slice image comparison depicts that aliasing effects

at this location are quite strong for thin structures reconstructed with only 242 projections and at least 400 projections should be and are used for the real serial inspection of this cylinder head at Nemak Linz.

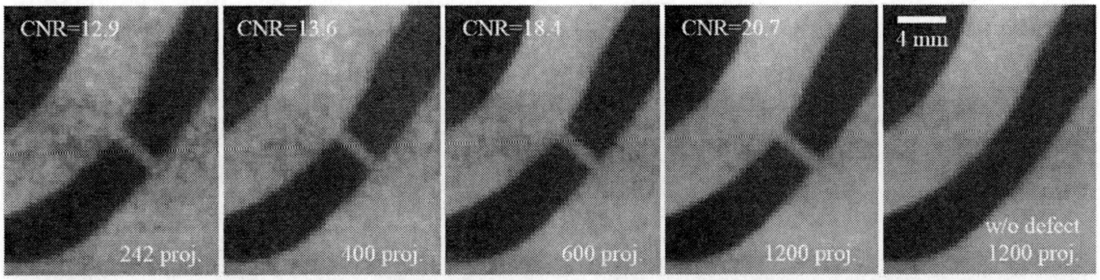

Fig. 6. Slice images at a water jacket position generated by simulation with varying number of projections. The thickness of the flash corresponds to the real artificial feature. The number of projections is proportional to the scan time.

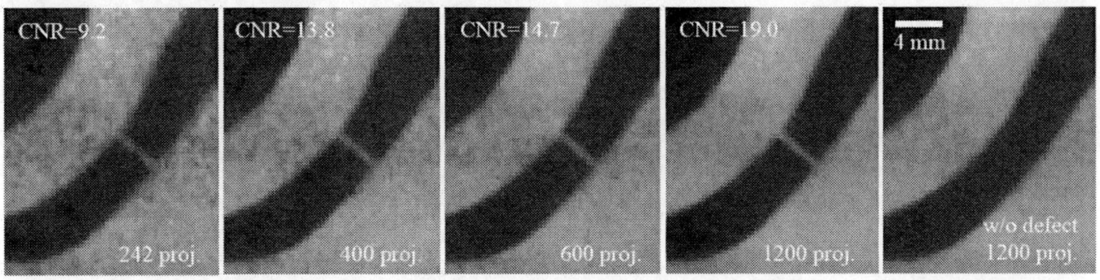

Fig. 7. Slice images at a water jacket position generated by simulation with varying number of projections. The thickness of the flash is two voxels.

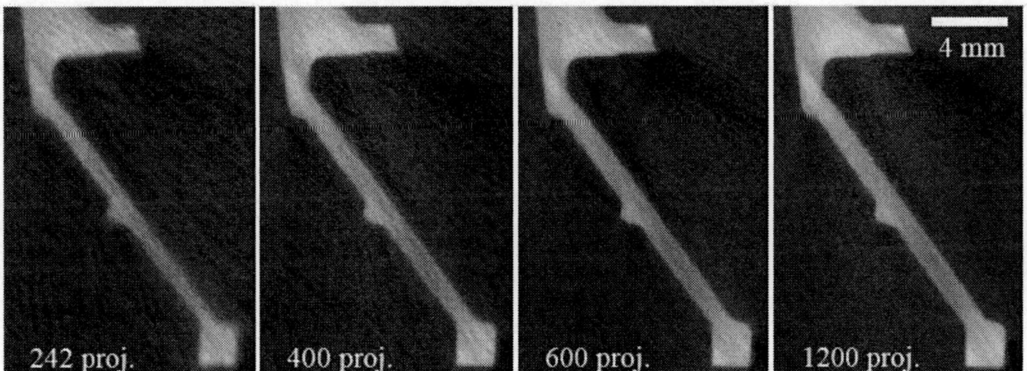

Fig. 8. Simulated slice images at thin walls that are prone to sampling artefacts with varying number of projections. The structure seems blurred due to aliasing.

3.3 Automated defect detection supported by numerical simulations

Fig. 9 shows three different ways of using the ZADD software at the defect of Position 2. The first row shows slice images of the mean golden part representation and the second row the XCT-data that should be analysed. The last row shows the segmented data, where the green and blue contours indicate the examination areas for the raw material respectively water jacket. The yellow line indicates areas with detected deviations and the light-red area the actual segmentation of the defect.

The left column shows the first standard way of using the ZADD, were 10 (typically up to 50) real scans of OK parts are used to generate the golden part. A real scan is tested against this golden part and the artificial defects are found. The middle column is a test run, where a virtual part is tested against a virtual golden part and the column on the right shows the inspection of real data against a virtual golden part. Although, there are still slight differences between real and simulated data, it seems to be possible to perform the initial training of the anomaly detection with simulated data.

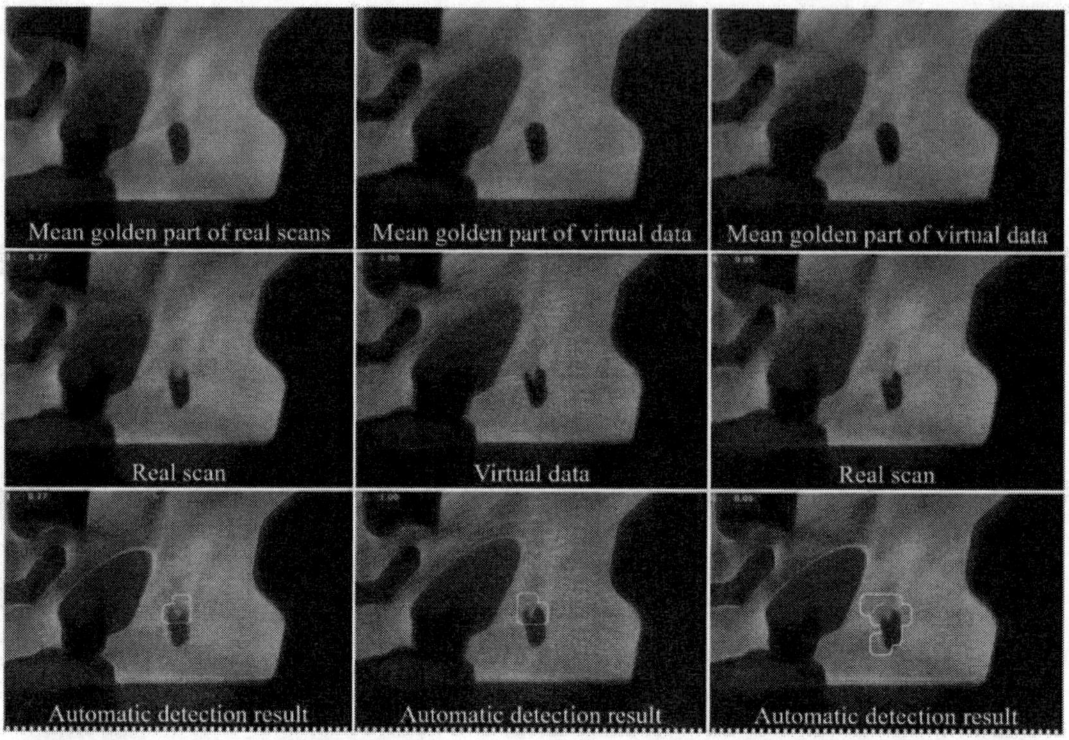

Fig. 9. The left column shows the scenario of a defect detection scenario with real golden part and real scans, the middle column shows a pure virtual detection scenario and the right column a combination of a real scan tested against a virtual golden part

In practice, after the first inspections of real parts are done, every time an OK part is identified, it can be appended to the stack of scans that is used to calculate the golden part. This means the golden part could be iteratively updated and consider more and more effects, which gradually improves the robustness over time. Optionally, virtual datasets could be completely replaced by real scans, when enough scans of manufactured OK parts are available.

In other words, the virtual datasets could be used to identify OK parts, in the setup phase of a serial inspection process, which can be one of the most challenging and time-consuming steps with the described approach.

4. Conclusion

At-line XCT has the potential to replace standard methods such as flow tests, light through methods or endoscopy to check the continuity of water jackets in cylinder heads. The presented XCT inspection scenario is fully automated, except in the case of a NOK part, where a decision of the operator is needed, can handle highly complex inner structures of cylinder heads and avoids manual and user dependent inspections.

This paper showed the successful implementation of fast serial inspection with XCT at Nemak Linz for one exemplary cylinder head. Additionally, the inspection was successfully

supported by numerical simulation to optimise scan parameters and on the other hand, initial tests have shown that SimCT can support XCT-operators during the teaching phase of the used ZADD software. Both a-priori studies and the teaching of the defect detection software can be performed even before an actual part has been casted.
Further, numerical simulations give insight into the detection limits of XCT inspections, since any kind of defect (size, shape and material) can be virtually positioned at any location in the specimen. Within this work one possible defect type has been physically created in a cylinder head and real XCT scans have been compared to virtual ones, which further increases the acceptance of numerical simulations. Nevertheless, additional efforts will be made to reduce existing deviations between numerical simulations and real XCT scans.

Acknowledgements

This work was supported by the K-Project "Photonic Sensing for Smarter Processes" financed by FFG and the government of Upper Austria and Styria. We further want to thank the XCT development team at Carl Zeiss Industrielle Messtechnik GmbH in Oberkochen for their support on modelling the Zeiss VoluMax in SimCT.

References

[1] P. Fuchs, T. Kröger, T. Dierig, C. S. Garbe, Generating Meaningful Synthetic Ground Truth for Pore Detection in Cast Aluminum Parts, Proceedings of Conference on Industrial Computed Tomography (iCT2019), Italy, Padua, 2019.
[2] M. Reiter, M. Erler, C. Kuhn, C. Gusenbauer, J. Kastner, SimCT: a simulation tool for X-ray imaging, Proceedings of Conference on Industrial Computed Tomography (iCT2016), Austria, Wels, 2016.
[3] M. Reiter, J. Kastner, Investigation towards simulation-based determination of measurement uncertainties for X-ray computed tomography, In Proceedings 4th International Conference on Image Formation in X-Ray Computed Tomography. Bamberg, Germany, July, 2016.

Automatic data analysis within X-ray inspection of welds

Erik LINDGREN [1], Thomas STOCKER [2], Christopher ZACH [3]
[1] Swerim, Nohabgatan 18A, SE- 461 53, Trollhättan, Sweden
[2] Fraunhofer-Development Center X-ray Technology EZRT, Fürth, Germany
[3] Chalmers University of Technology, Göteborg, Sweden

Contact e-mail: erik.lindgren@swerim.se

Abstract. In this paper, different approaches to automatic and semi-automatic data analysis within X-ray inspection of welds have been reviewed. The reason for the limited number of industrial implementations of such automatic data analysis solutions within the quality critical manufacturing industry segment (e.g. aerospace and oil and gas industry) has then been analyzed. A conceptual solution for semi-automatic data analysis, potentially more suitable for industrial implementation in such quality critical applications, has been outlined and proposed together with a set of identified research gaps that require attention. Our proposed solution is semi-automatic rather than fully automatic. It leverages recent developments in machine learning in order to minimize the human effort required to generate annotated training data. Thus, the method effectively selects an as small as possible dataset that needs to be analyzed (and annotated) by a human operator. A key ingredient of our proposed method is to utilize a conservative estimate of confidences, which is valuable feedback data for further analysis and also assists in training data generation.

1. Introduction

This paper is focused on the data analysis part of nondestructive evaluation (NDE) of welds with digital X-ray radiography. NDE with the aim of ensuring a certain level of material quality at manufacturing can be divided into three phases: planning, data collection, and data analysis. In the planning phase a hardware setup and overall procedure is derived for the specific application (e.g. the material defect to detect in what component geometry). In the data collection phase an object to be inspected is physically brought into an inspection setup and in the case of digital X-ray radiography a set of X-ray radiographs (1- or 2- dimensional images), one or more, is collected. In the third phase the collected data is analyzed, e.g. radiographs interpreted, and turned into quantitative information (e.g. positions and type of material defects) that can be utilized to take a decision on the quality of the inspected object.

For many applications, a high level of automatization of the data collection is engineering art rather than an open research problem. This is however, in general, not true for the data analysis part which then becomes the bottle neck when it comes to increasing the level of automatization of the whole inspection. Solutions with fully automatic analysis do exist for some applications like aluminum castings (e.g. automotive wheels) or electronic components. However, for X-ray inspection of welds it is still an open research question. The overall difficulty when trying to increase the level of automatization of the data analysis is for the computer algorithms to cope with variation. The variation can be for example variation in material defects (e.g. size, type, orientation) or the material and geometry of the products (e.g. weld shape and position).

In this paper different approaches to automatic and semi-automatic data analysis within X-ray inspection of welds have been reviewed. The reason for the limited industrial implementation of such automatic data analysis solutions within the quality critical manufacturing industry segment (e.g. aero industry and oil and gas) has then been analyzed. A conceptual solution has been proposed, in theory, together with a set of identified research gaps that require attention.

2. Review of scientific studies

The automatic and computer assisted analysis of X-ray images with the objective to identify material defects within the imaged object can be divided into (inspired by (1; 2)) three overlapping categories: generic minimum prior knowledge based, typically machine learning (in this context the learning data is not considered prior knowledge); reference based, requiring a lot of prior knowledge; statistical hypothesis testing based, requiring prior knowledge. Some examples of prior knowledge, information, or assumptions are: the weld position within the radiograph; ranges for the physical size of the object; typical size, type, and position of relevant material defects; information formation chain from the X-ray physical interactions with matter. In this section the current state of the art within each analysis approach as well as some important differences and similarities between them will be described.

All of the listed approaches typically include a preprocessing step. This step includes, apart from the previously described calibration and correction operations close to the detector hardware, noise reduction and contrast enhancement. Noise removal consists of removing short length scale noise (for example electronic and X-ray quantum noise) and medium or large length scale noise (for example intensity variations from the weld geometry or the overall geometry of the component) from the radiograph.

Reference based approaches subtract an ideal (prior knowledge) radiograph from the acquired radiograph and evaluates the differences (3; 4), whereas the generic approach might for example utilize low pass filters (5), (6), and a statistical hypothesis approach typically assume some mathematical function or base functions at the medium and long length scale noise background (1). For example, in (7) and in (4), welding defects are detected as deviations from a curve fit, where the curve fit represents the intensity variations originating from the weld geometry. In some cases, for example in most additive manufacturing methods there exists a digital CAD-model of the object, and mathematical models of the X-ray inspection, e.g. (8), could potentially be used for such generalized background subtraction. As a rule of thumb, geometrical noise can also be suppressed by avoiding it in the first place by an appropriate choice of inspection procedure, e.g. single wall imaging instead of multiple wall imaging.

The preprocessing step might also contain some contrast enhancement operation. Methods based on histogram equalization (3), where the mapping of intensity values into on-screen gray values is adjusted, are commonly utilized.

In statistical hypothesis testing, a test statistic is calculated inside a kernel for each spatial position in the radiograph (image). Utilizing a spatially invariant model (prior knowledge) of the expected statistics of pure noise (the physics of the X-ray interaction has an intrinsic randomness) standard hypothesis tests can check for the probability of the region containing an anomaly of not (9; 10; 1). The term anomaly is used to differentiate it from a defect, since the algorithms typically stop at this stage without further characterization of the anomaly indication.

After the preprocessing stage, the generic based and reference based approaches apply segmentation, feature extraction, and classification steps. At the segmentation step potential material defect indications are selected, which also may include the selection of the region of interest. In each segmented region a set of features is extracted (feature vector). In a final step classification of the defects is performed, where the classification labels may also include a "not a defect" label.

There exists a multitude of segmentation methods in literature. A comprehensive study comparing the applicability of utilizing different segmentation methods for detecting different defect types in radiographs was conducted in (11). The conclusion was as follows: (i) edge detection (for example Canny or Laplacian edge detection) was applicable for detecting slag-inclusion, incomplete penetration, and transverse cracks; (ii) region growing was applicable for detecting lack of root penetration, undercuts, and gas cavities; (iii) watershed segmentation was applicable for detecting wormhole type cavities, lack of fusion, slag inclusion, and slag line. An overview of segmentation approaches in radiographs can also be found in (5). For an in-depth description of all of these segmentation methods see for example (12). However, it is worth noticing that region growing segmentation, where regions are expanded starting from seed sites are often based on a measure of variation within the growing region (11; 13), and are therefore similar to the already described statistical hypothesis testing approach.

An alternative approach to the above segmentation methods is to use a model of the object of interest (for example a specific defect type) and match it against regions in the radiograph. Interesting results on detecting longitudinal cracks in welds with an advanced crack indication model has been reported (14). The matching was formulated as an optimization problem, and the formulated model included noise (similarity to statistical hypothesis testing) and parameters (similarity to adaptive segmentation methods). The algorithm was extended in (15) to include an estimation of the true- or false defect probability for each of the proposed defects, advancing it towards the classification and decision step as well as making it further similar to the statistical hypothesis testing approach.

The segmentation stage is usually followed by a feature extraction step, where scalars measuring some representative property of the region are extracted. Geometrical features are common as well as texture, see Table 1. These features are then utilized in the final classification step. One study (16) utilizes so called Haar-like features (17) which are closely related to the variation among neighboring pixel values and thus similar to the statistical hypothesis testing approach.

An overview of studies utilizing supervised machine learning algorithms for the classification step can be seen in Table 1. In supervised machine learning a classification algorithm is trained on data with known classification (e.g. training sets labeled by skilled humans) in order to make class predictions for future data, see for example (18) for an introduction. As can be seen in Table 1 many studies have used artificial neural networks and fuzzy logic systems. In addition, k-nearest neighbor and support vector machines have also been explored. It is worth noting also that (19) explores simulated defects, where defects are manually inserted into the radiographs, as training set with positive results. Since almost all studies are evaluated on different radiographs an objective comparison of the results is difficult, however the successful categorization rate ranges from 46 to 100 %. It is important to note that successful categorization depends highly on the variability of the input, which is in most cases different and unclear, hence these values should be taken rather as rough indications. It should be noted that since 2015 there is a publicly available database (GDXray) of relevant industrial radiographs (20), which some of the studies have utilized.

Recently, so-called Deep Learning (21) has shown great promise for many different data classification applications. Instead of explicit segmentation and feature selection, the image to be analyzed is instead divided in patches or sliding window/kernels within which features and implicit selected (e.g. with unsupervised learning) with minimal or no manual intervention; this is followed by a supervised learning phase. In (22) such a Deep Learning approach was explored for the weld X-ray inspection data interpretation task with the result of around 90 % accuracy on experimental data (GDXray).

Most studies in Table 1, as well as the deep learning study, focused on how to detect and correctly classify as many defects as possible. Only one study included the estimation of the confidence in each predicted classification. In (23) a confidence level, referred to as credibility level, is estimated for the defect proposals detected by a fuzzy logic system. The objective of the study was to minimize the false alarm (or false positive) rate, rather than maximizing the detection accuracy subject to a low false positive rate. We are not aware of any follow-up work on this study, which elaborates further on the confidence estimation.

The class of 3D point reconstruction algorithms can also be related to the approaches described above (generic, statistical, reference). In 3D point reconstruction (13) (24) the 3D positions of specific points (e.g. the material defect mass centers) are derived from a set of radiographs taken of the object at different views. It is similar to X-ray computed tomography but in contrast to tomography-based approaches it is not a volumetric reconstruction method. One might argue that the 3D point reconstruction algorithms include the decision step intrinsically (e.g. the threshold on the deviation from predicted measurement points, when deriving the 3D points of the potential material defects (13)). 3D point reconstruction-based approaches can also be interpreted as feature extraction methods, where the calculated measurement point deviations act as features; however, no study has been found connecting the 3D point reconstruction in this way to the statistical hypothesis testing or generic low prior knowledge classification approach. The concept of extracting indication features has however been included into the algorithm class (25) but not for defect characterization but for improving the robustness of the method.

Table 1. Overview of studies where some type of a supervised machine learning algorithm has been used for classification of defects in welds, the generic low prior knowledge approach. A plus-sign is used to indicate a combination of and a comma-sign is used to indicate "and".

Study	Segmentation	Feature vector	Classifier	Probability of detection/Correct classification percentage [%]
(16) (Dong, et. al. 2018)	Fixed patch/sliding window	Haar-like	Random forest	83
(26) (Kumar, et. al. 2014)	Edge, region growing, watershed	Texture, geometrical	ANN	87
(27) (Zahran et. al., 2013)	Morphological+ Wavelet+Region growing	Mel-Frequency Cepstral Coefficients (MFCCs) and Power density spectra	ANN	75-100
(28) Zapata, vilar, Ruiz, 2011	Region/Edge growing	Geometrical + PCA	Multiple Layer ANN, adaptive-network-based fuzzy inference system (ANFIS)	79-83

(29) (Valavanis et. Al., 2010)	Graph-based (region growing + connectivity similarities)	Geometrical, texture	SVM, ANN, k-NN	46-85
(30) (Vilar, et al., 2009)	Region/Edge growing	Geometrical + PCA	Multiple Layer ANN	100
(31) (Wang, et. Al., 2008)	Multiple thresholds, fixed regions	Geometrical, texture	SVM + Hough Tranform (Line defects)	96
(19) (2007)	Unclear thresholding	Geometrical	Multiple layer ANN	98
(23) (Kaftandjian et al, 2003)	Region based + reference + morphological	Geometrical	Fuzzy logic	80 % at 0 % false alarms
(32) Wang and Liao, 2002	Histogram thresholding	Geometrical, position in weld	Fuzzy k-NN, Multiple layer ANN	92
(33) (Lashkia, 2001)	Region growing after classification	Contrast, variance	Neuro-fuzzy system (Training set based)	98

3. Commercially available solutions & current implementations in industry

Fully automatic X-ray inspection systems are established for several applications like electronic components or aluminum castings. For these applications, it is possible to detect all currently relevant defects automatically. The main challenge for these systems is mostly to avoid false defect detects to be detected as real defects, resulting in unnecessary scrapping (or repairing) of parts. There are two common modes of operating these systems: either in a fully automatic mode or in a controlled automatic/semi-automatic mode. In the semi-automatic mode a human operator reviews all images with potential defect indications marked in the images by an automatic defect recognition (ADR) software. The purpose of this semi-automatic mode is to avoid scrapping parts with no real critical defects, to identify the false positives and thus decrease the false defect detection rate. In a typical foundry implementation, one person at one workplace can review the images from all X-ray inspection systems. The false detection rate for castings in such a setup usually is below 3 % of all inspected parts.

Currently there is no commercially available solution that provides a 100 % fully automatic defect recognition in welds. The only systems we know of, such as XPlus from VisiConsult (34) or ISAR from Fraunhofer EZRT (35), only assist an operator by highlighting potential defects. Those solutions do not guarantee to detect all typical critical defects within welds. The ISAR system is capable of detecting voids, cracks, foreign material and inhomogeneities of the welding. It also estimates the extension of inhomogeneities in beam direction. However, depending on the specific application it is not capable of reliably detecting all defect types.

The Fraunhofer ADR solution, ISAR, is based on a generic minimum prior knowledge (material thickness, type of material) approach. In a few pre-processing steps the orientation and extension of the weld is detected. Afterwards, with a standard image processing chain, special types of defects are detected with special filters; for example, median based algorithms are used to detect voids and pores and gradient based algorithms are used to detect insufficient weld penetration depths (36). Within a last step a classification of the defects in real and false defects is made with a support vector machine.

4. Proposed approach

If the impact of a quality control failure is critical, then identifying an unexpected scenario is just as important as correctly handling the expected one. Therefore, automatic analysis algorithms must correctly handle both situations, (i) previously unseen but in principle detectable defects, and (ii) outlier measurements that failed for some other reason (not expected defect detectability). Based on the review the following approach is proposed in order to minimize risks in the nondestructive quality control at the industrial implementation of automatic and semi-automatic analysis algorithms: 1) the analysis of the evaluation results and the corresponding final decision should be made by humans and computer algorithms in collaboration; 2) human effort should primarily focus on increasing the accuracy of the confidence estimation in the computer algorithm results and only secondary on maximizing the detection rate and correct classification rate of the algorithms; 3) the analysis algorithms should be capable of learning from corrections made by humans.

Most of the above points seem trivial. However, we are only aware of one scientific publication within the field of automatic interpretation/analysis of X-ray weld inspection data, that focused on confidence estimation in each analysis output by the algorithms (23), which—in contrast to our proposed framework—did not aim to achieve a semi-automatic system with a robust human-machine interaction. Broadening the literature search it was found that (37) explored similar ideas as listed above in connection to nondestructive evaluation in the context of ultrasonic testing. The focus shift as proposed in 2) (i.e. the emphasis on minimizing the false positive rate via human operator feedback) is also justified by the fact that in the existing literature AI-based systems have shown to be capable of detecting and classifying defect indications in the radiographs as long as new data is similar to training data. The framework of active learning, see for example (38), or semi-supervised learning could potentially be utilized. The overall objective of those algorithm classes is to maximize the classification accuracy and at the same time to minimize the required amount of labelled training data; e.g. only querying/asking a human operator to label data that would add maximum of accuracy to the prediction model.

Estimation of confidences in a machine learning-based system is most critical. A reliable estimate of confidences coupled with on-line learning utilizing new human labeled data (in cases when the AI would be low in confidence) will lead to a significant reduction in the fraction of analyses requiring human interpretation or annotation. The frequency of such difficult cases in need of human intervention will drop over time due to the successively improved coverage of the input domain. When the AI-based system has low confidence and requires human assistance, the labelled learning set grows, and new predictions for similar input ideally increase in their estimated confidence; in addition, outliers in the form of failed measurements will still not be accepted since the corresponding confidence level will remain low. The accuracy of the quality control should be comparable to an entirely manual approach based on human operators, as long as the confidence estimates are correct. It remains to define such a robust estimate of level of confidences. An often employed confidence measure is the posterior induced by the classification method (as used in (37)), which has been demonstrated to be largely over-confident especially on out-of-distribution data. One alternative is to estimate the so-called evidence of input data, which is intrinsically linked to probabilistic models and unsupervised learning. Our hypothesis is that it will be in particular crucial for the confidence estimator to correctly identify any out-of-distribution data.

5. Conclusions

A review of the last fifteen years of scientific studies on automatic analysis, interpretation, of image data in X-ray inspection of welds (with respect to material defects, also referred to as automatic defect recognition) has been conducted as well as a review of the commercial availability of such solutions for the industry. Even though the review of the scientific studies concluded that sophisticated state-of-the-art computer algorithms can achieve a high detection, accuracy, of the relevant critical weld material defects, we can find none of the solutions implemented in industry. The aim of the scientific studies was typically a fully automatic solution and the studies often focused on how to achieve an as high accuracy as possible on input data with a fixed and known variance. However, we believe this is not a correct aim and focus for such solutions to be industrially implementable in quality critical industries without risking the quality of the X-ray inspection. We propose instead to focus on a reliable estimate of confidence, in each computer analysis, also with respect to out of training distribution data; that the aim rather should be a semi-automatic approach. The framework of active learning machine learning together with on-line learning is proposed as a suitable starting point for such research.

Acknowledgements

Much of the background work in this manuscript was done with the financial support from Swedish Strategic Innovation Program Produktion2030 which is a joint venture by VINNOVA, FORMAS, and Swedish Energy Agency.

References

1. *Statistical detection of defects in radiographic images using an adaptive parametric model.* **Cogranne, R. och Retraint, F.** 2014, Vol. 96, ss. 173-189.
2. *A Review of Methods for Automated Recognition of Casting Defects.* **Mery, D., Jaeger, T. och Filbert, D.** 2002, J. Br. Inst. Non-Destruct. Test., Vol. 44, ss. 428-436.
3. *Automatic identification of different types of welding defects in radiographic images.* **Wang, G. och Liao, W.** 2002, NDT&E International, Vol. 35, ss. 519–528.
4. *A novel method for detecting weld defects accurately and reliably in radiographic images.* **Dang, C., o.a.** 2016, Insight-Non-Destructive Testing and Condition Monitoring, Vol. 58, ss. 28-34.
5. *State-of-the-Art of Weld Seam Inspection by Radiographic Testing: Part I – Image Processing.* **Silva, R.R och Mery, D.** 2007, Image processing. Materials Evaluation, Vol. 65, ss. 643–647.
6. *Detection, 3-D positioning, and sizing of small pore defects using digital radiography and tracking.* **Lindgren, E.** 2014, EURASIP Journal on Advances in Signal Processing, s. 9.
7. *An automated radiographic NDT system for weld inspection: Part II Flaw detection.* **Liao, W. och Li, Y.** 1998, NDT&E International, ss. 183-192.
8. *Radiographic Simulator aRTist: Version 2.* **Bellon, C., o.a.** 2012. Proceedings of 18th World Conference on Nondestructive Testing.
9. *Implementation of automated 3D defect detection for low signal-to noise features in NDE data.* **Grandin, R. och Gray, J.** u.o. : AIP, 2014. Review of progress in Quantitative Nondestructive Evaluation 2013. Vol. 1581, ss. 1840-1847.
10. *Statistical detection of defects in radiographic images in nondestructive testing.* **Kazantseva, I. G., o.a.** 2002, Signal Processing, Vol. 82, ss. 791-801.
11. *A comparative study of different segmentation techniques for detection of flaws in NDE weld images.* **Rathod, V.R. och Anand, R.S.** 2011, Journal of Nondestructive Evaluation, Vol. 31, ss. 1-16.
12. **Gonzalez, R.C.** *Digital Image Processing.* u.o. : Pearson Education, 2013.
13. *Detection and 3-D positioning of small defects using 3-D point reconstruction, tracking, and the radiographic magnification technique.* **Lindgren, E.** 2015, NDT&E International, Vol. 76, ss. 1-8.
14. **Alekseychuk, O.** *Detection of crack-like indications in digital radiography by global optimisation of a probabilistic estimation function.* u.o. : Technische Universität Dresden, 2005. PhD Thesis.

15. *Recent developments for longitudinal crack detection using a Continuous Density Hidden Markov Model (CDHMM).* **Alekseychuk, O. och Zscherpel, U.** 2007. International Symposium on Digital industrial Radiology and Computed Tomography.
16. *Automatic Inspection of Aerospace Welds Using X-ray images.* **Dong, X., Taylor, C. J. och Cootes, T. F.** 2018. 24th International Conference on Pattern Recognition (ICPR). ss. 2002-2007.
17. *Rapid object detection using a boosted cascade of simple features.* **Viola, P. och Jones, M.** 2001. Proceedings of the 2001 IEEE Computer Society Conference on Computer Vision and Pattern Recognition. CVPR 2001.
18. **Murphy, K.P.** *Machine Learning - a probabilistic perspective.* u.o. : Massachusetts Institute of Technology, 2012.
19. *Automatic classification of weld defects using simulated data and an MLP neural network.* **Lim, T.Y., Ratnam, M.M. och Khalid, M.A.** 2007, Insight: Non-Destr Test Cond Monitor, Vol. 49, ss. 154-159.
20. *GDXray: The database of X-ray images for nondestructive testing.* **Mery, D, o.a.** 2015, Journal of Nondestructive Evaluation, Vol. 34, ss. 1-12.
21. *Deep Learning.* **LeCun, Y., Bengio, Y. och Hinton, G.** 2015, Nature, Vol. 521, ss. 436-444.
22. *Automatic Detection of Welding Defects using Deep Neural Network.* **Hou, W., o.a.** 2017. IOP Conf. Series: Journal of Physics: Conf. Series 933.
23. *Uncertainty modelling using Dempster–Shafer theory for improving detection of weld defects.* **Kaftandjian, V., o.a.** 2003, Pattern Recognition Letters, Vol. 24, ss. 547-564.
24. **Mery, D.** *Computer Vision for X-ray Testing.* u.o. : Springer, 2015.
25. *Automatic multiple view inspection using geometrical tracking and feature analysis in aluminum wheels.* **Carrasco, M. och Mery, D.** 2011, Machine Vision and Applications, Vol. 22, ss. 157-170.
26. *Flaws Classification using ANN for Radiographic Weld Images.* **Kumar, J., Anand, R.S. och Srivastava, S.P.** 2014. International Conference on Signal Processing and Integrated Networks. ss. 145-150.
27. *Automatic weld defect identification from radiographic images.* **Zahran, O., o.a.** 2013, NDT&E International, Vol. 57, ss. 26-35.
28. ***Zapata, J., Vilar, R. och Ruiz, R.*** *2011, Expert Systems with Applications, Vol. 38, ss. 8812-8824.*
29. Multiclass defect detection and classification in weld radiographic images using geometric and texture features. ***Valavanis, I. och Kosmopoulos, D.*** *2010, Expert Systems with Applications, Vol. 37, ss. 7606-7614.*
30. An automatic system of classification of weld defects in radiographic images. ***Vilar, R., Zapata, J. och Ruiz, R.*** *2009, NDT&E International, Vol. 42, ss. 467-476.*
31. Detection of line weld defects based on multiple thresholds and support vector machine. ***Wang, Y., o.a.*** *2008, NDT&E International, Vol. 41, ss. 517-524.*
32. Automatic identification of different types of welding defects in radiographic images. ***Wang, G. och Liao, W.*** *2002, NDT&E International, Vol. 35, ss. 519-528.*
33. Defect detection in X-ray images using fuzzy reasoning. ***Lashkia, V.*** *2001, Image and Vision Computing, Vol. 19, ss. 261-269.*
34. https://visiconsult.de/products/ndt/xrhpipe/. [Online]
35. https://www.iis.fraunhofer.de/en/ff/zfp/products/isar.html. [Online]
36. Schweißnahtprüfung bei der Fertigung von Großrohren. ***Wenzel, T., o.a. u.o.*** *: Deutsche Gesellschaft für Zerstörungsfreie Prüfung e.V. -DGZfP-, 2007. Seminar Aktuelle Fragen der Durchstrahlungsprüfung und des Strahlenschutzes.*
37. An Incremental Learning Algorithm with Confidence Estimation for Automated Identification of NDE Signals. ***Polikar, R., o.a.*** *2004, IEEE Transactions on Ultrasonics, Ferroelectrics, and Frequency Control, Vol. 51, ss. 990-1001.*
38. ***Settles, B.*** *Active Learning Literature Survey. u.o. : University of Wisconsin-Madison, 2009. Computer Sciences Technical Report 1648.*

Mapping Performance of CT

Nick BRIERLEY [1]
[1] The Manufacturing Technology Centre (MTC), Ansty Business Park, Coventry, UK
Contact e-mail: nick.brierley@the-mtc.org

Abstract. One of the major hurdles preventing the more widespread deployment of X-ray Computed Tomography (CT) as a tool of industrial quality control for safety critical components is the limited understanding of an inspection's capability to detect defects of concern. In more traditional Non-Destructive Testing (NDT) the challenge of inspection performance demonstration and qualification is typically addressed using samples with "representative" defects. Whilst this approach can in principle be applied to CT inspection, the complexity of the geometries to which CT is frequently applied undermines the level of inspection verification, as the scope for extrapolation from a few example defects is limited. This is especially true when the spatially varying effects, including imaging artefacts for example, that will determine whether or not an indication is detectable in the collected data, are considered.

In practice, the voxel size of the scan is frequently quoted as a measure of the expected inspection performance, but this is an entirely inadequate metric when component stress and lifing calculations require the quantification of the largest defect that could have been missed. Here, a computational framework to address this challenge is presented. It merges a simulation capability with experimental scan data to compute a map of the detectability of defects across the component volume, which can then be used to compute probability of detection and false call rate inspection reliability metrics. The structure of the algorithm is described, as well as initial results.

Introduction

Cone-beam X-ray Computed Tomography (CT) is an inspection technology of increasing industrial importance [1], to a large extent due to its suitability for the inspection of components made by Additive Manufacturing (AM) methods [2]. To enable the technique to be used for sentencing safety-critical components, it is essential to understand the limitations, and hence reliability, of the inspection. Furthermore, given the role of CT as a validation tool to other Non-Destructive Testing (NDT) techniques, this is a matter of broad importance [3].

Whilst it is known that a wide range of factors affect the performance of a CT inspection [4], the collective consequences for the performance are poorly understood. Moreover, several of these factors are liable to induce spatial variations in inspection performance across the sample volume. Such factors include beam-hardening, scattering, focal spot blurring, detector pixel imperfections and misalignment [5-6]. Additionally, the decreasing radiographic contrast of a defect of a certain size with increasing material path length (due to *Beer-Lambert* attenuation) [7] will, in the presence of noise, itself give rise to spatial variations in defect contrast, given the range of material path lengths associated with different locations across the component volume. Note that these considerations all relate to the data acquisition, before any data analysis, which will further affect the overall performance of the inspection.

Without spatial variations in inspection performance, it would be possible to demonstrate the capability of the inspection using a single test defect, introduced on a test piece. However, in the presence of potentially severe spatial variations in defect detectability, a conservative experimental assessment of inspection performance becomes almost impossible – it is clearly not possible to introduce a known inspection target at every location in the volume.

This paper presents an experimental study of the spatial variations in detection performance, as well as a simulation-based route to CT inspection performance assessment and hence qualification / validation, in a manner similar to the ultrasonics-based work in [8], as a means of overcoming the described challenges. Note a comprehensive account of the work is provided by [9], which should be consulted for some details omitted here for brevity.

2. Method

2.1 Experimental

The results relate to a test artefact that was developed to allow a sample defect to be placed in multiple locations in a body liable to give rise to spatial variations in inspection performance. The artefact was made of 304 stainless steel by Electrical Discharge Machining (EDM), including a spare defect-free larger insert and two (well-separated) nominally hemispherical test defects on the other of the larger inserts. The larger of the test defects that is the focus here was measured on an Alicona G4 focus variation microscope [10] to be 1.08 mm across, with a depth of 402 μm. A photograph of the reconfigurable artefact is shown in Fig. 1. In total, the set-up allows the two inspection targets / test defects to be placed in a total of 16 locations each, with negligible differences between configurations beyond defect positions.

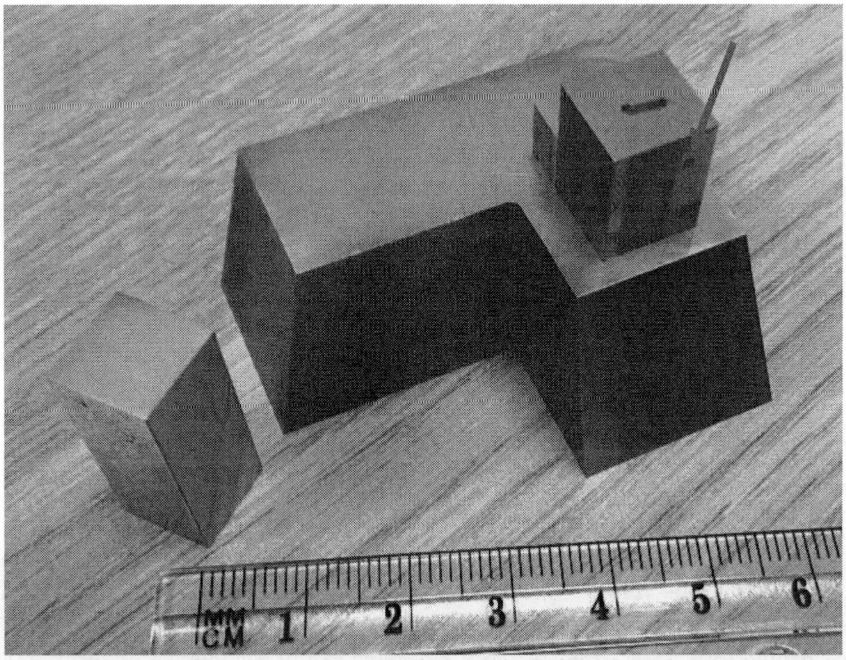

Fig. 1. Photograph of reconfigurable artefact, showing tightly-fitting sliding inserts. Note the test defect (arrow) visible on side of larger insert protruding from the main body of the artefact.

The artefact was scanned in 8 configurations (7 defect insert arrangements plus a scan with the blank insert) on a Nikon XTH 225 ST [11] CT system (fitted with a 2000 × 2000 pixel flat panel detector), otherwise maintaining identical scan parameters. The scan set-up was chosen to give a relatively fast scan (given the number required), with a data output that would not immediately be too overwhelming to process by means of the algorithm developed. Additionally, the intention was not to perform the highest quality scan possible, but rather to induce qualitatively appreciable differences in inspection performance across the set of scans. A summary of scan parameters is shown below in Table 1.

Table 1. CT scan parameters.

Parameter	**Value**
Source voltage (kV)	205
Source current (mA)	0.125
Exposure time (ms)	708
Number of projections	1200
Frames per projection	2
Detector binning	2 × 2
Filter material	Copper
Filter thickness (mm)	0.5
Approx. scan time (min)	29
Native voxel size (mm)	0.074

2.2 Computational

The operation of the algorithm developed is summarised in Fig. 2.

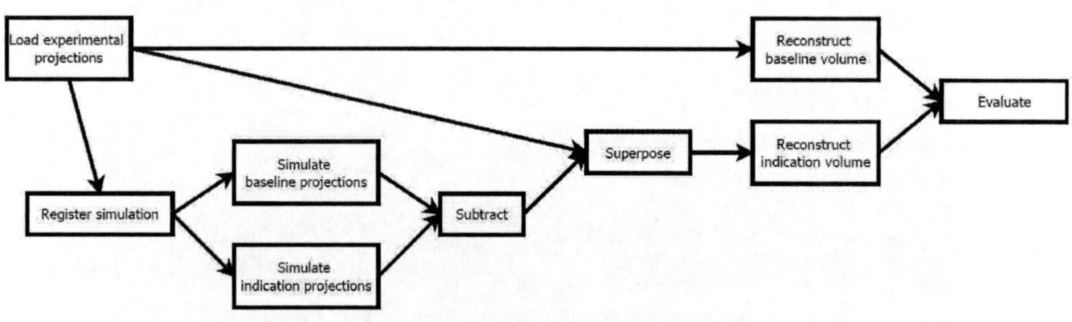

Fig. 2. Flow diagram representation of algorithm.

In slightly more detail:

- The experimental projection dataset forms the basis of calculation, allowing all nuances of the experimental hardware to be reflected.
- A simulation model of the inspection (using the aRTist simulation engine [12]) is registered to the experimental projection dataset based on the component CAD and the method described in [13].
- A set of projections corresponding to the experimental inspection is simulated.
- A second, similar, set of projections is simulated, but placing a hypothetical defect of interest into the component volume, in line with [14].
- Corresponding projections from the two sets are subtracted, to leave a set of difference images attributable to the presence of the hypothesised defect.

- The difference images are superposed into the corresponding experimental images, in a manner comparable to the approach in [15], to produce a hybridised projection set.
- Original experimental (baseline) and hybridised projections are reconstructed independently (using the highly efficient FDK [16] reconstruction implementation X-AID from MITOS [17]).
- The two reconstructed volumes are compared to allow the detectability of the hypothesised defect to be assessed.
- The process is then repeated as required for alternative hypothetical defects, sampling a range of locations across the component volume. Note that not all mentioned steps need to be repeated for subsequent defect evaluations.

3. Results

3.1 Experimental

The 7 locations in the component volume sampled with the larger test defect are shown in Fig. 3, marked on slices through reconstruction volumes of the artefact in its defect-free configuration. The images clearly show that the scan is subject to significant beam-hardening and scatter artefacts (no attempt was made to correct for these in the reconstruction), given the bright edges of the part in the images and the high greyscales recorded on the concave region exterior to the sample.

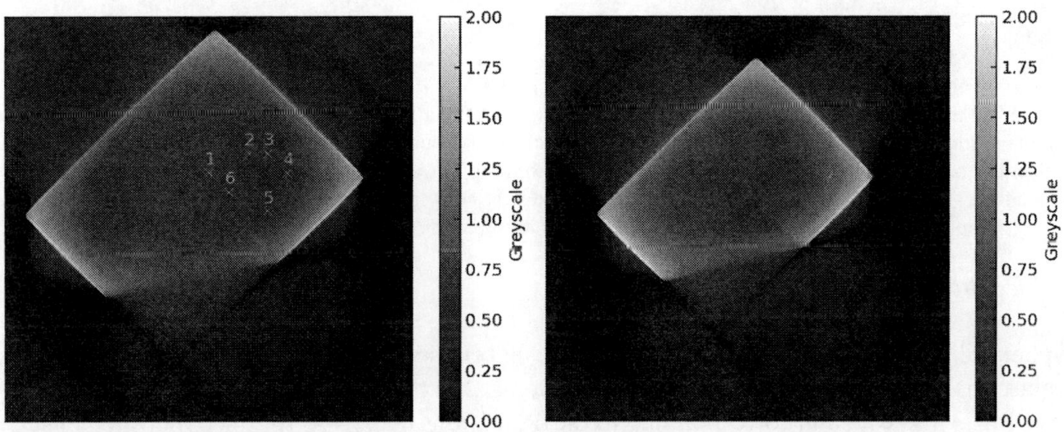

Fig. 3. Defect locations sampled experimentally with the larger test defect, marked on slices through the reconstructed (defect-free) volume. The left image corresponds to the height of the defect in the main body of the artefact when at the bottom of the insert, the right to the height of the defect when the insert is inverted.

Figure 4 shows slices through the indications obtained for 4 of the sampled locations – note more results can be found in [9]. The 4 images reveal obvious variations in the detectability of one and the same test defect, depending only on the position of the defect in the sample volume: despite being nearly 15 voxels across, the defect to all intents and purposes disappears in one of the locations illustrated. These results emphatically demonstrate the problem described in the Introduction, and highlight the extent to which voxel-based rules of thumb on inspection detectability in CT are flawed.

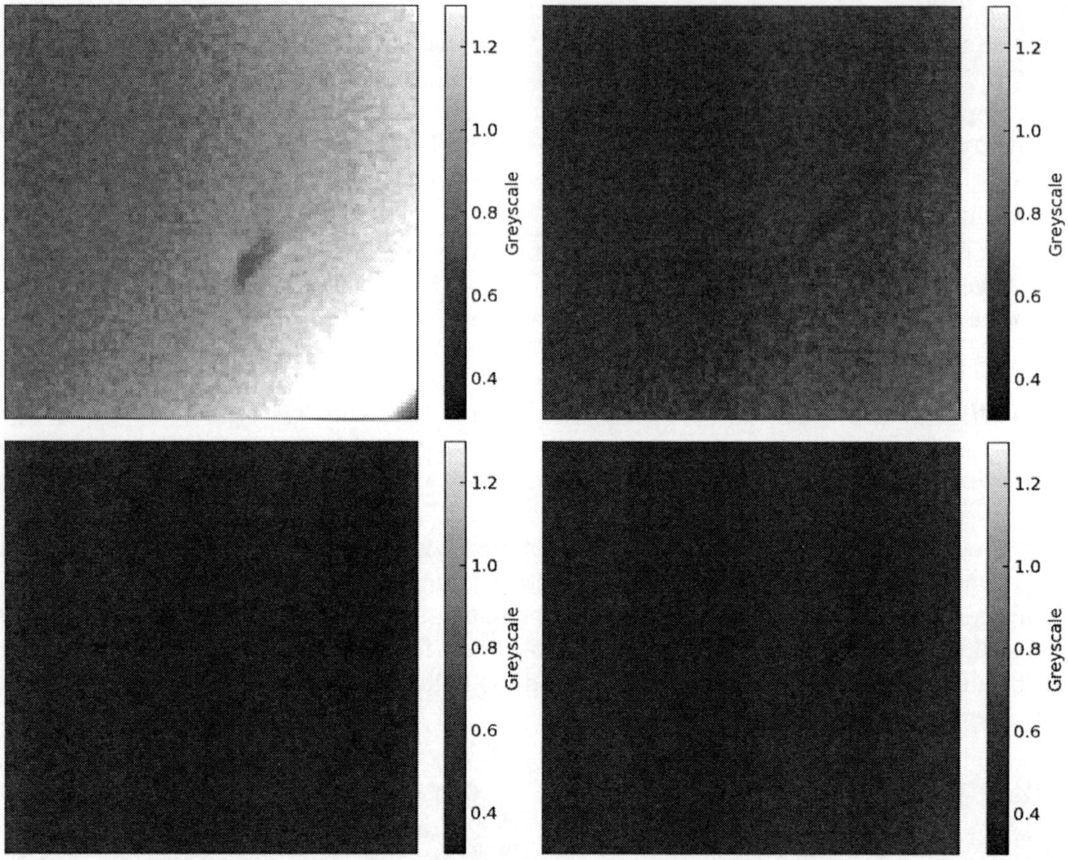

Fig. 4. Slices through the experimental volumes around the defect locations for some of the positions sampled in the artefact volume. The images are 100 × 100 pixels in size and shown with the same greyscale. They are also ordered by decreasing detectability, clockwise from top left, corresponding to positions marked 7, 5, 2, and 1, respectively, in Fig. 3. Note how the same defect goes from being entirely visible to essentially impossible to discern, and despite the slice location having been locally optimised for contrast.

3.2 Computational

Spherical defects of 0.5 mm diameter were hybridised into the defect-free experimental dataset using the algorithm described in Section 2.2 in the locations sampled experimentally (see Fig. 3). The results corresponding to the 4 locations displayed for the experimental scans in Fig. 4 are presented in Fig. 5.

Despite the differences in the size and morphology of the simulated defect compared against the experimental test defect, the slices from the hybridised volumes are quite similar to the experimental images for the same locations. Most importantly, broadly the same trend in qualitative detectability is observed – the correlation is somewhat less strong for the locations not displayed here (see [9]). Refinement of the simulation, to make the hypothesised defects more accurate should allow the results to be improved further.

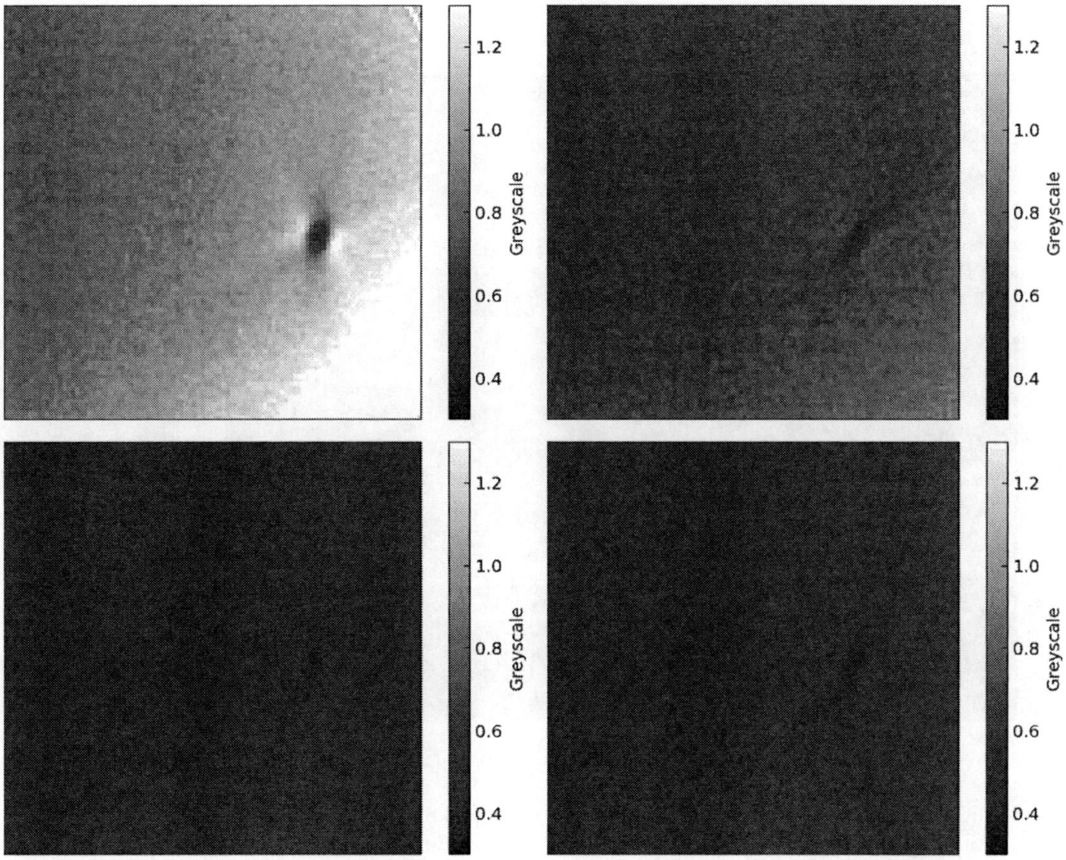

Fig. 5. Slices through the hybridised volumes corresponding to the experimental scans shown in Fig. 4. Clockwise from top left the locations again relate to positions marked 7, 5, 2, and 1, respectively, in Fig. 3. Note the similarity of the images here to those in the preceding figure and the reproduction of the same trend in detectability, despite the difference between physical & simulated defect morphologies.

Extending the computation to a set of defect locations across the cross-section of the artefact at the height of the larger defect allows the image in Fig. 6 to be built up. Here 44 hybridised indication volumes were superposed into the same volume for ease of visualisation. The image further reveals significant variations in the appearance of a test defect across the plane sampled, depending solely on the location of that defect. Clearly, metrics for detectability and size can be computed for this grid of indications, to quantitatively map out the inspection performance, taking into account the 3D nature of the indications. Such maps can then also be used to guide any experimental work with artificial test defects, ensuring that the experimental inspection performance assessment remains conservative.

As an aside, readers may be curious about the computing effort associated with producing such an image. Despite limited speed optimisation to date, the time per indication sampled was under 5 min (for 1200 projections of 1000 × 1000 pixels plus reconstruction and evaluation, excluding prior registration, on a custom workstation with AMD 1950x Threadripper CPU, twin NVIDIA Titan XP GPUs, 128 GB RAM and 1 TB NVMe SSD).

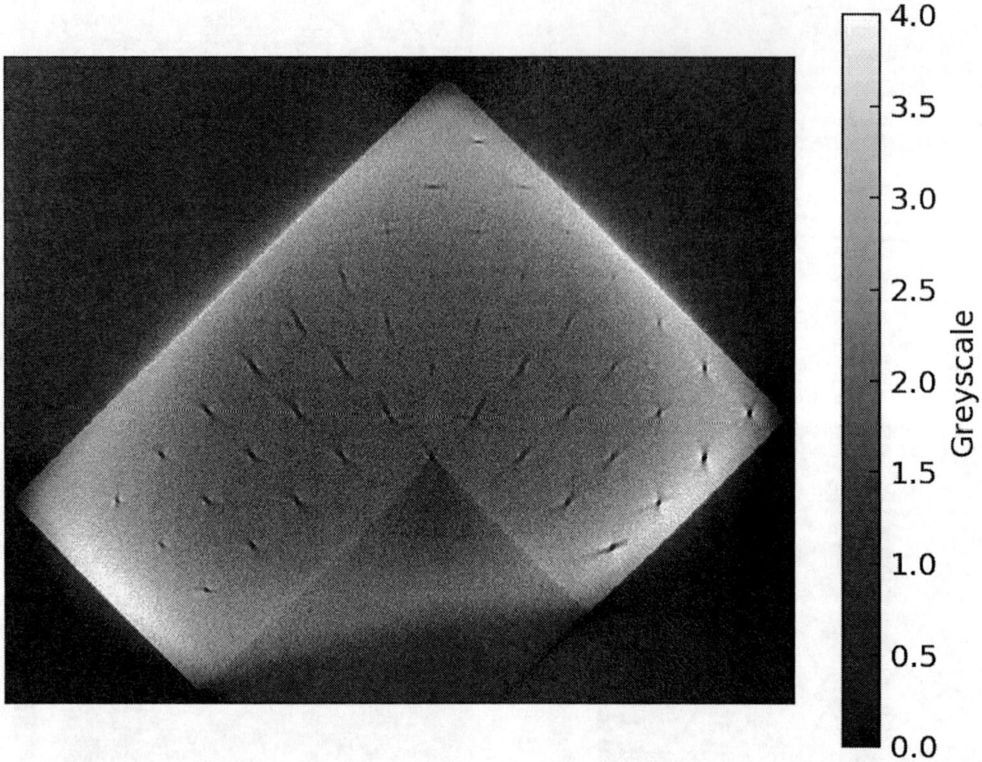

Fig. 6. Slice through a compound volume of the artefact formed by sampling a regular grid of defect locations with a 0.5 mm spherical void. Each defect was hybridised into the scan separately and the indication volumes formed in this way were superposed onto the scan of the defect-free artefact for visualisation. In contrast to previous slice images, some beam-hardening correction was applied in the reconstruction. Note the significant and complex variability in the appearance of one and the same defect in different locations in the scan.

4. Conclusions

This paper has described the challenge of spatial variations in CT inspection performance, supported by an experimental study starkly demonstrating the problem, and proposed an algorithm as a means of addressing the challenge, to enable the validation / qualification of CT inspections. Results obtained from the algorithm are promising, with reasonable correspondence to the experimental data. There is of course scope for further work, including on speed optimisation of the algorithm implementation, but even as it stands, it constitutes a significant advance in the understanding of the limitations of CT inspections.

Acknowledgements

The author thanks colleagues Bethan Nye and John McGuinness for taking on some supporting tasks. This work was completed as part of a Core Research Programme project at the MTC, and funded by industrial membership fees and the UK High Value Manufacturing Catapult. The author is grateful to Imperial College London and Trevor Tippetts, formerly of Imperial College London (now Los Alamos National Laboratory), for permitting the code base developed during the author's doctorate in collaboration with Trevor Tippetts to be reused and built on for the purposes of this project. The project is also indebted to all contributors to the open source code libraries that the program has been built on.

References

[1] L. D. Chiffre, S. Carmignato, J.-P. Kruth, R. Schmitt, and A.Weckenmann, Industrial applications of computed tomography, CIRP Annals, vol. 63, no. 2, pp. 655-677, 2014.
[2] E. Todorov, R. Spencer, S. Gleeson, M. Jamshidinia, and S. M. Kelly, America makes: National additive manufacturing innovation institute (NAMII) - project 1: Nondestructive evaluation (NDE) of complex metallic additive manufactured (AM) structures, June 2014.
[3] M. V. Felice, A. Velichko, P. D. Wilcox, T. Barden, and T. Dunhill, Optimization of ultrasonic array inspections using an efficient hybrid model and real crack shapes, AIP Conference Proceedings, vol. 1650, no. 1, p. 978, 2015.
[4] H. Villarraga-Gómez, C. Lee, and S. T. Smith, Dimensional metrology with X-ray CT: A comparison with CMM measurements on internal features and compliant structures, Precision Engineering, vol. 51, pp. 291-307, 2018.
[5] W. Dewulf, Y. Tan, and K. Kiekens, Sense and non-sense of beam hardening correction in CT metrology, CIRP Annals - Manufacturing Technology, vol. 61, pp. 495-498, 2012.
[6] M. Ferrucci, E. Ametova, S. Carmignato, and W. Dewulf, Evaluating the effects of detector angular misalignments on simulated computed tomography data, Precision Engineering, vol. 45, pp. 230-241, 2016.
[7] T. M. Bezug, Computed tomography: from photon statistics to modern cone-beam CT. Springer, 2010.
[8] A. Ballisat, P. Wilcox, R. Smith, and D. Hallam, Model assisted qualification of NDE techniques, AIP Conference Proceedings, vol. 1806, no. 1, p. 140005, 2017.
[9] N. Brierley, Bethan Nye, John McGuinness, Mapping the Spatial Performance Variability of an X-ray Computed Tomography Inspection, NDT&E Int., in review, submitted Mar' 19.
[10] Alicona InfiniteFocus, https://www.alicona.com/products/infinitefocus, 2018. [Accessed 27/11/18].
[11] Nikon XTH 225, https://www.nikonmetrology.com/en-gb/product/xt-h-225, 2018. [Accessed 27/11/18].
[12] aRTist - Analytical RT Inspection Simulation Tool, http://artist.bam.de, 2016. [Accessed 14/08/18].
[13] N. Brierley, Simulation-supported Analysis of Complex Radiographs, 10th International Symposium on NDT in Aerospace, Oct 2018.
[14] N. Brierley, C. Bellon, and B. Lazaro Toralles, Optimized multi-shot imaging inspection design, Proceedings of the Royal Society of London A: Mathematical, Physical and Engineering Sciences, vol. 474, no. 2216, 2018.
[15] C. Liu, J. Dobson, and P. Cawley, Efficient generation of receiver operating characteristics for the evaluation of damage detection in practical structural health monitoring applications, Proceedings of the Royal Society of London A: Mathematical, Physical and Engineering Sciences, vol. 473, no. 2199, 2017.
[16] L. A. Feldkamp, L. C. Davis, and J. W. Kress, Practical cone-beam algorithm, Journal of the Optical Society of America A, vol. 1, pp. 612-619, June 1984.
[17] MITOS X-AID, https://www.mitos.de/en/software.html, 2018. [Accessed 15/08/18].

Establishment of Acceptance Criteria for Tomography Inspection of Aluminium Alloy Castings

Abdel Rahman DAKAK [1,2], Pr. Valérie KAFTANDJIAN [2],
Dr. Philippe DUVAUCHELLE [2], Mr. Patrick BOUVET [1]
[1] CTIF, Sèvres, France
[2] INSA Lyon, Lyon, France

Contact E-Mail: dakak@ctif.com, abdelrahman.dakak@insa-lyon.fr

Abstract. In October 2018, the French company CTIF, in collaboration with the laboratory LVA of INSA Lyon, has launched a PhD thesis. The aim of this thesis is to establish a set of acceptance criteria for tomography inspection of aluminium alloy castings.

Currently, the agreement on conformity of aluminium castings inspected by 2D X-ray radiography is based on the ASTM reference images. However, when it comes to tomography inspection, such images have not yet been established in 3D. Consequently, tomography inspectors and their casting customers do not have an approved reference on which they could rely to create their own scope statement and evaluate the 3D casting defects and their criticality.

Based on the aforementioned motivation, the objective of our thesis is the establishment of such quality standards that will be applicable to the characterization of aluminium castings defects and their types by 3D tomography. The methodological approach proposed to reach our endeavour is as follows: (1) development of image-processing algorithms to establish automatic detection of casting defects in the tomographic slices, (2) a 3D reconstruction of the defects, and finally (3) automatic processing and measurement of features from the 3D volume of the defects to classify them and finally determine a 3D quality reference.

For the first step, which is currently in making, two classes of regions can be detected on a digital tomographic slice of an aluminium casting after segmentation: regions belonging to the geometric structures of the specimen or tomographic artefacts (false alarms), and those relating to real defects. The distinction between these two classes can be done by using pattern recognition techniques for object detection and classification. By using a defects library, a classifier can be trained to determine whether a segmented indication is a false alarm, or whether it belongs to a real defect.

Introduction

The usage of aluminium alloy casting parts by the automotive and aeronautics industries is constantly on the rise. These parts, such as cylinder housing, cylinder heads and car-to-ground connecting components, are subjected to high mechanical strain, and hence require serious non-destructive testing to ensure their integrity. The cost of such quality controls plays an important role in the price of the casting as well.
The internal conformity of the moulded parts is often inspected by radiography. This produces images of the projected volume on a plane. As moulded parts are complex shapes

by definition, it is often difficult to visualize the component sides' projections and to locate the discontinuities along the thickness variations. Such ambiguous results may mislead to reject a component with shrinkage cavity for example, which lies in the neutral axis and has no influence on its service life. On the other hand, small indications (possible defects) located in the superficial zone in accordance with customer specifications, can lead to a reduction of more than 30% of the fatigue strength. Tomography provides a better detection of discontinuities, and it helps locating them in the volume and gives access to the internal three-dimensional measurements of the specimen. With tomography, it is possible to know if the discontinuities will disappear during the machining or clog on the surface or if they are in a zone designated as crucial for the mechanical resistance of the part. Advances in computing technology and the increase in image reconstruction speeds are now making it possible to consider its use for the inspection of the casting production. In addition, some manufacturers, seduced by this technique and the quality of results, have invested in CT scanners, not only for the development of their parts, but also for the validation of their in-line production.
However, the manufacturers who use this technology do not have a set of acceptance criteria based on 3D tomographic images to decide the criticality of the detected defects. Currently, acceptance of parts inspected by radiography is based solely on ASTM reference (X-ray) images. There are many standards depending on the alloy / process families. For aluminium alloys, the standards of reference images are ASTM E155 [1] and E505 [2] for radiography with film and, ASTM E2422 [3] and 2973 [4] for radiography with digital detectors.
A qualified operator judges the choice of the criticality level of an indication manually after a comparison of the radiographic image of the defect with that of the reference image.
The disadvantages of radiography and comparison to standard images are major for manufacturers:

- The application of 2D criteria leads to eliminate parts that are safe for their service.
- The automation of RX detection is available, but systems of automatic sanctions sometimes lack reliability. In practice, manufacturers do not take the risk of using them.
- The progression of the reference images is not always "logical". Some levels are very close whereas sometimes a very important difference exists between two successive levels. This encourages the customers to ask for new levels of quality that manufacturers have trouble following.
- The comparison is a "little subjective", and differences between the operators' decisions, or between client and another are too often observed. This is the source of many disputes.

In conclusion, tomography is capable of providing much relevant information to judge the impact of an indication on the function of a casting. However, the customers and their foundry manufacturers do not have a quality reference on which they could rely to create their own specifications depending on the functional need of the part (mechanical resistance, appearance, tightness, dimensional details…). In addition, tomography would avoid all the drawbacks associated with the use of standard images.

Keywords: X-ray tomography, automatic analysis, defects detection, defects classification, extraction of characteristics.

1. Scientific and Industrial Objectives

Based on the aforementioned facts, our project therefore has for main industrial objectives:

- To help manufacturers (founders/customers) to transpose from 2D radiographic criteria to 3D tomographic ones.
- To define quality reference that will be relevant and applicable on the production line for the characterization of foundry parts by tomography.

- To enhance the non-destructive testing of the casting parts. That is, no longer rejecting parts whose defects are not critical and conversely not accepting parts with undetected critical defects.

To meet these industrial requirements, it is necessary to achieve the following scientific objectives:
- To measure and predict the differences between 2D inspection results and those provided by tomography depending on the geometry of the part.
- To build an algorithm according to the functional need to define a level of acceptability of the casting based on the characteristics that the tomography data provide. These data describe the dimension of the indication, its density (volume fraction), its form factor (acuity of the defect), and its position with respect to the finishing surface or to the zone most strained in practice.
- To foresee the transfer of these developed algorithms to the industries and virtually compare their contributions.

The development of tomography should allow the improvement of casting quality, a reduction of the risks and hazards, and creating relevant automatic sanctioning systems. This is particularly innovative because there are no quality criteria today for the acceptance of parts in tomography testing.

2. Scientific and Technological Obstacles

The main technological difficulties are:
- The high resolution necessary for a good detection of discontinuities. In tomography, it is necessary to compromise between the speed (and rate of control), the cost of a tomography machine (incorporating the amortization of the material), the resolution, the X-ray energy required and the volumes inspected. It is illusory to want a very fine resolution to inspect an important volume because the time and therefore the cost of control become very prohibitive. In practice, the resolution must be chosen according to the size of the indications to be detected. It is accepted that a minimum of three voxels per dimension (i.e. 9 voxels on a volume) is necessary to detect a discontinuity.
- The difficulty of crossing dense and thick parts. This is the main limitation of tomography. Dense materials such as steel, super alloys or cuprous are only controllable by tomography if their thickness is compatible with the power of installation. In practice, thicknesses greater than 10 mm will pose difficulties for conventional installations. In this project, the field of application is limited to light alloys even if it can be extended to dense parts of small thicknesses.
- Tomographic sections contain numerous artefacts (beam hardening, presence of concentric rings, blurs, noises,) that must be corrected before processing the images. For operators not familiar with tomography, it is sometimes difficult to realize that the images are of poor quality. To have reliable results, it is necessary to correct all these defects produced during the acquisition, to check the image quality by means of specific indicators, and to follow a validated procedure.
- Image processing is usually done upon binary images. To obtain these binary images, Thresholding is the delicate operation that can lead to deviant results if a validation procedure is not correctly applied.

The scientific obstacles to be overcome are:

- First of all, the development of relevant algorithms and integrating them in image processing software (Matlab at the first, then an industrial integration within an existing software in the future).
- Validation between the established criteria and the effective lifetimes of the parts.
- Consistency of objectives and results between the development control of the casting and its production. In production line, the time is a major parameter especially when the testing is done in real-time and on a high rate.

3. Proposed Segmentation Methodology

Before establishing the intended criteria, we have to develop some image processing algorithms for better visualizing the defects in the volume and characterize them. The aim of this step is to enhance the display and binarize the volume, i.e., isolate the defects from the background. There are two ways to achieve this step: we can either process the whole volume directly or take a slower approach and process each slice separately. Afterwards, the detected indications will be characterized in order to measure their characteristics (volume, boundary box, extent, etc…), based on which a pre-trained machine-learning algorithm will judge their types (true or false alarms). A finally, the criticality of these true alarms will be studied.

2.1 Slice-by-Slice Processing

Image segmentation is the division of an image into regions or categories, which correspond to different objects or parts of objects. Every pixel in an image is allocated to one of a number of these categories. A good segmentation is typically one in which:

- pixels in the same category have similar greyscale of multivariate values and form a connected region,
- Neighbouring pixels, which are in different categories, have dissimilar values.

There are three general approaches to segmentation [5]:

- In Thresholding, pixels are allocated to categories according to the range of intensities. In the case of global Thresholding, only one threshold is selected for the entire image, while in the adaptive case, local thresholds are selected for each groups of pixels.
- In edge-based segmentation, an edge filter is applied to the image, pixels are classified as edge or non-edge depending on the filter output, and pixels, which are not separated by an edge, are allocated to the same category.
- Finally, region-based segmentation algorithms operate iteratively by grouping together pixels, which are neighbours and have similar values and splitting groups of pixels, which are dissimilar in value.

Our approach (**Fig. 1**) is based on the second approach, which uses first order derivative in order to find the edges of the casting parts, as well as the contours of the defects.

- The first step of our approach is to denoise the image. This is done by using wavelet transformation [6], and it helps reduce the noise in the image without affecting the defects.
- Then, an image sharpening takes place. The aim of this step is to accentuate the defects as well as the geometrical structure of the casting part, and hence improve the efficiency of the edge operator during the segmentation.
- The third step consists of segmenting the image using the Thresholding approach [7]. This provides a binary image.
- After that, an emphasizing operation takes place by comparing the segmented image and the sharpened one. The resulting image highlights the casting and conceals the background.

- The fifth step consists of segmenting the resulting image from the previous step by using an edge operator [8].
- Then, the detected contours are dilated in order the overlay them on the original image.

Fig. 2 shows the result of the segmentation algorithm and its defects detection performance.

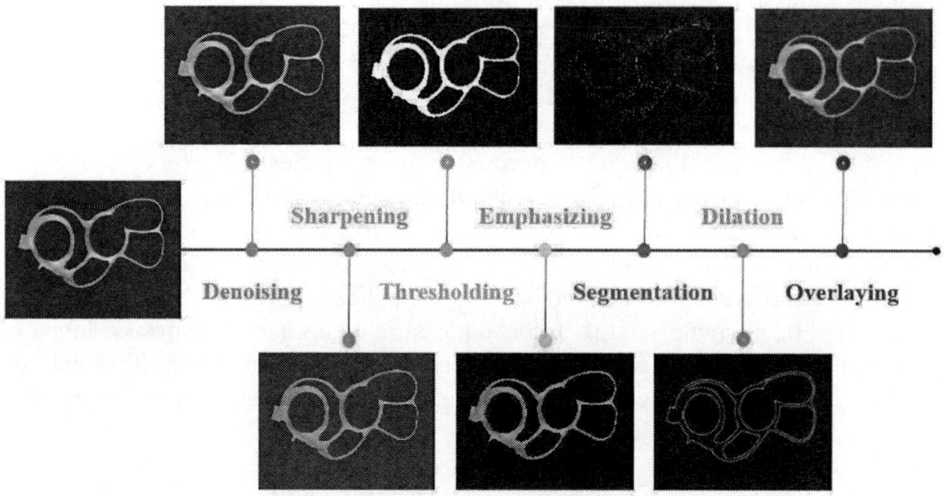

Fig. 1. Proposed 2D Segmentation Approach.

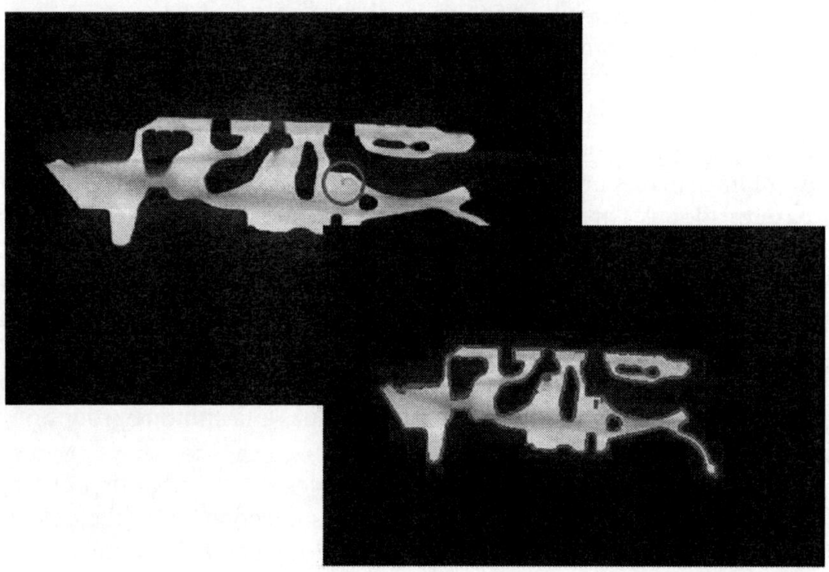

Fig. 2. Tomographic slice Segmentation using the proposed algorithm.

4. Conclusion

An edge-based segmentation technique has been established in order to highlight the indications in a tomographic slice of an aluminium casting. Such indications will be tested by a machine-learning algorithm in order to decide their types (false or true alarms). Finally, the true alarms (defects) will be characterized in 2D and in 3D as well in order to establish the acceptance criteria for tomography inspection of aluminium alloy castings.

5. Acknowledgement

We thank CTIF Company that funds the thesis, and LVA laboratory of INSA Lyon where the research is conducted. We thank the ANRT (Association Nationale de la Recherche et de la Technologie) for accepting to launch the thesis. In addition, we thank the company who participate in our project and provide us with the necessary data: Renault, Constellium Ussel, Montupet, Eurocast groupe GMD, SAB Montmerle.

References

[1] ASTM E155, Standard Reference Radiographs for Inspection of Aluminium and Magnesium Castings.

[2] ASTM E505, Standard Reference Radiographs for Inspection of Aluminum and Magnesium Die Castings.

[3] ASTM E2422, Standard Digital Reference Images for Inspection of Aluminium Castings.

[4] ASTM E2973, Standard Digital Reference Images for Inspection of Aluminium and Magnesium Die Castings.

[5] Cristina Stolojescu-Crisan, Holban StefanA Comparison of X-Ray Image Segmentation Techniques, 2013.

[6] Survey of Image Denoising Techniques Mukesh C. MotwaniMukesh C. GadiyaRakhi C. Motwani.

[7] C.A. Glasbey. An analysis of histogram-based thresholding algorithms. GMIP, 55(6), 199.

[8] Edge Detection Techniques for Image Segmentation: A Survey of Soft Computing Approaches N. Senthilkumaran and R. Rajesh.

Using Experiment and Simulation to Understand the Effects of Source Filtering on Scatter Reduction for Industrial X-ray Computed Tomography

Don ROTH [1], Alexander SUPPES [2]
[1] BHGE Inspection Technologies, Sharonville, OH USA
[2] BHGE Inspection Technologies, Wunstorf, Germany
Contact e-mail: don.roth@bhge.com
Contact e-mail: alexander.suppes@bhge.com

Abstract. X-ray tubes create polychromatic x-ray spectra from tungsten and other target materials. The lower and moderate energies, when left unfiltered, result in several undesirable artifacts in X-ray CT images (namely scatter and beam hardening artifacts). In this study, we performed experiments and simulation with different tube (source) filters to visualize the effects of filtration on reduction of scatter artifacts. A commercial CT system with the *Scatter Correction* method was also employed for comparison with the filtration scenarios. The simulations are used in a sense to validate the experiments (normally we do the opposite) and explore the broad range of simulation possible for x-ray CT. Finally, further simulations are performed to show the effects of filtration on the x-ray spectra themselves in order to understand the effects we see in the CT images.

Introduction

X-ray tubes generate a broadband of energy based on the applied kV level. It would be desirable to generate as narrow a band as energy as possible in order to avoid some of the artifacts associated with a broad energy band. The lower and moderate energy portion of the x-ray spectra for most materials is prone to being scattered when x-rays encounter a material [1].* One example of scatter is the haze seen at the concave surface of a turbine blade.

Source and detector filtering are used in industrial x-ray computed tomography (CT) in order to reduce scatter (and beam hardening) artifacts (as well as to gain latitude and avoid detector saturation). Filtering at the source will absorb lower energy prior to it impinging on the test object, reducing scattering (and beam hardening) artifacts within the test object. To the best of the author's knowledge, a robust method for selecting filtering type and thickness for specific scenarios does not exist - filtering tends to be somewhat of an art and is determined primarily empirically for each application. The goals of this study are to 1) help understand the effects of different source filtering materials and thicknesses on scatter artifact reduction and 2) utilize computational simulation software to aid in validating and understanding these effects as well as demonstrate CT simulation software capabilities. The software package utilized in this study was CIVA software developed by CEA (French Nuclear Agency) [2].

1. Effect of Source Filtration on Scatter Reduction: Experiment

The first set of experiments studied the effects of source filtering on scatter reduction in a nickel-based superalloy turbine blade scanned with x-ray CT. The system utilized was a 450 kV minifocus x-ray CT system. The main CT scan parameters were tungsten target, 400 µm minifocus focal spot size, voltage = 400 kV, current = 400 µA, 2x magnification, 100 µm voxel dimension, and no beam hardening artifact software correction during reconstruction. Filters selected were those commonly used in x-ray (aluminium [Al], copper [Cu], tin [Sn], and tungsten [W]), and thicknesses were those available commercially. For all experiments, it was desired to leave setup parameters the same while only varying filter. However, in some cases where minimum gray values (for a 14-bit detector) at the detector < 100 as filtering was increased, current was raised to obtain gray values ≥ 100. Frame averaging was used to equalize noise levels as filtering was increased.

Figure 1 shows the effect of various filters on scatter reduction on a CT slice at the center of the blade airfoil (~ center slice shown as dotted red line in picture of blade in figure 1). The histogram of gray values at the detector is included to show the distribution of gray values for the air and material. Brighter and darker areas of the histogram are considered as material and air, respectively. An edge detection / surface determination method in visualization software was used to provide a visualization of scatter extent. The gray value of the surface determination position in the image is determined from the average of the material and air peak value locations on the x-axis of the histogram. The "No Filter" and "4.0 mm Al filter" show the most scatter as shown by the surface determination lines. As the filter is changed from 2.0 mm Cu to 1.0 mm Sn to 0.6 mm W, scatter reduction is observed as visualized by the surface determination lines. Also note that the spread distance (contrast) between material and air distribution center locations in the histogram changes with different filtering (see dotted white lines on slices in figure 1).

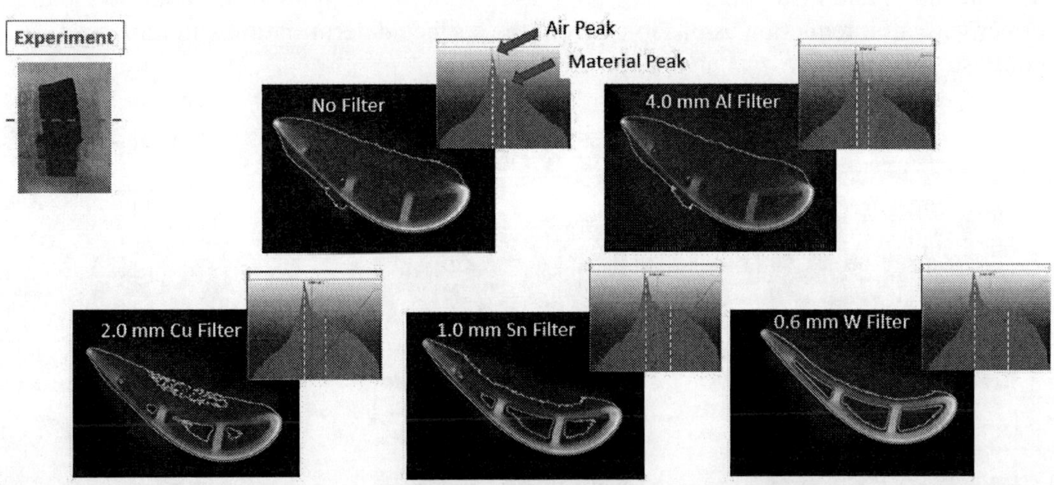

Figure 1. Effect of Various Filters on Scatter Reduction.

Figure 2 compares compound filters (those composed of more than one filter material) against the two most effective single material filters with regards to their effect on scatter reduction. Compound filters allow the combination of different attenuation characteristics of different materials. This can result in more effective filtering by reducing intensity of low energies over a wider band. Again, CT slices and their surface determinations and histograms are shown. The compound filter composed of 0.3 mm W, 0.5 mm Sn, 1.0 mm Cu, and 2.0 mm Al is the most effective of the three compound filters for scatter reduction as shown in terms of surface determination. The single filter 0.6 mm W appears from the surface determination to be about as effective as the most effective compound filter for scatter reduction. However, the larger thicknesses of W (≥ 0.3 mm) reduce signal-to-noise ratio (SNR) significantly. The reduced SNR must be compensated for by an increase in frame averaging, which will increase scan time (and this may be acceptable).

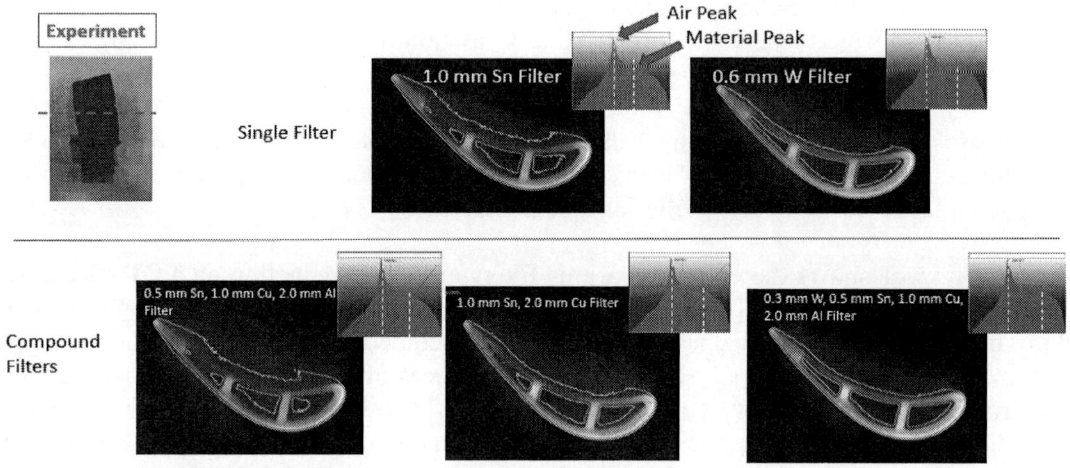

Figure 2. Effect of Compound Filters on Scatter Reduction.

Figure 3 introduces the effect of a patented scatter reduction solution ("Scatter Correction") provided by commercial vendor Baker-Hughes GE Inspection Technologies that incorporates a scatter grid in front of the detector [3]. Scatter correction alone with no source filtering provides a similar scatter reduction to the compound filter composed of 1.0 mm Sn and 2.0 mm Cu. Scatter correction used with the compound filter shows the most effective scatter reduction as observed from the surface determination and increased contrast.

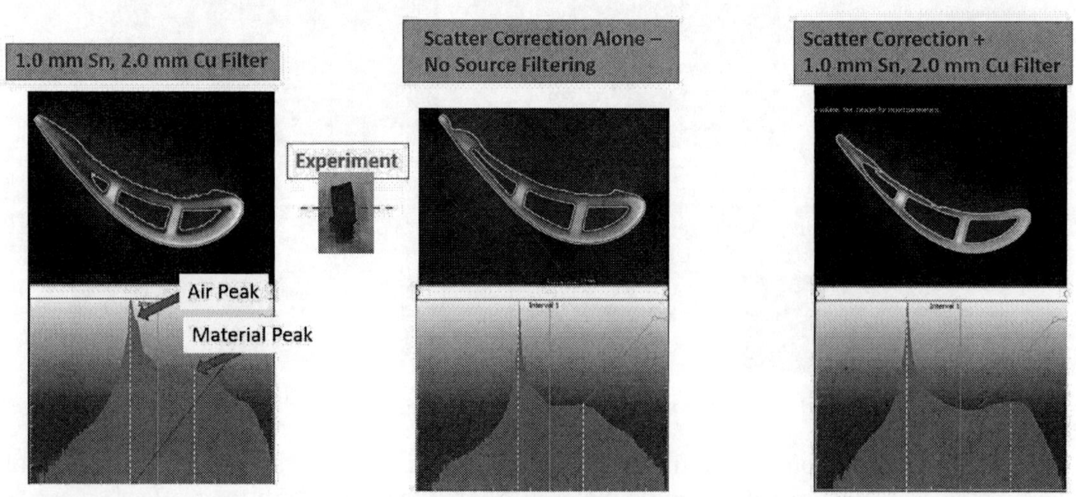

Figure 3. Effect of Patented Scatter Correction Solution on Scatter Reduction.

As previously noted and now observed in figures 1 – 3, the spread distance (contrast) between material and air distribution center locations in the histogram changes with different filtering. The spread distance can serve as a figure of merit for the effectiveness of scatter reduction. Figure 4 shows a bar chart of the normalized histogram spread for the various filter situations of figures 1 – 3. The x-axis values are normalized by dividing the spread distance into the spread distance for the scatter correction with the compound filter (which was the most effective filter for scatter reduction according to the surface determinations). The scatter reduction scenarios are ordered according to effectiveness with larger x-axis values illustrating more scatter reduction.

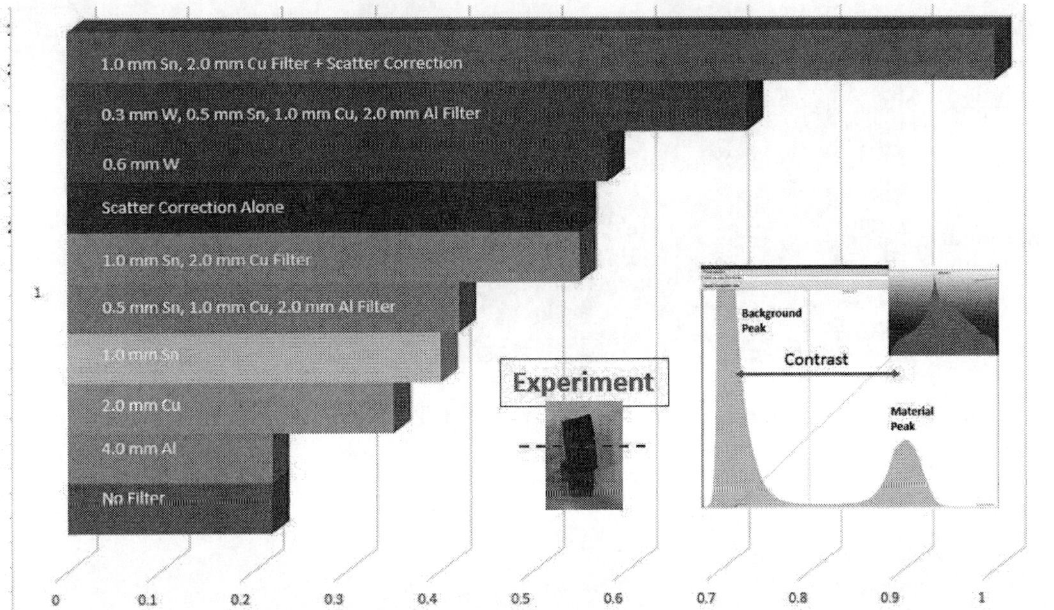

Figure 4. Normalized Histogram Spread Distance (Contrast) Between Material and Air Distribution Centers For the Filtering Experiments Shown in Figures 1 - 3.

It is seen that compound filtering was generally more effective than single filters for scatter reduction. However, the 0.6 mm tungsten filter was the third most effective of the filter scenarios (but will significantly reduce SNR as previously mentioned). 4.0 mm Al filtering does not show a measurable effect as compared to no filtering and thus would not be a wise choice to be used alone for filtering when used at these kV levels. However, Al filtering can be utilized to reduce the intensities of very low energies [1]. As will be discussed in section 3, the effectiveness of filtering for scatter reduction will be a combination of filter thickness and attenuation coefficient.

2. Effect of Source Filtration on Scatter Reduction: Simulation

A standard triangle language (STL) model was created from the CT data of the turbine blade scanned using scatter correction with compound filtering (the highest quality CT data). The STL model was imported into the simulation software and the simulations were set up to mimic the experimental setups as closely as possible. Figure 5 shows simulation results compared with those from the experiments for CT slices for three scenarios: no filter, 2.0 mm Cu filter, and 0.3 mm W, 0.5 mm Sn, 1.0 mm Cu, 2.0 mm Al compound filter.

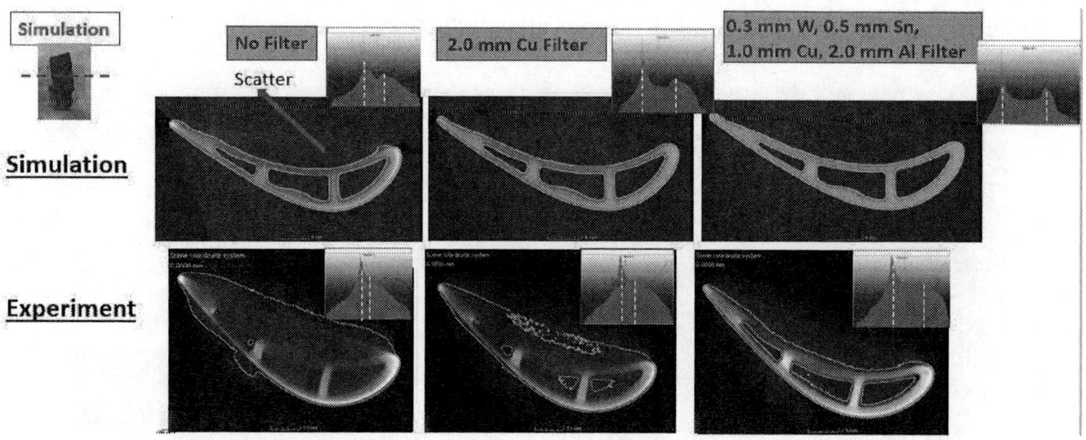

Figure 5. Simulation Compared to Experimental CT Results For Scatter Reduction.

The level of scatter shown for the simulations is significantly less for the simulations compared to the experiments as shown by the surface determination results. However, scatter haze is still easily visualized at the blade concave surface for the simulation with no filter, and is reduced as filters are applied. Additionally, it also seen that the surface determination gets tighter, and the histogram spread gets larger, as filters are applied for the simulations. Similar to figure 4, figure 6 shows a bar chart of the normalized histogram spread for the simulation and experiment filter situations of figures 5. Simulation shows a larger histogram spread as compared to experiment, but the trend of increased spread with increased filtering is clearly shown in both cases.

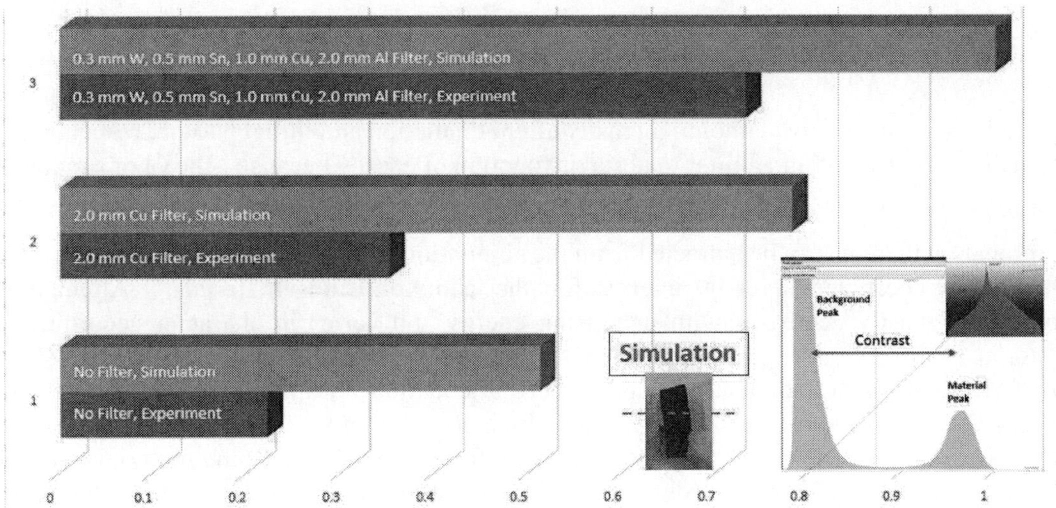

Figure 6. Normalized Histogram Spread Distance Between Material and Air Distribution Centers For the Simulation and Experiment Filter Situations of Figure 5.

3. Understanding Effect of Source Filtration on Scatter Reduction: Examining and Analyzing X-ray Spectras

This section will provide insight into how different source filter materials and thicknesses reduce low x-ray energies and thus reduce scatter artifacts. All of the plots in this section were derived from files generated by the simulation software. The linear attenuation coefficient (μ[keV]) and thickness (x) of filters impact source filtering effectiveness according to the Beer – Lambert law [1]

$$I(keV) = I_o e^{-\left(\frac{\mu(keV)}{\rho}\right)\rho x} \quad (1)$$

where I and Io are received photon intensity after attenuation and initial photon, respectively. Intensity decrease through a filter material is function of energy because μ[keV] of a material is function of energy. The mass attenuation coefficient ($\frac{\mu(keV)}{\rho}$) for any material (or combination thereof) can be extracted from the simulation software. Figure 7 shows the mass attenuation coefficients versus energy for the common filter materials. Attenuation coefficient mostly decreases with increasing energy and varies in character according to material. Absorption edges, highlighted for Al and Sn in the enlarged boxes, are sharp upward edges on the attenuation curve where attenuation abruptly increases. This spike occurs due to the photoelectric effect in which x-ray photons are absorbed. Note that absorption edges are located at different energies for each material. Some materials such as W have multiple absorption edges. Individual contributions to the total attenuation coefficient curve include absorption (usually dominant below 1 MeV), scattering (compton and coherent) (usually dominant below 1 MeV), and pair production (usually dominant above 1 MeV). Most major attenuation using the common x-ray filters shown in figure 7 takes place below 100 keV.

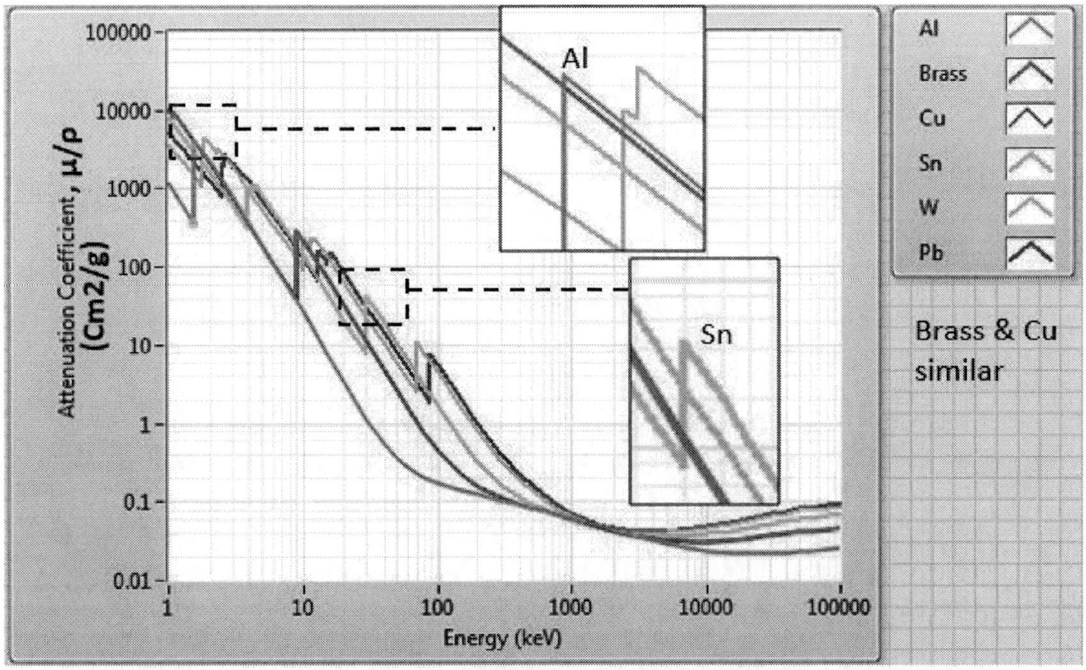

Figure 7. Mass Attenuation Coefficients versus Energy for the Common Filter Materials.

Figure 8 shows the tungsten (target) x-ray spectras for an applied voltage of 450 kV after application of common filter material types. The different filter types reduce different low energy portions of the spectrum due to the different attenuation coefficients and thicknesses. 0.6 mm W and 1.0 mm Sn reduce the intensity and extent of low energy band the most, while 4.0 mm Al reduces the least amount of intensity and extent of low energy band.

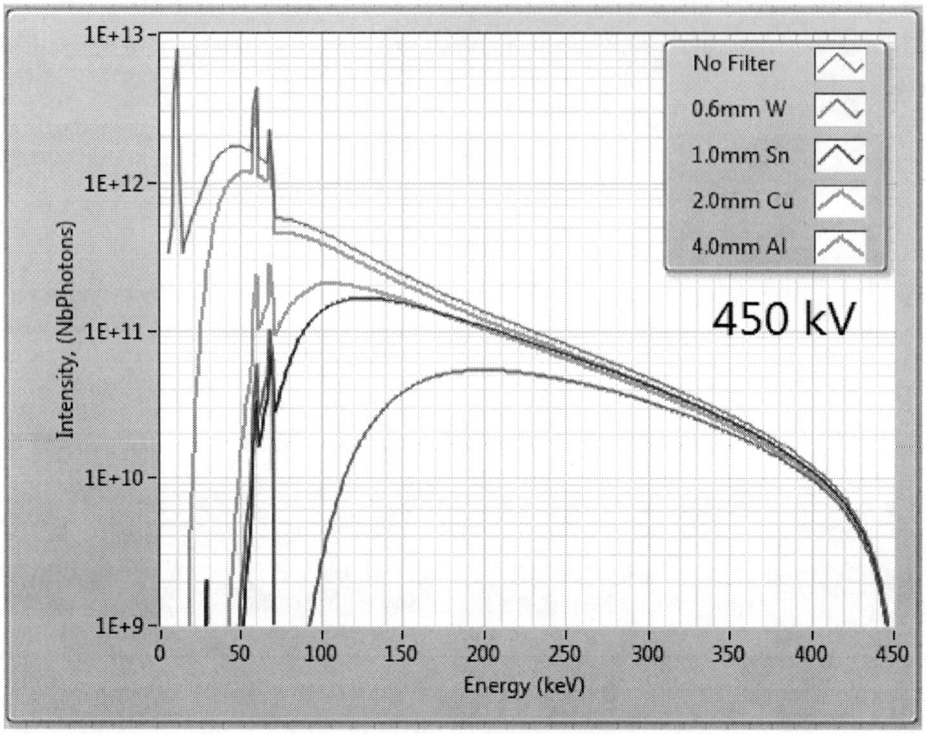

Figure 8. X-ray Intensity versus Energy For Different Common Filter Types and Thicknesses. NbPhotons = number of photons.

Figure 9 shows the tungsten (target) x-ray spectras for an applied voltage of 270 kV using no filtering and with Cu filter thicknesses of 0.1 mm, 0.3 mm, and 1 mm. It can be seen that the filter increasingly reduces the intensity and extent of low energy band as thickness is increased.

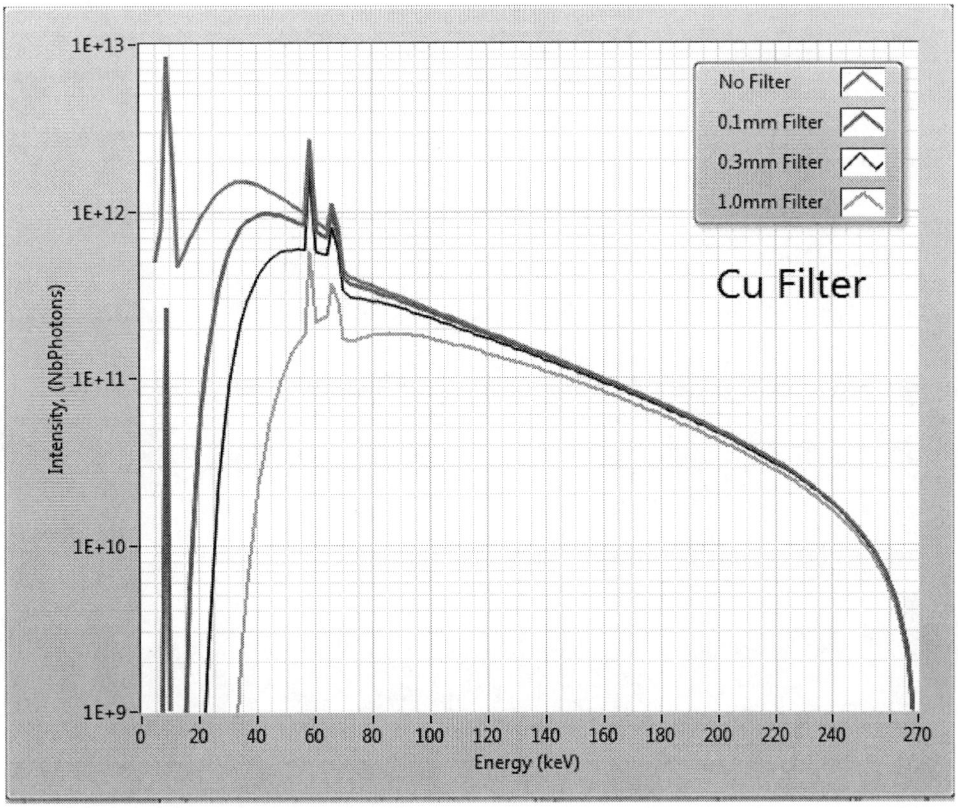

Figure 9. Intensity Versus Energy Spectra for Tungsten Target, 270 kV, for Various Cu Filter Thicknesses.

Figure 10 shows normalized intensity versus filter thickness at 200 keV for an applied voltage of 270 kV for the common filter types. The effect of increasing filter thickness on spectra intensity (at 200 keV) will be dependent upon the filter material because the mass attenuation coefficient is a function of energy. Al filter thickness change is seen to have the least impact on spectra intensity change, while Lead (Pb) and W filter thickness change are seen to have the most impact on spectra intensity change. One specific example is shown on this plot for 0.5 mm Sn versus 0.5 mm W. In this example, a filter of 0.5 mm thickness Sn shows $I/I_0 \sim 0.9$ (10% reduced intensity at 200 keV). In comparison, for 0.5 mm W thickness, $I/I_0 \sim 0.5$ (50% reduced intensity at 200 keV).

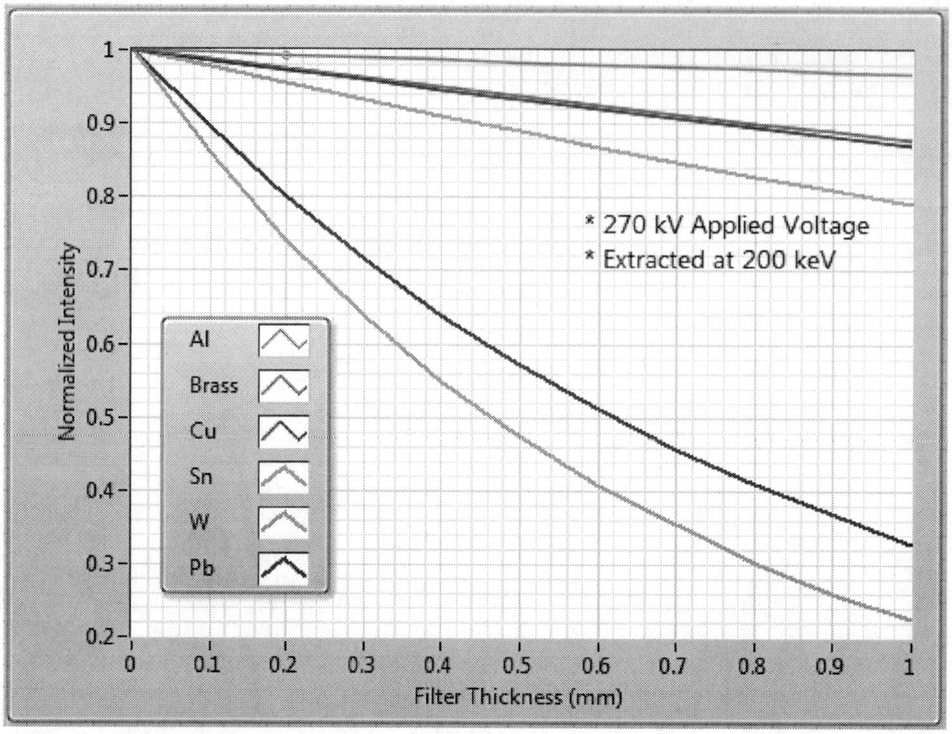

Figure 10. Normalized Intensity Versus Filter Thickness at 200 keV For Common Filter Types. Tungsten Target, 270 kV Applied Voltage.

Figure 11 shows the tungsten (target) x-ray spectras for an applied voltage of 450 kV for two similar compound filter types (identical except one with 0.6 mm W and one without 0.6 mm W). Adding W reduced intensity and filtered slightly more low energy width. However, the use of Al, Sn, and Cu compound filter without the use of W may be more beneficial in some circumstances to achieve a higher SNR and avoid irregular effects such as the large spike shown on the x-ray spectra.

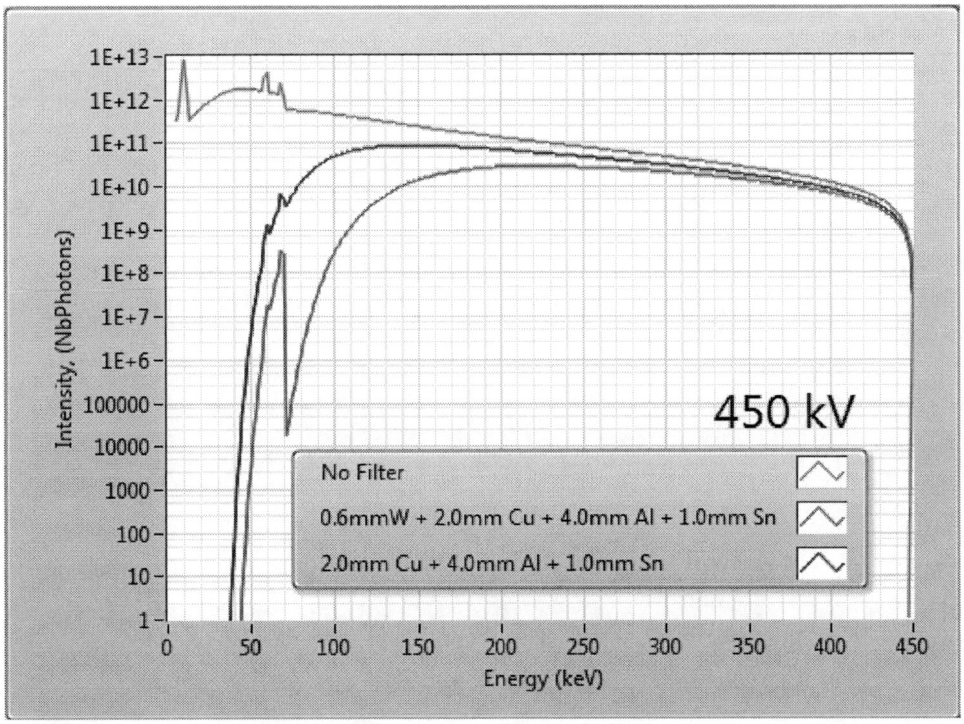

Figure 11. Intensity Versus Energy Spectra for Tungsten Target, 450 kV, for two Compound Filters.

4. Conclusions

The low and moderate energy bands are responsible for scattering artifacts in CT. Experimental and simulation CT tests were conducted to understand the effects of source filtering on scatter artifact reduction. CT slice images from experiment and simulation for different source filter scenarios showed similar trends for scatter artifact reduction. Further modeling demonstrated the reduction of the intensity and width of low energy band from the x-ray spectra. The effectiveness of scatter artifact reduction involves a combination of filter material and filter thickness. Compound filters allow the combination of different attenuation characteristics of different materials. This can result in more effective filtering by reducing intensity of low energies over a wider band. Simulation is a powerful tool that helped us "validate" experimental results and can perhaps help us design ideal filters for different industrial CT inspection scenarios.

5. Footnotes

* The lower energy portion of the x-ray spectra is also prone to being absorbed as x-rays travel into a material leading to beam hardening artifact. Beam hardening artifact causes the outer portion of a CT slice to show higher intensity as compared to the inner portion. The effect of filtration on reduction of beam hardening artifact will be discussed in a future article.

6. References

[1] W. J. Meredith and J. B. Massey; Fundamental Physics of Radiology, 3rd Edition. John Wright & Sons Ltd. 1977.
[2] M. Costin, D. Tisseur, C. Vienne, R. Guillamet, H. Banjak, N. Bhatia and R. Fernandez; *CIVA CT, an advanced simulation platform for NDT*, 6th Conference on Industrial Computed Tomography (iCT 2016), Wels, Austria.
[3] O. Brunke and S. Zahorodny; *CT for the Production Floor: Recent Developments in Hardware and Software for Industrial CT Systems*, ASNT Digital Imaging Conference, July 27-28, 2015. Warwick, RI.

On quality assessment of 3D printed sand mold using non-destructive X-ray μ-CT

Saptarshee MITRA [1], Mohamed EL MANSORI [1], Marius COSTIN [2],
Antonio RODRÍGUEZ DE CASTRO [3]
[1] Arts et Métiers ParisTech, Aix-en-Provence, France
[2] CEA, LIST, Department of Imaging & Simulation for Non-Destructive Testing, Gif-sur-Yvette, France
[3] Arts et Métiers ParisTech, Châlons-en-Champagne, France

Contact e-mail: saptarshee.mitra@ensam.eu

Abstract

One of the major concerns in sand molding using Three-Dimensional Printing (3DP) technology is the proper characterization of the local density and permeability of the mold for quality assessment without destruction. Molding sand compaction and liquid binder jetting bonding plays a crucial role on the mold quality and hence the quality of casting. The air in the mold cavity can escape through the interstices of the porous 3DP mold during metal casting, limiting casting defects caused by gas trapping in the melt. Density, permeability, porosity, grain size distribution, pore size distribution and average throat diameter are important properties for any kind of porous geometry. The determination of these mold properties is often based on destructive tests and carried out by trial-error methods, which is subjective, expensive and time consuming. This paper discusses the use of computer tomography a promising method for non-destructive testing and analyzing additively manufactured sand mold to quantitatively evaluate its local permeability and density values. For this reason X-ray μ-CT was used in generating a virtual structure of the 3DP mold, and then as an input for a detailed characterization of permeability and density. The present work focusses to perform both numerical simulation and experimental analysis from X-ray μ-CT to assess spatial distribution of the density and the permeability of 3DP sand mold under industrial conditions.

Keywords: Additive manufacturing; 3D-printed casting sand mold; Non-destructive testing; Permeability; X-ray μ-CT; Grain Size Distribution; Pore Size Distribution; Numerical simulations;

A Forward Projection Model for Computed Tomography Reconstruction Based on Photon Transport

Ammar ALSAFFAR [1], Sven SIMON [1]
[1] Institute for Parallel and Distributed System, Stuttgart, Germany
Contact e-mail: alsaffar@ipvs.uni-stuttgart.de

Abstract

A forward projection model based on photon transport has been implemented for the optimization of computed tomography reconstruction algorithms. This model is very accurate concerning the physics of the photon transport, including the photo-electric absorption, Compton and Rayleigh scattering. In addition it takes into consideration the important effects of the bound force of the atom so as the Doppler Effect which considers the non-static nature of the electron in the atom that might collide with the photons during their penetration through matter. The goal to apply this model is to improve the accuracy of the reconstruction algorithms for 3D computed tomography, as the results obtained using this model has been integrated with a CT scan data already exists in order to include the important effect of the scattered radiation. This model has been implemented based on several point detectors that scores the flux on the detector which support the image formation of the scanned object, the consideration of each pixel on the detector as a point detector reduces the statistical fluctuation of the result. Several statistical parameters check such as for example the estimated relative error, support the accuracy of this model. The results obtained using this method for different scanned objects formed using simple and complex geometries shows very fine agreement with the well-known X-Ray simulation tool EGSnrc even with fewer number of photons.

Damage characterisation of short glass fibre reinforced polyamide with different fibre content by an interrupted in-situ X-Ray computed tomography test

Julia MAURER[1], Christian HANNESSCHLÄGER [1], Bernhard PLANK [1], Johann KASTNER [1]
[1] UNIVERSITY OF APPLIED SCIENCES UPPER AUSTRIA, Wels, Austria

Contact e-mail: julia.maurer@fh-wels.at

Abstract. Short glass fibre reinforced polymers (SGFR) are increasingly used in automotive industry for the replacement of metals since they provide a significant reduction of weight. Parts with anisotropic fibre orientation were produced during the injection moulding process. Local microstructure parameters such as fibre length and orientation are decisive for the damage behaviour of the injection moulded parts.

Damage type and propagation investigation are necessary for better understanding of material behaviour under stress. X-ray computed tomography (XCT) is a suitable non-destructive method to characterise internal defects in SGFR specimens. With an interrupted in-situ XCT tensile test three dimensional damage propagation in SGFR can be observed.

In this study an interrupted in-situ computed tomography technique was used to quantify the main damage mechanism in polyamide specimens with a glass fibre content (PAGF) of 15 and 30 wt%. Double notched adapted miniature tensile specimens were cut from a plate in two orientations, 0° and 90° relatively to the melt flow. Therefore the tensile force was applied parallel and perpendicular to the expected main fibre orientation (0° respectively 90°). After every force step, a XCT scan with a voxel size of (2 μm)^3 was performed to detect the internal defects. Four different defect types were classified: matrix fractures, fibre pull-outs, fibre/matrix debondings and fibre fractures. Fibre pull-out was the dominating defect type. Defects were mainly induced in the shear region of the 0° specimens. The core region showed nearly no defects at the last load step before breakage.

Summing up the results show that visualisation and quantification of the defect volume and mechanism at certain load steps was possible with this method. Clear differences between damage induction of the 0° and 90° specimens were observed.

Introduction

Fibre reinforced polymers are of interest in many different industrial sectors. Especially short glass fibre reinforced polymers (SGFR) are increasingly used in automotive industry for the replacement of metals. Due to high material requirements, in terms of mechanical and physical properties, it is necessary to have a closer look on the microstructure of such SGFR polymers. The structural properties are mainly determined using three-dimensional fibre orientation and fibre length distribution. To understand the material behaviour under stress, a microstructural investigation of the damage type and damage propagation is necessary. X-ray computed tomography (XCT) is widely used for scientific and industrial applications. A good overview on XCT is given in the book of Carmignato et. al [1]. It is a suitable non-destructive method for qualitative and quantitative characterization of filled polymers. For the quantification of the damage mechanisms interrupted in-situ XCT technique can be used. Due to a quasi-static tensile test until fracture the defects occurring at each loading step can be detected and classified in four different types [2]: matrix fracture, fibre pull-out, fibre/matrix debonding and fibre fracture.

This work presents an approach for the quantification and visualisation of different defect types. Also the damage evolution can be investigated, due to stepwise increase of the loading. We report on a workflow for appropriate measurement, segmentation and characterization of such fibre filled materials. Furthermore a direct comparison of defect type and defect volume for different material systems is presented and discussed.

1. Experimental

1.1 Material and Specimen

Short fibre reinforced polymer specimen were investigated, differing in glass fibre content and fibre orientation. The matrix material was polyamide 6.6 reinforced with 15 and 30 wt% glass fibres (PAGF). Double notched adapted miniature tensile specimens were cut from an injection moulded plate in two orientations relatively to the melt flow respectively expected main fibre orientation (FO) (figure 1). Thus the tensile force was applied parallel to the main fibre orientation for a 0° test specimen and perpendicular for a specimen with expected main fibre orientation of 90°. The injection moulded plates are approximately 2 mm thick and were conditioned at 50 % moisture content.

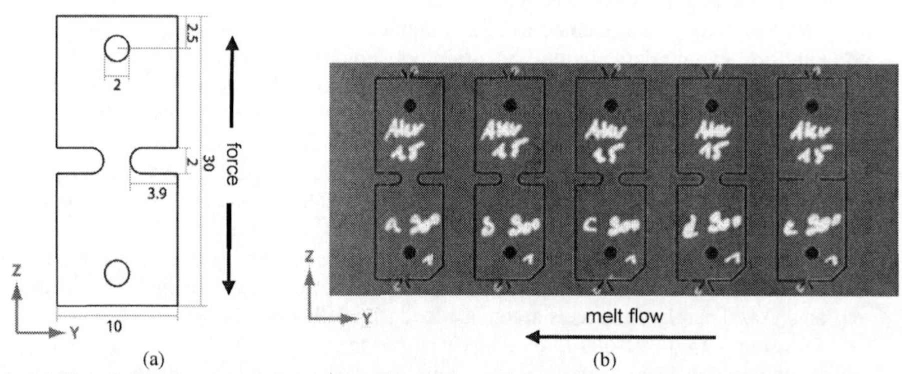

Fig. 1. (a) Sketch of adapted miniature tensile specimen (dimensions given in millimetre) (b) Part of an injection moulded plate with five miniature tensile specimen oriented 90° to the melt flow direction.

1.2 High Resolution XCT

The specimens were scanned at the Nanotom 180 NF XCT device (GE phoenix|x-ray, Germany) with a voxel size of (2 μm)³ at certain load steps. This system uses a sub-μ-focus X-ray tube and a 2304 x 2304 pixels flat panel detector (Hamamatsu). The scanning time was approximately 190 minutes per load step. Molybdenum was used as target material. The reconstruction of the data was done with the software phoenix datos|x 2 reconstruction (GE phoenix|x-ray, Germany).

1.3 In-situ tensile testing stage

The CT500 in-situ tensile stage (Deben, UK) was used to apply the tensile load (figure 2). The specimens were mounted mechanically by clamps. Main specifications of the tensile/compression stage can be found in table 1. All experiments were performed at room temperature.

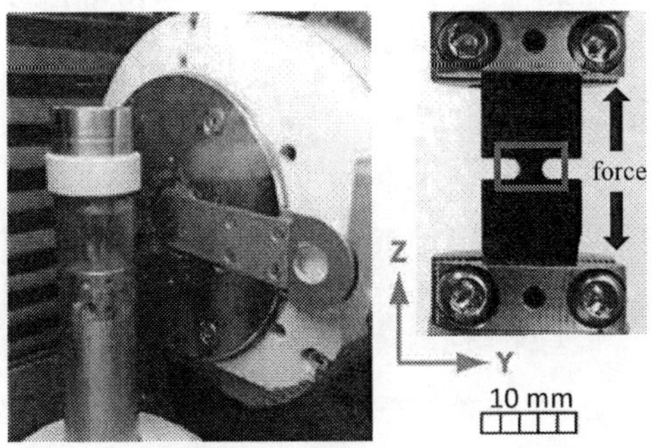

(a) (b)

Fig. 2. (a) Tensile stage in front of the X-ray source (b) mounted specimen with red rectangle indicating the measurement area

Table 1. Specification of tensile stage CT500 (Deben)

Specifications	
Max. load	500 N
Max. motor speed	1 mm/min
Modes	tensile/compression
Max. extension	10 mm
Accuracy	1 % of full scale range
Weight	approx. 1.0 kg

As preliminary tests show and as explained by Sause et al. [3] the loading of the specimen need to be performed in displacement-controlled mode, which implicates constant strain. A displacement of 0.2 mm/min was applied by the servo motor of the stage until the experimentally defined peak force was reached. Stable conditions, as well as no movements, during the scanning time are crucial for a reconstructed volume without artefacts and required for the segmentation of the defects. Therefore, an experimental determined delay time of approximately 30 minutes and for higher load levels of up to 90 minutes between the peak force and the start time of the scan was applied. During this, so called relaxation time, the force decreases significant. An example for such a loading curve is presented in figure 3.

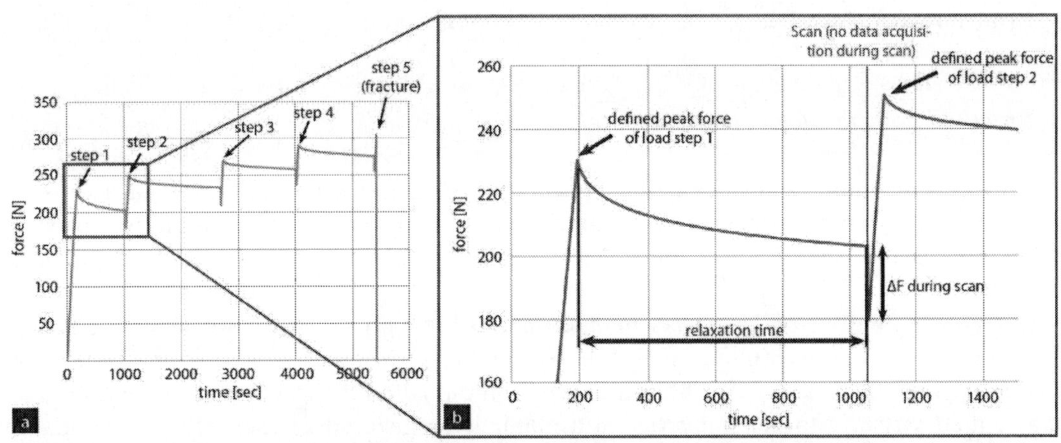

Fig. 3. Exemplary loading steps and relaxation of the test specimen (a) and detailed diagram for load step 1 (b). During the scan the data acquisition was stopped.

Two reference specimen of the injection moulded plate were tested without a CT scan to determine the quasi static load steps for the in-situ experiment. The maximum force was identified by a continuous increase of the force until fracture occurs. After evaluation of the peak force the second reference specimen was tested with interrupted force steps to evaluate the final scan points. These test scans revealed that first defects can be detected at approximately 70 % of F_{max}. Furthermore, the scan after the loading step of approximately 95 % of F_{max} shows that the specimen movement during the scan disables the possibility of defect quantification. Hence, the measurements started at approximately 70 % of F_{max}, defined by the preliminary tests and were increased in steps of 30 N for the PAGF specimen with glass fibre content of 30 wt% and main fibre orientation of 0° respectively 20 N for all other specimens. This resulted in the percentage load levels for the different material systems shown in table 2 to 5.

Table 2. Planned load steps for PAGF 15 wt% | FO 0°

Load step	Force applied	relative percentage to reference F_{max}
0	0 N	0 %
1	230 N	70 %
2	250 N	76 %
3	270 N	83 %
4	290 N	89 %
5 (fracture)	327 N (F_{max})	100 %

Table 3. Planned load steps for PAGF 15 wt% | FO 90°

Load step	Force applied	relative percentage to reference F_{max}
0	0 N	0 %
1	205 N	67 %
2	225 N	74 %
3	245 N	80 %
4	265 N	87 %
5 (fracture)	305 N (F_{max})	100 %

Table 4. Planned load steps for PAGF 30 wt% | FO 0°

Load step	Force applied	relative percentage to reference F_{max}
0	0 N	0 %
1	280 N	65 %
2	310 N	72 %
3	340 N	79 %
4	370 N	86 %
5 (fracture)	430 N (F_{max})	100 %

Table 5. Planned load steps for PAGF 30 wt% | FO 90°

Load step	Force applied	relative percentage to reference F_{max}
0	0 N	0 %
1	200 N	67 %
2	220 N	73 %
3	240 N	80 %
4	260 N	87%
5 (fracture)	300 N (F_{max})	100 %

1.4 Methodology for defect detection and classification

There are four major steps for defect detection and classification: XCT data acquisition, pre-processing, post-processing as well as result analysis and interpretation. The used methodology is based on the publication of Rao et al. [4] and is shown in the flow chart in figure 4.

The reconstructed 32 bit data was mapped to 16 bit with the software VG Studio MAX version 3.1 (Volume Graphics GmbH). Hence the grey level peak of air (10,000) and material (45,000) is always at the same grey value position. The volume of each load step was cutted virtually to the same size and manually registered to each other. The adaptive rectangular tool was used to get rid of the surrounding air. For the noise reduction, the in open_iA [5] implemented gradient anisotropic diffusion filter (GADF) was applied. In VG Studio MAX the advanced surface determination tool was used to create a defect mask. Therefore an experimentally determined ISO-65 threshold was used for pore segmentation. The defect classification tool was developed in-house by Amirkhanov et. al [6] and requires a labeled data set of the defects and the position and orientation of each fibre for the evaluation. Open_iA was used for labelling the binarised data and the in-house developed fibre characterization tool by Salaberger et. al [7, 8] was used to get the information of each fibre. The defects were classified as fibre pull-outs, fibre fractures, matrix fractures and fibre/matrix debondings. With increasing force the existing defects grow together. Therefore, defects bigger than 400 voxels respectively 3,200 μm^3 were classified as matrix fractures. This threshold was determined experimentally. A matlab script, provided by Salaberger, was used to create a defect classified 3D volume for the visualization.

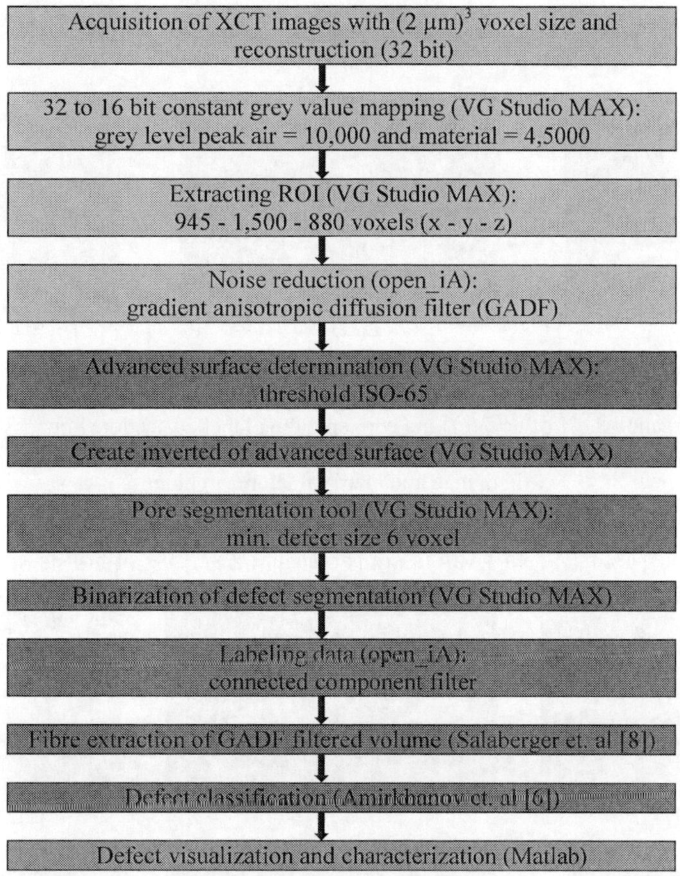

Fig. 4. Methodology flow chart; major steps color-coded: orange – data acquisition, green – pre-processing, red – post-processing, blue – result analysis and interpretation

2. Results

Each defect is assigned to one of the four defect classes and can be color-coded in the CT data. The defects are color-coded as follows: fibre fractures are shown in red, fibre/matrix debondings in green, fibre pull-outs in orange and matrix fractures in blue.

The XCT scans after each loading step enables the tracking of certain fibres respectively defects. An example of such a tracking sequence is shown with sectional images (unlabelled and labelled) in figure 5 and 6 for test specimens with 15 wt% glass fibres and a main fibre orientation of 0° respectively 90°.

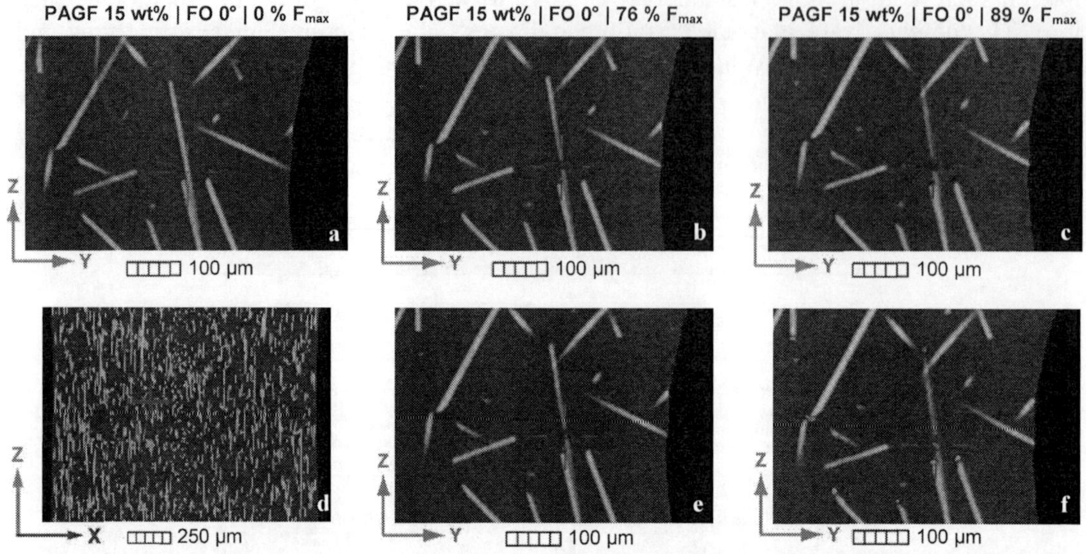

Fig. 5. Sectional images of 15 wt% specimen with an expected main fibre orientation of 0°. Sectional image over thickness (d) (30 µm thick slab), red cross marks the position (shear region) of the sectional images at 0 % (a), 76 % (b) and 89 % of F_{max} (c), and corresponding labelled sectional images for 76 % (e) and 89 % of F_{max} (f). Fibre fractures shown in red, fibre/matrix debondings in green, fibre pull-outs in orange and matrix fractures in blue.

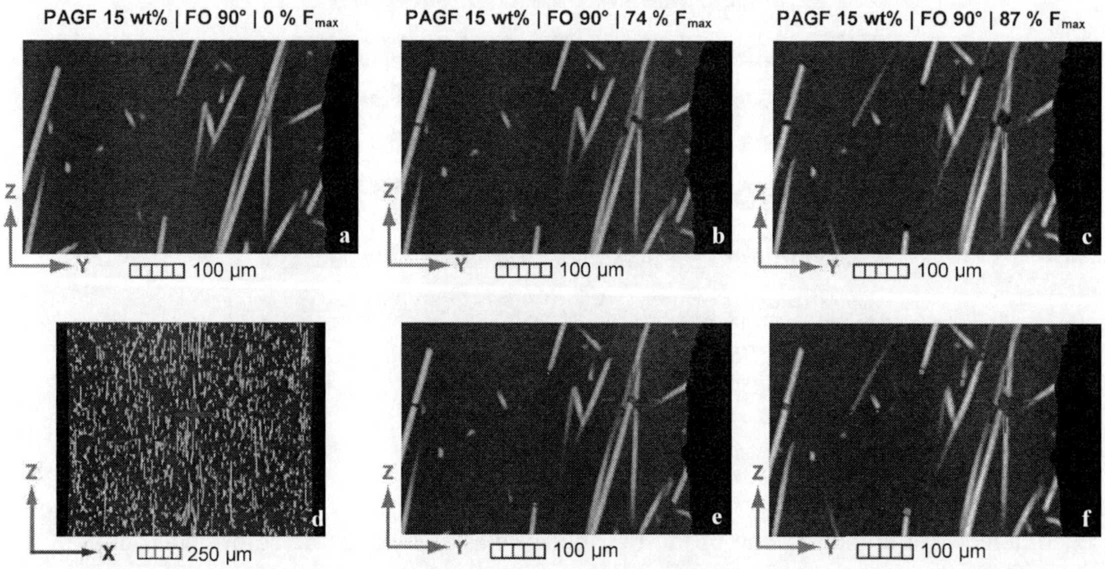

Fig. 6. Sectional images of 15 wt% specimen with an expected main fibre orientation of 90°. Sectional image over thickness (d) (30 µm thick slab), red cross marks the position (core region) of the sectional images at 0 % (a), 74 % (b) and 87 % of F_{max} (c), and corresponding labelled sectional images for F_{max} 74 % (e) and 87 % of F_{max} (f). Fibre fractures shown in red, fibre/matrix debondings in green, fibre pull-outs in orange and matrix fractures in blue.

Figure 7 show 3D rendered top views with labelled defects at the last loading step before breakage for all material systems (the material is shown transparent, the defects are displayed in the corresponding colors).

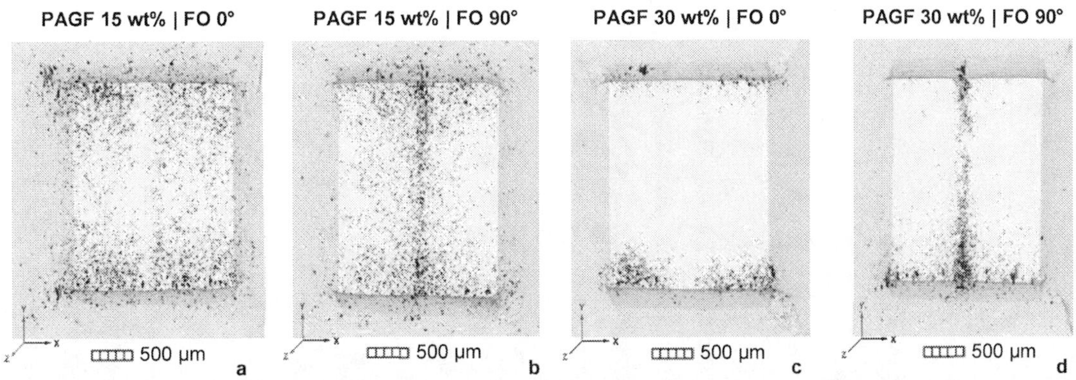

Fig. 7. 3D rendered top view at last loading step before breakage. Specimen with 15 wt% (a, b) and 30 wt% (c, d) glass fibre content with an expected main fibre orientation of 0° (a, c) respectively 90° (b, d). Fibre fractures shown in red, fibre/matrix debondings in green, fibre pull-outs in orange and matrix fractures in blue.

For each scan the number of defects as well as the number of voxels for each defect is provided and can be presented in many different ways. In figure 8 and 9 the number of defects and the corresponding relative defect volume per loading step are shown for the 15 wt% respectively the 30 wt% glass fibre filled polymers.

It should be noted that for each scan the same volume size of approximately 10 mm^3 was analysed. Therefore the in figure 8 and 9 presented numbers allow a direct comparison.

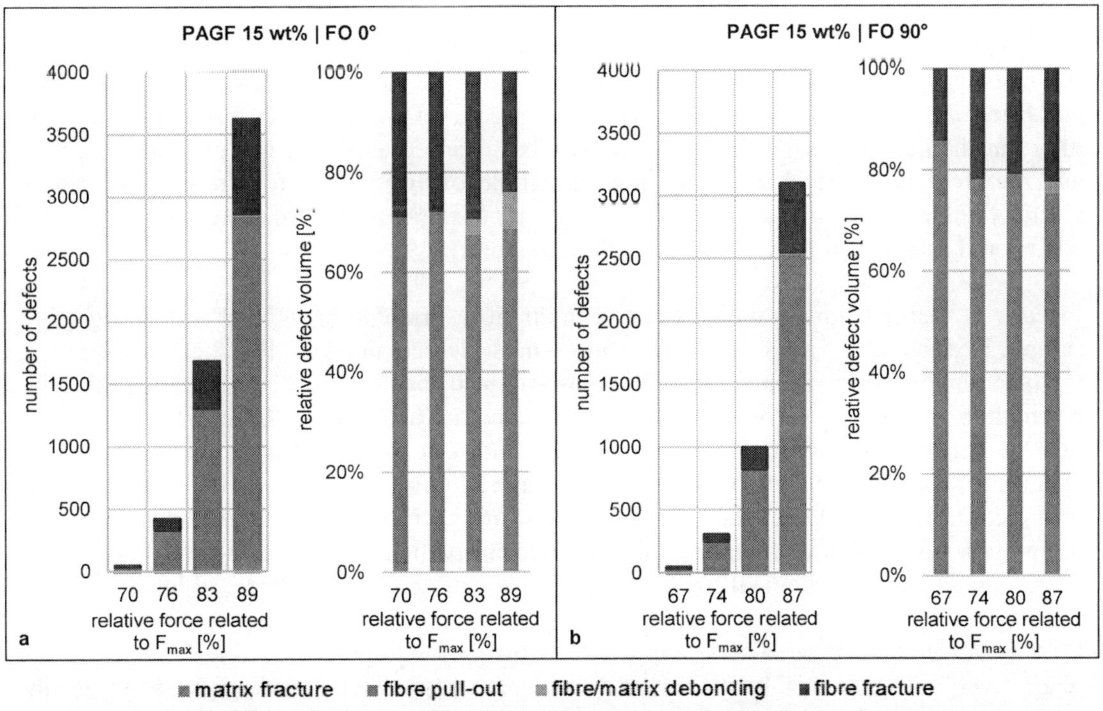

Fig. 8. Number of defects and relative defect volume for each loading step for specimens with a fibre content of 15 wt% and an expected main fibre orientation of 0° (a) and 90° (b).

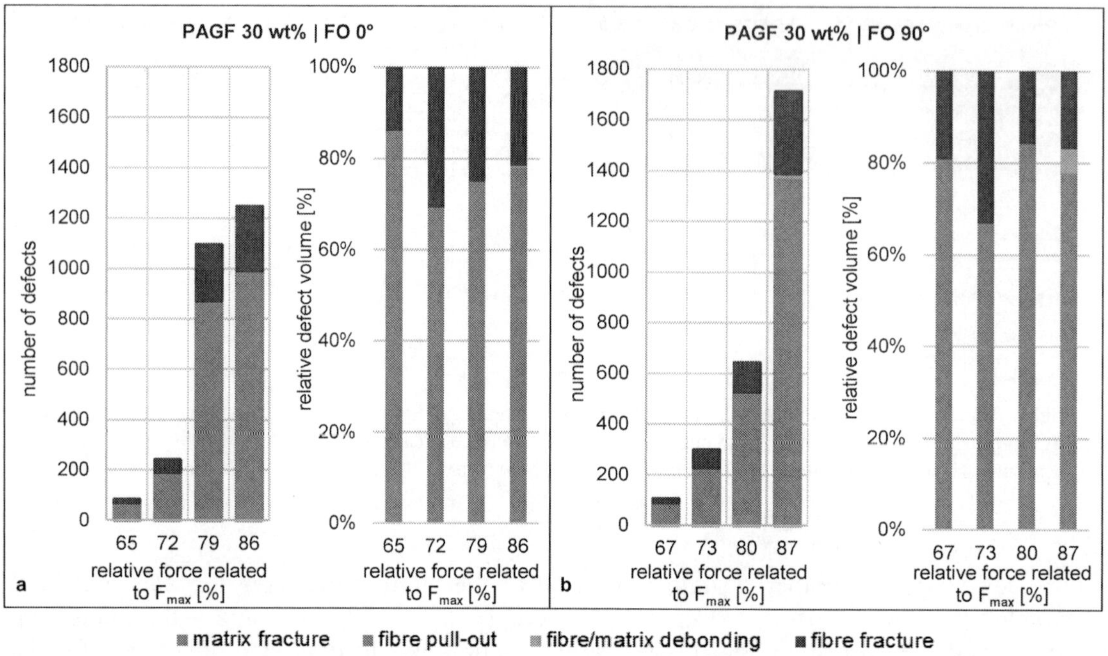

Fig. 9. Number of defects and relative defect volume for each loading step for specimens with a fibre content of 30 wt% and an expected main fibre orientation of 0° (a) and 90° (b).

3. Discussion

The results look very promising and a lot can be deduced, whereas some questions need to be investigated in detail in further work. For example the fibre fracture in figure 6e is classified as matrix fracture in 6f. This may seem to be a misclassification, but on a micromechanical view, the fibre fracture can be regarded as the origin of matrix fractures and therefore the experimentally defined threshold of 400 voxels for the classification of matrix fractures is confirmed. Despite random matrix fractures, also fibre pull-outs can be considered as a reason for damage growth in the matrix [9].

In case of test specimens with expected main fibre orientation 0°, rather loading of the sample in direction of main fibre orientation, most defects occur in the shear region of the test specimen (figure 7a and 7c). Whereas for test specimen with expected main fibre orientation of 90° respectively loading perpendicular to the main fibre orientation, most defects occur in the core region (figure 7b and 7d). Those regions of higher defect occurrence can be explained by the fibre orientation in those regions, which is in both cases in direction of load. Due to the fact that in case of specimens with an expected main fibre orientation of 0° most of the fibres are oriented in loading direction, those specimens can withstand a higher mechanical stress which is in accordance to the observed loading curves.

Usually fewer defects would be expected in lower filled specimens due to longer fibres respectively a higher weighted average fibre length [8], which was approved by the fibre characterization. The higher defect number for 15 wt% glass fibre filled specimens may possibly be explained by the lower fibre volume ratio. Especially for the 90° specimens, filled with 15 wt% (figure 7b) and 30 wt% (figure 7d) glass fibres, the maximum force is at the same range (table 3 respectively 5). It can be concluded that for a 15 wt% fibre filled material the force is distributed on less fibres in comparison to a 30 wt% fibre filled material. Therefore the force which needs to be absorbed by one single fibre is higher and causes fibre defects, mainly fibre fractures and fibre pull-outs. It should be noted that the fibre diameter (fibre paradoxon of Griffith [10]), the local fibre content and the occurrence of fibre fragments are not considered in this explanation, but may also influence the damage behaviour.

Figure 8 and 9 clearly illustrate that fibre pull-out is the dominating defect class, independent of the specimen configuration. In terms of the number of defects, the second most frequent defect class is fibre fracture. Only a few fibre-matrix debondings were detected at the last steps. These behaviours are also in accordance to other investigations [9]. For all material systems a strong increase in the number of defects, especially at the last loading steps before breakage, can be observed. The relative defect volume of the matrix fracture for the 30 wt% fibre filled specimens (figure 9) is higher for loading step one compared to step two. This can be explained by the lower total defect number which causes another defect volume distribution.

4. Conclusion and Outlook

With the presented method quantification and visualisation of the defect volume at certain load steps is possible. Based on sectional images and on 3D images as well as on the quantitative evaluation the defect propagation can be studied in detail. Clear differences between damage induction of specimens with an expected main fibre orientation of 0° and 90° can be observed. In specimens with expected main fibre orientation of 0° defects occur mainly in the shear region, whereas for specimens with expected main fibre orientation of 90° defects occur mainly in the core region. Also differences in the defect number for 15 wt% and 30 wt% glass fibre filled polymers can be seen. The analyses show that lower filled specimens have a higher total number of defects compared to higher filled specimens.

Depending on the material systems, other micromechanical processes may dominate and thus the measurement and characterisation parameters may need to be adapted. Further work should also include a detailed investigation of fibre diameter, local fibre content and ideally at least two test specimens per material systems, to get a more accurate statement on the micromechanical processes.

Acknowledgments

This work was supported by the project "Multimodal and in-situ characterisation of inhomogeneous materials" (MiCi) by the federal government of Upper Austria and the European Regional Development Fund (EFRE) in the framework of the EU-program IWB2020. It also received funding by the K-Project "Photonic Sensing for Smarter Processes" (PSSP), the COMET program of FFG and the federal government of Upper Austria and Styria.

References

[1] S. Carmignato, W. Dewulf and R. Leach, "Industrial X-Ray Computed Tomography", *Springer International Publsihing,* 2018.

[2] K. Friedrich, "Composite Materials Series, 6: Application of Fracture Mechanics to Composite Materials". *Elsevier, Science Publishers B. V.*, 1. ed., 1989.

[3] M. Sause, "In Situ Monitoring of Fibre-Reinforced Composites", *Springer International Publishing*. Switzerland, 2016.

[4] G. Rao, A. Amirkhanov, D. Salaberger, J. Kastner and C. Heinzl, "Damage characterization in SFRP using X-ray computed tomography after application of incremental and interrupted in situ quasi static tensile loading", in *Proceedings of iCT2017 Industrial Computed Tomography,* (Leuven, Belgium), February 2017.

[5] B. Fröhler, J. Weissenböck, M. Schiwarth, J. Kastner and C. Heinzl, "open_iA: A tool for processing and visual analysis of industrial computed tomography datasets", *Journal of Open Source Software*, 4(35), 1185, 2019. https://doi.org/10.21105/joss.01185.

[6] A. Amirkhanov, A. Amirkhanov, D. Salaberger, J. Kastner, E. Gröller and C. Heinzl, "Visual Analysis of Defects in Glass Fibre Reinforced Polymers for 4DCT Interrupted In situ Tests", *COMPUTER GRAPHICS FORUM*, Vol. 35(3), pp. 201-210, April 2016.

[7] D. Salaberger, K. A. Kannappan, J. Kastner, J. Reussner and T. Auinger. "Evaluation of Computed Tomography Data from Fibre Reinforced Polymers to Determine Fibre Length Distribution", *International Polymer Processing*: Vol. 26, No. 3, pp. 283-291, 2011.

[8] D. Salaberger, "Micro-structure of discontinuous fibre polymer matrix composites determined by X-ray Computed Tomography", PhD thesis Technical University Vienna, 2019. http://katalog.ub.tuwien.ac.at/AC15350300.

[9] H. Rolland, N. Saintier and G. Robert, "Damage mechanisms in short glass fibre reinforced thermoplastics during in situ microtomography tensile tests", *Journal of Composites Part B*. Vol 90, pp. 365-377, 2016.

[10] A. A. Griffith, "The Phenomena of Rupture and Flow in Solids", *Philosophical Transactions of the Royal Society of London. Series A, Containing Papers of a Mathematical or Physical Character*, Vol. 221, 163-198, 1921.

Robot accuracy and its impact on robot-CT reconstruction quality

Risheng KANG [1], Gabriel PROBST [1], Peter SLAETS [1], Wim DEWULF [1]
[1] KU Leuven, Department of Mechanical Engineering,
Leuven, Belgium

Contact e-mail: risheng.kang@kuleuven.be

Abstract

Nowadays, Robot CT systems are gaining more and more attention: they allow multiple scan geometries, highly customized scan trajectories, and the potential to make mobile or in-line platforms.

The quality of CT reconstruction is, however, highly dependent on the accuracy of the position and orientation information of the source and the detector, which pose a big challenge for current commercial serial industrial robots. In order to assess the obtainable reconstruction quality of Robot CT systems, this paper elaborates on a set of realistic simulations.

An industrial robot, ABB IRB120, together with an optical CMM, Nikon K600, are used to evaluate the robot path accuracy and repeatability. The K600 system is capable of measuring the 6D information with up to 1000Hz, which is quite suitable for this evaluation task. The volumetric accuracy of the K600 is 0.18 mm (2 sigma) and its repeatability is 0.015 mm (2 sigma), as evaluated using a Mitutoyo CMM with MPE = 4.2 μm + 5 μm/m.

Two different circular paths are defined to simulate the respective path of X-Ray source and detector of a twin-robot CT system. Repeatability, for both positioning and orientation, are expressed as 3 sigma. The measured mean path positioning accuracy is below 0.49 mm, and the mean path positioning repeatability is below 0.17 mm. The mean orientation accuracy is below 0.83 degrees, and the mean path orientation repeatability is below 0.02 degrees. The orientation error is described by axis angle instead of three Euler angles. The two circular paths under test with different radii showed similar error distributions. Configurations with both robot bases situated at the same side of the workpiece (referred to as in-phase error configuration) and both robots situated at opposite side of the workpiece (referred to as out-of-phase error configuration) have been simulated.

Based on the measured robot accuracy information, five groups of CT simulation are conducted with 6D positioning and orientation errors of source and detector. A cluster of spheres was simulated as a workpiece. Results showed that the realistic simulations yield sphere-to-sphere distance errors as well as sphere form errors of up to 200 μm. The radius errors are better and below 50 μm, which could be a result of averaging effect.

Both the robot orientation error and the robot positioning error contribute to the reconstruction errors to a similar extent, which implies both errors should be compensated to improve the reconstruction quality.

In this experiment, the reconstructions subjected to in-phase error and out-of-phase error do not show a significant difference to each other in terms of the radius, distance and form error.

Resolution Limits of the CRYCAM X-ray Microradiography Camera

Jan TOUS [1], Jiri PARIZEK [1], Karel BLAZEK [1], Miroslav SULC [2], Radek MELICH [2], Tomas VANEK [3]

[1] CRYTUR, spol. s r.o., Palackeho 175, 51101 Turnov, Czech Republic

[2] Academy of Sciences of the Czech Republic - Institute of Plasma Physics, Za Slovankou 1782/3, CZ182 00 Prague, Czech Republic

[3] Academy of Sciences of the Czech Republic - Institute of Physics, Cukrovarnicka 10, CZ16253 Prague, Czech Republic

Contact e-mail: tous@crytur.cz

Abstract. Recently, high resolution micrography became of great interest. One of the approaches are very thin scintillator layers of about 5-20 micrometers used to achieve spatial resolution below one micrometer. Such thin screens are mainly used in micro-CT and nano-CT systems with either micro-focus X-ray tubes or with synchrotron sources.

This work deals with a high resolution CMOS camera together with different optical systems and different single crystal scintillators in application for low energy X-ray micro-radiography. The light distribution on the screen is transferred by an optical system to a high-resolution chip. A theoretical modeling and practical comparison of several set-ups has been done. Thin screens used were prepared from different scintillators. The resolution is presented on test patterns.

Introduction

X-ray micro-radiography is a non-destructive method that has been receiving much interest in recent years. Recently, three-dimensional CT imaging has been experiencing growing interest.

Very thin single crystal scintillator imaging plates are radiation detectors that are used in low energy X-ray micro-radiography. In high resolution X-ray projection imaging, very thin single crystal scintillator plates of about 5-20 micrometres in thickness are used to achieve spatial resolution of about one micrometre. Such thin plates are mainly used in micro-CT and nano-CT systems with either micro-focus X-ray tubes or with synchrotron sources. The light distribution on the plate is transferred by an optical system to a high-resolution CCD or CMOS chip. For a specific optical collection system, there exists an optimum thickness for the scintillator plate. A thicker plate absorbs high-energy photons more efficiently but the image on the sensor becomes blurred [1]. On the other hand, an excessively thin scintillator does not provide enough absorption so that the integration time for an image is rather high.

1. Instrumentation

1.1 Scintillators

Single crystal thin plates are made from single crystal boles. The most used crystals are garnets doped with cerium. High quality industrial YAG:Ce, LuAG:Ce and recently GdGAG:Ce single crystals are prepared by the Czochralski method [2] at Crytur. These inorganic crystal scintillators are characterized by good mechanical and chemical stability, non-hygroscopicity, high scintillation efficiency and fast decay below 100 ns [3], [4]. The imaging scintillator screen made of such single crystal is optically transparent. The emission wavelength of Cerium doped garnets is around 550 nm. Imaging screens prepared from these crystals can be used in equipment for the detection of different kinds of radiation and particles such as UV, VUV, electrons or ions or their beams, X- or gamma-rays. Thin YAG:Ce or LuAG:Ce scintillator imaging screen were used in low-energy X-ray radiation radiography [5].

1.2 X-ray CMOS Camera

A high-resolution X-ray camera consists of a highly sensitive digital CCD or CMOS detector and a scintillator. The CRYCAM camera sensor size is of 1940 x 1460 pixels, 4,54 x 4,54 µm each. A Peltier thermoregulation system is used for cooling and temperature stabilization. The optical system uses objectives with magnification of 10 times or 20x. The pixel size together with the optics magnification determines one of the limits for system spatial resolution. Also, the maximum system spatial resolution is limited by the screen thickness and numerical aperture of the objective.

2. Experiments

2.1 Set-up

In the experimental setup, the highly sensitive CRYCAM camera and YAG:Ce single crystal thin plates were used for resolution evaluation. The thickness of the scintillator was 10 micrometers or 20 micrometers, to reach the highest possible resolution. The objective was focused into the plane inside the scintillator where the absorption image has the best contrast. The camera was in a sideway position to avoid direct impact of the X-rays on camera electronics.

The X-ray CMOS camera resolution was tested using the JIMA RT RC02 resolution chart. The position of the object was very close to the scintillator plate, so almost no projection magnification was used. A Cu anode microfocus X-ray source of X-RayWorx was used with intensity maximum at about 8 keV.

2.2 Results

The best resolution was achieved using the combination of 10 micrometers YAG and 20-times magnification optics, see Fig. 1. The resolution is then limited by the spot of the x-ray source. The contrast of the JIMA pattern image is 27%.

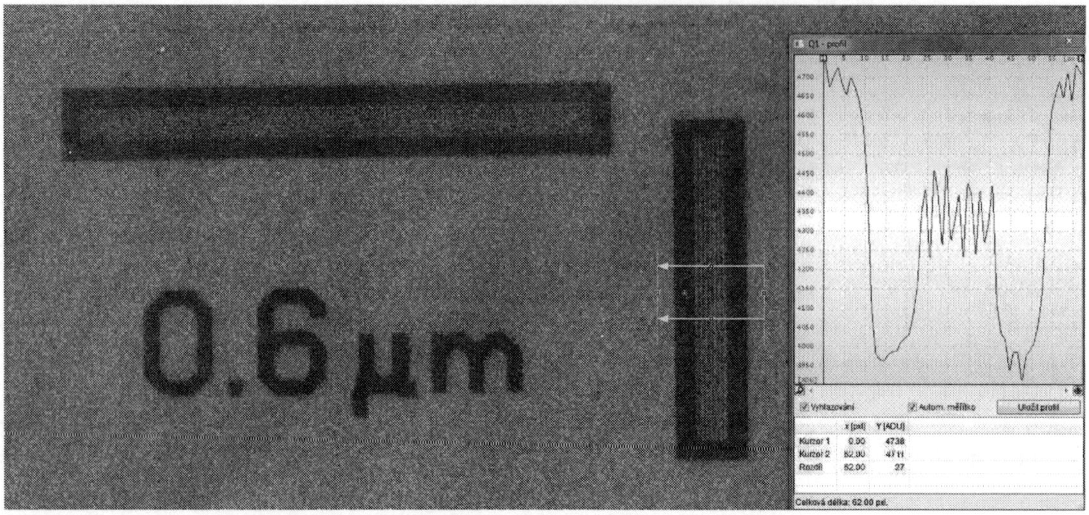

Fig. 1. JIMA pattern presents the resolution of the current system.

4. Discussion

The resolution of a scintillator-based system depends on several factors, mainly on the X-ray absorption process, screen geometry (mainly thickness), and the optical system.

An X-ray photon is absorbed and produces scintillation photons in a volume which is energy dependent. In [1] the authors report Monte Carlo simulations for YAG:Ce which show that the great majority of scintillation photons is generated in a volume with dimensions of less than 100 nm. This is less than the diffraction limited resolution of any optical system, thus the scintillation material and X-ray absorption process itself are not a limiting factor at all. The resolution is limited practically only by geometry and optical system issues.

Moreover, there is a fundamental limit for the resolution of any optical system - diffraction limit. Thus, for the typical scintillator emission wavelength of 550 nm, the highest submicron resolution has probably already been reached in a synchrotron laboratory [6].

The mean absorption depth of X-ray radiation in the scintillator depends on photon energy and the material. The YAG:Ce screen is optically transparent so the image of interaction points is easily transferred to the CMOS sensor. However, the advantage of material transparency decreases with increasing imaging plate thickness. If the scintillator is thinner, the mean absorption depth is lower and the image created is sharper due to less blurring of the image caused by lateral spread of the scintillation photons. Hence, the thinner the imaging plate is, the better the resolution achieved in the image. On the other hand, the detection efficiency decreases with scintillator thickness.

The efficiency of scintillation screens depends on the absorption of the X-ray energy spectrum in the material. The highest 2D-spatial resolution achieved in the images using the system presented was about 0.6 micron utilizing a standard microfocus X-ray source.

Acknowledgement

The support of Czech TA CR Project No. TH04010525 is gratefully acknowledged.

References

[1] Koch, A., Raven, C., Spanne, P., and Snigirev, A., X-ray imaging with submicrometer resolution employing transparent luminescent screens, J. Opt. Soc. Am. A 15, 1940-1951 (1998).
[2] Brandle, C.D., Czochralski growth of oxides, J. Crystal Growth 264 (2004) 593-604.
[3] Nikl, M., Scintillation detectors for X-rays. Meas. Sci. Technol. 17, R37-R54 (2006).
[4] J.A. Mares, A. Beitlerova, M. Nikl, N. Solovieva, C. D'Ambrosio, K. Blazek, P. Maly, K. Nejezchleb and F. De Notaristefani, Rad. Meas. 38, 353-356 (2004).
[5] Tous, J., Horvath, M., Pina, L., Blazek, K. and Sopko, B., High-resolution application of YAG:Ce and LuAG:Ce imaging detectors with a CCD X-ray camera. Nucl. Instr. Meth. Phys. Research A, 591 264-267 (2008).
[6] T. Martin et al., Recent developments in X-ray imaging with micrometer spatial resolution, J. Synchrotron Rad. 13, 180-194 (2006)

pyXIT: Python GUI Software for CT Imaging, Learning and Experimental Algorithms; Example: Multi Material Phase Retrieval

Maximilian ULLHERR [1], Simon ZABLER [2], Jonas GRÄTZ [1], Randolf HANKE [3]
[1] Universität Würzburg, Würzburg, Germany
[2] Fraunhofer EZRT, Würzburg, Germany
[3] Fraunhofer EZRT, Fürth, Germany

Contact e-mail: maximilian.ullherr@physik.uni-wuerzburg.de

Abstract

The Python X-Ray Imaging Tool (pyXIT) is a software for CT reconstruction and image processing. It is open and free software (BSD license) [1]. pyXIT is easily extendable with plugins written in Python and includes a basic CT reconstruction algorithm (cone FBP). The processing speed is comparable to commercial software and it is designed to process many (> 20) and large data sets (4K) comfortably. The software runs in a Python environment (≥ v3.6) with the usual modules for scientific computing installed and has no other dependencies.

pyXIT is not meant to replace commercial software, but it is rather an environment for learning CT and using own or experimental code. It is intended to be used by an experienced user. The original use case is for reconstructing high-resolution CTs at the chair for X-Ray microscopy of the University Würzburg. Some features developed for this use case are: phase retrieval, field of view extension, drift correction (from rewind images), lens distortion correction.

Plugins for image processing can easily be integrated into pyXIT and the preview allows to test/debug or parameterize an algorithm. One example for such a plugin is our implementation of a multi material phase retrieval algorithm [2]. It is a volume filter which applies phase retrieval for inline phase contrast in multiple material samples.

As a framework, pyXIT provides many features which simplify plugin development or can be used for independent projects. The most important are: GPU computing with OpenCl (own API layer), multiprocessing, Fourier filters, file access, preview.

[1] physik.uni-wuerzburg.de/lrm/forschung/software/

[2] Ullherr, M, and Zabler, S.. "Correcting multi material artifacts from single material phase retrieved holo-tomograms with a simple 3D Fourier method." Optics express (2015), DOI: 10.1364/OE.23.032718

SNR Spectra as a Theoretical Model for Understanding (Polychromatic) X-Ray Image Quality and as a Measurable Quantity

Maximilian ULLHERR [1], Simon ZABLER [2]
[1] Universität Würzburg, Würzburg, Germany
[2] Fraunhofer EZRT, Würzburg, Germany

Contact e-mail: maximilian.ullherr@physik.uni-wuerzburg.de

Abstract

We use the known quantity of SNR spectra as a way to better understand which effects significantly influence the image quality of a x-ray imaging measurement. Commonly used models for optimizing image quality often face difficulties when applied to real setups e.g. from neglecting noise properties (temporal SNR) or assuming monochromaticity (DQE).

One important factor for image quality is the sample. It has an energy-dependent signal strength and determines which x-ray energies can be used for imaging. We show by modeling the image formation why the SNR spectra model requires less approximations and how this leads to both a quantitatively and qualitatively better description of image quality.

Another reason to use SNR spectra instead of e.g. a combination of (temporal) SNR and MTF is that a unified measure allows a reliable optimization. In the case of phase contrast imaging, SNR spectra are the only known way to quantitatively describe image quality [1].

SNR spectra not only allow a better theoretical understanding of image quality, they can also be directly calculated from a series of measured projection images in a real setup [1]. If comparability of measurements is required, standardized test objects need to be used, for which we will discuss several possibilities. Another application is the automatic optimization of a measurement setup to a specific sample.

[1] Ullherr, M., et al. "Using measurements of the spatial SNR to optimize phase contrast X-ray imaging." NIM A (2018), DOI: 10.1016/j.nima.2017.09.044

AUTHOR INDEX

Deutsche Gesellschaft Fuer Zerstoerungsfreie Pruefung (DGZfP)
Max-Planck-Str. 6
12489 Berlin
Germany

ISBN 978-1-7138-3709-1

CURRAN ASSOCIATES INC.
proceedings
.com